The Metagenomics Revolution

The Metagenomics Revolution: Unlocking the Potential Microbial Communities in Diagnostics and Beyond serves as a comprehensive resource covering the fundamental principles, methodologies, and cutting-edge advancements in metagenomics. It bridges the gap between theoretical understanding and practical applications, making it an essential guide for researchers, clinicians, engineers, and industry professionals. By integrating concepts from microbiology, bioinformatics, biotechnology, and clinical diagnostics, this book offers a multidisciplinary approach to metagenomic analysis and its real-world applications. The primary aim is to equip readers with a thorough understanding of metagenomics, from sample collection and sequencing technologies to computational tools and data interpretation. The book explores how metagenomic sequencing can be leveraged for diverse applications, including clinical diagnostics, antimicrobial resistance surveillance, environmental monitoring, agricultural microbiome research, and industrial biotechnology. It provides a structured framework for analyzing complex microbial datasets, highlighting best practices in bioinformatics pipelines, multi-omics integration, and AI-driven microbial analysis.

The Metagenomics Revolution
Unlocking the Potential Microbial Communities in Diagnostics and Beyond

Edited by Vivek Kumar and Amit Chaudhary

CRC Press
Taylor & Francis Group
Boca Raton London New York

CRC Press is an imprint of the
Taylor & Francis Group, an **informa** business

Designed cover image: Vivek Kumar

First edition published 2026
by CRC Press
2385 NW Executive Center Drive, Suite 320, Boca Raton FL 33431

and by CRC Press
4 Park Square, Milton Park, Abingdon, Oxon, OX14 4RN

CRC Press is an imprint of Taylor & Francis Group, LLC

ISBN: 978-1-041-13514-2
ISBN: 978-1-041-13515-9
ISBN: 978-1-003-67015-5

DOI: 10.1201/9781003670155

Typeset in Times
by Apex CoVantage, LLC

Contents

Preface

Metagenomics has transformed microbiology by enabling direct, sequence-based access to the diversity, function, and evolution of the vast uncultivated majority of microbes that shape ecosystems and human health. By moving beyond culture-dependent discovery to community-level genomics, the field now links who is present with what they can do, offering a coherent lens on ecology, biotechnology, and medicine alike. This volume is organized to guide readers from foundational concepts to cutting-edge applications and open challenges, balancing experimental practice, computational rigor, and translational impact across 15 chapters. Early chapters introduce core definitions, study design, and the rationale for metagenomics, establishing how careful sampling, metadata, and statistical planning underpin credible inference and reproducibility. Technical chapters survey tools and techniques in metagenomic analysis, ranging from nucleic acid extraction and library construction to marker-based and assembly-first strategies, while highlighting the trade-offs in sensitivity, specificity, and scalability. Data processing pipelines are presented end-to-end—from quality control through taxonomic profiling, binning, and functional annotation—emphasizing best practices, software ecosystems, and pitfalls to avoid. Applications in human microbiome research illustrate how metagenomics resolves strain-level variation, gene catalogs, and community dynamics across body sites and life stages, and how these insights inform health, disease, and therapeutic development. Complementary chapters examine diagnostics and disease detection, showcasing pathogen discovery, outbreak tracing, and host–microbe signatures that expand the reach of clinical microbiology. Viromics receives dedicated treatment, addressing the unique challenges of viral community profiling in the absence of universal markers, as well as the opportunities to map viral diversity, host interactions, and ecological roles across DNA and RNA viromes. Environmental metagenomics spans oceans, soils, and engineered systems, connecting community structure to biogeochemical cycles and ecosystem services at local to global scales. Agriculture and food science applications demonstrate how metagenomics advances crop resilience, livestock health, fermentation quality, and food safety by resolving microbes and their functions along production chains. Chapters on bioremediation and bioengineering show how metabolic potential mined from communities can be harnessed to degrade pollutants, valorize waste, and prototype synthetic consortia. Given the global threat of antimicrobial resistance, a focused chapter outlines metagenomic strategies for surveilling resistomes, tracking mobile genetic elements, and evaluating interventions across clinical and environmental reservoirs. Computational challenges and solutions are treated candidly, covering reference biases, compositionality, contamination, and scalable algorithms for assembly, binning, and statistical inference. Emerging modalities—single-cell and long-read metagenomics—bridge the contiguity and resolution gaps, enabling the linkage of genes to genomes and mobile elements to hosts, while improving the recovery of complex repeats and structural variants. Multi-omics integration expands beyond DNA to transcriptomes, proteomes, and metabolomes, illuminating activity, regulation, and phenotype in context-rich community models. The closing chapters address challenges, ethics, and policy, including data stewardship, privacy, equitable access, and standards that sustain open, reproducible, and responsible science across research and clinical domains. Looking ahead, metagenomics will increasingly fuse robust experimental design, FAIR data principles, and interoperable analytics to deliver actionable insights for planetary health and precision medicine.

Editor Bios

Dr. Vivek Kumar is a postdoctoral researcher with a rich academic and professional background in the field of biotechnology and multi-omics research. He earned both his bachelor's and master's degrees in biotechnology, followed by a Ph.D. in biotechnology from the Motilal Nehru National Institute of Technology (MNNIT), Allahabad, in 2022. His doctoral research focused on the role of microRNAs (miRNA), liquid biopsy techniques, biomarkers, and epigenetic mechanisms in ovarian cancer, contributing to the advancement of early detection and diagnostic methodologies. Following his Ph.D., Dr. Kumar joined the Department of Laboratory Medicine/Pathology at the All India Institute of Medical Sciences (AIIMS), Patna, as a postdoctoral fellow under a Department of Science and Technology (DST) project. Here, his work centered on breast cancer research, where he applied his expertise in molecular diagnostics and cancer biology to explore new avenues for understanding and diagnosing breast cancer. Dr. Kumar's pursuit of excellence then led him to the United States, where he joined the University of South Florida (USF) Microbiome Institute within the Center of Excellence in Aging and Brain Repair as a postdoctoral researcher. His current research delves into the complex interplay between the gut microbiome and the brain, specifically examining how epigenetic modifications influence the gut–brain axis in cognitively impaired patients. By integrating multi-omics approaches, Dr. Kumar aims to uncover the molecular mechanisms underlying cognitive disorders and identify potential biomarkers for early diagnosis and therapeutic intervention. Dr. Vivek Kumar's work stands at the cutting edge of biotechnology, epigenetics, and neurodegenerative disease research, contributing valuable insights into cancer diagnostics and brain health through the lens of microbiome science and molecular biology.

Dr. Amit Chaudhary is currently a postdoctoral scholar in the Department of Cell, Developmental, and Integrative Biology at The University of Alabama at Birmingham, Alabama, USA. Prior to this, Dr. Chaudhary worked as a postdoctoral fellow in the Polymer Simulation and Characterization Lab in the Department of Metallurgical Engineering and Materials Science at the Indian Institute of Technology Bombay, Mumbai, India. He completed his Ph.D. in biotechnology in the Department of Biotechnology, Motilal Nehru National Institute of Technology, Allahabad, India. Dr. Chaudhary also holds an M.Tech. in bioinformatics from the Indian Institute of Information Technology, Allahabad, India, and a Bachelor of Pharmacy from the United Institute of Pharmacy, Allahabad, India. Dr. Chaudhary's research interests span a broad range of topics, including transcriptomics analysis, molecular modeling, molecular docking, molecular dynamics simulations, phylogenetic analysis, protein-protein/gene-gene interaction, and functional and pathways analysis of proteins/genes.

1 Introduction to Metagenomics

Diksha Singh[1], *Madhuri Tiwari*[1#], *Meghna Agrawal*[1#], *and Ashutosh Mani*[1*]

[1] Department of Biotechnology, Motilal Nehru National Institute of Technology, Allahabad

*Corresponding author

[#] Equally contributed author

1.1 MICROBES SHAPE OUR WORLD

Microorganisms are fundamental to an ecosystem's function and host biology; these microbes play a crucial role in every organism's life, including humans (Gilbert & Neufeld, 2014). Microbes regulate Earth's atmosphere by driving geochemical cycles of carbon, nitrogen, oxygen, and sulfur, and by controlling the production and consumption of greenhouse gases (Ahn et al., 2023; Suvarnalatha Devi, 2024). In humans, gut and oral microbiota extract energy from complex dietary components and defend against pathogens. Additionally, through fermentation, microbial enzymes convert raw substrates into widely consumed foods such as yogurt, cheese, and bread, thereby keeping us healthy (Hou et al., 2022; Valentino et al., 2024). All animals and plants are closely associated with microbes that mobilize nutrients, solubilize metals, and synthesize essential vitamins (Hnini et al., 2024). Plant growth is supported by soil and rhizosphere bacteria that supply nutrients to roots, while microbial biocontrol agents protect plants from disease. Microbial communities play an immense role in crop health and productivity worldwide (El-Saadony et al., 2022).

Microbes also clean the environment by decomposing organic waste, carrying out bioremediation of naturally occurring toxins, transforming toxic compounds—including man-made pollutants, spilled oil, and industrial chemicals—into harmless substances, and recycling nutrients. Consequently, soil and water detoxification largely relies on microbial activity (Maglione et al., 2024). Influencing geochemical and greenhouse-gas cycles, supporting human and plant health, and cleansing the environment are just a few examples of microbial community functions. Their collective actions profoundly shape the biosphere. The strength of these communities lies in the diverse metabolic capabilities of their constituent species; by working together, these microbial communities carry out the majority of matter and energy transformations on Earth. Complex microbial assemblages are integrated, balanced, and capable of rapid adaptation to environmental change (Maglione et al., 2024). The role of microbes in different aspects is depicted in Figure 1.1.

DOI: 10.1201/9781003670155-1

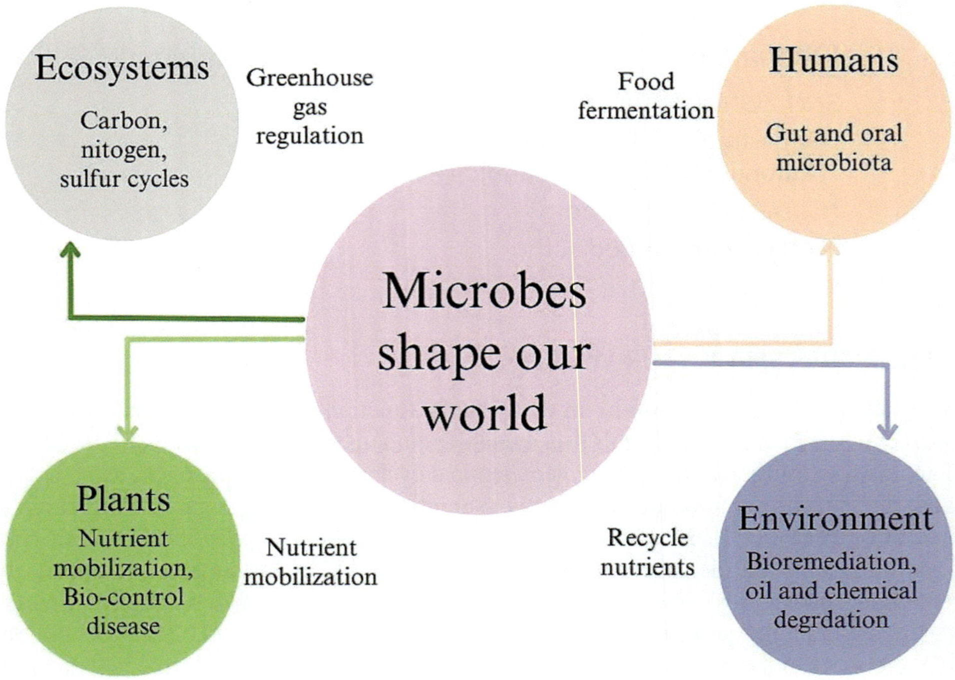

FIGURE 1.1 Roles of microbes in different systems for various purposes, such as in ecosystems for greenhouse gas regulation, in humans for food fermentation, in plants for nutrient mobilization, and in environments for recycling of nutrients.

1.2 EVOLUTION OF MICROBIAL STUDY: FROM ISOLATES TO COMPLEX CONSORTIA

Historically, studying microbes in their native habitats was difficult, so microbiologists relied on growing individual species in pure culture in the laboratory. As a result, understanding of the microbial community lagged behind knowledge of isolated organisms. For instance, the planktonic state was long considered the inherent condition of microorganisms, whereas biofilms were regarded as deviations primarily studied under laboratory conditions (Lund et al., 2020). Now, it is understood that microbes in nature are more than free-floating entities, and their community composition is far from random. Instead, microbes function as multicellular, often multispecies consortia that are metabolically interdependent and sometimes physically connected. Some communities thrive without direct cell-to-cell contact, interacting through shared environments (e.g., water or soil) by exchanging nutrients, biochemical products, and chemical signals. Others grow on surfaces (e.g., pipe walls, teeth), where they exist in close proximity to both their own species and others. Biofilms are structured clusters of microbial cells embedded in a matrix of extracellular polysaccharides, enabling physical connectivity and coordinated activity. Microbes have evolved to grow in communities and undertake collective metabolic activities that reflect physical organization and division of labor in microbial communities (Geesink et al., 2024). In the pure-culture paradigm, the coexistence of multiple microbial species in the same culture medium was considered as "contamination." Microbes that grew well as single-cell suspensions in liquid media or formed discrete colonies on Petri dishes were considered ideal study models. This perspective made it difficult to identify and characterize microbial species that relied on the metabolic by-products of other organisms for growth (Wan et al., 2023).

1.3 REVOLUTION OF METAGENOMICS

All microbes cannot be cultivated under standard laboratory conditions because different microbes within an environment interact with each other. Moreover, microbes that grow alone often behave differently compared with when they are part of a multispecies community. Hundreds, thousands, and millions of microbial species can be present in a handful of soil, in water, and in the human gut (Yan et al., 2023). The development of metagenomics has completely changed the way microbial ecology is studied by allowing scientists to analyze all microbial genes from an environment without culturing them in the laboratory (Pérez-Cobas et al., 2020). Metagenomics is the study of methodologies and techniques for the collection and analysis of DNA from all organisms in an environment in a culture-independent manner (Kumar et al., 2024). It enables researchers to reconstruct the genetic potential of entire microbial communities by identifying the genomic composition of a community, detecting previously uncultured or unknown species, and uncovering genes that drive complex metabolic networks. Metagenomics has also revealed alternative metabolic pathways, antibiotic resistance, and microbial adaptation that have occurred through horizontal gene transfer (Stewart, 2012).

Amplicon sequencing and shotgun sequencing are two major sequencing methods used in metagenomics. Amplicon sequencing targets specific marker genes such as bacterial 16S rRNA, whereas shotgun sequencing fragments and sequences all DNA in a sample. Although the sequencing strategies are complementary, together they provide complete information on the abundance, diversity, and interactions of microbes within a sample (Pérez-Cobas et al., 2020).

1.4 MILESTONES IN METAGENOMICS

Since the Human Genome Project, metagenomics projects have provided insights into diverse microbial communities across the globe. The Human Genome Project (HGP) was started in 1990 and completed in 2003. The cost of this project was approximately $3 billion, and the sequencing method used was Sanger sequencing. Since then, NGS has made genome sequencing much faster and cheaper. Now, sequencing can be completed in less than a day for a few hundred dollars. The decline in sequencing costs has transformed large-scale analyses of multiple genomes. Today, metagenomics is one of the most powerful tools for interpreting complex microbial communities (Kim et al., 2024).

- ***Global Ocean Sampling Expedition (GOS):*** Led by J. Craig Venter, this project involved collecting and sequencing microbial DNA from marine environments worldwide. It was the first global-scale metagenomics project. It revealed that oceanic microbial composition depends on depth, temperature, and nutrients; discovered new microbial species involved in photosynthesis, geochemical cycles, and nutrient cycling; and identified previously unknown genes (Rusch et al., 2007).
- ***Human Microbiome Project (HMP):*** Launched by the NIH in 2008, this project analyzed microbiomes from various body sites of hundreds of healthy individuals. Major sites included the gut, mouth, skin, nose, and vagina. The project revealed that healthy individuals can harbor distinct microbial communities, that most microbes are not harmful, and that dysbiosis is associated with disease. It generated extensive datasets that enabled the development of microbiome-based therapies and precision medicine (Turnbaugh et al., 2007).
- ***Earth Microbiome Project (EMP):*** Initiated in 2010, this project was not focused on specific ecosystems or humans like GOS and HMP, it was an international effort in which microbial samples were collected and sequenced from diverse environments including soil, oceans, and animals. Standardized protocols enabled comparison of approximately 200,000 samples. The project revealed that most microbes remain unknown, community composition depends on pH, temperature, and salinity, and soil harbors the greatest microbial diversity (Gilbert et al., 2014).
- ***MetaSUB Project:*** This project focuses on microbiomes in urban environments such as subways, railings, seats, buses, and air across cities worldwide. It revealed that urban

microbiomes vary between cities and influence air quality, human health, and disease. The project also identified previously unknown genes and demonstrated that some microbes contribute to pollutant degradation (*The metagenomics and metadesign of the subways and urban biomes (MetaSUB) international consortium inaugural meeting report*, 2016, June 3).

Together, these milestones illustrate how metagenomics has evolved from sequencing a single human genome to exploring the vast diversity of microbial life across humans, oceans, soils, and urban environments. Metagenomics has revolutionized the understanding of diseases in which complex microbial communities play critical role in disease progression. Metagenomic sequencing helps elucidate the characteristics of disease-associated microbes and their impact. Several studies have reported association between the gut microbiome and diseases such as breast cancer, colorectal cancer, and Alzheimer's disease (Madigan et al., 2025; Nair et al., 2024; Prajapati et al., 2024). Dysbiosis of microbiomes is correlated with immune dysfunction, metabolic-based inflammation, cognitive decline, and cancer progression (Mishra et al., 2024; Sarfraz et al., 2024; James et al., 2024). These studies reveal the potential of metagenomics in modern biology to discover previously unknown microbial communities and genes. The complete milestone events in metagenomics are depicted in Figure 1.2.

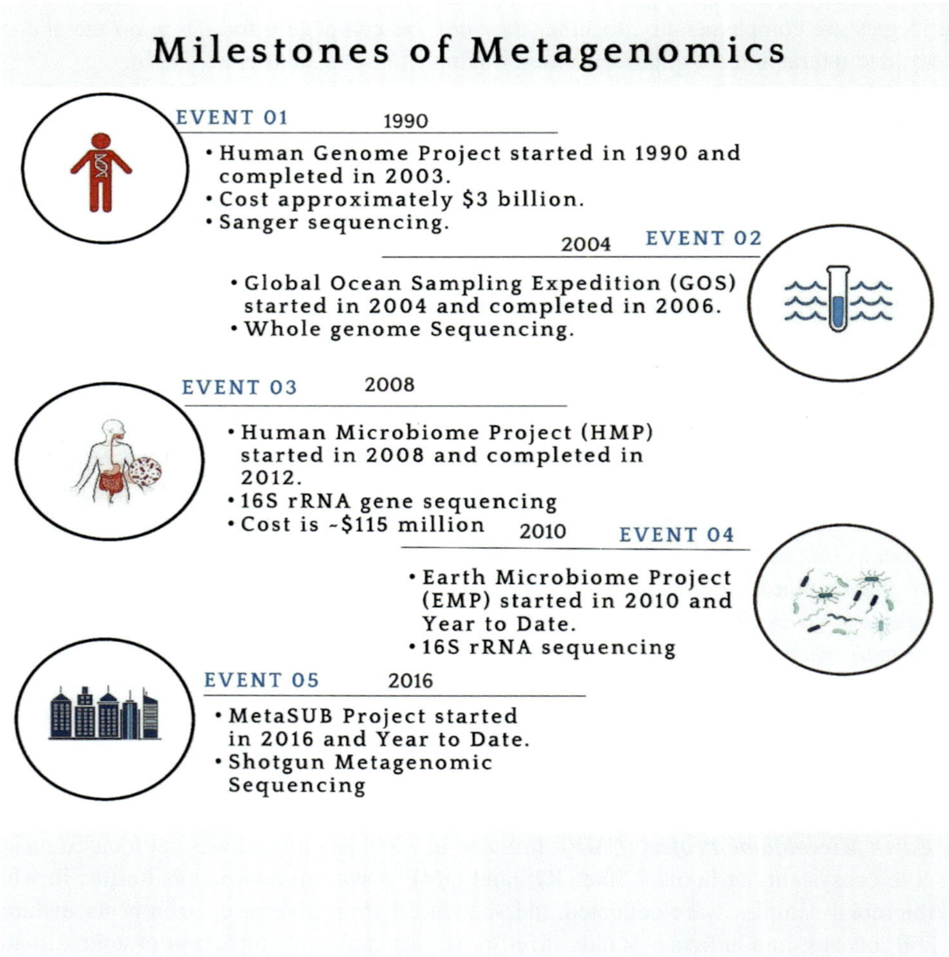

FIGURE 1.2 Major events in metagenomics milestones: event 1 (1990)—the human genome project; event 2 (2004)—the global ocean sampling expedition; event 3 (2008)—the human microbiome project; event 4 (2010)—the earth microbiome project; and event 5 (2016)—the MetaSUB project.

1.5 METAGENOMIC APPROACH

The introduction of the term "metagenome" by Handelsman et al. in 1998 set the stage for modern microbiology, emphasizing the importance of studying the genetic makeup of entire microbial communities rather than focusing solely on cultivable isolates. Over the years, metagenomic approaches have evolved into two main strategies: amplicon metagenomics and shotgun metagenomics (Nam et al., 2023).

1.5.1 Amplicon Metagenomics

1.5.1.1 Principle and Workflow

Amplicon sequencing includes the amplification of a marker gene, such as the 16S rRNA gene in bacteria. This gene is 1,500 base pairs long and has nine hypervariable regions that are important in detecting evolutionary changes between microbes. The role of this sequencing approach is to utilize it for taxonomic classification in the analysis of microbial communities. This sequencing approach, also known as targeted sequencing, is still considered the primary approach for the analysis of community structure (Buetas et al., 2024). Sequencing of the 16S rRNA gene involves several stages that are performed in sequential order (Gupta et al., 2019). The complete procedure is depicted in Figure 1.3.

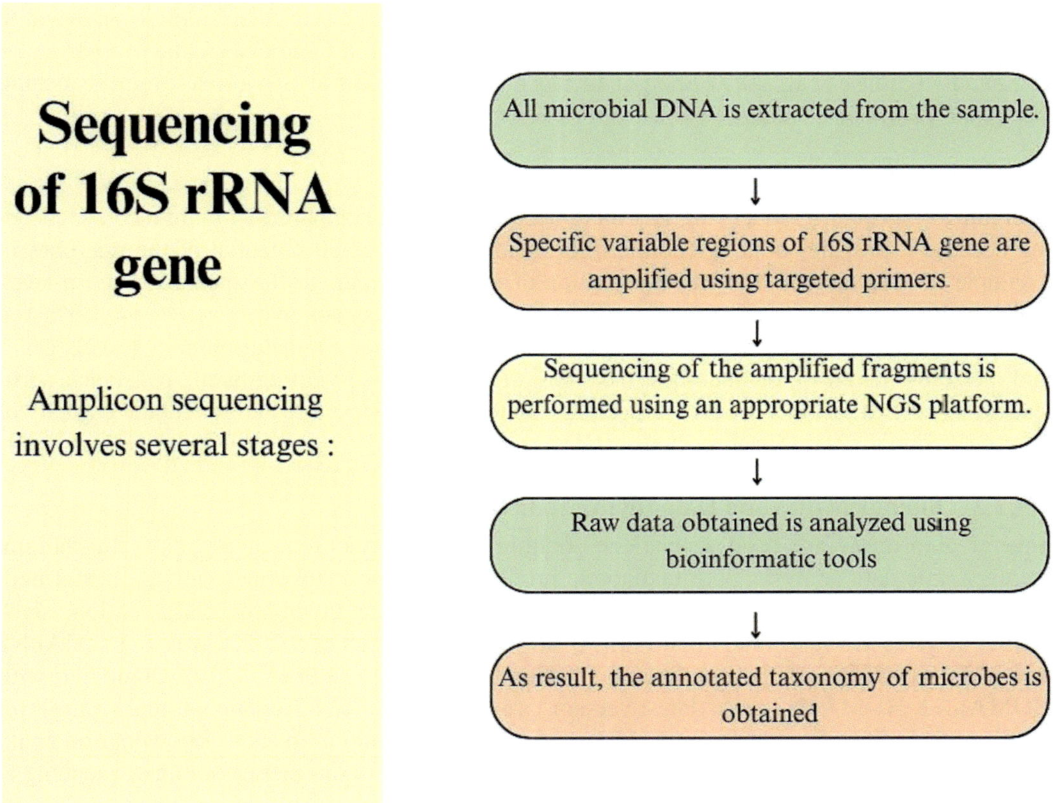

FIGURE 1.3 The sequential stages of 16S rRNA gene amplicon sequencing.

1.5.1.2 Applications

16S rRNA gene sequencing has been used for the analysis of microbial diversity due to its low price and effectiveness (Zhang et al., 2023). This method is useful mainly for low-abundance or complex samples, such as upper respiratory swabs or sputum samples. However, careful DNA extraction and proper storage are essential to provide accurate results (Claassen-Weitz et al., 2020).

1.5.1.3 Bioinformatics and Data Interpretation

Several algorithms and specific computational tools are required to convert raw amplicon reads into functionally informative data, such as microbial profiling and the abundance of each microbe. Raw data are first preprocessed to remove low-quality or duplicate reads, and the processed reads are then analyzed using bioinformatic pipelines. Several bioinformatic pipelines exist for the analysis of 16S rRNA gene sequencing data, but there is currently no universally accepted standard workflow. At the final stage of analysis, sequences are assigned taxonomy using reference databases such as Greengenes, SILVA, and RDP (Ibal et al., 2022; Prodan et al., 2020).

The choice of pipeline, database, and settings can affect how taxa are identified, sometimes introducing bias. Different pipelines use different methods for clustering, cleaning, and classification, so results from the same dataset may not always match (Abellan-Schneyder et al., 2021).

1.5.2 Shotgun Metagenomics

1.5.2.1 Principle and Workflow

Shotgun metagenomic sequencing is another method used for studying microbes, in contrast to amplicon sequencing. In this sequencing method, all the DNA present in a sample is analyzed to characterize the taxonomy as well as the functional properties of all microbes found in a particular community (Quince et al., 2017). Shotgun metagenomic sequencing also involves several sequential steps (Sharpton, 2014; Kumar, Shukla, et al., 2024).

1.5.2.2 Applications

Shotgun metagenomic sequencing is a more accurate method compared with amplicon sequencing because it provides both taxonomic information and functional potential of the microbes in a sample. In this method, the characterization of microbes occurs at the species or strain level, which makes it easier to identify rare or novel organisms that may not be detected by marker-based sequencing methods (Sanguineti et al., 2024). Furthermore, shotgun metagenomics reveals genes and metabolic pathways of microbes, including those associated with antibiotic resistance, virulence, and other traits, thereby providing insights into both the structure and functional capacity of microbial communities (Sanabria et al., 2021).

1.5.2.3 Bioinformatics and Data Interpretation

Several algorithms are used to analyze shotgun metagenomic sequencing data. In shotgun metagenomic data analysis, both taxonomic profiling and functional annotation are performed, and several algorithms are available for these purposes. For taxonomic classification, algorithms such as Kraken2 and MetaPhlAn3 are used (Portik et al., 2022), while HUMAnN2 and HUMAnN3 are applied for functional annotation (Franzosa et al., 2018). Compared with HUMAnN2, HUMAnN3 is the latest version, providing taxonomic, functional, and strain-level profiling and offering more accurate results for microbial functions and activities (Beghini et al., 2021). The comparison between both amplicon-based and shotgun metagenomic sequencing is depicted in Table 1.1.

TABLE 1.1
Comparative Workflow: Amplicon versus Shotgun Metagenomic Sequencing

Step	Amplicon Sequencing (Targeted)	Shotgun Metagenomic Sequencing (Untargeted)
1. Sample Collection	Environmental or host sample (soil, water, gut, etc.)	Environmental or host sample (soil, water, gut, etc.)
2. DNA Extraction	Total microbial DNA isolated	Total DNA from all organisms (bacteria, archaea, fungi, viruses, eukaryotes)
3. Gene Targeting	PCR amplification of marker genes (16S rRNA, 18S rRNA, ITS)	Random fragmentation of entire DNA
4. Library Preparation	Not required (PCR products directly sequenced)	Required—adapters ligated to DNA fragments
5. Sequencing	NGS of specific marker gene regions	NGS of all DNA fragments (whole community genomes)
6. Bioinformatics Analysis	Quality filtering to taxonomic assignment	Quality filtering to Assembly, binning, taxonomic classification, functional annotation
7. Output	Community composition	Community composition and functional potential

TABLE 1.2
Comparative Studies of Strengths and Limitations of Both Sequencing Methods

Features	16S rRNA Amplicon Sequencing	Shotgun Metagenomic Sequencing
Primary functions	Taxonomy classification and community structure	Comprehensive taxonomic, functional classification, and community structure
Key strength	Lower sequencing cost, reproducible results with few reads	Enhanced species-level resolution, access to functional genes
Major limitations	Biases from amplification, short reads, incomplete databases, poor resolution of closely related taxa	High sequencing cost, technical demands, require millions of reads for coverage/depth
Expertise required	Standard sequencing and bioinformatics	Specialized laboratory and complex bioinformatic pipelines

1.5.3 Limitations and Strengths of Metagenomics

Amplicon sequencing is the primary taxonomic classification method, but it has several limitations. To overcome these limitations, researchers are using full-length 16S sequencing with third-generation technologies such as PacBio to enhance species-level resolution and selectively analyzing specific variable regions, such as V1–V3, to improve short-read classification. In contrast, shotgun metagenomic sequencing offers comprehensive functional and taxonomic classification; however, it requires millions of high-quality reads and specialized expertise, which leads to significantly higher costs (Usyk et al., 2023; Zhang et al., 2023). Detailed strengths and limitations are discussed in Table 1.2.

1.6 METHODOLOGIES IN METAGENOMICS

Metagenomic methodologies include both experimental and computational workflows to study microbial communities directly from environmental or host-associated samples. Raw metagenomic

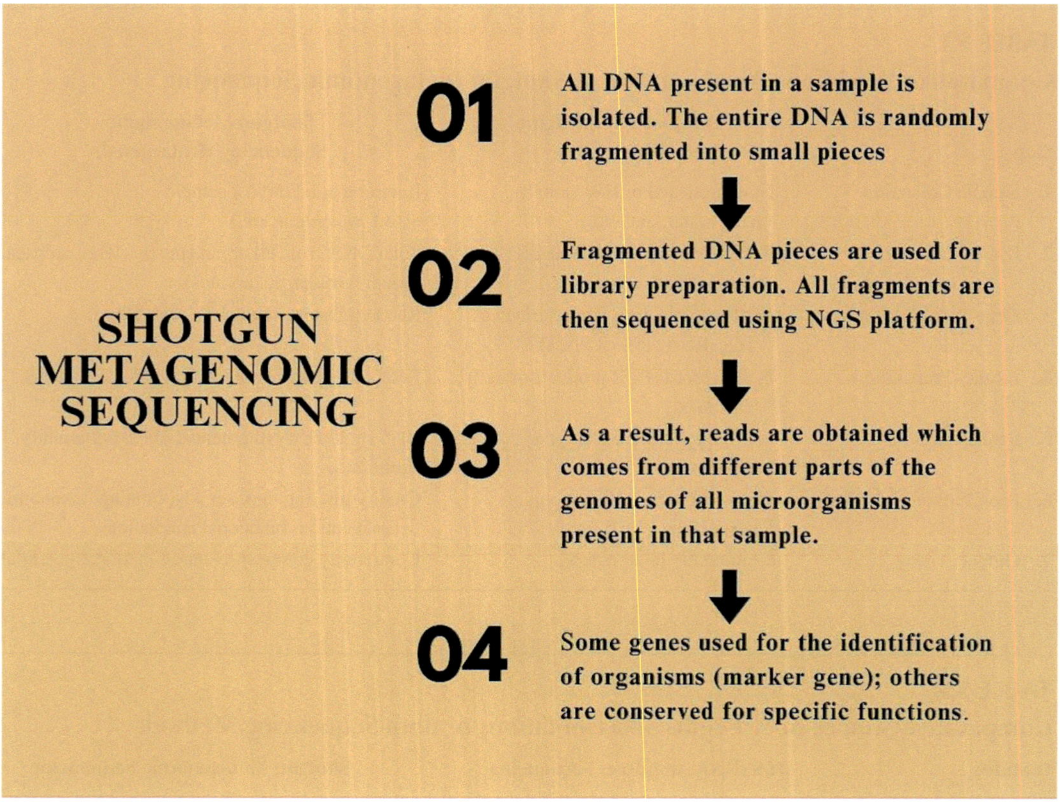

FIGURE 1.4 The sequential steps of shotgun metagenomic sequencing.

data obtained from sequencing technologies, such as 16S rRNA gene sequencing or shotgun metagenomics, are decoded using bioinformatic pipelines (Liu et al., 2021). The methodologies for metagenomic research are a stepwise process; they begin with sample collection and continue through sequencing and advanced data analysis, and this scheme is applicable to both amplicon and shotgun metagenomic data (Lindner et al., 2024). The comprehensive metagenomics workflow is depicted in Figure 1.4.

1.6.1 Sample Collection and DNA Extraction

- *Sample collection:* Sample collection is the first step in microbiome studies, and samples can come from various sources such as the environment (soil and water) or biological sites (human gut or oral cavity) (Kuruppu et al., 2024; Pollock et al., 2018). To obtain accurate results, researchers need to follow certain standardization procedures at every step in metagenomic studies. Sample type not only affects the results, but the sampling site, collection method, and storage conditions—including temperature, duration, and the use of preservatives—can also affect downstream analysis. Therefore, consistency in methodology should be prioritized (Pollock et al., 2018).
- *DNA extraction:* After sample collection, DNA is extracted from all microbial cells present in the sample. The choice of extraction method is necessary to obtain pure DNA free from contamination. Therefore, to avoid problems in downstream analysis processes, extraction methods should be chosen carefully. DNA purity and quantity should be checked before library preparation, which is the next step in metagenomic studies (Buck et al., 2021).

TABLE 1.3
Comparison of Amplicon versus Shotgun Sequencing Library Preparation

Feature	Amplicon Sequencing	Shotgun Sequencing	References
Platforms used	Illumina (MiSeq), Roche 454, Ion Torrent, Nanopore, PacBio	Illumina (NovaSeq), other high-throughput WGS platforms	Dong et al. (2017); Kampmann et al. (2025)
Process	Uses PCR to amplify specific gene regions (e.g., 16S rRNA in bacteria)	Three main steps: DNA fragmentation → adaptor ligation → amplification (commercial kits such as Nextera which combine fragmentation and adaptors ligation into a single steps, called tagmentation)	Ni et al. (2023); Feng et al. (2018)
Traditional protocol	Repeated PCR amplification with risk of contamination	DNA fragmentation done by enzymatic/physical method, adapters ligated in separate steps, and then PCR amplification	Ni et al. (2023); Feng et al. (2018)
Advantages	Simple, cost-effective, focuses on targeted genes	Provide whole-genome coverage, high-resolution data	Ni et al. (2023); Feng et al. (2018)
Limitations	Can be biased due to PCR amplification	More expensive, requires more sequencing depth and resources	Ni et al. (2023); Kampmann et al. (2025)

1.6.2 Library Preparation and Sequencing

Further development of next-generation sequencing (NGS) technology is responsible for transforming microbiome research by allowing microbial communities to be analyzed through faster and cheaper methods. Unlike traditional Sanger sequencing, which provides only a few long DNA reads at a time, NGS can generate millions of short DNA fragments simultaneously, providing the depth and detail that are required for accurate characterization of complex microbiota (Hodkinson & Grice, 2015). These technologies now form the foundation of modern microbiome studies, supporting both taxonomic and functional analyses across diverse environments (Yi et al., 2024). Differences in library preparation between both methodologies are discussed in Table 1.3.

1.6.3 Data Processing and Quality Control

Quality control of the data is essential for both amplicon and shotgun metagenomic workflows. Certain common processing steps are applied to both types of data:

- *Demultiplexed:* This is the first step of data processing, in which reads are sorted according to their sample-specific barcodes (Oulas et al., 2015).
- *Quality filtering or preprocessing:* Raw sequencing data, whether obtained from amplicon or shotgun sequencing, often contain sequencing errors, adapter sequences, and poor-quality reads. Therefore, it is important to remove these unwanted portions (Hoff, 2009; Pérez-Cobas et al., 2020). Several tools are used for this purpose; the most commonly used tools include Cutadapt (Martin, 2011), Trimmomatic (Bolger et al., 2014), and fastp (Chen et al., 2018). This processing step ensures that the data are reliable and accurate for downstream analyses (Zhou et al., 2014) (see Figure 1.5).

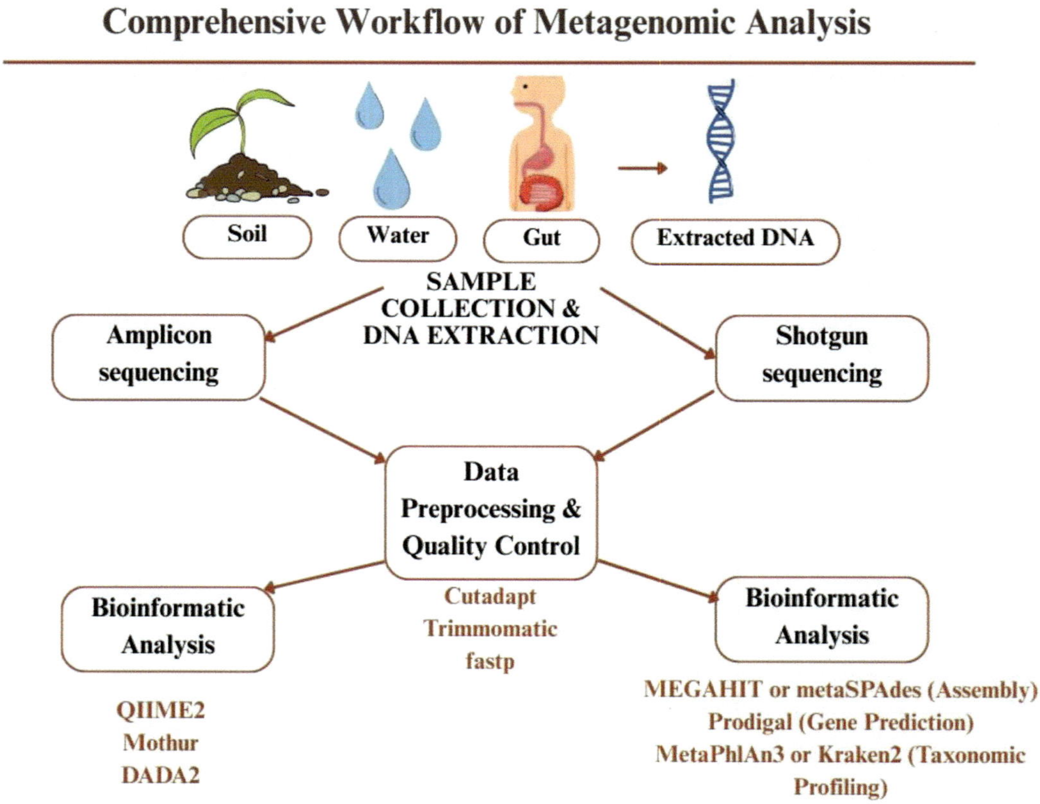

FIGURE 1.5 Comparative bioinformatic analysis workflow of amplicon and shotgun sequencing data.

1.6.4 Bioinformatic Analysis of Amplicon and Shotgun Metagenomic Data

The bioinformatic analysis of amplicon and shotgun metagenomic data involves several steps to obtain meaningful biological information, which can be used for essential processes in human health, sustainable energy production, or other applications that benefit society. For amplicon data, workflows include taxonomic assignment and statistical analysis, mainly to measure microbial diversity and differential abundance. Shotgun metagenomic analysis is more complex and typically includes the assembly of reads into contigs, binning into metagenome-assembled genomes (MAGs), gene prediction and functional annotation, taxonomic profiling, and visualization of results. Various specialized pipelines and tools are available for each step to provide accurate and reliable biological information. The tools and pipelines applicable for analysis of both methodologies are given in Table 1.4.

TABLE 1.4
Analysis of Amplicon and Shotgun Metagenomic Data

Data Type	Analysis Step	Description	Tools/Pipelines	References
Amplicon (16S rRNA)	Taxonomic Classification	Processing of sequences including denoising, chimera removal, and extraction of representative sequences before taxonomic assignment.	QIIME 2 (DADA2 plugin)	Marizzoni et al. (2020)
		In mothur, several processing steps also performed before taxonomic assignment like dereplication, merging of reads.	Mothur	
	Statistical Analysis	Alpha diversity (within-sample) and beta diversity (between samples); identification of differentially abundant taxa.	General statistical methods	Cassol et al. (2025)
Shotgun Metagenomic	Assembly	Quality-controlled reads are assembled into longer contigs; assemblies may be fragmented due to presence of repeated sequences, low coverage for certain species, or closely related strains.	MEGAHIT, metaSPAdes	Li et al. (2015); Nurk et al. (2017)
	Binning	Grouping of contigs based on sequence composition or coverage patterns to reconstruct metagenome assemble genomes (MAGs) to avoid fragmented assemblies drawback.		Benoit et al. (2024)
	Gene Prediction	Identification of ORFs and functional annotation to understand metabolic potential; Gene prediction can follow two main approaches; assembly-based prediction (contigs are created) and read-based prediction (directly from short reads).	Prodigal, FragGeneScan (FGS)	Hyatt et al. (2010); Van der Jeugt et al. (2022)
	Functional Annotation	After gene prediction, identified genes are compared against comprehensive reference databases to understand metabolic potential and functional capabilities of the microbial community.	Reference databases such as: KEGG (Kyoto Encyclopedia of Genes and Genomes), COG (clusters of Orthologous Groups), Pfam, and SEED	Kanehisa et al. (2015); Galperin et al. (2024); Mistry et al. (2020); Overbeek et al. (2014)
	Taxonomic Profiling	For short-read datasets: classify reads based on sequence composition or mapping reads against specific marker gene.	Kraken2 and Centrifuge (sequence composition based profiler)	Wood et al. (2019); Kim et al. (2016)
		For long-read datasets: use nucleotide or protein alignments combined with lowest-common-ancestor (LCA) approaches	MetaPhlAn3 and mOTUs2 (marker-based profiler)	Beghini et al. (2021); Portik et al. (2022); Ruscheweyh et al. (2022)
	Visualization	Interactive exploration of taxonomic and functional profiles.	Krona; KEGG Mapper	Ondov et al. (2011); Kanehisa and Sato (2020)

1.7 APPLICATIONS OF METAGENOMICS

Metagenomics has become a powerful tool to study microorganisms directly from their environment. Unlike traditional methods that require growing microbes in the lab, metagenomics allows researchers to observe the full diversity of microbial communities and understand their roles. This approach is useful in many fields, including human health, agriculture, environmental science, and biotechnology.

1.7.1 Human Health and Microbiome

Microbiome studies are used for mapping microbial signatures, pathogen detection, and personalized medicine.

- *Disease Mapping:* Identification of microbial patterns linked to major diseases, including cancer, obesity, diabetes, and inflammatory bowel disease (IBD) (Jin et al., 2024).
- *Pathogen Detection:* Rapid detection of rare, novel, and emerging pathogens with high sensitivity. For example, mNGS identified pathogens with 78% sensitivity (Zhang, Zhou, et al., 2024).
- *Personalized medicine:* Analysis of the microbiome, including the human gut virome, allows researchers to create predictive models for patient prognosis and to tailor treatment responses (Lathakumari et al., 2025).

1.7.2 Environmental Science and Conservation

Metagenomic studies reveal ecosystem function, the role of microbes in bioremediation, and the use of extreme-environment microbes in industries.

- ***Ecosystem Function:*** Metagenomic studies explain the essential role of microbes in soil and marine environments, specifically their influence on global nutrient cycling (carbon and nitrogen) and carbon storage (Sunagawa et al., 2015).
- ***Bioremediation:*** These studies identify and characterize specific microbes and their unique properties in the natural degradation processes of harmful materials such as industrial chemicals and plastics (Yoshida et al., 2016; Kumar et al., 2025).
- ***Extreme Environments:*** Metagenomic studies of specialized microbes that live in harsh conditions, such as high heat or high salinity, are useful for identifying novel enzymes (extremoenzymes) for industrial applications (Cowan et al., 2015).

1.7.3 Agriculture and Soil Health

Soil microbiome research reveals interactions of microbes with plants, their role in soil health management, and their contribution to crop protection.

- ***Plant-Microbe Interaction:*** Metagenomic studies characterize microbial communities that directly benefit plants by enhancing nutrient uptake, increasing stress tolerance, and providing natural protection against disease (Babalola et al., 2025).
- ***Soil Health Management:*** Analysis of soil microbes explains how microbes decompose organic matter and fix nitrogen, thereby supporting microbial diversity and overall soil health (Thakur et al., 2025).
- ***Crop Protection:*** Microbiome studies allow researchers to rapidly detect specific crop pathogens and identify microbial genes suitable for use as biocontrol agents, helping to reduce reliance on chemical pesticides (Zhang, Wang, et al., 2024).

1.8 EMERGING TRENDS AND FUTURE PERSPECTIVES

Metagenomics is becoming more advanced and is moving toward new techniques while improving computational tools. Currently, metagenomics and other omics technologies are being combined, with AI and machine learning increasingly used for data evaluation. As a result, this information can be applied in various fields, such as clinical applications, while also addressing ethical and data-sharing limitations. The emerging trends are depicted in Table 1.5.

TABLE 1.5
Current Trends in Metagenomics and Other Omics Technologies Include the Application of AI and Machine Learning

Trend/ Perspective	Description	Application	Example	Reference
Integration with multi-omics	Combining metagenomics with meta-transcriptomics, meta-proteomics, and metabolomics to understand microbial functions in detail	Understanding disease mechanisms and microbial functions in host	Linking gut microbes to metabolic disorders	Almeida et al. (2021); Quince et al. (2017)
AI and machine learning	Using ML/AI to analyze large and complex datasets, identify biomarkers, and classify taxa accurately	Biomarker discovery, functional prediction, clinical diagnostics	Predicting host-microbiome interactions and detecting resistance genes	de Abreu et al. (2021)
Clinical translation and microbiome therapeutics	Applying metagenomic insights to design therapies	Treatment of IBD, cancer, and other microbiome-related diseases	Fecal microbiota transplantation (FMT), engineered microbial consortia	Manrique et al. (2017); Yadav and Chauhan (2021)
Ethical and data-sharing challenges	Addressing privacy, consent, reproducibility, and data accessibility	Ensuring ethical and responsible use of metagenomic research	Development of standardized pipelines and international data-sharing frameworks	Filardo et al. (2024)

1.9 CONCLUSION

Metagenomics provides knowledge of the microbial world in much greater detail than traditional cultivation methods, overcoming their limitations by obtaining information directly from the environment. Microbiome studies are conducted using two main advanced sequencing technologies: amplicon and shotgun metagenomic sequencing. Each of these technologies has its own strengths, limitations, and applications across various fields. Several pipelines and algorithms have been developed for the analysis of each type of sequencing data; while the workflows are connected in some ways, they differ in others. The final results of metagenomic analyses reveal microbial diversity, microbial abundance, functional potential, and metabolic pathways. This information helps identify the key roles of particular microbes in specific fields. Metagenomic studies provide insights into the benefits of microbes for society in areas such as medicine, sustainable agriculture, environmental protection, and the early diagnosis of life-threatening diseases, as well as ecosystem functions. Emerging tools such as AI and machine learning are expanding the potential of sequencing and bioinformatics, and multi-omics approaches have further enhanced understanding of microbial interactions in metagenomics for industrial innovation and personalized medicine. Overall, metagenomics provides a detailed and rapid platform for detecting and exploring the microbial world, enabling the utilization of microbes for global impact.

REFERENCES

Abellan-Schneyder, I., Matchado, M. S., Reitmeier, S., Sommer, A., Sewald, Z., Baumbach, J., . . . Neuhaus, K. (2021). Primer, pipelines, parameters: issues in 16S rRNA gene sequencing. *mSphere*, *6*(1), 1202–1220.

Ahn, Y. J., Lee, J. A., Choi, K. R., Bang, J., & Lee, S. Y. (2023). Can microbes be harnessed to reduce atmospheric loads of greenhouse gases? *Environmental Microbiology*, *25*(1), 17–25.

Almeida, A., Nayfach, S., Boland, M., Strozzi, F., Beracochea, M., Shi, Z. J., . . . Finn, R. D. (2021). A unified catalog of 204,938 reference genomes from the human gut microbiome. *Nature Biotechnology*, *39*(1), 105–114.

Babalola, O. O., Adebayo, A. A., & Enagbonma, B. J. (2025). Shotgun metagenomics dataset of the core rhizomicrobiome of monoculture and soybean-precedent carrot. *BMC Genomic Data*, *26*(1), 26. https://doi.org/10.1186/s12863-025-01320-7

Beghini, F., McIver, L. J., Blanco-Míguez, A., Dubois, L., Asnicar, F., Maharjan, S., . . . Segata, N. (2021). Integrating taxonomic, functional, and strain-level profiling of diverse microbial communities with bioBakery 3. *Elife*, *4*(10), 65088.

Benoit, G., Raguideau, S., James, R., Phillippy, A. M., Chikhi, R., & Quince, C. (2024). High-quality metagenome assembly from long accurate reads with metaMDBG. *Nature Biotechnology*, *42*(9), 1378–1383. https://doi.org/10.1038/s41587-023-01983-6

Bolger, A. M., Lohse, M., & Usadel, B. (2014). Trimmomatic: a flexible trimmer for Illumina sequence data. *Bioinformatics*, *30*(15), 2114–2120. https://doi.org/10.1093/bioinformatics/btu170

Buck, M., Garcia, S. L., Fernandez, L., Martin, G., Martinez-Rodriguez, G. A., Saarenheimo, J., . . . Peura, S. (2021). Comprehensive dataset of shotgun metagenomes from oxygen stratified freshwater lakes and ponds. *Scientific Data*, *8*(1), 131. https://doi.org/10.1038/s41597-021-00910-1

Buetas, E., Jordán-López, M., López-Roldán, A., D'Auria, G., Martínez-Priego, L., De Marco, G., . . . Mira, A. (2024). Full-length 16S rRNA gene sequencing by PacBio improves taxonomic resolution in human microbiome samples. *BMC Genomics*, *25*(1), 310. https://doi.org/10.1186/s12864-024-10213-5

Cassol, I., Ibañez, M., & Bustamante, J. P. (2025). Key features and guidelines for the application of microbial alpha diversity metrics. *Scientific Reports*, *15*(1), 622. https://doi.org/10.1038/s41598-024-77864-y

Chen, S., Zhou, Y., Chen, Y., & Gu, J. (2018). fastp: an ultra-fast all-in-one FASTQ preprocessor. *Bioinformatics*, *34*(17), i884–i890.

Claassen-Weitz, S., Gardner-Lubbe, S., Mwaikono, K. S., du Toit, E., Zar, H. J., & Nicol, M. P. (2020). Optimizing 16S rRNA gene profile analysis from low biomass nasopharyngeal and induced sputum specimens. *BMC Microbiology*, *20*(1), 113. https://doi.org/10.1186/s12866-020-01795-7

Cowan, D. A., Ramond, J. B., Makhalanyane, T. P., & De Maayer, P. (2015). Metagenomics of extreme environments. *Current Opinion in Microbiology*, *25*, 97–102.

de Abreu, V. A. C., Perdigão, J., & Almeida, S. (2021). Metagenomic approaches to analyze antimicrobial resistance: an overview. *Frontiers in Genetics*, *11*(575592).

Dong, X., Kleiner, M., Sharp, C. E., Thorson, E., Li, C., Liu, D., & Strous, M. (2017). Fast and simple analysis of MiSeq amplicon sequencing data with MetaAmp. *Frontiers in Microbiology*, *8*(1461).

El-Saadony, M. T., Saad, A. M., Soliman, S. M., Salem, H. M., Ahmed, A. I., Mahmood, M., . . . AbuQamar, S. F. (2022). Plant growth-promoting microorganisms as biocontrol agents of plant diseases: mechanisms, challenges and future perspectives. *Frontiers in Plant Science*, *13*. https://doi.org/10.3389/fpls.2022.923880

Feng, K., Costa, J., & Edwards, J. S. (2018). Next-generation sequencing library construction on a surface. *BMC Genomics*, *19*(1), 416. https://doi.org/10.1186/s12864-018-4797-4

Filardo, S., Di Pietro, M., & Sessa, R. (2024). Current progresses and challenges for microbiome research in human health: a perspective. *Frontiers in Cellular and Infection Microbiology*, *14*(1377012).

Franzosa, E. A., McIver, L. J., Rahnavard, G., Thompson, L. R., Schirmer, M., Weingart, G., . . . Huttenhower, C. (2018). Species-level functional profiling of metagenomes and metatranscriptomes. *Nature Methods*, *15*(11), 962–968.

Galperin, M. Y., Vera Alvarez, R., Karamycheva, S., Makarova, K. S., Wolf, Y. I., Landsman, D., & Koonin, E. V. (2024). COG database update 2024. *Nucleic Acids Research*, *53*(D1), D356–D363. https://doi.org/10.1093/nar/gkae983

Geesink, P., Ter Horst, J., & Ettema, T. J. G. (2024). More than the sum of its parts: uncovering emerging effects of microbial interactions in complex communities. *FEMS Microbiology Ecology*, *100*(4).

Gilbert, J. A., Jansson, J. K., & Knight, R. (2014). The Earth microbiome project: successes and aspirations. *BMC Biology*, *12*(69), 14–69.

Gilbert, J. A., & Neufeld, J. D. (2014). Life in a world without microbes. *PLoS Biology*, *12*(12).

Gupta, S., Mortensen, M. S., Schjørring, S., Trivedi, U., Vestergaard, G., Stokholm, J., . . . Sørensen, S. J. (2019). Amplicon sequencing provides more accurate microbiome information in healthy children compared to culturing. *Communications Biology*, *2*(1), 291. https://doi.org/10.1038/s42003-019-0540-1

Hnini, M., Rabeh, K., & Oubohssaine, M. (2024). Interactions between beneficial soil microorganisms (PGPR and AMF) and host plants for environmental restoration: a systematic review. *Plant Stress*, *11*, 100391. https://doi.org/10.1016/j.stress.2024.100391

Hodkinson, B. P., & Grice, E. A. (2015). Next-generation sequencing: a review of technologies and tools for wound microbiome research. *Advances in Wound Care*, *4*(1), 50–58.

Hoff, K. J. (2009). The effect of sequencing errors on metagenomic gene prediction. *BMC Genomics*, *10*(1), 520. https://doi.org/10.1186/1471-2164-10-520

Hou, K., Wu, Z. X., Chen, X. Y., Wang, J. Q., Zhang, D., Xiao, C., . . . Chen, Z. S. (2022). Microbiota in health and diseases. *Signal Transduction and Targeted Therapy*, *7*(1), 22–974.

Hyatt, D., Chen, G.-L., LoCascio, P. F., Land, M. L., Larimer, F. W., & Hauser, L. J. (2010). Prodigal: prokaryotic gene recognition and translation initiation site identification. *BMC Bioinformatics*, *11*(1), 119. https://doi.org/10.1186/1471-2105-11-119

Ibal, J. C., Park, Y.-J., Park, M.-K., Lee, J., Kim, M.-C., & Shin, J.-H. (2022). Review of the current state of freely accessible web tools for the analysis of 16S rRNA sequencing of the gut microbiome. *International Journal of Molecular Sciences*, *23*(18), 10865.

James, A. S., Adil, N. A., Goltz, D., Tangudu, D., Chaudhari, D. S., Shukla, R., . . . Yadav, H. (2024). Abnormalities in gut virome signatures linked with cognitive impairment in older adults. *Gut Microbes*, *16*(1), 2431648. https://doi.org/10.1080/19490976.2024.2431648

Jin, D.-M., Morton, J. T., & Bonneau, R. (2024). Meta-analysis of the human gut microbiome uncovers shared and distinct microbial signatures between diseases. *mSystems*, *9*(8), e00295.24. https://doi.org/10.1128/msystems.00295-24

Kampmann, M.-L., Børsting, C., Jepsen, A. H., Andersen, M. M., Aagreen, C. I. V., Poggiali, B., . . . Andersen, J. D. (2025). Preparing for shotgun sequencing in forensic genetics: evaluation of DNA extraction and library building methods. *Forensic Science International: Genetics*, *76*, 103234. https://doi.org/10.1016/j.fsigen.2025.103234

Kanehisa, M., & Sato, Y. (2020). KEGG Mapper for inferring cellular functions from protein sequences. *Protein Science*, *29*(1), 28–35.

Kanehisa, M., Sato, Y., Kawashima, M., Furumichi, M., & Tanabe, M. (2015). KEGG as a reference resource for gene and protein annotation. *Nucleic Acids Research*, *44*(D1), D457–D462. https://doi.org/10.1093/nar/gkv1070

Kim, D., Song, L., Breitwieser, F. P., & Salzberg, S. L. (2016). Centrifuge: rapid and sensitive classification of metagenomic sequences. *Genome Research*, *26*(12), 1721–1729.

Kim, N., Ma, J., Kim, W., Kim, J., Belenky, P., & Lee, I. (2024). Genome-resolved metagenomics: a game changer for microbiome medicine. *Experimental & Molecular Medicine*, *56*(7), 1501–1512. https://doi.org/10.1038/s12276-024-01262-7

Kumar, V., Gangani, S., Shukla, R., Prajapati, S. K., Shekhar, H., & Shukla, V. (2024). Next-generation sequencing and genomic data analysis. In A. Chaudhary, S. K. Sethi, & A. Verma (Eds.), *Unraveling New Frontiers and Advances in Bioinformatics* (pp. 49–76). Springer Nature. https://doi.org/10.1007/978-981-97-7123-3_3

Kumar, V., Sharma, J. K., Kumar, S., Gulia, J., Singh, N. P., & Santal, A. R. (2025). Metagenomic and proteomic approach for bioremediation of environmental pollutants. In M. P. Shah (Ed.), *Development in Waste Water Treatment Research and Processes*. Elsevier. https://doi.org/10.1016/B978-0-443-13615-3.00020-0

Kumar, V., Shukla, R., Gangani, S., Shekhar, H., & Shukla, V. (2024). Metagenomics and microbiome analysis. In A. Chaudhary, S. K. Sethi, & A. Verma (Eds.), *Unraveling New Frontiers and Advances in Bioinformatics* (pp. 77–92). Springer Nature. https://doi.org/10.1007/978-981-97-7123-3_4

Kuruppu, M., Ling, K. L., Ding, P., Ahmad, K., Ali, A., & Siddiqui, Y. (2024). Decoding the fruit microbiome: a climate smart strategy to manage postharvest decays. *Horticultural Plant Journal*, *10*(5), 1061–1072. https://doi.org/10.1016/j.hpj.2023.02.017

Lathakumari, R. H., Vajravelu, L. K., Gopinathan, A., Vimala, P. B., Panneerselvam, V., Ravi, S. S. S., & Thulukanam, J. (2025). The gut virome and human health: from diversity to personalized medicine. *Engineering Microbiology*, *5*(1).

Li, D., Liu, C.-M., Luo, R., Sadakane, K., & Lam, T.-W. (2015). MEGAHIT: an ultra-fast single-node solution for large and complex metagenomics assembly via succinct de Bruijn graph. *Bioinformatics*, *31*(10), 1674–1676. https://doi.org/10.1093/bioinformatics/btv033

Lindner, B. G., Gerhardt, K., Feistel, D. J., Rodriguez, R. L., Hatt, J. K., & Konstantinidis, K. T. (2024). A user's guide to the bioinformatic analysis of shotgun metagenomic sequence data for bacterial pathogen detection. *International Journal of Food Microbiology*, *410*(110488), 17.

Liu, Y. X., Qin, Y., Chen, T., Lu, M., Qian, X., Guo, X., & Bai, Y. (2021). A practical guide to amplicon and metagenomic analysis of microbiome data. *Protein Cell*, *12*(5), 315–330.

Lund, P. A., De Biase, D., Liran, O., Scheler, O., Mira, N. P., Cetecioglu, Z., . . . O'Byrne, C. (2020). Understanding how microorganisms respond to acid pH is central to their control and successful exploitation. *Frontiers in Microbiology*, *11*(556140).

Madigan, K. E., Maffei, V. J., Shukla, R., Kumar, V., Yadav, H., & Weinberg, R. B. (2025). Fecal microbiome richness and evenness are reduced and the relative abundance of firmicutes species is increased in patients with intestinal methanogen overgrowth. *Gastroenterology*, *169*(S1), S1263. https://doi.org/10.1016/S0016-5085(25)03787-4

Maglione, G., Zinno, P., Tropea, A., Mussagy, C. U., Dufossé, L., Giuffrida, D., & Mondello, A. (2024). Microbes' role in environmental pollution and remediation: a bioeconomy focus approach. *AIMS Microbiology*, *10*(3), 723–755.

Manrique, P., Dills, M., & Young, M. J. (2017). The human gut phage community and its implications for health and disease. *Viruses*, *9*(6).

Marizzoni, M., Gurry, T., Provasi, S., Greub, G., Lopizzo, N., Ribaldi, F., . . . Cattaneo, A. (2020). Comparison of bioinformatics pipelines and operating systems for the analyses of 16S rRNA gene amplicon sequences in human fecal samples. *Frontiers in Microbiology*, *11*(1262).

Martin, M. (2011). Cutadapt removes adapter sequences from high-throughput sequencing reads. *EMBnet Journal*, *17*, 10–12.

The MetaSUB International Consortium. (2016, June 3). The metagenomics and metadesign of the subways and urban biomes (MetaSUB) international Consortium inaugural meeting report. *Microbiome*, *4*(1), 24. https://doi.org/10.1186/s40168-016-0168-z

Mishra, S., Buddendorff, L., Hoover, J., Shukla, R., Kumar, V., Jain, S., & Yadav, H. (2024). microbiome role in cognitive impairment: MIAGB cohort insights. *Innovation in Aging*, *8*(S1), 811. https://doi.org/10.1093/geroni/igae098.2629

Mistry, J., Chuguransky, S., Williams, L., Qureshi, M., Salazar, G. A., Sonnhammer, E. L. L., . . . Bateman, A. (2020). Pfam: the protein families database in 2021. *Nucleic Acids Research*, *49*(D1), D412–D419. https://doi.org/10.1093/nar/gkaa913

Nair, S. U., Lin, Y., Jain, S., Chaudhari, D. S., Shukla, R., Kumar, V., . . . MiaGB Consortium. (2024). Unique gut microbiome signatures associated with physical and cognitive function. *Alzheimer's & Dementia*, *20*(S2), e092044. https://doi.org/10.1002/alz.092044

Nam, N. N., Do, H. D. K., Loan Trinh, K. T., & Lee, N. Y. (2023). Metagenomics: an effective approach for exploring microbial diversity and functions. *Foods*, *12*(11).

Ni, J., Pan, J., Wang, Y., Chen, T., Feng, X., Li, Y., . . . Li, W. (2023). An integrative protocol for one-step PCR amplicon library construction and accurate demultiplexing of pooled sequencing data. *Marine Life Science and Technology*, *5*(4), 564–572.

Nurk, S., Meleshko, D., Korobeynikov, A., & Pevzner, P. A. (2017). metaSPAdes: a new versatile metagenomic assembler. *Genome Research*, *27*(5), 824–834.

Ondov, B. D., Bergman, N. H., & Phillippy, A. M. (2011). Interactive metagenomic visualization in a web browser. *BMC Bioinformatics*, *12*(1), 385. https://doi.org/10.1186/1471-2105-12-385

Oulas, A., Pavloudi, C., Polymenakou, P., Pavlopoulos, G. A., Papanikolaou, N., Kotoulas, G., . . . Iliopoulos, I. (2015). Metagenomics: tools and insights for analyzing next-generation sequencing data derived from biodiversity studies. *Bioinformatics and Biology Insights*, *9*, 75–88.

Overbeek, R., Olson, R., Pusch, G. D., Olsen, G. J., Davis, J. J., Disz, T., . . . Stevens, R. (2014). The SEED and the rapid annotation of microbial genomes using subsystems technology (RAST). *Nucleic Acids Research*, *42*(Database issue), 29.

Pérez-Cobas, A. E., Gomez-Valero, L., & Buchrieser, C. (2020). Metagenomic approaches in microbial ecology: an update on whole-genome and marker gene sequencing analyses. *Microbial Genomics*, *6*(8), 24.

Pollock, J., Glendinning, L., Wisedchanwet, T., & Watson, M. (2018). The madness of microbiome: attempting to find consensus "best practice" for 16S microbiome studies. *Applied and Environmental Microbiology*, *84*(7), 2617–2627.

Portik, D. M., Brown, C. T., & Pierce-Ward, N. T. (2022). Evaluation of taxonomic classification and profiling methods for long-read shotgun metagenomic sequencing datasets. *BMC Bioinformatics*, *23*(1), 541. https://doi.org/10.1186/s12859-022-05103-0

Prajapati, S. K., Kumar, V., Shukla, R., Szekeres, C., Ma, Y., Jain, S., & Yadav, H. (2024). Microbiome-immune interplay in aging brains: a pilot investigation from the MiaGB cohort. *Alzheimer's & Dementia*, *20*(S2), e090734. https://doi.org/10.1002/alz.090734

Prodan, A., Tremaroli, V., Brolin, H., Zwinderman, A. H., Nieuwdorp, M., & Levin, E. (2020). Comparing bioinformatic pipelines for microbial 16S rRNA amplicon sequencing. *PLoS One*, *15*(1).

Quince, C., Walker, A. W., Simpson, J. T., Loman, N. J., & Segata, N. (2017). Shotgun metagenomics, from sampling to analysis. *Nature Biotechnology*, *35*(9), 833–844.

Rusch, D. B., Halpern, A. L., Sutton, G., Heidelberg, K. B., Williamson, S., Yooseph, S., . . . Venter, J. C. (2007). The Sorcerer II global ocean sampling expedition: Northwest Atlantic through Eastern tropical Pacific. *PLoS Biology*, *5*(3), 0050077.

Ruscheweyh, H.-J., Milanese, A., Paoli, L., Karcher, N., Clayssen, Q., Keller, M. I., . . . Sunagawa, S. (2022). Cultivation-independent genomes greatly expand taxonomic-profiling capabilities of mOTUs across various environments. *Microbiome*, *10*(1), 212. https://doi.org/10.1186/s40168-022-01410-z

Sanabria, A. M., Janice, J., Hjerde, E., Simonsen, G. S., & Hanssen, A.-M. (2021). Shotgun-metagenomics based prediction of antibiotic resistance and virulence determinants in Staphylococcus aureus from periprosthetic tissue on blood culture bottles. *Scientific Reports*, *11*(1), 20848. https://doi.org/10.1038/s41598-021-00383-7

Sanguineti, D., Zampieri, G., Treu, L., & Campanaro, S. (2024). Metapresence: a tool for accurate species detection in metagenomics based on the genome-wide distribution of mapping reads. *mSystems*, *9*(8), e00213.24. https://doi.org/10.1128/msystems.00213-24

Sarfraz, H., Czerniecki, B. J., Stankiewicz, K., Jaglal, M. V., Jain, S., Yadav, H., . . . Bari, S. (2024). Evaluating gut microbiome as a biomarker of pathological complete response in patients with early locally advanced triple negative breast cancer (LA TNBC): a pilot study. *Journal of Clinical Oncology*, *42*(S16), 593. https://doi.org/10.1200/JCO.2024.42.16_suppl.593

Sharpton, T. J. (2014). An introduction to the analysis of shotgun metagenomic data. *Frontiers in Plant Science*, *5*(209).

Stewart, E. J. (2012). Growing unculturable bacteria. *Journal of Bacteriology*, *194*(16), 4151–4160.

Sunagawa, S., Coelho, L. P., Chaffron, S., Kultima, J. R., Labadie, K., Salazar, G., . . . Bork, P. (2015). Ocean plankton: structure and function of the global ocean microbiome. *Science*, *348*(6237), 1261359.

Suvarnalatha Devi, K. R. P., Gnaneswari, K., & Rajaswi Devi, M. (2024). Microbial influence on climate change: drivers, mediators, and mitigators of global environmental shifts. *Indian Journal of Ecology*, *51*(6), 1464–1470. https://doi.org/10.55362/IJE/2024/4425

Thakur, S. S., González, S. S., Suravajhala, P., Jain, S. K., Yadav, S., Narayan, K. S., . . . Ratnasari, A. (2025). Metagenomic insights of microbial functions under conventional and conservation agriculture. *World Journal of Microbiology and Biotechnology*, *41*(3), 25–4312.

Turnbaugh, P. J., Ley, R. E., Hamady, M., Fraser-Liggett, C. M., Knight, R., & Gordon, J. I. (2007). The human microbiome project. *Nature*, *449*(7164), 804–810. https://doi.org/10.1038/nature06244

Usyk, M., Peters, B. A., Karthikeyan, S., McDonald, D., Sollecito, C. C., Vazquez-Baeza, Y., . . . Burk, R. D. (2023). Comprehensive evaluation of shotgun metagenomics, amplicon sequencing, and harmonization of these platforms for epidemiological studies. *Cell Reports Methods*, *3*(1), 23.

Valentino, V., Magliulo, R., Farsi, D., Cotter, P. D., O'Sullivan, O., Ercolini, D., & De Filippis, F. (2024). Fermented foods, their microbiome and its potential in boosting human health. *Microbial Biotechnology*, *17*(2), 1751–7915.

Van der Jeugt, F., Dawyndt, P., & Mesuere, B. (2022). FragGeneScanRs: faster gene prediction for short reads. *BMC Bioinformatics*, *23*(1), 198. https://doi.org/10.1186/s12859-022-04736-5

Wan, X., Yang, Q., Wang, X., Bai, Y., & Liu, Z. (2023). Isolation and cultivation of human gut microorganisms: a review. *Microorganisms*, *11*(4).

Wood, D. E., Lu, J., & Langmead, B. (2019). Improved metagenomic analysis with Kraken 2. *Genome Biology*, *20*(1), 257. https://doi.org/10.1186/s13059-019-1891-0

Yadav, M., & Chauhan, N. S. (2021). Microbiome therapeutics: exploring the present scenario and challenges. *Gastroenterology Report*, *15*(10).

Yan, C., Owen, J. S., Seo, E. Y., Jung, D., & He, S. (2023). Microbial interaction is among the key factors for isolation of previous uncultured microbes. *Journal of Microbiology*, *61*(7), 655–662.

Yi, X., Lu, H., Liu, X., He, J., Li, B., Wang, Z., . . . Yu, X. (2024). Unravelling the enigma of the human microbiome: evolution and selection of sequencing technologies. *Microbial Biotechnology*, *17*(1), 1751–7915.

Yoshida, S., Hiraga, K., Takehana, T., Taniguchi, I., Yamaji, H., Maeda, Y., . . . Oda, K. (2016). A bacterium that degrades and assimilates poly(ethylene terephthalate). *Science*, *351*(6278), 1196–1199.

Zhang, H., Wang, X., Chen, A., Li, S., Tao, R., Chen, K., . . . Zhang, S. (2024). Comparison of the full-length sequence and sub-regions of 16S rRNA gene for skin microbiome profiling. *mSystems*, *9*(7), 00399.24.

Zhang, H., Zhou, F., Liu, X., & Huang, J. (2024). Clinical application of metagenomic next-generation sequencing in patients with different organ system infection: a retrospective observational study. *Medicine*, *103*(4), 36745.

Zhang, W., Fan, X., Shi, H., Li, J., Zhang, M., Zhao, J., & Su, X. (2023). Comprehensive assessment of 16S rRNA gene amplicon sequencing for microbiome profiling across multiple habitats. *Microbiology Spectrum*, *11*(3), e00563.23. https://doi.org/10.1128/spectrum.00563-23

Zhou, Q., Su, X., & Ning, K. (2014). Assessment of quality control approaches for metagenomic data analysis. *Scientific Reports*, *4*(1), 6957. https://doi.org/10.1038/srep06957

2 Tools and Techniques for Metagenomic Analysis

Rohit Shukla[1*], *Arvind Kumar Yadav*[2], *Vivek Kumar*[1], *Anshuman Chandra*[3], *and Nidhi Awasthi*[4]

[1]USF Center for Microbiome Research, Microbiomes Institute, University of South Florida, Tampa, FL, USA

[2]Functional Genomics & Bioinformatics Laboratory, Department of Animal Science and Technology, Chung-Ang University, Anseong, Gyeonggi-do, Republic of Korea

[3]School of Physical Sciences, Jawaharlal Nehru University, New Delhi, India

[4]Department of Chemistry, IIT Kanpur, Kanpur, India

*Corresponding author

2.1 INTRODUCTION

Metagenomics has revolutionized and transformed the microbial field from one generation to the next by enabling the identification of novel microbial communities without the need for cultivation. Researchers can gain insight into the diversity and functional potential of microbial communities from different samples, such as soil, water, saliva, and stool. Metagenomics not only provides taxonomic-level information, but it also provides functional-level information, which can reveal what these microorganisms are doing in an ecosystem. However, metagenomics also requires carefully designed workflows. The first and most critical step is sample collection, where each type of sample comes with distinct challenges related to heterogeneity, preservation, and host DNA contamination. For example, saliva samples often consist of high proportions of host DNA, which makes it difficult to identify microbial communities. The depletion of host DNA in saliva samples by combining osmotic lysis with propidium monoazide (lyPMA) treatment has improved the enrichment of microbial sequences compared with conventional kits (Marotz et al., 2018). After sample collection, scientists should focus on the nucleic acid extraction process, which must balance purity, yield, and taxonomic breadth. For samples with low biomass or complex compositionality, several integrated methods in recent kits, such as bead beating, viability staining, and liquid–liquid extraction, can outperform conventional approaches and provide greater reliability and representation of microbial diversity (Shen et al., 2022). In sequencing, Illumina short-read platforms are still the dominant choice because of their exceptional accuracy and depth (Kumar, Gangani, et al., 2024; Kumar, Shukla, et al., 2024). However, their limited read length makes it complicated to achieve fine-scale genome assembly resolution in complex communities. In contrast, Oxford Nanopore and PacBio are long-read sequencing technologies that generate substantially longer sequences and enable more contiguous assemblies, facilitating the recovery of complete metagenome-assembled genomes (MAGs), albeit with higher error rates (Eisenhofer et al., n.d.; Patin & Goodwin, 2022). Increasingly, hybrid sequencing approaches that combine Illumina's base-level accuracy with the long-read continuity of Nanopore or PacBio can generate more complete genomes from metagenomic data across ecosystems, including the human microbiome (Eisenhofer et al., n.d.;

DOI: 10.1201/9781003670155-2

C. Kim et al., 2024; N. Kim et al., 2024). In metagenomics, reproducibility is a critical outcome. Reusable pipelines with transparent code have been developed using workflow managers such as Nextflow and Snakemake. Community platforms, such as Galaxy and MG-RAST, are also available for analysis. The selection of suitable tools is equally important and depends on research goals. For example, amplicon sequencing (e.g., 16S rRNA) is well suited for broad taxonomic surveys, whereas shotgun metagenomics is used to obtain deeper insights into taxonomy and functional potential. The depth of shotgun metagenomics can provide a better understanding of pathogen surveillance, antimicrobial resistance monitoring, and genome assembly construction (Davis et al., 2023). Validation and benchmarking of analyses further ensure result reliability. The use of mock communities and synthetic datasets allows systematic assessment of pipeline specificity, sensitivity, and error rates. The continuous emergence of new approaches in this field has redefined the potential of metagenomics. Long-read sequencing is continuously increasing taxonomic resolution and enabling the integration of metagenomic data (Chen et al., 2022; Han et al., 2024) with several other data types, such as metaproteomics, metabolomics, and metatranscriptomics. The use of machine learning also opens new avenues for linking genetic potential (Shukla et al., 2024; Yadav et al., 2021) to realized functions and phenotypes in dynamic microbial systems. Collectively, in this chapter, we provide details on sample collection, sequencing, preprocessing, computational workflows, and tool selection, with an emphasis on reproducibility, benchmarking, and emerging frontiers in metagenomic research.

2.2 SAMPLE COLLECTION AND PREPARATION

Sample collection, preservation, and processing are key steps in metagenomics research. Sample collection defines the microbial communities captured for downstream analysis. Improper collection or inadequate sample preparation can reduce DNA yield, introduce contaminants, and distort microbial diversity, which ultimately compromises result accuracy.

2.3 SAMPLE TYPES (SOIL, STOOL, SALIVA, WATER, ETC.)

The selection of sample type depends on the research goal, and each type has its own advantages and disadvantages.

- **Soil samples** are among the most diverse sample types compared with others, containing microorganisms embedded within complex matrices of organic and inorganic material. Due to high spatial heterogeneity, multiple replicates from different locations and depths are recommended to minimize sampling bias (Sessitsch & Mitter, 2015). Mechanical disruption, such as bead beating, is often required to release DNA from resistant spores and biofilms.
- **Stool samples** are mainly used to characterize the gut microbiome. They provide a snapshot of distal intestinal microbial populations, although they may underrepresent mucosal populations. Stool collection kits generally include preservative solutions that prevent the need for immediate freezing and make samples more suitable for collection and transport (Gemmell et al., 2024).
- **Saliva samples** are a key noninvasive sample type for microbiome analysis. However, they often contain high levels of host contamination, which can hamper true microbiome signals. To address this issue, recent approaches have employed selective lysis or propidium monoazide treatment to enrich microbial DNA prior to sequencing (Bang et al., 2023; Marotz et al., 2018).

- **Water samples** are a major resource for studying environmental microbiology but typically have low microbial biomass. This necessitates large-volume collection followed by filtration to concentrate microbial cells. The choice of filtration pore size (commonly 0.22 μm) is critical for balancing microbial recovery with processing efficiency (Chang et al., 2020; Moghaddam et al., 2024).

All sample types require careful handling and appropriate preservation, typically storage at −80 °C within a few hours. For RNA preservation, additional stabilizing reagents, such as DNA/RNA Shield, are required and play an important role in downstream processing.

2.4 DNA/RNA EXTRACTION PROTOCOLS

Microbial taxa differ in cell wall structure, which can make them resistant to lysis; therefore, the choice of extraction protocol has a strong influence on metagenomics outcomes (Shaffer et al., n.d.). A well-optimized method should maximize nucleic acid yield, minimize impurities, and avoid introducing taxonomic bias.

- **Mechanical lysis methods** (e.g., bead beating, sonication) are highly effective at disrupting resistant bacterial and fungal cell walls. However, the application of strong physical forces often shears DNA and produces shorter fragments, which are less suitable for long-read sequencing.
- **Chemical and enzymatic lysis methods** (e.g., SDS, lysozyme, proteinase K) generally preserve DNA integrity but may fail to lyse particularly resilient microorganisms, leading to underrepresentation of certain taxa.

Several commercial DNA extraction kits combine chemical, mechanical, and enzymatic approaches to balance yield, fragment length, and taxonomic representation. Comparative studies demonstrate

TABLE 2.1
Sample Types and DNA/RNA Extraction Considerations in Metagenomics

Sample Type	Major Challenges	Extraction Considerations/Best Practices	Representative DOIs
Soil	Highly heterogeneous; DNA bound to organic/inorganic matrices; presence of humic acids inhibiting PCR	Mechanical disruption (e.g., bead beating) required to lyse spores and biofilms; replicate sampling across depths reduces bias	Child et al. (2024); Wydro (2022)
Stool	Underrepresentation of mucosal populations; inhibitors like bile salts	Stabilizing buffers preserve DNA during transport; widely used in gut microbiome studies	Gemmell et al. (2024)
Saliva	High host DNA background masking microbial signals	Selective lysis and propidium monoazide (PMA) treatment enrich microbial DNA; osmotic lysis + PMA (lyPMA) enhances signal	Bang et al. (2023); Marotz et al. (2018)
Water	Low microbial biomass; dilution effect in large volumes	Large-volume collection followed by filtration (typically 0.22 μm pore size) concentrates cells; careful pore-size choice balances capture versus processing time	Chang et al. (2020)
RNA (all sources)	Intrinsic instability; rapid degradation by RNases	Rigorous RNase-free handling; stabilizing reagents essential; rRNA depletion improves metatranscriptomic resolution	Ojala et al. (2023); Pereira-Marques et al. (n.d.)

that extraction methods significantly shape observed community profiles, underscoring the importance of methodological standardization within studies (Willner et al., 2012). The inherent instability of RNA molecules makes RNA extraction more challenging. Therefore, during sample collection, adding preservatives that inhibit RNase activity and stabilize sample integrity is essential. For metatranscriptomic applications, workflows typically include simultaneous extraction of total RNA and ribosomal RNA (rRNA) depletion, thereby enriching messenger RNA (mRNA) and enhancing the resolution of functional studies (Barber et al., 2024; Kumar et al., 2025; Shakya et al., 2019). A detailed summary of challenges and sample types is shown in Table 2.1.

2.5 SEQUENCING TECHNOLOGIES

Advances in sequencing techniques are the primary component in metagenomics analysis. The advancement of sequencing technologies has now transformed into portable, real-time, long-read platforms compared with earlier clone libraries and Sanger sequencing, which transformed the way microbial communities were investigated. These technologies have defined a new path for various analyses, such as genome reconstruction, structural variation analysis, and evolutionary studies (Vasudeva et al., 2023). In this section, we describe the two most used approaches—short-read and long-read sequencing—and compare their advantages, limitations, and applications in metagenomic research (C. Kim et al., 2024; Qayyum et al., 2025).

2.6 SHORT-READ PLATFORMS (ILLUMINA AND ITS APPLICATIONS)

Illumina sequencing is a widely used platform, and it is based on the sequencing-by-synthesis (SBS) principle. Fluorescently labeled nucleotides are incorporated sequentially into immobilized DNA fragments on a flow cell, with each incorporation detected via high-resolution imaging. After imaging, the next cycle is enabled by cleaving the dye. This massively parallel process generates billions of clusters altogether, yielding exceptional throughput at a relatively low cost. Typical read lengths range from 50 to 300 base pairs, depending on the platform (e.g., MiSeq, HiSeq, NovaSeq) (Ayling et al., 2019).

2.7 LONG-READ PLATFORMS (OXFORD NANOPORE, PACBIO, AND ITS APPLICATIONS)

2.7.1 Pacific Biosciences (PacBio)

PacBio sequencing is based on single-molecule real-time (SMRT) technology. DNA polymerase is immobilized at the bottom of nanostructures known as zero-mode waveguides, where fluorescently labeled nucleotides are incorporated and detected continuously in real time. The latest innovation, HiFi reads, generates long reads (10–25 kb) with extremely high accuracy (>99.9%) by repeatedly sequencing the same DNA molecule (Han et al., 2024).

Applications: PacBio HiFi sequencing is now a premier method for genome-resolved metagenomics. A landmark study demonstrated its ability to recover near-complete microbial genomes from complex soil communities with unprecedented accuracy (Nayfach et al., 2021).

2.7.2 Oxford Nanopore Technologies (ONT)

ONT sequencing operates by threading single DNA molecules through protein nanopores embedded in membranes. As nucleotides pass through the pore, base-specific disruptions in ionic current are measured to infer sequence identity. Unlike other platforms, ONT generates reads ranging from hundreds of kilobases to several megabases, far surpassing Illumina or PacBio in length.

ONT sequencing works by threading single DNA molecules using embedded protein nanopores in the membrane.

- **Portability**: ONT's compact devices, such as the MinION, enable sequencing in remote and resource-limited environments. They have been deployed in outbreak settings, including Ebola, for rapid pathogen detection (Hoenen et al., 2016).
- **Rapid diagnostics**: Real-time data acquisition allows near-instantaneous analysis, making ONT particularly useful for clinical microbiology and urgent diagnostic applications.
- **Structural resolution**: Nanopore sequencing is particularly powerful for identifying plasmids, phages, and other mobile elements, offering unique advantages in structural genomics (González-Escalona et al., 2019).

2.7.3 Limitations of Long-Read Sequencing

Although highly valuable, PacBio and ONT have historically exhibited higher error rates compared with Illumina. ONT's per-base error rates (~2–5%) are steadily improving with better basecalling algorithms, while PacBio HiFi reads now deliver near-perfect accuracy, although at a higher cost (Wenger et al., 2019). Additional limitations include:

- **DNA input requirements**: Both platforms require substantial amounts of high-molecular-weight DNA, posing challenges for low-biomass samples.
- **Cost**: The cost per gigabase remains higher than Illumina, limiting feasibility for large cohort studies.
- **Computational demands**: Long-read data requires significant computational resources for error correction and assembly of complex genomes.

2.8 COMPARATIVE ADVANTAGES, LIMITATIONS, AND USE-CASES

Choosing between different sequencing platforms should be based on technological features and research objectives. One or more types of sequencing can be performed depending on the research objective. A detailed comparison of sequencing types is shown in Table 2.2.

Use-case examples

- **Short-read dominance**: Illumina is still the most widely used and cost-effective option for conducting in-depth and reasonably priced microbiome surveys of thousands of individuals worldwide (Thompson et al., 2017).
- **Long-read advantage**: PacBio HiFi reads offer higher resolution for the reconstruction of nearly complete genomes from heterogeneous soil samples (Kovaka et al., 2025).
- **Field diagnostics**: As evidenced during the Ebola and COVID-19 pandemics, ONT's portability and speed demonstrated strong efficiency for real-time outbreak investigations (Bull et al., 2020).
- **Hybrid approaches**: Hybrid approaches combine two or more sequencing methods to take advantage of depth, precision, and contiguity. Researchers are increasingly combining Illumina and long-read sequencing to enable highly precise assemblies and functional annotation (Chen et al., 2022).

2.9 LIBRARY CONSTRUCTION AND PREPROCESSING

In metagenomic research, library preparation and data preprocessing are critical steps that directly shape the depth, accuracy, and interpretability of downstream analyses. The sequencing library determines how effectively the diverse range of microbial genomes within a sample is represented, while preprocessing ensures that only high-quality and biologically meaningful reads are retained.

TABLE 2.2
Comparative Overview of Sequencing Platforms for Metagenomics

Feature	Illumina (Short-Read)	PacBio (Long-Read, HiFi)	Oxford Nanopore (Long-Read)
Read length	50–300 bp	10–25 kb (HiFi)	Hundreds of kb to >1 Mb
Accuracy	>99.9%	>99.9% (HiFi consensus)	95–98% (improving with basecalling)
Throughput	Very high (billions of reads/runs)	Moderate	Variable, depends on device (e.g., MinION versus PromethION)
Cost per base	Lowest	Higher	Moderate, but decreasing
DNA input	Low	High-molecular-weight DNA required	High-molecular-weight DNA preferred
Assembly power	Limited (fragmented assemblies, repeats unresolved)	Excellent (high contiguity, near-complete MAG recovery)	Excellent (captures plasmids, phages, structural variants)
Portability	No	No	Yes (field-deployable, e.g., MinION)
Use-case examples	Diversity surveys, resistome studies, large cohorts	Genome-resolved metagenomics, complex community assemblies	Rapid outbreak diagnostics, structural genomics, mobile element detection

Metagenomic libraries must capture a broad spectrum of microbial genomes, often within a background dominated by host DNA and environmental contaminants. Preprocessing steps, such as adapter trimming, contaminant removal, and host DNA depletion, can play a central role in reducing noise and preventing misleading results. By carefully implementing these procedures, researchers can ensure that data advancing to taxonomic classification and functional annotation reflect the true biological signal rather than technical artifacts (Eisenhofer et al., 2019).

2.10 BIOINFORMATICS TOOLS AND WORKFLOWS

Sequencing technologies generate a large amount of raw data, but these data do not have meaning without preprocessing through a standard computational workflow. Bioinformatics tools transform these data into interpretable information for identifying microbial communities present in a sample, the genes they encode, and their biological and ecological functions (Raja et al., 2025; Kumar et al., 2023). Metagenomics analysis generally consists of four key steps: taxonomic classification, functional annotation, assembly and binning, followed by data visualization. Each stage requires careful tool selection, and it directly influences accuracy, reproducibility, and the overall relevance of results to the research question (Breitwieser et al., 2019).

2.11 TAXONOMIC CLASSIFICATION (KRAKEN2, METAPHLAN, ETC.)

The first step in metagenomics analysis is taxonomic classification, which answers the basic question of who is present in a microbial community. Two main types of methods are used for microbial identification.

- **Alignment-based classifiers**: These techniques require substantial computational power and map sequencing reads to reference genomes. Based on precise matching, they provide taxonomic classification.
- **Marker gene-based approaches**: These tools provide higher specificity and a lower risk of false positives by inferring community composition using clade-specific genes. They use the clade-specific genes for the identification of community composition and reduce the risk of false positives due to more specificity.

Kraken2 is an alignment-based tool that is widely used for shotgun metagenomics. By using exact k-mer matching against large reference databases, Kraken2 provides exceptional speed and scalability and can classify millions of reads within minutes. However, its performance is highly dependent on the completeness of the reference database (Wood et al., 2019, p. 2).

In contrast to Kraken2, MetaPhlAn uses a marker gene-based strategy that relies on curated clade-specific markers derived from reference genomes. This approach enhances specificity and reduces misclassification from shared genomic regions that can occur in alignment-based tools, making it particularly valuable for studies of human-associated microbiomes (Blanco-Míguez et al., 2023, p. 4).

Several benchmark studies suggest that no single classifier can consistently outperform others; however, combining outputs from different tools can make results more reliable.

2.12 FUNCTIONAL ANNOTATION (HUMANN, PROKKA, EGGNOG, ETC.)

While taxonomic profiling identifies the composition of microbial communities, understanding their functional potential is essential for interpreting ecological and biomedical relevance. Functional annotation links genomic sequences to antimicrobial resistance genes, metabolic pathways, and other traits that define microbial activity and interactions.

- **HUMAnN (The HMP Unified Metabolic Analysis Network)** is a widely accepted pipeline for the functional analysis of metagenomic reads. It uses alignment-based tools that align reads against several databases, such as UniRef and MetaCyc. HUMAnN then reconstructs microbial pathways and gene families, enabling community-level metabolic characterization (Beghini et al., n.d., p. 3).
- **PROKKA** was originally designed for bacterial isolate genomes, but it is now a frequently used tool to annotate contigs generated from metagenomic assemblies. It reliably identifies and predicts coding sequences, rRNAs, and tRNAs and also assigns putative functions through curated databases, thereby supporting the annotation of draft microbial genomes (Seemann, 2014).
- **eggNOG mapper** improves functional interpretation by adding one of the largest orthologous group databases. It is particularly useful for poorly characterized organisms because it assigns gene ontologies, functional categories, and KEGG pathways where homology-based annotation alone may be insufficient (Huerta-Cepas et al., 2019).

Other databases, such as CARD and VFDB, also contribute to the identification of virulence factors and mobile genetic elements and expand resolution for detecting antimicrobial resistance genes and virulence factors (Alcock et al., 2020). Collectively, these tools transform raw sequencing data into biologically meaningful insights, addressing the central question: *"What can these microbes do?"*

2.13 ASSEMBLY AND BINNING (MEGAHIT, METABAT, ETC.)

The genomic representation of a microorganism can be achieved from fragmented short reads using assembly and binning processes, which provide genome-resolved insights into microbial communities (Saroj et al., 2023).

Assembly involves reconstructing contiguous sequences (contigs) from overlapping reads.

- **MEGAHIT** is one of the most widely used assemblers for reconstructing genomes from large-scale metagenomic datasets and bacterial isolates. It uses succinct de Bruijn graph algorithms and can efficiently handle billions of reads while maintaining relatively modest memory requirements (Li et al., 2015).
- **metaSPAdes** is another assembler designed for complex communities and uneven sequencing depth. It uses algorithms that are particularly well suited for reconstructing genomes from diverse and heterogeneous datasets (Nurk et al., 2017).

Binning uses assembled contigs and transforms them into draft genomes, which are called metagenome-assembled genomes (MAGs).

- **MetaBAT** applies probabilistic distance measures based on tetranucleotide frequencies and coverage profiles to achieve highly accurate binning (Kang et al., 2015).
- Alternative tools, such as **MaxBin** (Wu et al., 2016) and **CONCOCT** (Alneberg et al., 2014), use distinct algorithmic strategies and are frequently combined in ensemble workflows to maximize genome recovery.

  The MAGs recovery has revolutionized the microbes identification by enabling the discovery of thousands of novel microbial lineages that were not identified previously using culture based methods (Thompson et al., 2017; Kumar et al., 2019). These methods have greatly increased our understanding of the microbial tree of life and continue to provide crucial insights into the diversity, evolution, and ecological roles of uncultured microorganisms.

2.14 VISUALIZATION PLATFORMS (KRONA, PAVIAN, ETC.)

Data visualization fills the gap between the complexity of metagenomic outputs and their biological interpretation. Effective visualization tools transform raw computational reads into intuitive representations and enable researchers to identify patterns, detect abnormalities, and communicate findings to both scientific and non-scientific audiences.

- **Krona** generates interactive plots that allow users to explore relative abundances of species at different taxonomic levels. Its zooming feature makes it particularly effective for examining large and complex microbial communities (Ondov et al., 2011).
- **Pavian** is used to compare results from different metagenomic classifiers, such as Centrifuge and Kraken2. It provides detailed exploration of classification results using interactive taxonomic trees, heatmaps, and bar plots (Toro-Domínguez et al., 2019).

These visualization tools are not merely supplementary; they are essential for identifying outliers, detecting contamination, and ensuring that complex microbial community structures are accurately and reproducibly characterized (see Table 2.3). A comprehensive workflow is shown in Figure 2.1.

2.15 CLOUD-BASED PLATFORMS (MG-RAST, GALAXY, ETC.)

Cloud-based platforms offer effective solutions for laboratories that do not have access to local high-performance computing (HPC) infrastructure. Several user-friendly platforms, such as MG-RAST and Galaxy, provide preconfigured metagenomic workflows that lower technical barriers and broaden accessibility.

- **MG-RAST (Metagenomics Rapid Annotation using Subsystem Technology)** is a widely used web server for metagenomics analysis, including data submission, annotation, and analysis. It provides automated pipelines for quality control, taxonomic classification, and functional analysis, making it suitable for non-specialists. Since its launch, it has processed billions of reads and remains widely used for community-level comparative studies (Keegan et al., 2016).
- **Galaxy** offers a graphical user interface that enables researchers to run bioinformatics tools without requiring command-line expertise. It is highly flexible for both research and education, supports metagenomic workflows, and accommodates other omics data (Afgan et al., 2018). Galaxy is a widely used web-based server which enables the researchers to analyze their data without the need of command-line expertise. It is a very useful tool for both research and education, and also supports metagenomic workflows while accommodating other omics data too (Afgan et al., 2018).

TABLE 2.3
Key Bioinformatics Tools in Metagenomic Workflows

Workflow Stage	Representative Tools	Key Functions	Notes/Considerations
Taxonomic classification	Kraken2, MetaPhlAn, Centrifuge, mOTUs	Assign reads to microbial taxa using k-mers or marker genes	Kraken2 is very fast but database dependent; MetaPhlAn is highly specific (human microbiome studies)
Functional annotation	HUMAnN, PROKKA, eggNOG mapper, CARD (RGI)	Links genes to metabolic pathways, resistance determinants, and ecological functions	HUMAnN excels in pathway-level analysis; PROKKA and eggNOG useful for novel assemblies
Assembly	MEGAHIT, metaSPAdes	Construct contigs from overlapping reads	MEGAHIT is scalable for large datasets; metaSPAdes handles uneven depth well
Binning	MetaBAT, MaxBin, CONCOCT	Group contigs into metagenome-assembled genomes (MAGs)	Ensemble approaches often improve recovery of high-quality genomes
Visualization	Krona, Pavian, Anvi'o	Generate interactive taxonomic charts, heatmaps, and genome-binning interfaces	Visualization critical for detecting contamination, outliers, and communication

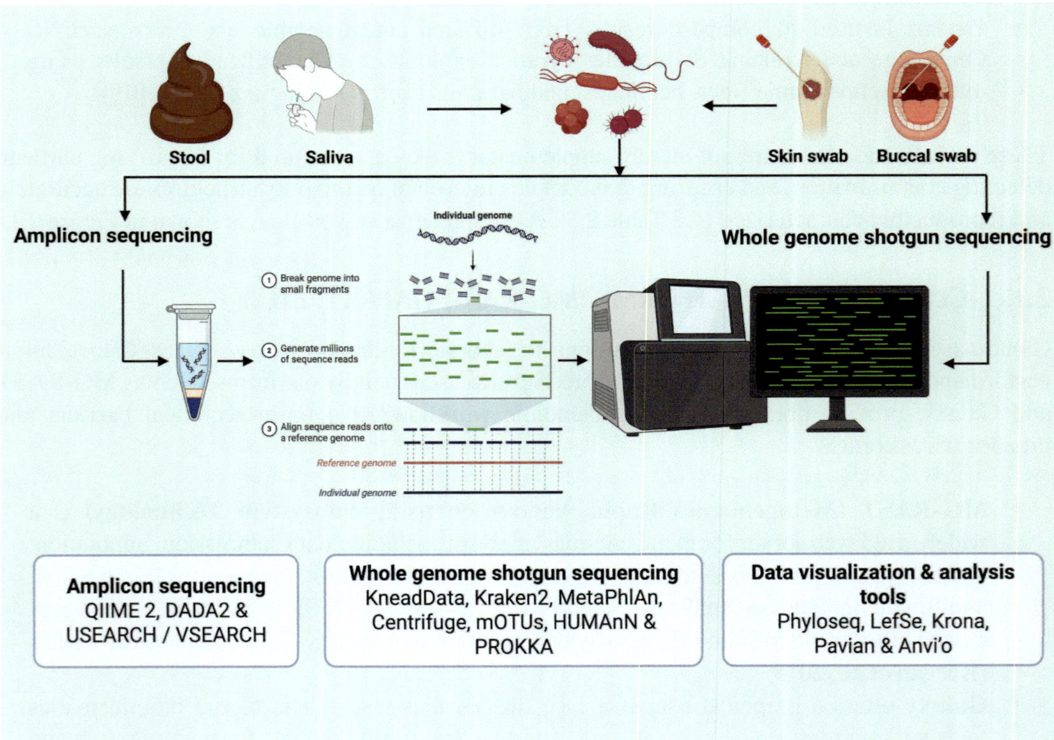

FIGURE 2.1 A comprehensive workflow for metagenomics analysis.

Cloud solutions offer several advantages:

- **Accessibility**: Researchers without HPC access can still perform large-scale analyses.
- **Reproducibility**: Standardized pipelines ensure consistency across users.
- **Collaboration:** Cloud-based data sharing fosters community-driven projects and meta-analyses.

Although it has several challenges too. Analysis of large datasets on cloud platforms is not allowed and if allowed then it can charge a substantial amount of computational costs and data security is also a critical concern, particularly when handling clinical samples. Additionally, it totally depends on the centralized databases which can introduce the vulnerabilities, as pipeline updates or service outages may compromise reproducibility and continuity of analyses.

2.16 BENCHMARKING AND BEST PRACTICES

The rapid proliferation of metagenomic tools has created both opportunities and challenges. With dozens of classifiers, assemblers, and annotation pipelines available, outcomes can differ substantially depending on software choice, parameter settings, and reference databases. To ensure reliability and reproducibility, benchmarking and adherence to best practices have become critical components of metagenomic research. This process involves systematic tool evaluation, pipeline validation, and compliance with data-reporting standards that support transparency and comparability across studies.

2.17 BENCHMARKING TOOLS AND DATASETS

Benchmarking is a key technique that assesses the performance of bioinformatics tools against controlled or semi-controlled datasets in a structured manner. This assessment highlights each tool's strengths, inherent biases, and weaknesses, enabling researchers to select approaches that best suit their scientific objectives.

- **Synthetic and mock communities:** Defined microbial mixtures serve as gold-standard criteria for benchmarking. The Critical Assessment of Metagenome Interpretation (CAMI) initiative provides standardized datasets that are widely adopted for comparing assemblers, classifiers, and annotation pipelines (Sinha et al., 2017).
- **Simulated datasets:** In this approach, reads are generated *in silico* from reference genomes, providing controlled datasets. These datasets allow systematic evaluation of tool performance under varying sequencing depths, error rates, and community complexities (Ghaffaari & Marschall, 2019).
- **Real-world datasets:** Tool assessment using realistic datasets generated from complex samples, such as soil or human gut microbiomes, provides performance evaluation under real conditions. However, the lack of absolute ground truth in such datasets limits the precision of performance assessment.

Comparative studies emphasize key trade-offs. Tools such as Kraken2 offer excellent speed and scalability, but they can introduce false positives when reference databases are incomplete or of low quality. MetaPhlAn, which uses a marker-based approach, provides higher specificity but may under-detect rare taxa (Blanco-Míguez et al., 2023). The MEGAHIT and metaSPAdes are preferred choices for short-read assembly, whereas long-read assemblers, such as Flye and Canu, are increasingly adopted in hybrid sequencing projects, offering improvements in genome contiguity and completeness (Li et al., 2015).

2.18 PIPELINE VALIDATION STRATEGIES

To find whether the metagenomic analysis is predicting true signals instead of artifacts arising from software design, database variability, or parameter selection, pipeline validation is essential. Laborious validation practices can enhance both reproducibility and reliability of findings.

- **Cross-validation with multiple tools:** Running the sequencing data using more than one classifier or assembler can generate true signals instead of tool-specific artifacts.
- **Use of internal controls:** Incorporating spike-in standards or mock communities within sequencing runs also enables the assessment of specificity, sensitivity, and contamination, thereby strengthening quality assurance (Sinha et al., 2017).
- **Version tracking:** Recording the log of software versions, database releases, and their parameter settings also ensures that computational pipelines can be accurately reproduced and updated as methodologies evolve.
- **Independent replication:** Repeating analyses across different datasets, laboratories, or sequencing runs provides an additional layer of validation, enhancing confidence in conclusions.

Community-driven initiatives, such as the PrecisionFDA challenges and the CAMI, have been instrumental in shaping validation practices (Sinha et al., 2017). These collaborative efforts provide standardized evaluation frameworks, improve transparency, and establish widely accepted benchmarks for the metagenomics community (see Table 2.4).

TABLE 2.4
Benchmarking and Best Practices in Metagenomic Analysis

Category	Key Approaches/Standards	Purpose/Benefits	Representative DOIs
Benchmarking datasets	*Synthetic/mock communities* • *Simulated datasets* (in silico reads) • *Real-world datasets* (soil, gut microbiomes)	• Establish gold standards for evaluation • Test performance under controlled complexity • Assess tool robustness in realistic scenarios	Hardwick et al. (2018)
Pipeline validation	• Cross-validation with multiple tools • Internal controls (spike-ins, blanks) • Version tracking of software/databases • Independent replication across datasets/labs	• Distinguish true biological signals from artifacts • Improve sensitivity/specificity assessment • Ensure reproducibility across platforms and time	Sczyrba et al. (2017)
Reporting standards	• MIxS (Minimum Information about any (x) Sequence) guidelines • FAIR data principles	• Guarantee essential metadata inclusion (sample type, platform) • Promote findable, accessible, interoperable, and reusable datasets	Wilkinson et al. (2016)
Data repositories	• NCBI SRA • EBI-ENA • MG-RAST	• Enable long-term archiving • Facilitate re-analysis and meta-studies • Support transparency and community sharing	Kodama et al. (2012)
Workflow sharing	• GitHub/GitLab repositories • Containerization (Docker, Singularity) • Workflow managers (Nextflow, Snakemake, CWL)	• Ensure reproducibility and portability • Provide modular, transparent pipelines • Enhance collaboration across research groups	Di Tommaso et al. (2017)

2.19 CONCLUSION

Metagenomics is now an advanced and progressive field that has evolved from a primarily descriptive discipline to a data-driven science with several applications in biotechnology, environmental research, agriculture, and medicine. In this chapter, we have outlined the essential components of metagenomics analysis, such as sample collection, nucleic acid extraction, sequencing strategies, library preparation, preprocessing, and bioinformatics workflows. The metagenomics field now stands at the threshold of a new era marked by innovation. Multi-omics integration of different omics data using advanced tools has enabled the capture of more holistic views of microbial ecosystems (Shukla et al., 2022; Yadav et al., 2022). The development of artificial intelligence and machine learning can transform raw data into knowledgeful data, which can help in predictive modeling and diagnostic applications. These advancements are transforming the metagenomics field beyond cataloging microbial diversity toward generating actionable insights, which can help in ecosystem management, promote agricultural sustainability, and guide medical interventions. With the advancement in sequencing technologies, computational infrastructure is also becoming more advanced and more accessible to all. By combining robust technical foundations with emerging innovations, metagenomics is set to drive the next generation of discoveries, which can link microbial communities to health, disease, and environmental change in several ways that will shape the life sciences of the 21st century.

REFERENCES

Afgan, E., Baker, D., Batut, B., van den Beek, M., Bouvier, D., Čech, M., Chilton, J., Clements, D., Coraor, N., Grüning, B. A., Guerler, A., Hillman-Jackson, J., Hiltemann, S., Jalili, V., Rasche, H., Soranzo, N., Goecks, J., Taylor, J., Nekrutenko, A., & Blankenberg, D. (2018). The galaxy platform for accessible, reproducible and collaborative biomedical analyses: 2018 update. *Nucleic Acids Research*, *46*(W1), W537–W544. https://doi.org/10.1093/nar/gky379

Alcock, B. P., Raphenya, A. R., Lau, T. T. Y., Tsang, K. K., Bouchard, M., Edalatmand, A., Huynh, W., Nguyen, A.-L. V., Cheng, A. A., Liu, S., Min, S. Y., Miroshnichenko, A., Tran, H.-K., Werfalli, R. E., Nasir, J. A., Oloni, M., Speicher, D. J., Florescu, A., Singh, B., . . . McArthur, A. G. (2020). CARD 2020: Antibiotic resistome surveillance with the comprehensive antibiotic resistance database. *Nucleic Acids Research*, *48*(D1), D517–D525. https://doi.org/10.1093/nar/gkz935

Alneberg, J., Bjarnason, B. S., de Bruijn, I., Schirmer, M., Quick, J., Ijaz, U. Z., Lahti, L., Loman, N. J., Andersson, A. F., & Quince, C. (2014). Binning metagenomic contigs by coverage and composition. *Nature Methods*, *11*(11), 1144–1146. https://doi.org/10.1038/nmeth.3103

Ayling, M., Clark, M. D., & Leggett, R. M. (2019). New approaches for metagenome assembly with short reads. *Briefings in Bioinformatics*, *21*(2), 584–594. https://doi.org/10.1093/bib/bbz020

Bang, E., Oh, S., Ju, U., Chang, H. E., Hong, J.-S., Baek, H.-J., Kim, K.-S., Lee, H.-J., & Park, K U. (2023). Factors influencing oral microbiome analysis: From saliva sampling methods to next-generation sequencing platforms. *Scientific Reports*, *13*(1), 10086. https://doi.org/10.1038/s41598-023-37246-2

Barber, D. G., Davies, C. A., Hartley, I. P., & Tennant, R. K. (2024). Evaluation of commercial RNA extraction kits for long-read metatranscriptomics in soil. *Microbial Genomics*, *10*(9), 1298. https://doi.org/10.1099/mgen.0.001298

Beghini, F., McIver, L. J., Blanco-Míguez, A., Dubois, L., Asnicar, F., Maharjan, S., Mailyan, A., Manghi, P., Scholz, M., Thomas, A. M., Valles-Colomer, M., Weingart, G., Zhang, Y., Zolfo, M., Huttenhower, C., Franzosa, E. A., & Segata, N. (n.d.). Integrating taxonomic, functional, and strain-level profiling of diverse microbial communities with bioBakery 3. *eLife*, *10*, e65088. https://doi.org/10.7554/eLife.65088

Blanco-Míguez, A., Beghini, F., Cumbo, F., McIver, L. J., Thompson, K. N., Zolfo, M., Manghi, P., Dubois, L., Huang, K. D., Thomas, A. M., Nickols, W. A., Piccinno, G., Piperni, E., Punčochář, M., Valles-Colomer, M., Tett, A., Giordano, F., Davies, R., Wolf, J., . . . Segata, N. (2023). Extending and improving metagenomic taxonomic profiling with uncharacterized species using MetaPhlAn 4. *Nature Biotechnology*, *41*(11), 1633–1644. https://doi.org/10.1038/s41587-023-01688-w

Breitwieser, F. P., Lu, J., & Salzberg, S. L. (2019). A review of methods and databases for metagenomic classification and assembly. *Briefings in Bioinformatics*, *20*(4), 1125–1136. https://doi.org/10.1093/bib/bbx120

Bull, R. A., Adikari, T. N., Ferguson, J. M., Hammond, J. M., Stevanovski, I., Beukers, A. G., Naing, Z., Yeang, M., Verich, A., Gamaarachchi, H., Kim, K. W., Luciani, F., Stelzer-Braid, S., Eden, J.-S., Rawlinson, W. D., van Hal, S. J., & Deveson, I. W. (2020). Analytical validity of nanopore sequencing for rapid SARS-CoV-2 genome analysis. *Nature Communications*, *11*(1), 6272. https://doi.org/10.1038/s41467-020-20075-6

Chang, C.-J., Huang, C.-P., Chen, C.-Y., & Wang, G.-S. (2020). Assessing the potential effect of extreme weather on water quality and disinfection by-product formation using laboratory simulation. *Water Research*, *170*, 115296. https://doi.org/10.1016/j.watres.2019.115296

Chen, L., Zhao, N., Cao, J., Liu, X., Xu, J., Ma, Y., Yu, Y., Zhang, X., Zhang, W., Guan, X., Yu, X., Liu, Z., Fan, Y., Wang, Y., Liang, F., Wang, D., Zhao, L., Song, M., & Wang, J. (2022). Short- and long-read metagenomics expand individualized structural variations in gut microbiomes. *Nature Communications*, *13*(1), 3175. https://doi.org/10.1038/s41467-022-30857-9

Child, H. T., Wierzbicki, L., Joslin, G. R., & Tennant, R. K. (2024). Comparative evaluation of soil DNA extraction kits for long read metagenomic sequencing. *Access Microbiology*, *6*(9), 000868.v3. https://doi.org/10.1099/acmi.0.000868.v3

Davis, B. C., Brown, C., Gupta, S., Calarco, J., Liguori, K., Milligan, E., Harwood, V. J., Pruden, A., & Keenum, I. (2023). Recommendations for the use of metagenomics for routine monitoring of antibiotic resistance in wastewater and impacted aquatic environments. *Critical Reviews in Environmental Science and Technology*, *53*(19), 1731–1756. https://doi.org/10.1080/10643389.2023.2181620

Di Tommaso, P., Chatzou, M., Floden, E. W., Barja, P. P., Palumbo, E., & Notredame, C. (2017). Nextflow enables reproducible computational workflows. *Nature Biotechnology*, *35*(4), 316–319. https://doi.org/10.1038/nbt.3820

Eisenhofer, R., Minich, J. J., Marotz, C., Cooper, A., Knight, R., & Weyrich, L. S. (2019). Contamination in low microbial biomass microbiome studies: Issues and recommendations. *Trends in Microbiology*, *27*(2), 105–117. https://doi.org/10.1016/j.tim.2018.11.003

Eisenhofer, R., Nesme, J., Santos-Bay, L., Koziol, A., Sørensen, S. J., Alberdi, A., & Aizpurua, O. (n.d.). A comparison of short-read, HiFi long-read, and hybrid strategies for genome-resolved metagenomics. *Microbiology Spectrum*, *12*(4), e03590.23. https://doi.org/10.1128/spectrum.03590-23

Gemmell, M. R., Jayawardana, T., Koentgen, S., Brooks, E., Kennedy, N., Berry, S., Lees, C., & Hold, G. L. (2024). Optimised human stool sample collection for multi-omic microbiota analysis. *Scientific Reports*, *14*(1), 16816. https://doi.org/10.1038/s41598-024-67499-4

Ghaffaari, A., & Marschall, T. (2019). Fully-sensitive seed finding in sequence graphs using a hybrid index. *Bioinformatics*, *35*(14), i81–i89. https://doi.org/10.1093/bioinformatics/btz341

González-Escalona, N., Allard, M. A., Brown, E. W., Sharma, S., & Hoffmann, M. (2019). Nanopore sequencing for fast determination of plasmids, phages, virulence markers, and antimicrobial resistance genes in Shiga toxin-producing Escherichia coli. *PLoS One*, *14*(7), e0220494. https://doi.org/10.1371/journal.pone.0220494

Han, Y., He, J., Li, M., Peng, Y., Jiang, H., Zhao, J., Li, Y., & Deng, F. (2024). Unlocking the potential of metagenomics with the PacBio high-fidelity sequencing technology. *Microorganisms*, *12*(12), 2482. https://doi.org/10.3390/microorganisms12122482

Hardwick, S. A., Chen, W. Y., Wong, T., Kanakamedala, B. S., Deveson, I. W., Ongley, S. E., Santini, N. S., Marcellin, E., Smith, M. A., Nielsen, L. K., Lovelock, C. E., Neilan, B. A., & Mercer, T. R. (2018). Synthetic microbe communities provide internal reference standards for metagenome sequencing and analysis. *Nature Communications*, *9*(1), 3096. https://doi.org/10.1038/s41467-018-05555-0

Hoenen, T., Groseth, A., Rosenke, K., Fischer, R. J., Hoenen, A., Judson, S. D., Martellaro, C., Falzarano, D., Marzi, A., Squires, R. B., Wollenberg, K. R., de Wit, E., Prescott, J., Safronetz, D., van Doremalen, N., Bushmaker, T., Feldmann, F., McNally, K., Bolay, F. K., . . . Feldmann, H. (2016). Nanopore sequencing as a rapidly deployable ebola outbreak tool. *Emerging Infectious Diseases*, *22*(2), 331–334. https://doi.org/10.3201/eid2202.151796

Huerta-Cepas, J., Szklarczyk, D., Heller, D., Hernández-Plaza, A., Forslund, S. K., Cook, H., Mende, D. R., Letunic, I., Rattei, T., Jensen, L. J., von Mering, C., & Bork, P. (2019). eggNOG 5.0: A hierarchical, functionally and phylogenetically annotated orthology resource based on 5090 organisms and 2502 viruses. *Nucleic Acids Research*, *47*(D1), D309–D314. https://doi.org/10.1093/nar/gky1085

Kang, D. D., Froula, J., Egan, R., & Wang, Z. (2015). MetaBAT, an efficient tool for accurately reconstructing single genomes from complex microbial communities. *PeerJ*, *3*, e1165. https://doi.org/10.7717/peerj.1165

Keegan, K. P., Glass, E. M., & Meyer, F. (2016). MG-RAST, a metagenomics service for analysis of microbial community structure and function. *Methods in Molecular Biology (Clifton, N.J.)*, *1399*, 207–233. https://doi.org/10.1007/978-1-4939-3369-3_13

Kim, C., Pongpanich, M., & Porntaveetus, T. (2024). Unraveling metagenomics through long-read sequencing: A comprehensive review. *Journal of Translational Medicine*, *22*, 111. https://doi.org/10.1186/s12967-024-04917-1

Kim, N., Ma, J., Kim, W., Kim, J., Belenky, P., & Lee, I. (2024). Genome-resolved metagenomics: A game changer for microbiome medicine. *Experimental & Molecular Medicine*, *56*(7), 1501–1512. https://doi.org/10.1038/s12276-024-01262-7

Kodama, Y., Shumway, M., Leinonen, R., & On Behalf of the International Nucleotide Sequence Database Collaboration. (2012). The sequence read archive: Explosive growth of sequencing data. *Nucleic Acids Research*, *40*(D1), D54–D56. https://doi.org/10.1093/nar/gkr854

Kovaka, S., Hook, P. W., Jenike, K. M., Shivakumar, V., Morina, L. B., Razaghi, R., Timp, W., & Schatz, M. C. (2025). Uncalled4 improves nanopore DNA and RNA modification detection via fast and accurate signal alignment. *Nature Methods*, *22*(4), 681–691. https://doi.org/10.1038/s41592-025-02631-4

Kumar, V., Gangani, S., Shukla, R., Prajapati, S. K., Shekhar, H., & Shukla, V. (2024). Next-generation sequencing and genomic data analysis. In A. Chaudhary, S. K. Sethi, & A. Verma (Eds.), *Unraveling New Frontiers and Advances in Bioinformatics* (pp. 49–76). Springer Nature. https://doi.org/10.1007/978-981-97-7123-3_3

Kumar, V., Pandey, A., Arora, A., Gautam, P., Bisht, D., Gupta, S., Chaurasia, A., & Sachan, M. (2023). Diagnostics and therapeutic potential of miR-205 and miR-34a in ovarian cancer management: A miRNA-target-based analysis. *DNA and Cell Biology*, *42*(3), 151–162. https://doi.org/10.1089/dna.2022.0487

Kumar, V., Sharma, J. K., Kumar, S., Gulia, J., Singh, N. P., & Santal, A. R. (2025). Metagenomic and proteomic approach for bioremediation of environmental pollutants. In M. P. Shah (Ed.), *Development in Waste Water Treatment Research and Processes*. Elsevier. https://doi.org/10.1016/B978-0-443-13615-3.00020-0

Kumar, V., Shukla, R., Gangani, S., Shekhar, H., & Shukla, V. (2024). Metagenomics and microbiome analysis. In A. Chaudhary, S. K. Sethi, & A. Verma (Eds.), *Unraveling New Frontiers and Advances in Bioinformatics* (pp. 77–92). Springer Nature. https://doi.org/10.1007/978-981-97-7123-3_4

Kumar, V., Singh, A., Gautam, P., & Sachan, M. (2019). Techniques/tools to study epigenetic biomarkers in human cancer detection. In S. Paul (Ed.), *Biomedical Engineering and Its Applications in Healthcare* (pp. 327–351). Springer. https://doi.org/10.1007/978-981-13-3705-5_14

Li, D., Liu, C.-M., Luo, R., Sadakane, K., & Lam, T.-W. (2015). MEGAHIT: An ultra-fast single-node solution for large and complex metagenomics assembly via succinct de Bruijn graph. *Bioinformatics*, *31*(10), 1674–1676. https://doi.org/10.1093/bioinformatics/btv033

Marotz, C. A., Sanders, J. G., Zuniga, C., Zaramela, L. S., Knight, R., & Zengler, K. (2018). Improving saliva shotgun metagenomics by chemical host DNA depletion. *Microbiome*, *6*(1), 42. https://doi.org/10.1186/s40168-018-0426-3

Moghaddam, H. S., Abkar, L., & Fowler, S. J. (2024). Making waves: From tap to gut-exploring the impact of drinking water on gut microbiota. *Water Research*, *267*, 122503. https://doi.org/10.1016/j.watres.2024.122503

Nayfach, S., Roux, S., Seshadri, R., Udwary, D., Varghese, N., Schulz, F., Wu, D., Paez-Espino, D., Chen, I.-M., Huntemann, M., Palaniappan, K., Ladau, J., Mukherjee, S., Reddy, T. B. K., Nielsen, T., Kirton, E., Faria, J. P., Edirisinghe, J. N., Henry, C. S., . . . Eloe-Fadrosh, E. A. (2021). A genomic catalog of Earth's microbiomes. *Nature Biotechnology*, *39*(4), 499–509. https://doi.org/10.1038/s41587-020-0718-6

Nurk, S., Meleshko, D., Korobeynikov, A., & Pevzner, P. A. (2017). metaSPAdes: A new versatile metagenomic assembler. *Genome Research*, *27*(5), 824–834. https://doi.org/10.1101/gr.213959.116

Ojala, T., Häkkinen, A.-E., Kankuri, E., & Kankainen, M. (2023). Current concepts, advances, and challenges in deciphering the human microbiota with metatranscriptomics. *Trends in Genetics*, *39*(9), 686–702. https://doi.org/10.1016/j.tig.2023.05.004

Ondov, B. D., Bergman, N. H., & Phillippy, A. M. (2011). Interactive metagenomic visualization in a web browser. *BMC Bioinformatics*, *12*(1), 385. https://doi.org/10.1186/1471-2105-12-385

Patin, N. V., & Goodwin, K. D. (2022). Long-read sequencing improves recovery of picoeukaryotic genomes and zooplankton marker genes from marine metagenomes. *mSystems*, *7*(6), e00595.22. https://doi.org/10.1128/msystems.00595-22

Pereira-Marques, J., Ferreira, R. M., & Figueiredo, C. (n.d.). A metatranscriptomics strategy for efficient characterization of the microbiome in human tissues with low microbial biomass. *Gut Microbes*, *16*(1), 2323235. https://doi.org/10.1080/19490976.2024.2323235

Qayyum, H., Ishaq, Z., Ali, A., Kayani, M. U. R., & Huang, L. (2025). Genome-resolved metagenomics from short-read sequencing data in the era of artificial intelligence. *Functional & Integrative Genomics*, *25*(1), 124. https://doi.org/10.1007/s10142-025-01625-x

Raja, A., Shukla, R., Vannier, B., Moussa, A., Kant, A., & Singh, T. R. (2025). Technologies, computations, and data analysis for next-generation sequencing. In *Next-Generation Sequencing Standard Operating Procedures and Applications*. CRC Press. ISBN: 9781003354062.

Saroj, D. B., Ahire, J. J., & Shukla, R. (2023). Genetic and phenotypic assessments for the safety of probiotic Bacillus clausii 088AE. *3 Biotech*, *13*(7), 238. https://doi.org/10.1007/s13205-023-03662-4

Sczyrba, A., Hofmann, P., Belmann, P., Koslicki, D., Janssen, S., Dröge, J., Gregor, I., Majda, S., Fiedler, J., Dahms, E., Bremges, A., Fritz, A., Garrido-Oter, R., Jørgensen, T. S., Shapiro, N., Blood, P. D., Gurevich, A., Bai, Y., Turaev, D., . . . McHardy, A. C. (2017). Critical assessment of metagenome interpretation—a benchmark of metagenomics software. *Nature Methods*, *14*(11), 1063–1071. https://doi.org/10.1038/nmeth.4458

Seemann, T. (2014). Prokka: Rapid prokaryotic genome annotation. *Bioinformatics*, *30*(14), 2068–2069. https://doi.org/10.1093/bioinformatics/btu153

Sessitsch, A., & Mitter, B. (2015). 21st century agriculture: Integration of plant microbiomes for improved crop production and food security. *Microbial Biotechnology*, *8*(1), 32–33. https://doi.org/10.1111/1751-7915.12180

Shaffer, J. P., Marotz, C., Belda-Ferre, P., Martino, C., Wandro, S., Estaki, M., Salido, R. A., Carpenter, C. S., Zaramela, L. S., Minich, J. J., Bryant, M., Sanders, K., Fraraccio, S., Ackermann, G., Humphrey, G., Swafford, A. D., Miller-Montgomery, S., & Knight, R. (n.d.). A comparison of DNA/RNA extraction protocols for high-throughput sequencing of microbial communities. *Biotechniques*, *70*(3), 149–159. https://doi.org/10.2144/btn-2020-0153

Shakya, M., Lo, C.-C., & Chain, P. S. G. (2019). Advances and challenges in metatranscriptomic analysis. *Frontiers in Genetics*, *10*. https://doi.org/10.3389/fgene.2019.00904

Shen, J., McFarland, A. G., Blaustein, R. A., Rose, L. J., Perry-Dow, K. A., Moghadam, A. A., Hayden, M. K., Young, V. B., & Hartmann, E. M. (2022). An improved workflow for accurate and robust healthcare environmental surveillance using metagenomics. *Microbiome*, *10*(1), 206. https://doi.org/10.1186/s40168-022-01412-x

Shukla, R., Yadav, A. K., & Singh, T. R. (2024). Application of deep learning in biological big data analysis. In *Research Anthology on Bioinformatics, Genomics, and Computational Biology* (pp. 225–250). IGI Global Scientific Publishing. https://doi.org/10.4018/979-8-3693-3026-5.ch011

Shukla, R., Yadav, A. K., Sote, W. O., Junior, M. C., & Singh, T. R. (2022). Chapter 25: Systems biology and big data analytics. In D. B. Singh & R. K. Pathak (Eds.), *Bioinformatics* (pp. 425–442). Academic Press. https://doi.org/10.1016/B978-0-323-89775-4.00005-5

Sinha, R., Abu-Ali, G., Vogtmann, E., Fodor, A. A., Ren, B., Amir, A., Schwager, E., Crabtree, J., Ma, S., Abnet, C. C., Knight, R., White, O., & Huttenhower, C. (2017). Assessment of variation in microbial community amplicon sequencing by the microbiome quality control (MBQC) project consortium. *Nature Biotechnology*, *35*(11), 1077–1086. https://doi.org/10.1038/nbt.3981

Thompson, L. R., Sanders, J. G., McDonald, D., Amir, A., Ladau, J., Locey, K. J., Prill, R. J., Tripathi, A., Gibbons, S. M., Ackermann, G., Navas-Molina, J. A., Janssen, S., Kopylova, E., Vázquez-Baeza, Y., González, A., Morton, J. T., Mirarab, S., Zech Xu, Z., Jiang, L., . . . Knight, R. (2017). A communal catalogue reveals Earth's multiscale microbial diversity. *Nature*, *551*(7681), 457–463. https://doi.org/10.1038/nature24621

Toro-Domínguez, D., Martorell-Marugán, J., López-Domínguez, R., García-Moreno, A., González-Rumayor, V., Alarcón-Riquelme, M. E., & Carmona-Sáez, P. (2019). ImaGEO: Integrative gene expression meta-analysis from GEO database. *Bioinformatics*, *35*(5), 880–882. https://doi.org/10.1093/bioinformatics/bty721

Vasudeva, K., Kaur, P., & Munshi, A. (2023). Chapter 28: High-throughput sequencing technologies in metagenomics. In V. Kumar, M. Bilal, S. K. Shahi, & V. K. Garg (Eds.), *Metagenomics to Bioremediation* (pp. 685–708). Academic Press. https://doi.org/10.1016/B978-0-323-96113-4.00005-6

Wenger, A. M., Peluso, P., Rowell, W. J., Chang, P.-C., Hall, R. J., Concepcion, G. T., Ebler, J., Fungtammasan, A., Kolesnikov, A., Olson, N. D., Töpfer, A., Alonge, M., Mahmoud, M., Qian, Y., Chin, C.-S., Phillippy, A. M., Schatz, M. C., Myers, G., DePristo, M. A., . . . Hunkapiller, M. W. (2019). Accurate circular consensus long-read sequencing improves variant detection and assembly of a human genome. *Nature Biotechnology*, *37*(10), 1155–1162. https://doi.org/10.1038/s41587-019-0217-9

Wilkinson, M. D., Dumontier, M., Aalbersberg, Ij. J., Appleton, G., Axton, M., Baak, A., Blomberg, N., Boiten, J.-W., da Silva Santos, L. B., Bourne, P. E., Bouwman, J., Brookes, A. J., Clark, T., Crosas, M., Dillo, I., Dumon, O., Edmunds, S., Evelo, C. T., Finkers, R., . . . Mons, B. (2016). The FAIR guiding principles for scientific data management and stewardship. *Scientific Data*, *3*(1), 160018. https://doi.org/10.1038/sdata.2016.18

Willner, D., Daly, J., Whiley, D., Grimwood, K., Wainwright, C. E., & Hugenholtz, P. (2012). Comparison of DNA extraction methods for microbial community profiling with an application to pediatric bronchoalveolar lavage samples. *PLoS One*, *7*(4), e34605. https://doi.org/10.1371/journal.pone.0034605

Wood, D. E., Lu, J., & Langmead, B. (2019). Improved metagenomic analysis with Kraken 2. *Genome Biology*, *20*(1), 257. https://doi.org/10.1186/s13059-019-1891-0

Wu, Y.-W., Simmons, B. A., & Singer, S. W. (2016). MaxBin 2.0: An automated binning algorithm to recover genomes from multiple metagenomic datasets. *Bioinformatics*, *32*(4), 605–607. https://doi.org/10.1093/bioinformatics/btv638

Wydro, U. (2022). Soil microbiome study based on DNA extraction: A review. *Water*, *14*(24), 3999. https://doi.org/10.3390/w14243999

Yadav, A. K., Shukla, R., & Singh, T. R. (2021). Chapter 11: Machine learning in expert systems for disease diagnostics in human healthcare. In P. Kumar, Y. Kumar, & M. A. Tawhid (Eds.), *Machine Learning, Big Data, and IoT for Medical Informatics* (pp. 179–200). Academic Press. https://doi.org/10.1016/B978-0-12-821777-1.00022-7

Yadav, A. K., Shukla, R., & Singh, T. R. (2022). Chapter 22: Topological parameters, patterns, and motifs in biological networks. In D. B. Singh & R. K. Pathak (Eds.), *Bioinformatics* (pp. 367–380). Academic Press. https://doi.org/10.1016/B978-0-323-89775-4.00012-2

3 Data Processing Pipelines

Kavita Patel[1] *and Ashutosh Mani*[1*]
[1]Department of Biotechnology, Motilal Nehru National Institute of Technology, Prayagraj
*Corresponding author

3.1 INTRODUCTION

Metagenomics is a science that applies high-throughput sequencing technology and bioinformatics to study the functional and taxonomic composition of microbial communities without the isolation of individual microorganisms. The complexity and diversity of the microbiome are vast and include bacteria and archaea present in a number of samples, and the study of new and known species represents a significant advancement in microbiology. Metagenomics has been used extensively worldwide for microbiological research due to the affordability and popularity of next-generation sequencing, which bypasses the limitations of conventional image-based microscopy techniques for identifying and naming new species from various sample types. Sequencing techniques can generate high-throughput short reads for multiple samples present in a library, providing adequate information for functional annotation or taxonomic classification, depending on the specific metagenomic strategy used. Sequencing is more capable than microarrays for screening samples, with the ability to observe unknown species (Hugerth & Andersson, 2017, Kumar, Gangani, et al., 2024).

Amplicon sequencing uses polymerase chain reaction (PCR) to construct amplicons by targeting regions of interest in marker genes and generating reads to analyze taxonomic information. rRNA genes have been used as marker genes for two reasons: first, they are highly conserved genes that exist in all bacterial genomes; second, the conserved regions' structural patterns are interspersed with variable regions within the rRNA gene, which are used as universal primers to capture fragments and to identify sequence differences among species. Notably, 16S rRNAs have been utilized as marker genes for prokaryotes, whereas 18S rRNAs have been employed for the taxonomic analysis of eukaryotes, such as protists. The aim of amplicon sequencing primarily focuses on reading all sequences targeted by paired primers and measuring the diversity and abundance of microbial species by assigning the taxonomic classification to every high-quality sequence (Gołębiewski & Tretyn, 2020; Kumar, Shukla, et al., 2024; Kumar et al., 2023, 2019). This entire procedure is referred to as taxonomic profiling. Additionally, amplicon sequencing may not be susceptible to contaminants present in samples, and it cannot provide information on protein-coding genes from the metagenome that are involved in metabolic pathways beyond the marker gene. Therefore, amplicon sequencing is used to analyze microbial species across interdisciplinary areas, such as plant–microbe–animal interactions, the role of pathogens in infectious diseases, and environmental microbial ecology (Bharti & Grimm, 2021).

Whole-genome shotgun sequencing (WGS) is an alternative approach based on constructing a random sample of total DNA, which allows the detection of all DNA genomes. Limitations of shotgun metagenomics include cost, laboratory and bioinformatics technical requirements, and the need for millions of high-quality sequence reads. Transforming these raw reads into interpretable outputs requires a series of steps, including quality control, filtering, host DNA removal, denoising, taxonomic and functional classification, genome assembly, binning, and statistical analysis.

 DOI: 10.1201/9781003670155-3

The modern pipeline integrates specialized tools such as QIIME2 (Bolyen et al., 2019), DADA2 (Callahan et al., 2016), Mothur (amplicon data) (Schloss et al., 2009), and Kneaddata (*KneadData—The Huttenhower Lab*, n.d.), Kraken2 (*Improved Metagenomic Analysis with Kraken 2 | Genome Biology | Full Text*, n.d.), MetaPhlAn (Blanco-Míguez et al., 2023), HUMaN (*Humann3—The Huttenhower Lab*, n.d.), and MEGAHIT (*MEGAHIT: An Ultra-Fast Single-Node Solution for Large and Complex Metagenomics Assembly via Succinct de Bruijn Graph | Bioinformatics | Oxford Academic, n.d.*) (WGS) for processing and ensuring reproducibility. The selection of tools and parameters at each stage significantly influences the biological interpretation of results. This chapter systematically presents end-to-end workflows for both amplicon and shotgun metagenomics data, including step-by-step protocols, tool comparisons, commands, and best practices to aid researchers in effectively using metagenomics datasets for scientific discovery.

3.2 AMPLICON SEQUENCING PIPELINES

Raw amplicon sequencing data are generated in FASTQ files by sequencing platforms. It is recommended that these analyses be conducted on a Linux operating system to handle files efficiently and manageably. The software used in this pipeline can be installed individually from respective websites or via a package manager such as Conda, which simplifies installation. For Conda installation and downloads, visit https://bioconda.github.io/ and run the Miniconda installer, and set up channels using the provided instructions. Follow the steps below to set up the Conda environment and install the required packages (see Figure 3.1, Table 3.1).

1. Create a new Conda environment named metagenomics_data:
   conda create —name metagenomics_data
2. Activate Conda metagenomics environment:
   conda activate metagenomics_data

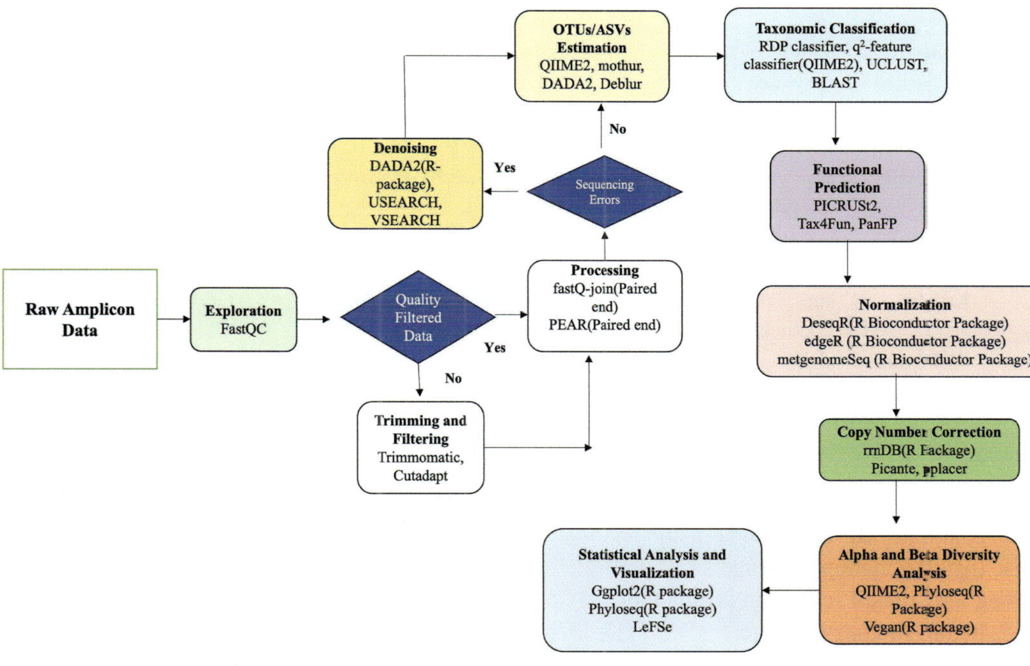

FIGURE 3.1 Flowchart of the amplicon data analysis pipeline, starting from raw amplicon reads to taxonomic classification, functional annotation, normalization, and diversity analysis. Key tools and R packages used at each step are highlighted to ensure accurate microbial profiling.

TABLE 3.1
List of Software for Amplicon Analysis and Whole Shotgun Sequencing Analysis

S.No.	Name	Link	Description	Reference
1.	FastQC	https://www.bioinformatics.babraham.ac.uk/projects/fastqc/	Sequencing reads quality evaluation (GC content, sequence length distribution, quality scores)	de Sena Brandine & Smith (2021)
2.	Cutadapt	https://cutadapt.readthedocs.io/en/stable/installation.html	Cutadapt finds and removes the adapter sequence, poly A tails, Primers and other different type of sequences from high throughput sequencing reads	Martin (2011)
3.	Trimmomatic	https://github.com/timflutre/trimmomatic	Remove adapters, primers, low quality bases	Bolger et al. (2014)
4.	Bowtie2	http://bowtie-bio.sourceforge.net/bowtie2	Used for rapid alignment and to remove host contamination	Langmead and Salzberg (2012)
5.	DADA2	https://benjjneb.github.io/dada2/	Chimeric sequence identification and removal of chimeric sequences	Callahan et al. (2016)
6.	Mothur	https://github.com/mothur/mothur/releases/tag/v1.48.3	Assign taxonomy using reference databases	Schloss et al. (2009)
7.	MetaPhlAn2	https://bitbucket.org/biobakery/metaphlan2	Taxonomic profiling tool and the output is relative abundance strains	Truong et al. (2015)
8.	QIIME2	https://library.qiime2.org/quickstart/amplicon	Assign taxonomy using reference databases and provide alpha and beta diversity metrics, ordination and statistical tests	Estaki et al. (2020)
7.	Phyloseq	https://joey711.github.io/phyloseq/	Generate bar plots, ordination plots, heatmaps	McMurdie & Holmes (2013)
8.	KneadData	https://github.com/biobakery/kneaddata	Remove host DNA and filters low quality reads	(*KneadData—The Huttenhower Lab*, n.d.)
9.	MEGAHIT	https://github.com/voutcn/megahit	Used to assemble shotgun reads into contigs	Li et al. (2015)
10.	MetaBAT2	https://anaconda.org/bioconda/metabat2	Used to group contigs into bins	(Kang et al., 2019)
11.	Kraken2	https://github.com/DerrickWood/kraken2	Classify contigs to taxonomic group	(Lu et al., 2022)
12.	HUMAnN3	https://github.com/biobakery/humann	Profiles the abundance of microbial metabolic pathways	(Beghini et al., n.d.)

3. Install FastQC package:
   conda install -c bioconda fastqc
4. Install Cutadapt package:
   conda install -c bioconda cutadapt
5. Install mothur package:
   conda install -c bioconda mothur
6. Install DADA2 package:
   conda install -c bioconda bioconductor-dada2
7. Install phyloseq package:
   conda install -c bioconda bioconductor-phyloseq

3.2.1 Preprocessing of Data

FASTQ data contain information about read names, sequences, and base quality scores. The following analysis applies to paired-end raw data that have already been demultiplexed. In paired-end sequencing, two FASTQ files are generated per sample—one containing forward reads and the other containing reverse reads (Cock et al., 2010).

FastQC (https://www.bioinformatics.babraham.ac.uk/projects/fastqc/) is used to examine the quality of individual FASTQ files.

```
fastqc<FASTQ_file_name>
fastqc<*.FASTQ_file>
```

, where "*" is used for processing multiple FASTQ input files.

The FastQC report provides several informative quality metrics, such as adapter content and per-base sequence quality, which guide downstream analysis strategies and parameter selection.

Cutadapt (Martin, 2011) is used to trim adapter sequences from the 5′ and/or 3′ ends of reads.

```
cutadapt -a AGATCGGAAGAGC -A AGATCGGAAGAGC -q 20 -m 30 -o trimmed_read1.
    fastq.gz -p trimmed_read2.fastq.gz read1.fastq.gz read2.fastq.gz
```

where -a AGATCGGAAGAGC: Adapter sequence to trim from **read1** (3′ end), -A AGATCGGAAGAGC: Adapter sequence to trim from **read2** (3′ end), -q 20: Trim low-quality bases (phred score < 20) from both ends, -m 30: Discard reads shorter than 30 bp after trimming, -o trimmed_read1.fastq.gz: Output file for trimmed read1, -p trimmed_read2.fastq.gz: Output file for trimmed read2, read1.fastq.gz: Input file for read1, read2.fastq.gz: Input file for read2.

3.2.2 Data Analysis Pipelines for 16S rRNA Amplicon Sequencing by Mothur

Mothur is a bioinformatics tool designed to provide a comprehensive suite of tools for amplicon sequencing data analysis. Its primary advantage is its ability to perform all key analyses within a unified framework using standardized file formats that allow seamless integration of commands. The tool is based on several principles: minimizing file size, discarding redundant data, reducing computation time, and retaining all key information. The following sections describe the fundamental steps of 16S rRNA amplicon sequencing analysis using Mothur for MiSeq data (*MiSeq SOP*, n.d.; Schloss et al., 2009).

3.2.2.1 Download the Reference Data

Two reference files are required: (1) Silva reference files (https://mothur.org/wiki/silva_reference_files/), used by align.seqs for sequence alignment and quality control, and (2) RDP reference file (https://mothur.org/wiki/rdp_refer ence_files/), used by classify.seqs for advanced quality control.

3.2.2.2 Paired Read Assembly and Quality Control

To start the mothur, the Conda environment should be activated, and the following command should be run to launch mothur:

```
mothur
```

The first thing to tell Mothur which fastq files go together, and this can be done by make.file command.

```
mothur > make.file(inputdir=., type=fastq, prefix=stability)
```

where make.file: command generates a "stability.files" file which list all FASTQ files to be processed, inputdir=.: Specifies the current directory (.) as the location of your FASTQ files, type=fastq: Indicates that the files are in FASTQ format, prefix=stability: the output file will be named *stability.files*.

Then, we combine our sets of reads for each sample and combine the data from all samples. This will be done by making.contigs commands that require stability.files as input. This command extracts the sequence and quality score data from fastq files and will create the reverse complement of the reverse read and then join the reads into contigs. The algorithm aligns the pairs of sequences and looks across the alignment for any positions where the two reads disagree.

```
mothur > make.contigs(file=stability.files)
```

This command will produce the stability.trim.contigs.fasta and stability.contigs.count_table; these files contain the sequence data and group identity for each sequence. The stability.contigs.report files contain the data about contig assembly for each read.

If the reads didn't assemble well and found ambiguous base calls or repetition in reads, the following command will remove any sequences with ambiguous bases and sequences longer than 275bp.

```
mothur > screen.seqs(fasta=stability.trim.contigs.fasta, count=stability.contigs.count_table,
   maxambig=0, maxlength=275, maxhomop=8)
```

3.2.2.3 Processing of Improved Sequences

To remove the duplication of sequences, the following command unique.seqs will be used:

```
mothur > unique.seqs(fasta=stability.trim.contigs.good.fasta, count=stability.contigs.good.
   count_table)
```

After removing the duplication, we need to align the sequences to the reference alignment. To align with the reference database (silva.bacteria.fasta), the following command should be followed.

```
mothur > align.seqs(fasta=stability.trim.contigs.good.unique.fasta, reference=silva.v4.fasta)
```

The overhang sequences at both ends, which are not aligned, should be removed after performing alignment. Also, there are many columns in alignment that contain gap characters, which can be pulled out without losing any information, using filter.seqs:

```
mothur > filter.seqs(fasta=stability.trim.contigs.good.unique.good.align, vertical=T, trump=.)
```

The terminal gap characters of the sequences were removed using trump=. and vertical gap characters were removed using vertical=T. There is a possibility of redundancy creation across the sequences by end trimming, so the unique.seqs can be re-run:

```
mothur>unique.seqs(fasta=stability.trim.contigs.good.unique.good.filter.fasta, count=stability.
   trim.contigs.good.good.count_table)
```

where unique.seqs finds unique sequences by combining identical sequences and creating a mapping file showing how many times each sequence occurs, and fasta=stability.trim.contigs.good.unique.

good.filter.fasta: input FASTA file containing quality-filtered, aligned sequences, count=stability.trim.contigs.good.good.count_table: input count table that tracks how often each sequence is observed in each sample.

3.2.2.4 Advanced Quality Control

This section addresses sequencing errors, contamination, and chimeras and generates aligned sequence, taxonomy, and count files.

3.2.2.4.1 De-Noising of Sequences

pre.cluster command identifies the identical sequences having mismatches due to sequencing errors and generates unique sequences based on search results. This command splits the sequences by group and then sorts the sequences by abundance from most to least abundant; this allows 1 difference for 100 bp of sequence.

```
mothur > pre.cluster(fasta=stability.trim.contigs.good.unique.good.filter.unique.fasta,
   count=stability.trim.contigs.good.unique.good.filter.count_table, diffs=2)
```

where, fasta=stability.trim.contigs.good.unique.good.filter.unique.fasta: the filtered and unique sequences file to be clustered, count=stability.trim.contigs.good.unique.good.filter.count_table: the count table that records how many times each sequence was observed, diffs=2: the maximum number of differences allowed when merging similar sequences (e.g., errors introduced during sequencing).

Notes: This step helps in removing the error by merging sequences that differ by only a few nucleotides.

3.2.2.4.2 Chimera Removal

After removing the sequencing error, the chimeras should be removed by chimera.vsearch command, this command again splits the data by sample and checks for chimeras. If sequence got flagged as chimeric in one sample, the default (dereplicate=F) will remove it from all samples.

```
mothur > chimera.vsearch(fasta=stability.trim.contigs.good.unique.good.filter.unique.pre-
   cluster.fasta, count=stability.trim.contigs.good.unique.good.filter.unique.precluster.count_
   table, dereplicate=t)
```

classify.seqs command used further to classify the sequences using Bayesian classifier:

```
mothur > classify.seqs(fasta=stability.trim.contigs.good.unique.good.filter.unique.precluster.
   denovo.vsearch.fasta, count=stability.trim.contigs.good.unique.good.filter.unique.preclus-
   ter.denovo.vsearch.count_table, reference=trainset.pds.fasta, taxonomy=trainset.pds.tax)
```

After classifying, we remove undesirables, which will be done by remove.lineage command:

```
mothur > remove.lineage(fasta=stability.trim.contigs.good.unique.good.filter.unique.pre-
   cluster.denovo.vsearch.fasta, count=stability.trim.contigs.good.unique.good.filter.
   unique.precluster.denovo.vsearch.count_table, taxonomy=stability.trim.contigs.
   good.unique.good.filter.unique.precluster.denovo.vsearch.pds.wang.taxonomy,
   taxon=Chloroplast-Mitochondria-unknown-Archaea-Eukaryota)
```

summary.tax command will reflect the updated taxonomy summary file.

```
mothur > summary.tax(taxonomy=current, count=current)
```

3.2.2.5 Analysis of Taxonomy

In order to create a shared OTU file and a taxonomy summary file, the analysis in this section aims to group representative sequences into operational taxonomic units (OTUs) at a 97% identity threshold for the species level and to categorize the OTUs or sequences into taxonomic annotations.

The *dist.seqs* command produces a matrix file containing distances between aligned sequences, and the *cluster* command assigns sequences to OTUs.

```
mothur > dist.seqs(fasta=final.fasta, cutoff=0.03)
```

This *.dist* file contains the names of pairwise representative sequences and their distance values when those values fall below the cutoff.

```
mothur > cluster(column=final.dist, count=final.count_table)
```

3.2.2.5.1 OTUs

At this stage, a tabular list file containing the names of representative sequences for each OTU group and the total number of OTUs is generated.

Next, the *make.shared* command integrates the list file with the count file to create a new shared file that contains a shared OTU table with the number of sequences in each sample for each OTU.

```
mothur > make.shared(list=final.opti_mcc.list, count=final.count_table, label=0.03)
```

Then, the *classify.otu* command generates the consensus taxonomy for each OTU.

```
mothur > classify.otu(list=final.opti_mcc.list, count=final.count_table, taxonomy=final.taxonomy, label=0.03)
```

3.2.2.5.2 ASVs

The *pre.cluster* method is used to identify ASVs in sample sequences. In this step, the *make.shared* command is used to form OTUs into a shared file.

```
mothur > make.shared(count=final.count_table)
```

The results in the shared file can be used to form OTUs or phylotypes, and the list file is used to generate consensus taxonomy for each ASV.

```
mothur > classify.otu(list=final.asv.list, count=final.count_table, taxonomy=final.taxonomy, label=ASV)
```

3.2.2.5.3 Phylotypes

For some analyses, sequences are classified into phylotypes according to their taxonomic classification, and this can be achieved using the *phylotype* command:

```
mothur > phylotype(taxonomy=final.taxonomy)
```

To create the genus-level shared file of these classified sequences:

```
mothur > make.shared(list=final.tx.list, count=final.count_table, label=1)
```

The *classify.otu* command classifies sequences into OTUs:

```
classify.otu(list=final.tx.list, count=final.count_table, taxonomy=final.taxonomy, label=1)
```

Here, *classify.otu* assigns taxonomy to OTUs by summarizing taxonomic information from individual sequences. The parameter *list=final.tx.list* specifies the input file grouping sequences into OTUs, *count=final.count_table* specifies the count table recording the frequency of each sequence in each sample, *taxonomy=final.taxonomy* specifies the taxonomy file, and *label=1* controls how taxonomy is summarized for each OTU. In this case, "1" indicates that taxonomy is assigned using majority-rule or similar aggregation methods.

3.2.2.5.4 Phylogenetic

Phylogenetic diversity calculation depends on a phylogenetic tree and can be calculated using the *dist.seqs* and *clearcut* commands:

```
mothur > dist.seqs(fasta=final.fasta, output=lt)
mothur > clearcut(phylip=final.phylip.dist)
```

3.2.2.6 OTU-Based Analysis

3.2.2.6.1 Alpha Diversity

This analysis measures alpha diversity for individual samples and provides diversity metrics for all samples, as well as scores for each dimension used in drawing PCoA plots. The *summary.single* command calculates the Inverse Simpson diversity and Shannon diversity indices for samples.

```
mothur > summary.single(shared=final.opti_mcc.shared, calc=nseqs-coverage-sobs-invsimpson, subsample=T)
mothur > summary.single(shared=final.opti_mcc.shared, calc=nseqs-coverage-sobs-shanon, subsample=T)
```

3.2.2.6.2 Beta Diversity

To compare microbial community composition across all samples, an OTU-based beta diversity approach is employed. The *dist.shared* command is used to calculate pairwise community dissimilarities based on Bray–Curtis and Jaccard classic distance metrics.

```
mothur > dist.shared(shared=final.opti_mcc.shared, calc=braycurtis-jclass, subsample=t)
```

3.2.2.6.2.1 Principal Coordinates Analysis (PCoA)

PCoA is used to visualize patterns in community dissimilarity by reducing high-dimensional data to two principal axes. This process uses the Bray–Curtis distance matrix:

```
mothur > pcoa(phylip=final.opti_mcc.braycurtis.0.03.lt.ave.dist)
```

3.2.2.6.2.2 Non-Metric Multidimensional Scaling (NMDS)

NMDS also uses the Bray–Curtis distance matrix to visualize community differences without assuming a linear relationship.

```
mothur > nmds(phylip=final.opti_mcc.braycurtis.0.03.lt.ave.dist, mindim=3, maxdim=3)
```

3.2.3 Phylogeny-Based Diversity Analysis

OTU- and phylotype-based approaches rely on binning sequences into taxonomic units, whereas phylogeny-based methods use evolutionary relationships to assess community diversity. These methods use a phylogenetic tree as input rather than a shared OTU table, enabling an accurate assessment of genetic relatedness among community members.

3.2.3.1 Alpha Diversity

Alpha diversity is quantified as the sum of unique branch lengths in a phylogenetic tree that connects all taxa present within a sample. The *phylo.diversity* command computes metrics reflecting richness and evolutionary distinctiveness of microbial communities.

```
mothur > phylo.diversity(tree=final.phylip.tre, count=final.count_table, rarefy=T)
```

The output is a rarefied table representing phylogenetic diversity values across samples.

3.2.3.2 Beta Diversity (UniFrac Distance Metrics)

UniFrac-based beta diversity metrics are used to compare microbial community composition between samples and incorporate phylogenetic tree structure and sequence abundance data. Unweighted UniFrac considers only the presence or absence of lineages, whereas weighted UniFrac incorporates relative abundances.

```
mothur > unifrac.unweighted(tree=final.phylip.tre, count=final.count_table, distance=lt,
   random=F, subsample=t)
mothur > unifrac.weighted(tree=final.phylip.tre, count=final.count_table, distance=lt,
   random=F, subsample=t)
```

These commands produce output files such as *final.phylip.1.weighted.ave.dist*, which can be further analyzed using ordination methods (PCoA, NMDS) and statistical tools (AMOVA, PERMANOVA).

3.2.3.2.1 Statistical Validation of Group Differences

Ordination methods allow visual assessment of sample clustering, while statistical methods determine whether observed differences are significant.

3.2.3.2.1.1 AMOVA (Analysis of Molecular Variance)

AMOVA evaluates whether the average distance between groups is significantly greater than the average distance within groups by partitioning variance.

```
mothur > amova(distance=final.opti_mcc.braycurtis.0.03.lt.ave.dist, design=your_design_
   file.design)
```

where distance=final.opti_mcc.braycurtis.0.03.lt.ave.dist: specifies the distance matrix file you created (e.g., Bray-Curtis distance at 3% clustering threshold).

design=your_design_file.design: refers to the experimental design file, which describes how samples are grouped (e.g., treatment vs control, or other metadata).

Note: Replace *your_design_file.design* with the actual design file name, and ensure that all required files (*.dist*, *.design*) are present in the working directory.

3.2.3.2.1.2 PERMANOVA (Permutation Multivariate ANOVA)

PERMANOVA assesses group separation using permutation tests and serves as a nonparametric alternative to AMOVA. It is particularly useful when community variance is not homogeneous.

```
mothur > permanova(phylip=final.opti_mcc.braycurtis.0.03.lt.ave.dist, design=your_design_
   file.design)
```

where phylip=final.opti_mcc.braycurtis.0.03.lt.ave.dist specifies the distance matrix file (in PHYLIP format) you created, design=your_design_file.design: refers to the design file that groups your samples according to treatments, conditions, or other metadata.

3.2.4 Data Analysis Pipelines for 16S rRNA Amplicon Sequencing by DADA2

DADA2 is an R package that provides a workflow for processing raw amplicon sequencing data into high-resolution amplicon sequence variants (ASVs). ASVs offer finer taxonomic resolution and improved reproducibility compared with traditional OTU clustering.

3.2.4.1 Download the Reference Data

The DADA2 website provides DADA2-formatted taxonomic reference files required by the *assignTaxonomy* function: https://benjjneb.github.io/dada2/training.html.

3.2.4.2 Standard Sequence Denoising Procedure

The R interactive environment must be launched before initializing the DADA2 pipeline by typing *R* at the command line. The standard workflow includes loading required libraries, learning error models, dereplicating sequences, and performing denoising steps. This process generates *derep*-class and *dada*-class objects, which are intermediate outputs in transforming raw reads into high-quality ASVs.

3.2.4.2.1 Load R Packages

To initialize the analysis, the DADA2 and phyloseq packages must be loaded:

```
library(dada2)
library(phyloseq)
```

3.2.4.2.2 Error Rate Learning

To denoise the sequences using the DADA2 algorithm, the error profiles of sequencing runs must be estimated using the learnErrors() function from filtered FASTQ files.

```
error_rate_forward <- learnErrors(filtFs, multithread=TRUE)
error_rate_reverse <- learnErrors(filtRs, multithread=TRUE)
```

This function models the probabilities of nucleotide transitions (e.g., A→C, G→T) over different quality scores and returns a list of error matrices containing transition rates for all 16 possible base substitutions, stratified by quality scores.

3.2.4.2.3 Dereplication of Sequences

This step condenses identical reads into unique sequences while preserving abundance and quality information.

```
unique_forward <- derepFastq(filtFs, verbose=TRUE)
unique_reverse <- derepFastq(filtRs, verbose=TRUE)
```

The derepFastq() function generates derep-class objects, each representing the unique sequences from a sample, their abundance, and a matrix of average quality scores by position, which are required inputs for the core denoising algorithm.

3.2.4.2.4 Denoising Sequencing Data

The actual denoising of sequences is performed using the dada() function, which distinguishes true biological variants from sequencing error by applying a statistical model.

```
dada_forward <- dada(unique_forward, err=error_rate_forward, multithread=TRUE)
dada_reverse <- dada(unique_reverse, err=error_rate_reverse, multithread=TRUE)
```

This step creates DADA-class objects, and each object includes inferred ASVs, the number of reads supporting each ASV, and quality scores associated with each base position.

3.2.4.3 Assembly of Paired-End Reads and Generation of the ASV Table

After denoising and error correction of forward and reverse reads, the next phase of the pipeline involves assembling paired-end reads, constructing the ASV table, and removing chimeric sequences. This results in a high-resolution matrix that accurately represents the abundance of true biological sequences across all samples.

3.2.4.3.1 Merging Paired-End Reads

Based on the overlap regions, the denoised forward and reverse reads were combined into full-length sequences.

```
merged_sequences <- mergePairs(dada_forward, unique_forward, dada_reverse, unique_
  reverse, verbose=TRUE)
```

The mergePairs() function aligns forward and reverse reads by identifying the overlapping region and returns a list of data frame objects having assembled sequences, number of reads supporting each merged sequence, and number of matches and mismatches. This is an essential step to reconstruct the full-length amplicon from paired-end reads while verifying sequence consistency.

3.2.4.3.2 ASV Table Construction

Matrix generation of ASVs (rows) by samples (columns), with each cell representing the number of reads assigned to a specific ASV in a specific sample.

```
sequence_table <- makeSequenceTable(merged_sequences)
```

This table is the core output of DADA2 and forms the basis for diversity analysis, statistical comparison, and taxonomic classification. Each sequence is treated as ASV.

3.2.4.3.3 Chimera Removal

The removeBimerDenovo() function uses a consensus-based approach to identify chimeric ASVs sequences that are formed erroneously by joining two or more parent sequences.

```
sequence_table_remove_chimeras <- removeBimeraDenovo(sequence_table, method=
  “consensus”, multithread=TRUE)
```

The output of chimera removal is a refined ASV table containing only non-chimeric sequences.

3.2.4.4 Taxonomic Classification of ASVs

This assigns biological identity to each unique sequence using a reference database, enabling ecological interpretation and downstream comparative analysis.

```
taxonomy_table <- assignTaxonomy(sequence_table_remove_chimeras, “path/to/silva_nr_
  v138_train_set.fa.gz”, multithread=TRUE)
```

3.2.4.5 Diversity Analysis

The phyloseq is commonly used for downstream ecological analyses, including alpha and beta diversity assessments after generating the amplicon sequence variant (ASV).

3.2.4.5.1 Constructing a Phyloseq Experiment-Level Object

The phyloseq() function is used to make a phyloseq-class object, which acts as a container for microbiome data, integrating ASV (or OUT) table, taxonomy, sample metadata, and phylogenetic tree.

```
phyloseq_object <- phyloseq(otu_table(asv_table, taxa_are_rows = FALSE))
```

The output generates phyloseq-class object containing an out_table component can be accessed by:

```
otu_table(phyloseq_object)
```

3.2.4.5.2 Alpha Diversity Estimation

Alpha diversity represents the richness of species within individual samples. The estimate_richness() function calculates multiple diversity metrics, including Shannon, Simpson, and Observed species.

```
shannon_table <- estimate_richness(phyloseq_object, measures = "Shannon")
```

The output generated data.frame containing Shanon diversity index per sample can be exported and visualized using ggplot2 or plot_richness().

```
plot_richness(ps, x="Day", measures=c("Shannon", "Simpson"), color="When")
```

3.2.4.5.3 Beta Diversity and Ordination

Beta diversity measures differences in microbial community composition between all samples. The ordinate() function from phyloseq performs ordination techniques like Principal Coordinates Analysis (PCoA).

```
pcoa_result <- ordinate(phyloseq_object, method = "PCoA", distance = "jaccard")
```

The output generates a list object containing eigenvalues and scores for each sample along multiple principal coordinates and can be visualized by the following command:

```
plot_ordination(phyloseq_object, pcoa_result, color = "SampleType")
```

This plot enables the interpretation of similarity or dissimilarity across experimental groups.

3.2.4.5.4 NMDS Using Bray-Curtis

The transform_sample_counts() function from the phyloseq package was used to normalize the ASV count data to relative abundance to ensure the comparability across samples with varying sequencing depths.

```
ps.prop <- transform_sample_counts(ps, function(otu) otu/sum(otu))
```

Here, ps is the original phyloseq object containing raw counts and ps.drop is the transformed object with relative abundances.

NMDS is performed using the ordinate() function with Bray-Curtis dissimilarity:

```
ord.nmds.bray <- ordinate(ps.prop, method="NMDS", distance="bray")
```

This function computes the NMDS ordination using Bray-Curtis distance matrix which is derived from normalized abundance data.

Ordination results can be visualized by plot_ordination() function.

```
plot_ordination(ps.prop, ord.nmds.bray, color="When", title="Bray NMDS")
```

This command produces a two-dimensional scatter plot where each point denotes a sample, and proximity reflects compositional similarity.

3.2.5 Data Analysis Pipelines for 16S rRNA Amplicon Sequencing by QIIME 2

The pipeline of QIIME2 processes raw amplicon sequencing data into actionable biological insights, and the pipeline includes data import, quality control, denoising, taxonomic classification, diversity analysis, and data visualization (Bolyen et al., 2019; Estaki et al., 2020; *QIIME 2 Homepage*, n.d.; *QIIME2 Enhances Multi-Amplicon Sequencing Data Analysis: A Standardized and Validated Open-Source Pipeline for Comprehensive 16S rRNA Gene Profiling | Microbiology Spectrum, n.d.*).

3.2.5.1 Raw Data Import in QIIME2

Samples are converted to.qza format from raw fastq files, to be used internally by QIIME2. The samples are already demultiplexed and imported into QIIME2 by using the command:

```
qiime tools import \ &mdash;type 'SampleData[PairedEndSequencesWithQuality]' \
&mdash;input-path manifest.csv \
&mdash;output-path demux-paired-end.qza \
&mdash;input-format PairedEndFastqManifestPhred33
```

The input format defines here that Phred for positional quality scores is 33.

3.2.5.2 Summarize and Visualize Sequence Quality

Run the following command to visualize the data to evaluate per-base quality scores to know trimming and filtering thresholds.

```
qiime demux summarize \ &mdash;i-data demux-paired-end.qza \ &mdash;o-visualization
   demux-summary.qzv
```

demux-summary.qzv will show the interactive summary report can be accessible on the QIIME 2 viewer (https://view.qiime2.org).

3.2.5.3 Denoising and Construction of Feature Table DADA2

Denoising is used to correct sequencing errors, remove low-quality reads, remove chimeras, and generate ASVs. DADA2 corrects errors without clustering and allows higher resolution microbial profiling than traditional OTUs (Ansorge et al., 2021; Callahan et al., 2016).

```
qiime dada2 denoise-paired \ &mdash;i-demultiplexed-seqs Sample_man_new.qza \ &mdash;
   o-representative-sequences 16S_rep_seqs.qza \ &mdash;o-table 16S_table.qza \ &mdash;p-
   trunc-len-f 220 \ &mdash;p-trunc-len-r 220 \ &mdash;o-denoising-stats 16S_stats.txt
   &mdash;verbose \ &mdash;p-n-threads 0
```

The output of the denoising step is two.qza files, and can be used to create a summary statistic to review the denoising process:

```
qiime feature-table summarize \ &mdash;i-table 16S_table.qza \ &mdash;o-visualization
   16S_table.qzv \ &mdash;m-sample-metadata-file 16S_mapping.txt
```

3.2.5.4 RDP Classifier Training

The latest version of rdp in QIIME 2 acceptable format can be downloaded from here: https://forum.qiime2.org/t/rdp-reference-data base-in-qiime2-format/15146

The rdp.fa file should be converted to uppercase and this can be done by using r studio by following the command:

```
Fasta = readLines("/path/to/input.fasta"")
new = toupper(fasta)
writeLines(new, "path/to/output_uppercase.fasta")
```

3.2.5.4.1 Import the FASTA Sequences as Feature Data Using Command

```
qiime tools import \
&mdash;type 'FeatureData[Sequence]' \
&mdash;input-path /path/to/input_sequences.fasta \
&mdash;output-path /path/to/output_sequences.qza
```

3.2.5.4.2 Import Taxonomy as Feature Data Using Command

```
qiime tools import \
&mdash;type 'FeatureData[Taxonomy]' \
&mdash;input-format HeaderlessTSVTaxonomyFormat \
&mdash;input-path /path/to/input_taxonomy.tsv \
&mdash;output-path /path/to/output_taxonomy.qza
```

QIIME2 recommends training a classifier on sequences that have been extracted using a primer but remain untrimmed for classification using paired-end reads.

```
qiime feature-classifier extract-reads \
&mdash;i-sequences /path/to/ /rdp/ rdp_otus.qza\
&mdash;p-f-primer FORWARD_PRIMER_SEQUENCE \
&mdash;p-r-primer REVERSE_PRIMER_SEQUENCE \
&mdash;o-reads /path/to/rdp/ref-seqs.qza
```

The input is the rdp_otus.qza file, and the output is the ref-seqs.qza file.

3.2.5.4.3 After Classification, the Classifier Is Trained Using The Naïve Bayes Classifier

```
qiime feature-classifier fit-classifier-naive-bayes \ &mdash;i-reference-reads /rdp/ref-seqs.qza \ &mdash;i-reference-taxonomy /rdp/rdp-taxonomy.qza \ &mdash;o-classifier /rdp/classifier.qza
```

This command uses reference sequences and taxonomic artifacts as input and outputs a new classifier based on the RDP reference dataset that will be specific to the used primers.

3.2.5.5 Assign Taxonomy

Ensure RDP classifier "classifier.qza" saved in the same directory as representative sequences of interest and then use the following command to assign taxonomy and create a QIIME2 feature table:

```
qiime feature-classifier classify-sklearn \
&mdash;i-classifier /path/to/classifier.qza \
&mdash;i-reads /path/to/representative_sequences.qza \
&mdash;o-classification /path/to/output_taxonomy.qza
```

Input of command is classifier.qza: pre-trained Naïve Bayes classifier (e.g., RDP), and representative_sequences.qza: ASVs or OTUs from DADA2 or clustering.

Output is output_taxonomy.qza, which is the taxonomic classification of each feature.

3.2.5.6 Data Analysis

3.2.5.6.1 Kmer-Based Diversity Analysis

QIIME2 provides a bootstrapped plugin that performs bootstrapped diversity estimation using k-mer composition instead of alignment-based phylogenetic trees (Jenike et al., 2025; Moeckel et al., 2024). The command implements a bootstrapping framework to estimate alpha and beta diversity metrics.

```
qiime boots kmer-diversity \
&mdash;i-table asv-table.qza \
&mdash;i-sequences asv-seqs.qza \
&mdash;m-metadata-file sample-metadata.tsv \
&mdash;p-sampling-depth 10000 \
&mdash;p-n 100 \
&mdash;p-replacement \
&mdash;p-alpha-average-method median \
&mdash;p-beta-average-method medoid \
&mdash;output-dir boots-kmer-diversity
```

Output alpha-diversity.qzv used to plot sample diversity (e.g., Shannon, Simpson indices), beta-diversity.qzv used for ordination plots (e.g., NMDS, PCoA) based on Bray-Curtis or Jaccard distances.

3.2.5.6.2 Alpha Rarefaction Analysis

This analysis is used to evaluate whether sequencing depth is sufficient to capture the diversity of microbial communities within samples, and it generates rarefaction curves that observe diversity metrics as a function of sequencing depth (Wang et al., 2021; Willis, 2019). The qime diversity alpha-rarefaction command subsamples the feature table at multiple sequencing depths and calculates alpha diversity metrics for each subsample.

```
qiime diversity alpha-rarefaction \
&mdash;i-table asv-table.qza \
&mdash;p-max-depth 260 \
&mdash;m-metadata-file sample-metadata.tsv \
&mdash;o-visualization alpha-rarefaction.qzv
```

The alpha-rarefaction.qzv provides a plot where the x-axis represents sequencing depth, the y-axis represents an alpha diversity metric (e.g., Observed ASVs, Shannon index), and each line corresponds to a sample.

3.2.5.6.3 Bar Plot Analysis

Next, to prepare a metadata file to create bar plots of various taxa per sample by using taxonomic output from above, and combined with the abundance table.

```
qiime taxa barplot \
&mdash;i-table /path/to/feature_table.qza \
&mdash;i-taxonomy /path/to/output_taxonomy.qza \
&mdash;m-metadata-file /path/to/sample_metadata.tsv \
&mdash;o-visualization /path/to/output_taxa_barplot.qzv
```

Input is feature_table.qza: feature (ASV/OTU) table from denoising step (Fasolo et al., 2024; Jeske & Gallert, 2022), output_taxonomy.qza: taxonomic classifications, sample_metadata.tsv: sample metadata file in tab-delimited format. Output is output_taxa_barplot.qzv: interactive bar plots showing taxa composition per sample.

The output_taxa_barplot.qzv can be viewed on the QIIME 2 viewer (https://view.qiime2.org/).

3.2.5.6.4 Differential Abundance Analysis with ANCOM-BC2

The ANCOM-BC2 plugin used to test differences in feature abundance between groups while accounting for sample metadata variables (Hossine et al., 2025; Leech et al., 2025).

```
qiime feature-table filter-samples \
&mdash;i-table /path/to/input_feature_table.qza \
&mdash;m-metadata-file /path/to/sample_metadata.tsv \
&mdash;p-where '[MetadataColumn] IN ("Group1", "Group2", "Group3")' \
&mdash;o-filtered-table /path/to/output_filtered_table.qza
```

The next command performs differential abundance testing using ANCOM-BC2 method which corrects compositional biases and provides statistically robust identification of taxa differing between groups.

```
qiime composition ancombc2 \
&mdash;i-table /path/to/filtered_feature_table.qza \
&mdash;m-metadata-file /path/to/sample_metadata.tsv \
&mdash;p-fixed-effects-formula MetadataColumn \
&mdash;p-reference-levels 'MetadataColumn::ReferenceGroup' \
&mdash;o-ancombc2-output /path/to/ancombc2_results.qza
```

The output ancombc2-results.qza contains statistical test results for differentially abundant taxa.

ANCOM-BC2 visualization step integrates taxonomic information with statistical results to produce interpretable plots.

```
qiime composition ancombc2-visualizer \
&mdash;i-data /path/to/ancombc2_results.qza \
&mdash;i-taxonomy /path/to/taxonomy.qza \
&mdash;o-visualization /path/to/ancombc2_barplot.qzv
```

Barplot (.qzv) illustrates differential abundance patterns, which allows visual exploration of results. The complete amplicon pipeline is illustrated in Figure 3.1 for a clearer understanding of the pipeline, including all steps and the software used at each stage.

All software used in Amplicon and whole shotgun analysis is summarized in Table 3.1 to facilitate better understanding and provide downloading links.

3.3 SHOTGUN METAGENOMICS DATA ANALYSIS PIPELINES

3.3.1 Pre-Processing of Metagenomic Reads Using Kneaddata

Kneaddata is a widely used tool designed to perform trimming of low-quality reads, quality filtering, and host-derived sequence removal to generate high-quality microbial reads (*KneadData—The Huttenhower Lab, n.d.*) (see Figure 3.2). The workflow selection process for shotgun metagenomic analysis is illustrated in Figure 3.2, which shows a decision tree outlining the logical progression of tool selection and analysis strategies.

3.3.1.1 Installation

Install kneadata by using the following command:

```
pip install kneaddata
```

Download the reference database:

```
kneaddata_database &mdash;download human_genome bowtie2 $DIR
```

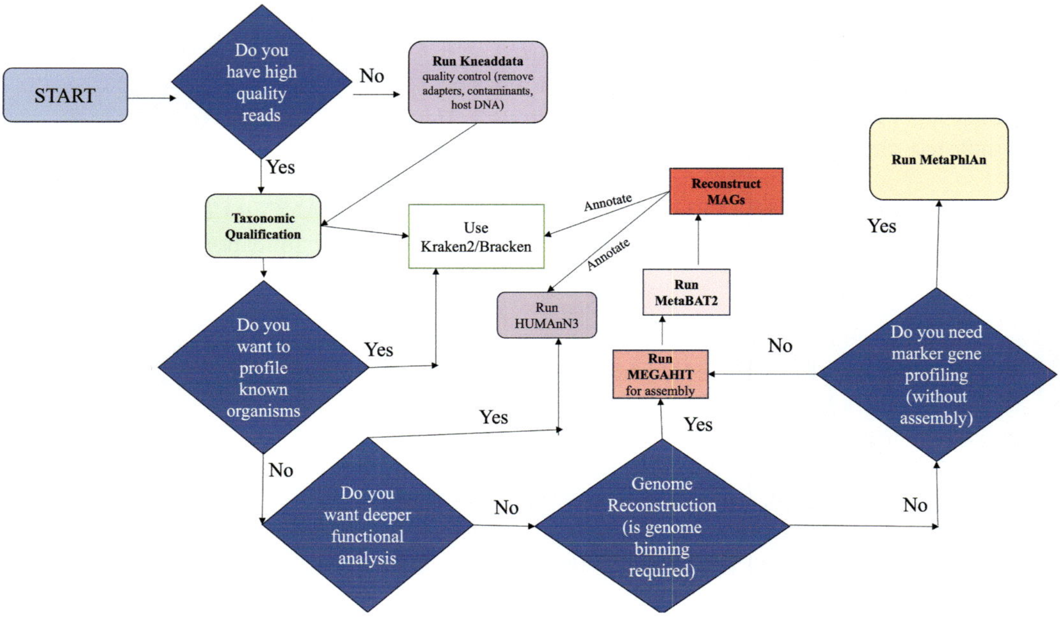

FIGURE 3.2 Decision tree to analyze shotgun metagenomic data based on their sample type, analytic needs, and research objectives. The diagram specifies when to use Kneaddata for quality checking, MEGAHIT for assembly, MetaBAT2 for binning, Kraken2/Bracken for taxonomic classification, and HUMAnN3 for functional profiling.

3.3.1.2 Preparation of Database

Kneaddata requires a Bowtie2 database of host genome for contamination removal, it use Bowtie2 to align reads against a reference host genome and discards matching reads.

```
kneaddata_database &mdash;download reference_genome bowtie2 /path/to/kneaddata_db/
reference_genome: for example human_genome.
```

The input files should ensure that raw reads are in FASTQ format (single or paired-end) from sequencing, and a pre-built Bowtie2 database.

3.3.1.3 Processing of Raw Reads

3.3.1.3.1 Single-End Reads

For single-end data, the unpaired flag is used to specify the input file; the command trimms low-quality reads and host sequences by using the specified reference database.

```
kneaddata \&mdash;unpaired /path/to/input_singleEnd.fastq \ &mdash;reference-db /path/
   to/host_database/ \ &mdash;output /path/to/output_dir/
```

Input of reference-db used to give the path to Bowtie2 database used for host sequence removal.

```
kneaddata_read_count_table \&mdash;input /path/to/kneaddata_output/ \&mdash;output /
   path/to/kneaddata_read_count_table.tsv
```

After kneaddata preprocessing, the number of reads can be summarized using the kneaddata_read_count_table utility, which will produce a table summarizing read counts for all processed samples.

3.3.1.3.2 Kneaddata with FASTQC

The following command runs Kneaddata on a single-end FASTQ file, generating FastQC reports both before and after trimming, and this provides a visual quality assessment at raw data and processed data.

```
kneaddata \ &mdash;unpaired /path/to/input_singleEnd.fastq \ &mdash;reference-db /
  path/to/host_database/ \ &mdash;output /path/to/output_dir/ \ &mdash;run-fastqc-start \
  &mdash;run-fastqc-end
```

3.3.1.3.3 Paired-End Reads

Kneaddata uses input1(forward read) and input2(reverse read) to specify the forward and reverse reads, respectively. The following command processes paired reads by trimming, quality filtering, and removing host-derived sequences.

```
kneaddata\&mdash;input1/path/to/sample_R1.fastq\&mdash;input2/path/to/sample_R2.fastq\
  &mdash;reference-db /path/to/kneaddata_db/ \ &mdash;output /path/to/output_dir/
```

Kneaddata efficiently performs quality trimming and host contamination removal, which produces high-quality microbial reads. The next step involves downstream processing, including taxonomic profiling (e.g., Kraken2, MetaPhlAn) and functional annotation (HUMAnN).

3.3.2 Metagenome Assembly Using MEGAHIT

MEGAHIT is a highly efficient assemble used for assembling large and complex metagenomic

Datasets, and it supports both single end and paired end reads. It is widely used due to its high speed performance and low memory footprint (*MEGAHIT: An Ultra-Fast Single-Node Solution for Large and Complex Metagenomics Assembly via Succinct de Bruijn Graph | Bioinformatics | Oxford Academic, n.d.*).

3.3.2.1 Downloading and Extraction of MEGAHIT

MEGAHIT can directly be download from GitHub and extracted for use by using following commands:

```
wget https://github.com/voutcn/megahit/releases/download/v1.2.9/MEGAHIT–1.2.9-Linux-
  x86_64-static.tar.gz tar zvxf MEGAHIT–1.2.9-Linux-x86_64-static.tar.gz
cd MEGAHIT–1.2.9-Linux-x86_64-static/bin/
```

and after downloading, the verification of installation can be done by using built in test:

```
./megahit—test
```

The above command will use the dataset provided with the package to ensure that MEGAHIT is functioning as expected.

MEGHIT assembles metagenomes from various read library formats, including paired-end, interleaved, and single-end reads.

3.3.2.2 Assembling a Single Paired-End Library

```
megahit -1 pe_1.fq -2 pe_2.fq -o out
```

In the above command, -1 denotes a forward read file, -2 denotes a reverse read file, and -o is the output directory where the assembled contigs will be saved, resulting in out/final.contigs.fa along with log and intermediate files.

3.3.2.3 Assembling an Interleaved Paired-End Library

For interleaved reads where forward reads and reverse reads are in the same file:

```
megahit&mdash;12 interleaved.fq -o out
```

In command, —12 specifies an interleaved FASTQ file containing both forward and reverse reads.

3.3.2.4 Assembling Multiple Paired-End Libraries and Single-End Libraries

MEGAHIT also allows combining multiple paired-end libraries and single-end libraries with a single assembly run command:

```
megahit -1 a1.fq,b1.fq,c1.fq -2 a2.fq,b2.fq,c2.fq -r se1.fq,se2.fq -o out
```

In the above command, -1 is comma separated list of forward reads form multiple libraries, -2 is comma separated list of reverse reads corresponding to previous (-1) libraries and -r is comma separated list of single-end reads files.

3.3.2.5 Extraction of FASTG Format from Intermediate Contigs

MEGAHIT stores intermediate contigs for different k-mer sizes in its output and megahit_core contig2fastg command converts those intermediate contigs into FASTG format which is used for the visualization of assembly graphs.

```
megahit_core contig2fastg 119 out/intermediate_contigs/k119.contig.fa > k119.fastg
```

Command 119 represents the k-mer size for intermediate contigs, out/intermediate_contigs/k119.contig.fa is the intermediate contigs file for k=119, and k119.fastg is output assembly graph in FASTG format.

3.3.3 Binning by MetaBAT2

Binning is the process of grouping assembled metagenomic contigs into bins and each bin represents a potential genome which is called as Metagenome-Assembled Genome (MAG). It helps in reconstructing genomes from complex microbial communities and relies on features like sequence composition (like tetranucleotide frequency) and coverage patterns across samples. MetaBAT2 is a widely used tool for binning contigs from metagenomic assemblies into MAGs.

Before running MetaBAT2, two inputs are required, 1. Assembled contigs (FASTA file) like final.contigs.fa (output from MEGAHIT) and 2. Read mapping coverage for each contig (BAM) file which is generated by aligning reads back to contigs (Kang et al., 2019).

3.3.3.1 Indexing of Contigs

This step indexes the assembled contigs to allow alignment reads using BWA and below command prepare the contig FASTA for alignment.

```
bwa index megahit_out/final.contigs.fa
```

3.3.3.2 Align Reads to Contigs

This step align paired-end reads to contigs using BWA and output a BAM file and map reads back to the assembled contigs to calculate coverage depth.

```
bwa mem -t 8 megahit_out/final.contigs.fa pe_1.fq pe_2.fq | samtools view -Sb&mdash;> aln.bam
```

3.3.3.3 Sorting and Indexing of BAM File

Meta BAT2 requires sorted and indexed BAM files to calculate depth per contig, the following command will sort and index the BAM file using samtools.

```
samtools sort aln.bam -o aln.sorted.bam
samtools index aln.sorted.bam
```

3.3.3.4 Depth File Generation

The following command will generate a depth table containing coverage information for each contig, and jgi_summarize_bam_contig_depths will generate a depth file.

```
jgi_summarize_bam_contig_depths &mdash;outputDepth depth.txt aln.sorted.bam
```

3.3.3.5 Meta BAT2 Run for Binning

MetaBAT2 with contig FASTA and depth file. This command performs binning, generates multiple. fa files (bins), and each file potentially represents a MAG.

```
metabat2 -i megahit_out/final.contigs.fa -a depth.txt -o bins/bin
```

3.3.4 Data Analysis by Kraken2

Kraken2 is a taxonomic sequence classifier that is used to rapidly assign DNA sequences to their corresponding taxonomic labels. The classification is based on exact k-mer matching strategy combined with lowest common ancestor mapping (*Improved Metagenomic Analysis with Kraken 2 | Genome Biology | Full Text*, n.d.).

3.3.4.1 Kraken2 Installation

Kraken 2 has two primary scripts: kraken2 (for classification) and kraken2-build (for database construction), and several supporting programs and smaller scripts. After downloading the source package from the official website, run the following code to install:

```
./install_kraken2.sh /path/to/kraken2_dir/
```

3.3.4.2 Kraken2 Build

The command k2 build is the core utility used to construct Kraken2 databases for standard or custom. This command enables the customization of k-mer parameters, minimizers and database size which makes it flexible for different metagenomic analysis needs.

```
k2 build &mdash;db /path/to/kraken2_db/
```

To build standard database, run following command:

```
k2 build &mdash;standard &mdash;db /path/to/kraken2_db/ &mdash;threads 16
```

3.3.4.3 Kraken2 Inspect

The k2 inspect command is used to examine the contents of the Kraken2 database, used to provide a database composition summary and optionally list the count for all taxa included in the database. This tool is used for validating database construction and understanding taxonomic structure prior to classification.

```
k2 inspect &mdash;db /path/to/kraken2_db/
```

The following command saves inspection results to a file:

```
k2 inspect &mdash;db /path/to/kraken2_db/ &mdash;output db_inspection.txt
```

The following command displays taxonomy on MetaPhlAn Style:

```
k2 inspect &mdash;db /path/to/kraken2_db/ &mdash;use-mpa-style &mdash;output db_
  mpa_format.txt
```

3.3.4.4 Standard Kraken2 Database

To create a standard Kraken2 database, the following command is used and make sure that your disk has 100 gb of space during creation. This command will download the NCBI taxonomic information and complete genomes in RefSeq for bacterial, archeal, and viral domains along with the human genome.

```
kraken2-build &mdash;standard &mdash;db $DBNAME
```

After downloading the taxonomic data, the build command is used. This command passes the content of the reference library to estimate the number of minimizers present in the reference library and conversion of taxonomy to an internal format.

```
kraken2-build &mdash;standard &mdash;threads 24 &mdash;db $DBNAME
```

$DBNAME change this according to your preferred database name/location.

3.3.4.4.1 Classification

The kraken2 command is used for taxonomic classification of sequencing reads against the Kraken2 database. It processes input reads like single end or paired end and matches k-mers to the database, then assigns taxonomic labels using Lowest Common Ancestor (LCA) approach.

```
kraken2 &mdash;db /path/to/kraken2_db/ &mdash;threads 16 seqs.fa
```

and the following command is used to save separate files of classified and unclassified reads:

```
kraken2 &mdash;db /path/to/kraken2_db/ \&mdash;classified-out classified_reads.fq \
  &mdash;unclassified-out unclassified_reads.fq \ seqs.fastq
```

for Paired-End Reads handling use following command:

```
kraken2 \ &mdash;db /path/to/kraken2_db/ \&mdash;paired \ clean_R1.fastq clean_R2.fastq
  \ &mdash;output paired_output.txt \ &mdash;report paired_report.txt
```

in output, classified_reads_1.fq denotes Forward reads and classified_reads_2.fq denotes Reverse reads.

3.3.4.4.2 Kraken2 Output Formats

The tool creates two primary types of outputs; first is standard read-level classification output, and second is sample-wide taxonomic reports.

3.3.4.4.2.1 Standard Kraken2 Output Format

Kraken2 produces a single line of classification having five tab-delimited fields. The first field indicates classification status where "C" denotes classified reads and "U" denotes unclassified reads. The second field contains the sequence ID, the third field lists the NCBI taxonomy ID assigned to the read, with 0 indicating an unclassified read. The fourth field displays the lengths of the sequences and finally, fifth field is a space-separated list detailing the Lowest Common Ancestor

(LCA) mapping of each k-mer in the read. When the—ise-names option is used in command, Kraken2 replaces numeric taxonomic IDs with scientific names and their associated taxID in parentheses.

3.3.4.4.2.2 Sample Report Output Format

Kraken2 creates a sample wide taxonomic report when the report flag is specified. The report is a tab-delimited file where each line represents a taxon and summarizes abundance in sample. These fields include percentage of fragments classified under clade rooted at taxon, total number of fragments in that clade, number of reads directly assigned to taxon, rank code denoting its taxonomic level (U for unclassified, D for domain, K for kingdom, P for Phylum, C for class. O for order, F for family, G for genus, and S for species). The taxa with no reads assigned to them are excluded from the report, so enabling the report-zero-counts option ensures that all taxa, even with zero counts are displayed which is useful for comparative analyses across multiple samples. And use-mpa-style option will generate the MetaPhlAn-compatible report format.

3.3.4.5 Custom Database

The standard databases may not be suitable for everyone's needs, so kraken2 also allows creation of customized databases.

3.3.4.5.1 Build Custom Database

First, the NCBI taxonomy is downloaded using the following command. This default NCBI taxonomy ensures compatibility with most standard workflows, while advanced users can modify taxonomy files for custom classification needs.

```
kraken2-build &mdash;download-taxonomy &mdash;db $DBNAME
```

This command retrieves the access number to taxon ID maps, taxonomic names, taxonomic tree structure. Those who wish to customize taxonomy (add or reassign specific taxa) can edit files like names.dmp to contain taxonomic names, nodes.dmp defines taxonomic relationships in tree structure and *.accession2taxid for maps sequence accessions to taxonomic IDs. After these modifications, rebuild the database to apply changes.

3.3.4.5.2 Reference Libraries Installation in Kraken2

The next step after installing taxonomy is database preparation, downloading and integrating reference libraries. Kraken2 provides different preconfigured libraries consisting complete genomes or proteins from NCBI RefSeq and different other sources. These reference datasets are the basis for classifying sequences during metagenomic analysis.

The following command is used to download single library, for example, bacterial genomes:

```
kraken2-build &mdash;download-library bacteria &mdash;db /path/to/kraken2_db/
```

Kraken2 also supports the addition of custom sequences, the following command is used to add custom FASTA files:

```
kraken2-build &mdash;add-to-library chr1.fa &mdash;db /path/to/kraken2_db/
kraken2-build &mdash;add-to-library chr2.fa &mdash;db /path/to/kraken2_db/
```

Or add multiple files by using a loop:

```
for file in chr*.fa; do
kraken2-build &mdash;add-to-library $file &mdash;db /path/to/kraken2_db/
done
```

Or, you can add all FASTA files from a directory:

```
find genomes/ -name '*.fa' -print0 | xargs -0 -I{} -n1 kraken2-build &mdash;add-to-library {}
    &mdash;db /path/to/kraken2_db/
```

after adding libraries or custom sequences, build and finalize the database:

```
kraken2-build &mdash;build &mdash;db $DBNAME &mdash;threads 16
```

After building the database, intermediate files can be removed to reduced disk space, this command deletes temporary build files while preserving the final database required for classification.

```
kraken2-build &mdash;clean &mdash;db /path/to/kraken2_db/
```

After building the database, the classification process is similar to the standard database pipeline.

3.3.5 Abundance Estimation by Bracken

Bracken (Bayesian Reestimation of Abundance with Kraken) is a complementary tool for Kraken2 which is designed to provide more accurate species or genus level abundance in metagenomics samples. Kraken2 assigns reads to Lowest Common Ancestor (LCA) and Bracken uses Bayesian probability and k-mer distribution models to redistribute reads from higher to lower taxonomic levels (Lu et al., 2017).

3.3.5.1 Bracken Build

Bracken requires a k-mer distribution file generated by Kraken classifiers kmers for each genome in the database.

```
./bracken-build \ -d /path/to/kraken2_db/ \ -t 16 \ -k 35 \ -l 100 \-x /usr/local/bin/kraken2
```

—d is for Kraken database directory, -t is for Number of threads, -k is for K-mer length used to build the database (default 35 for Kraken2), -l is for Read length for modeling (default 100 bp), -x is for Optional path to Kraken executable.

3.3.5.2 Classification

After building the Bracken, classify all library input sequences:

```
kraken2 \ &mdash;db /path/to/kraken2_db/ \ &mdash;threads 16 \ <(find /path/to/kraken2_
    db/library -name "*.fa*" -exec cat {} +) \ > database.kraken
```

Compute the classifications for each perfect read

```
./kmer2read_distr\&mdash;seqid2taxid/path/to/kraken2_db/seqid2taxid.map\&mdash;taxonomy
    /path/to/kraken2_db/taxonomy \ &mdash;kraken database.kraken \ &mdash;output databas-
    e100mers.kraken \
-k 35 \ -l 100 \ -t 16
```

After that, generate kmer distribution file:

```
python generate_kmer_distribution.py \-i database100mers.kraken \ -o database100mers.
    kmer_distrib
```

Bracken requires Kraken report (.kreport file) as input. Run the following command using Kraken2:

```
kraken2 \ &mdash;db /path/to/kraken2_db/ \&mdash;threads 16 \ &mdash;report sample.
    kreport2 \ sample.fastq \ > sample.kraken2
```

3.3.5.3 Abundance Estimation by Bracken

Abundance estimation can be run using bracken shell script or est_abundance python script. The shell script checks the database for its necessary files and the python script requires specification of the exact path to the database files.

Following command is used bracken by shell script:

```
bracken \-d /path/to/kraken2_db/ \ -i sample.kreport2 \ -o sample.bracken \ -r 100 \ -l S \ -t 10
```

where -r is used for read length in classification, -l is used for taxonomic level, and -t is used for the minimum read threshold for abundance reallocation.

The following command is used Bracken by python script:

```
python estimate_abundance.py \-i sample.kreport2 \-k /path/to/kraken2_db/database100mers.
    kmer_distrib \-o sample.bracken \ -l S \ -t 10
```

The bracken output file contains taxon name, taxonomy ID level (S=species, G=genus, etc.), Kraken assigned reads, added reads (after Bayesian redistribution), total reads (after abundance estimation) and fraction of total reads.

Bracken enhances the accuracy of taxonomic profiling by redistributing reads which are assigned by Kraken2 and specifically, taxa based on read-length-aware k-mer distributions. So this step makes Bracken a crucial downstream step for studying quantitative microbiome profiling at species or genus resolution.

3.3.6 Visualization by Krona

Krona tool is used for generating interactive hierarchical pie charts which is used to visualize the metagenomic classification results, this can integrate taxonomic data from multiple tools, including Kraken2, Bracken and other and display taxonomic composition interactively in Krona web browser (Ondov et al., 2011).

Krona requires two column input file having read ID and taxon name (or taxID), this can extract from Kraken2 classification output using following command:

```
cut -f2,3 sample.kraken > krona_input.txt
```

Also, Bracken output can be converted into Krona-compatible file by extracting taxonomic names and associated abundance values:

```
awk '{print $1"\t"$2}' sample.bracken > krona_bracken_input.txt
```

The ktImportTaxonomy is used to generate an interactive HTML visualization using following command:

```
ktImportTaxonomy krona_bracken_input.txt -o krona_bracken.html
```

and for Bracken outputs:

```
ktImportTaxonomy krona_bracken_input.txt -o krona_bracken.html
```

Krona can combine multiple datasets for comparative visualization:

```
ktImportTaxonomy -o krona_combined.html sample1_krona.txt sample2_krona.txt
```

and each dataset is displayed as a separate layer in the final chart.

3.3.7 MetaPhlAn3 for Taxonomic Profiling

MetaPhlAn3 (METAgenomic Phylogenetic Analysis) is a widely used bioinformatics tool to profile the composition of microbial communities from metagenomic shotgun sequencing data. The MetaPhlAn uses a curated database of clade-specific marker genes to achieve accurate species-level classification with minimal false positives (Blanco-Míguez et al., 2023; *MetaPhlAn4—The Huttenhower Lab, n.d.*).

```
metaphlan INPUT_FILE OUTPUT_FILE \ &mdash;input_type {fastq,fasta,bowtie2out,sam}
  \ &mdash;bowtie2db /path/to/metaphlan_db \ -x mpa_v30_CHOCOPhlAn_201901 \
  &mdash;nproc 8
```

In input -x is the index name of database (e.g., mpa_v30_CHOCOPhlAn_201901)

3.3.7.1 For Example If Have FASTQ Reads and Want Taxonomic Profiling

```
metaphlan sample.fastq sample_profile.txt \&mdash;input_type fastq \&mdash;bowtie2db /
  path/to/metaphlan_db \-x mpa_v30_CHOCOPhlAn_201901 \ &mdash;nproc 8
```

3.3.7.2 for Taxonomic Profiling from Pre-Aligned Bowtie2 Output, Use Following Command

```
metaphlan sample.bowtie2out sample_profile.txt \ &mdash;input_type bowtie2out \ &mdash;
  bowtie2db /path/to/metaphlan_db \ -x mpa_v30_CHOCOPhlAn_201901 \ &mdash;nproc 8
```

heatmap generated by hclust2, which is a convenience script, uses tabular abundance results from MetaPhlAn, and this can be downloaded by using the following command:

```
conda install -c biobakery hclust2
```

now to generate species only abundance table:

```
grep -E “s__|clade” merged_abundance_table.txt | sed ‘s/^.*s__//g’\ | cut -f1,3–8 | sed -e ‘s/
  clade_name/body_site/g’ > merged_abundance_table_species.txt
```

Now next step is to generate species only heatmap by running following command:

```
hclust2.py -i merged_abundance_table_species.txt -o abundance_heatmap_species.png
  &mdash;f_dist_f braycurtis &mdash;s_dist_f braycurtis &mdash;cell_aspect_ratio
  0.5 -l &mdash;flabel_size 10 &mdash;slabel_size 10 &mdash;max_flabel_len 100
  &mdash;max_slabel_len 100 &mdash;minv 0.1 &mdash;dpi 300
```

MetaPhlAn output can be directly imported into HUMAnN3 for functional profiling, Phyloseq (R) for diversity analyses and LEfSe for biomarker discovery.

3.3.8 Abundance of Microbial Metabolic Pathways Analysis by HUMAnN3

HUMAnN (the HMP Unified Metabolic Analysis Network) is a method used to profile the abundance of microbial metabolic pathways and other molecular functions from metagenomic sequencing data or metatranscriptomic sequencing data (*Humann3—The Huttenhower Lab, n.d.*).

```
humann \ &mdash;input sample.fastq \ &mdash;output humann_results/ \ &mdash;threads 8
```

the above command will remove the blank lines and identify community species (with MetaPhlAn, it will map reads to community pangenomes with bowtie2 and then align unmapped reads to protein databases with DIAMOND. And in the output, this produces Gene family abundance tables(output_genefamilies.tsv), Pathway abundance tables(output_pathabundance.tsv), Pathway coverage tables(output_pathcoverage.tsv).

3.3.8.1 HUMAnN Output Table Manipulation

3.3.8.1.1 RPKs Normalization to Relative Abundance

It is important to normalize the HUMAnN's default RPK values prior to performing statistical analyses to facilitate comparisons between samples with different sequencing depths and sum normalization to relative abundance or copes per million (cpm) units are common approaches for this. These methods will be implemented in the human_renorm_table script (with considerations for HUMAnN's output formats, e.g., feature stratification). The following command will normalize gene abundance into CPM units:

```
humann_renorm_table \ &mdash;input <path_to_genefamilies.tsv > \ &mdash;output <path_
   to_normalized_genefamilies-cpm.tsv > \ &mdash;units cpm &mdash;update-snames
```

The output files will be updated from output_Abundance-RPK to output_Abundance-CPM courtesy of the —update-snames flag. After normalization, the output includes the gene family abundance table. To facilitate the functional interpretation, these gene families can be regrouped into higher level functional categories (e.g. KEGG orthologs, MetaCyc pathways).

```
humann_regroup_table \ &mdash;input <path_to_input_genefamilies.tsv> \ &mdash;output
   <path_to_output_regrouped.tsv> \ &mdash;groups uniref90_rxn
```

3.3.8.1.2 Renaming Functional Features

After normalization and regrouping, the output resulting table may still contain cryptic identifiers. The human_rename_table command converts IDs into human-relatable names using built-in mapping files.

```
humann_rename_table\&mdash;input/project/pathway_abundance-cpm.tsv\&mdash;output
   /project/pathway_abundance_named.tsv \ &mdash;names metacyc-rxn
```

The—input is the HUMAnN3 table where features need renaming (gene families or pathways), and—output is the file with human-readable feature names. —names specify the mapping to use for renaming IDs. Common options: metacyc-rxn maps MetaCyc reaction IDs to descriptive reaction names, uniref90 → Maps UniRef90 IDs to UniProt protein names, uniref50 → Maps UniRef50 IDs to UniProt protein names, ko → Maps KEGG Orthology IDs to KEGG pathway descriptions, custom mapping file (two-column tab-delimited: ID → Name).

3.3.8.2 Multiple Sample Processing by HUMAnN3

For analyzing the multiple metagenomic samples, the common way to execute HUMAnN on each sample individually and then merge results into a combined table for downstream analysis.

Following command will process sample individually by specifying the input file with -I and the output directory with -o:

```
for f in *.fasta; do humann -i $f -o hmp_subset; done
```

this command will generate per sample outputs: gene family abundance tables, Pathway abundance tables, Pathway coverage tables

3.3.8.2.1 Merging Gene Family Tables

After processing all samples, merge the individual gene family outputs into a single multi-sample table:

```
humann_join_tables \ -i hmp_subset \ -o hmp_subset_genefamilies.tsv \ &mdash;file_name genefamilies
```

3.3.8.2.2 Normalization of Merged Data

The output raw RPK (Reads per Kilobase) abundances are sensitive to sequencing depth and should be normalized before comparison into CPM.

```
humann_renorm_table \ -i hmp_subset_genefamilies.tsv \ -o hmp_subset_genefamilies-cpm.tsv \ &mdash;units cpm
```

3.3.8.2.3 Stratified Function Exploration

Stratified outputs like pathway abundances with taxonomic contributions can be explored to identify taxa contributing to each pathway. Following example hmp_pathabund.pcl file will inspect the table:

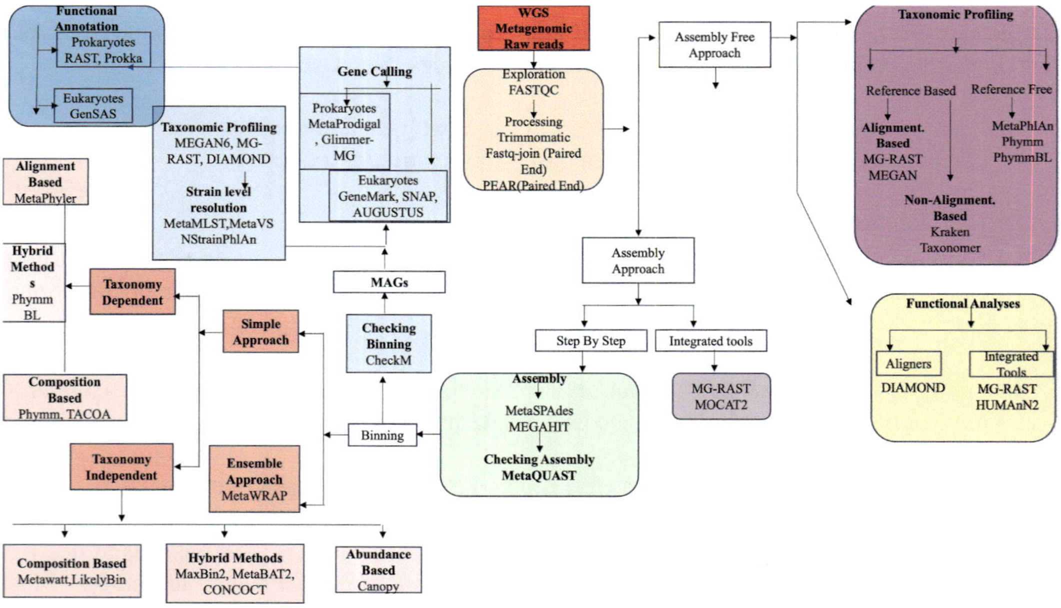

FIGURE 3.3 Workflow of whole genome shotgun (WGS) metagenomic analysis, which includes taxonomic profiling, gene calling, assembly, data exploration, and functional annotation. Alignment-based, reference-based, and hybrid methodologies are among the tools and techniques that are highlighted to facilitate precise analysis and interpretation of the microbial population.

```
less -S hmp_pathabund.pcl
```

if you want to extract stratified entries for a specific pathway, for example, L-homoserine and L-methionine biosynthesis (METSYN-PWY):

```
grep 'METSYN-PWY' hmp_pathabund.pcl | less -S
```

This command will reveal which taxa contribute to this pathway across samples.

2.8.2.4. Visualization of HUMAnN outputs

After normalizing and processing HUMAnN pathway abundance tables, the visualization is an important step to interpret functional differences across sample groups. HUMAnN provides the human_barplot utility to generate publication-ready barplots (see Figure 3.3).

```
humann_barplot \&mdash;input <path_to_abundance_table.pcl> \ &mdash;focal-metadata
  <metadata_column> \ &mdash;last-metadata <metadata_column> \ &mdash;output <out-
  put_plot.png> \ &mdash;focal-feature <feature_ID>
```

3.4 CONCLUSION

The study of metagenomic data offers a potent paradigm for thoroughly describing microbial populations in a variety of environments, including the human body. The process usually starts with raw sequencing read quality control, then moves on to taxonomic profiling using Kraken2 and refinement with Bracken, which allows precise microbial taxonomic identification and quantification. Community composition and metabolic capacity are further linked through functional profiling using pipelines like HUMAnN3, and interpretable insights into microbial functions are provided by downstream processes including regrouping, renormalization, and visualization (e.g., barplots and Krona charts). For assembly-based methods, binning algorithms like MetaBAT2 enable the reconstruction of almost entire metagenome-assembled genomes (MAGs), whereas programs like MEGAHIT reconstruct contigs from metagenomic data. Every stage adds to a multifaceted understanding of the structure, dynamics, and function of the microbiome in health and illness, from processing raw data to interpreting it functionally and ecologically. Thus, the most thorough insights are obtained by combining taxonomic and functional analysis, opening the door for translational applications in microbiome engineering, diagnostics, and treatments.

3.5 CONSENT FOR PUBLICATION

All authors of this chapter have read and approved the final manuscript and give their consent for publication. The images included in this article are originally generated and data are included with their proper citations or following ethical guidelines.

3.6 CONFLICT OF INTEREST

The authors of this chapter declare that they have no competing interests that could influence the interpretations presented in this study.

ACKNOWLEDGMENT

The authors of this chapter would like to express their sincere gratitude to all the individuals and institutions that contributed to this work.

REFERENCES

Ansorge, R., Birolo, G., James, S. A., & Telatin, A. (2021). Dadaist2: A toolkit to automate and simplify statistical analysis and plotting of metabarcoding experiments. *International Journal of Molecular Sciences*, *22*(10), 5309. https://doi.org/10.3390/ijms22105309

Beghini, F., McIver, L. J., Blanco-Míguez, A., Dubois, L., Asnicar, F., Maharjan, S., Mailyan, A., Manghi, P., Scholz, M., Thomas, A. M., Valles-Colomer, M., Weingart, G., Zhang, Y., Zolfo, M., Huttenhower, C., Franzosa, E. A., & Segata, N. (n.d.). Integrating taxonomic, functional, and strain-level profiling of diverse microbial communities with bioBakery 3. *eLife*, *10*, e65088. https://doi.org/10.7554/eLife.65088

Bharti, R., & Grimm, D. G. (2021). Current challenges and best-practice protocols for microbiome analysis. *Briefings in Bioinformatics*, *22*(1), 178–193. https://doi.org/10.1093/bib/bbz155

Blanco-Míguez, A., Beghini, F., Cumbo, F., McIver, L. J., Thompson, K. N., Zolfo, M., Manghi, P., Dubois, L., Huang, K. D., Thomas, A. M., Nickols, W. A., Piccinno, G., Piperni, E., Punčochář, M., Valles-Colomer, M., Tett, A., Giordano, F., Davies, R., Wolf, J., . . . Segata, N. (2023). Extending and improving metagenomic taxonomic profiling with uncharacterized species using MetaPhlAn 4. *Nature Biotechnology*, *41*(11), 1633–1644. https://doi.org/10.1038/s41587-023-01688-w

Bolger, A., Lohse, M., & Usadel, B. (2014). Trimmomatic: A flexible trimmer for Illumina sequence data. *Bioinformatics (Oxford, England)*, *30*. https://doi.org/10.1093/bioinformatics/btu170

Bolyen, E., Rideout, J. R., Dillon, M. R., Bokulich, N. A., Abnet, C. C., Al-Ghalith, G. A., Alexander, H., Alm, E. J., Arumugam, M., Asnicar, F., Bai, Y., Bisanz, J. E., Bittinger, K., Brejnrod, A., Brislawn, C. J., Brown, C. T., Callahan, B. J., Caraballo-Rodríguez, A. M., Chase, J., . . . Caporaso, J. G. (2019). Reproducible, interactive, scalable and extensible microbiome data science using QIIME 2. *Nature Biotechnology*, *37*(8), 852–857. https://doi.org/10.1038/s41587-019-0209-9

Callahan, B. J., McMurdie, P. J., Rosen, M. J., Han, A. W., Johnson, A. J. A., & Holmes, S. P. (2016). DADA2: High-resolution sample inference from Illumina amplicon data. *Nature Methods*, *13*(7), 581–583. https://doi.org/10.1038/nmeth.3869

Cock, P. J. A., Fields, C. J., Goto, N., Heuer, M. L., & Rice, P. M. (2010). The Sanger FASTQ file format for sequences with quality scores, and the Solexa/Illumina FASTQ variants. *Nucleic Acids Research*, *38*(6), 1767–1771. https://doi.org/10.1093/nar/gkp1137

de Sena Brandine, G., & Smith, A. D. (2021). Falco: High-speed FastQC emulation for quality control of sequencing data. *F1000Research*, *8*, 1874. https://doi.org/10.12688/f1000research.21142.2

Estaki, M., Jiang, L., Bokulich, N. A., McDonald, D., González, A., Kosciolek, T., Martino, C., Zhu, Q., Birmingham, A., Vázquez-Baeza, Y., Dillon, M. R., Bolyen, E., Caporaso, J. G., & Knight, R. (2020). QIIME 2 enables comprehensive end-to-end analysis of diverse microbiome data and comparative studies with publicly available data. *Current Protocols in Bioinformatics*, *70*(1), e100. https://doi.org/10.1002/cpbi.100

Fasolo, A., Deb, S., Stevanato, P., Concheri, G., & Squartini, A. (2024). ASV vs OTUs clustering: Effects on alpha, beta, and gamma diversities in microbiome metabarcoding studies. *PLoS One*, *19*(10), e0309065. https://doi.org/10.1371/journal.pone.0309065

Gołębiewski, M., & Tretyn, A. (2020). Generating amplicon reads for microbial community assessment with next-generation sequencing. *Journal of Applied Microbiology*, *128*(2), 330–354. https://doi.org/10.1111/jam.14380

Hossine, Z., Towers, I. N., & Kaehler, B. D. (2025). Network based differential abundance analysis: Bridging community interactions and host microbiome dynamics. *Network Modeling Analysis in Health Informatics and Bioinformatics*, *14*(1), 14. https://doi.org/10.1007/s13721-025-00506-4

Hugerth, L. W., & Andersson, A. F. (2017). Analysing microbial community composition through amplicon sequencing: From sampling to hypothesis testing. *Frontiers in Microbiology*, *8*, 1561. https://doi.org/10.3389/fmicb.2017.01561

Humann3—The Huttenhower Lab. (n.d.). Retrieved 26 August 2025, from https://huttenhower.sph.harvard.edu/humann/

Improved Metagenomic Analysis with Kraken 2 | Genome Biology | Full Text. (n.d.). Retrieved 26 August 2025, from https://genomebiology.biomedcentral.com/articles/10.1186/s13059-019-1891-0

Jenike, K. M., Campos-Domínguez, L., Boddé, M., Cerca, J., Hodson, C. N., Schatz, M. C., & Jaron, K. S. (2025). K-mer approaches for biodiversity genomics. *Genome Research*, *35*(2), 219–230. https://doi.org/10.1101/gr.279452.124

Jeske, J. T., & Gallert, C. (2022). Microbiome analysis via OTU and ASV-based pipelines—a comparative interpretation of ecological data in WWTP systems. *Bioengineering*, *9*(4), 146. https://doi.org/10.3390/bioengineering9040146

Kang, D. D., Li, F., Kirton, E., Thomas, A., Egan, R., An, H., & Wang, Z. (2019). MetaBAT 2: An adaptive binning algorithm for robust and efficient genome reconstruction from metagenome assemblies. *PeerJ*, *7*, e7359. https://doi.org/10.7717/peerj.7359

KneadData—The Huttenhower Lab. (n.d.). Retrieved 26 August 2025, from https://huttenhower.sph.harvard.edu/kneaddata/

Kumar, V., Gangani, S., Shukla, R., Prajapati, S. K., Shekhar, H., & Shukla, V. (2024). Next-generation sequencing and genomic data analysis. In A. Chaudhary, S. K. Sethi, & A. Verma (Eds.), *Unraveling New Frontiers and Advances in Bioinformatics* (pp. 49–76). Springer Nature. https://doi.org/10.1007/978-981-97-7123-3_3

Kumar, V., Pandey, A., Arora, A., Gautam, P., Bisht, D., Gupta, S., Chaurasia, A., & Sachan, M. (2023). Diagnostics and therapeutic potential of miR-205 and miR-34a in ovarian cancer management: A miRNA-target-based analysis. *DNA and Cell Biology*, *42*(3), 151–162. https://doi.org/10.1089/dna.2022.0487

Kumar, V., Shukla, R., Gangani, S., Shekhar, H., & Shukla, V. (2024). Metagenomics and microbiome analysis. In A. Chaudhary, S. K. Sethi, & A. Verma (Eds.), *Unraveling New Frontiers and Advances in Bioinformatics* (pp. 77–92). Springer Nature. https://doi.org/10.1007/978-981-97-7123-3_4

Kumar, V., Singh, A., Gautam, P., & Sachan, M. (2019). Techniques/tools to study epigenetic biomarkers in human cancer detection. In S. Paul (Ed.), *Biomedical Engineering and Its Applications in Healthcare* (pp. 327–351). Springer. https://doi.org/10.1007/978-981-13-3705-5_14

Langmead, B., & Salzberg, S. L. (2012). Fast gapped-read alignment with Bowtie 2. *Nature Methods*, *9*(4), 357–359. https://doi.org/10.1038/nmeth.1923

Leech, S. M., Barrett, H. L., Dorey, E. S., Mullins, T., Laurie, J., & Nitert, M. D. (2025). Consensus approach to differential abundance analysis detects few differences in the oral microbiome of pregnant women due to pre-existing type 2 diabetes mellitus. *Microbial Genomics*, *11*(4), 001385. https://doi.org/10.1099/mgen.0.001385

Li, D., Liu, C.-M., Luo, R., Sadakane, K., & Lam, T.-W. (2015). MEGAHIT: An ultra-fast single-node solution for large and complex metagenomics assembly via succinct de Bruijn graph. *Bioinformatics*, *31*(10), 1674–1676. https://doi.org/10.1093/bioinformatics/btv033

Lu, J., Breitwieser, F. P., Thielen, P., & Salzberg, S. L. (2017). Bracken: Estimating species abundance in metagenomics data. *PeerJ Computer Science*, *3*, e104. https://doi.org/10.7717/peerj-cs.104

Lu, J., Rincon, N., Wood, D. E., Breitwieser, F., Pockrandt, C., Langmead, B., Salzberg, S. L., & Steinegger, M. (2022). Metagenome analysis using the Kraken software suite. *Nature Protocols*, *17*(12), 2815–2839. https://doi.org/10.1038/s41596-022-00738-y

Martin, M. (2011). CUTADAPT removes adapter sequences from high-throughput sequencing reads. *EMBnet. Journal*, *17*(1), 10–12. https://doi.org/10.14806/ej.17.1.200

McMurdie, P. J., & Holmes, S. (2013). phyloseq: An R package for reproducible interactive analysis and graphics of microbiome census data. *PLoS One*, *8*(4), e61217. https://doi.org/10.1371/journal.pone.0061217

MEGAHIT: An Ultra-Fast Single-Node Solution for Large and Complex Metagenomics Assembly via Succinct de Bruijn Graph | Bioinformatics | Oxford Academic. (n.d.). Retrieved 26 August 2025, from https://academic.oup.com/bioinformatics/article/31/10/1674/177884

MetaPhlAn4—The Huttenhower Lab. (n.d.). Retrieved 17 September 2025, from https://huttenhower.sph.harvard.edu/metaphlan/

MiSeq SOP. (n.d.). Retrieved 16 July 2025, from https://mothur.org

Moeckel, C., Mareboina, M., Konnaris, M. A., Chan, C. S. Y., Mouratidis, I., Montgomery, A., Chantzi, N., Pavlopoulos, G. A., & Georgakopoulos-Soares, I. (2024). A survey of k-mer methods and applications in bioinformatics. *Computational and Structural Biotechnology Journal*, *23*, 2289–2303. https://doi.org/10.1016/j.csbj.2024.05.025

Ondov, B., Bergman, N., & Phillippy, A. (2011). Krona: Interactive metagenomic visualization in a web browser. *BMC Bioinformatics*, *12*, 385. https://doi.org/10.1186/1471-2105-12-385

QIIME2 Enhances Multi-Amplicon Sequencing Data Analysis: A Standardized and Validated Open-Source Pipeline for Comprehensive 16S rRNA Gene Profiling | Microbiology Spectrum. (n.d.). Retrieved 17 September 2025, from https://journals.asm.org/doi/10.1128/spectrum.01673-25

QIIME 2 Homepage. (n.d.). QIIME 2. Retrieved 26 August 2025, from https://qiime2.org

Schloss, P. D., Westcott, S. L., Ryabin, T., Hall, J. R., Hartmann, M., Hollister, E. B., Lesniewski, R. A., Oakley, B. B., Parks, D. H., Robinson, C. J., Sahl, J. W., Stres, B., Thallinger, G. G., Van Horn, D. J., & Weber, C. F. (2009). Introducing mothur: Open-Source, Platform-Independent, Community-Supported Software for Describing and Comparing Microbial Communities. *Applied and Environmental Microbiology*, *75*(23), 7537–7541. https://doi.org/10.1128/AEM.01541-09

Truong, D. T., Franzosa, E. A., Tickle, T. L., Scholz, M., Weingart, G., Pasolli, E., Tett, A., Huttenhower, C., & Segata, N. (2015). MetaPhlAn2 for enhanced metagenomic taxonomic profiling. *Nature Methods*, *12*(10), 902–903. https://doi.org/10.1038/nmeth.3589

Wang, J., Lu, H., Yu, L., Cheng, W., Yan, W., & Jing, X. (2021). Aggravation of airway inflammation in RSV-infected asthmatic mice following infection-induced alteration of gut microbiota. *Annals of Palliative Medicine*, *10*, 27. https://doi.org/10.21037/apm-20-2052

Willis, A. D. (2019). Rarefaction, alpha diversity, and statistics. *Frontiers in Microbiology*, *10*, 2407. https://doi.org/10.3389/fmicb.2019.02407

4 Human Microbiome Studies

*Diksha Singh[1], Kavita Patel[1], and Ashutosh Mani[1]**
[1] Department of Biotechnology, Motilal Nehru National Institute of Technology Allahabad, Prayagraj, India
*Corresponding author

4.1 INTRODUCTION

The human body is a microenvironment for various types of microorganisms that live in different parts of the body, and the genomes of these diverse microbes are collectively known as the human microbiome (Eloe-Fadrosh and Rasko 2013; Kumar, Shukla et al. 2024). These various types of microorganisms, including viruses, bacteria, fungi, and others, play a critical role in human physiology (Belkaid and Hand 2014). In terms of microbial population, the microbes in the human gut are so numerous that they outnumber host cells, and their aggregated gene content exceeds the human genome by more than a hundredfold (Bull and Plummer 2014).

The human microbiome has been recognized as a key contributor to human health and disease. These microbial communities influence digestion and metabolism, regulate immune system development, and provide defense against pathogenic organisms (Ma et al. 2024). Advances in metagenomics, high-throughput sequencing, and bioinformatics have provided remarkable insights into microbial composition and function. The NIH Common Fund launched the Human Microbiome Project to map healthy microbiomes across five body sites (skin, oral cavity, airways, gastrointestinal tract, and vagina) and provide these data to researchers, establishing a reference to study disease-related microbial changes (Gevers et al. 2012). In this chapter, we discuss the human microbiome studies carried out in recent years.

4.2 HUMAN MICROBIOME COMPOSITION AND DIVERSITY ACROSS BODY SITES

Humans are thought of as "holobionts," or superorganisms, because of the human microbiome, which is made up of trillions of symbiotic microorganisms and is now understood to be essential to proper physiology across several body locations. Every bodily environment, such as the skin, urogenital tract, respiratory system, and gastrointestinal tract, is home to unique microbial consortia, the formation of which is controlled by ecological concepts including drift, diversification, selection, and dispersal (2012; Vellend 2010). Microbes from maternal, environmental, or community sources are introduced by dispersal, and the species that flourish are determined by selection acting through habitat filters such as microbial competition, pH, host immune pressures, and nutritional availability (Costello et al. 2012). Diversification, through mutation, recombination, and horizontal gene transfer, quickly produces adaptive variations within communities, whereas drift creates random fluctuations that disproportionately harm low-abundance taxa, particularly following perturbations like antibiotic usage (Fodor et al. 2012; Nelson et al. 2010; Wylie et al. 2012).

DOI: 10.1201/9781003670155-4

4.2.1 Gut Microbiome

The gastrointestinal tract (GIT) is an integrated system of digestion, absorption, defense, and communication with the body. It contains the most abundant microbial ecosystems. The oral cavity has diverse microbial biofilms. The esophagus has transient taxa originating from the mouth. The stomach has low microbial loads, yet supports acid-tolerant *Helicobacter pylori.*

Its facultative anaerobes are adapted to bile acids and rapid transit (Rinninella et al. 2019). The ileum supports facultative and obligate anaerobes, primes host defenses, and sustains higher microbial densities with distinct immune functions. In contrast, the colon represents the densest microbial habitat, where anaerobic fermenters degrade complex polysaccharides into short-chain fatty acids, notably butyrate, which provide energy to colonocytes and maintain immune-metabolic balance (Schnorr et al. 2014). The composition of colonic microbes is strongly influenced by host immunity, bile acids, mucosal glycans, and diet. Their high diversity and resilience arise from niche partitioning. Disruption of these ecosystems can trigger inflammatory or metabolic disorders, whereas balanced communities support immunity, metabolism, and overall health (James et al. 2020).

4.2.2 Skin

The skin is one of the most dynamic microbial habitats since it is the largest external organ and is continuously exposed to changing environmental factors and host activities. Although regional and local variations in moisture, temperature, gland density, and topological features like hair follicles, sweat glands, and sebaceous glands generate unique biological niches, it is typically dry, cool, aerobic, and acidic (Chaudhari et al. 2020). The skin microbiome is the least stable of all body locations because of its low dispersal constraint, frequent disturbances, and high diversity and variability over time and across individuals (high beta diversity). An estimated 10^{11} microbial cells are supported by the skin's structurally diverse keratinized squamous epithelial layers, which are punctuated by microbial-rich follicles and glands that provide physiological gradients in salt, acidity, and moisture. While dry exposed regions, despite lower biomass, harbor diverse communities from Actinobacteria, Proteobacteria, Firmicutes, and Bacteroidetes, sebaceous sites like the face and back favor lipophilic taxa like *Propionibacterium acnes* and *Malassezia*, and moist areas like the axilla attract *Corynebacteria* and *Staphylococci* (Ruuskanen et al. 2022; Gupta et al. 2025). By promoting innate immunity (AMPs, β-defensins, complement), modifying adaptive immune responses (e.g., IL-1 production), and offering colonization resistance through direct inhibition of pathogens like *Staphylococcus aureus* and Group A *Streptococcus*, skin microbes promote host defense, even though many of them are commensal rather than obligate mutualists. To maintain its variety and defense capabilities, the skin microbiome exhibits a complex combination of dispersal, environmental filtration, stochasticity, and host–microbe interactions (Huang et al. 2020; Ho et al. 2019; Sarfraz et al. 2024; Prajapati et al. 2024; Nair et al. 2024; James et al. 2024; Mishra et al. 2024).

4.2.3 Urogenital Tract

Microbes find the urinary system to be a challenging environment because of factors such as low pH, antimicrobial chemicals released by the host, and frequent urine flushing. Despite this, *Lactobacillus* species play an important role in maintaining an acidic environment in females by producing lactic acid. This activity acts as a barrier against pathogens (Valenti et al. 2018). In males, communities are more diverse, including *Corynebacterium*, *Streptococcus*, *Staphylococcus*, and *Veillonella*, contributing to urinary tract colonization. These bacteria have protective functions by producing antimicrobial metabolites and competitively excluding other microorganisms. Crucially, the urogenital tract microbiome in both sexes demonstrates resilience, as it frequently re-establishes a stable baseline community, even though it may momentarily vary in response to factors like sexual activity, hormonal changes, antibiotic usage, or urine flow dynamics. However, in this environment, dysbiosis—whether due to overgrowth of opportunistic bacteria in males or loss of *Lactobacillus* dominance in females—has been linked to a higher risk of inflammation and infections (Di Paola et al. 2017).

4.2.4 Respiratory Tract

It is now known that the respiratory tract microbiome is a diverse microbial ecosystem that spans the upper (nasal cavity, oral cavity, and pharynx) and lower (trachea, bronchi, bronchioles, and alveoli) respiratory tracts. Previously believed to be sterile in its lower regions, culture-independent sequencing has revealed a low-biomass but significant microbial community in the lungs. Distinct regional communities are found in the upper respiratory tract, which is influenced by anatomical and physiological gradients. Skin-like taxa like *Corynebacterium*, *Staphylococcus*, and *Propionibacterium* are found in the nasal vestibule, whereas *Actinobacteria* and *Proteobacteria* dominate the warmer, mucus-lined nasal mucosa (Kang and Kang 2021; Teo et al. 2015; Kumar et al. 2023).Communities dominated by *Prevotella*, *Veillonella*, and *Streptococcus* are found in the lower respiratory system, despite being aerobic and nutrient-poor. These communities appear to reflect dissemination from the upper respiratory tract rather than strong local selection pressures, as demonstrated by bronchoalveolar lavage studies. The lung microbiome is described by ecological models, particularly island biogeography theory, as a balance between immigration from the upper tract and extinction through clearance mechanisms like coughing, mucociliary sweeping, and immune responses, with diversity declining farther from the upper respiratory tract source community. This equilibrium, however, shifts in disease states such as cystic fibrosis and COPD, where pathogen proliferation and spatial heterogeneity are driven by selection forces including inflammation, mucus accumulation, and anaerobic habitats (Stewart et al. 2017; Bisht et al. 2025). Overall, the respiratory microbiome provides new insights into respiratory health and therapeutic manipulation by illustrating how dispersal, selection, and stochastic processes interact to shape microbial communities and drive the transition from neutral to deterministic assembly forces in disease (Copeland et al. 2018). The major microbial taxa identified in different body parts are presented in Table 4.1.

TABLE 4.1
Major Microbial Taxa Identified Across Different Human Body Sites

S.No.	Body Site	Main Taxa	References
1.	Skin	*Staphylococcus, Streptococcus, Corynebacterium*, and *Propionibacterium*	Huang et al. (2020); Ho et al. (2019); Chang et al. (2018); Paller et al. (2019); Bay et al. (2020)
2.	Vagina	*Lactobacillus crispatus, L. jensenii Streptococcus, L. iners, L. gasseri*, and *Bifidobacterium*	Di Paola et al. (2017); Zhou et al. (2007); Felten et al. (1999)
3.	Ear	*Corynebacterium, Propionibacterium*, and *Staphylococcus*	Kolbe et al. (2019); Jervis-Bardy et al. (2019); Lappan et al. (2018)
4.	Eye	*Staphylococcus, Pseudomonas*, and *Propionibacterium*	Asao et al. (2019); Li et al. (2019); Suzuki et al. (2020)
5.	Oral	*Gemella, Streptococcus, Prevotella, Neisseria, Granulicatella, Rothia*, and *Abiotrophia*	Dewhirst et al. (2010); Crielaard et al. (2011); Huang et al. (2011); Kennedy et al. (2019); Sulyanto et al. (2019)

(Continued)

TABLE 4.1 *(Continued)*
Major Microbial Taxa Identified Across Different Human Body Sites

S.No.	Body Site	Main Taxa	References
6.	Nasopharyngeal tract	*Staphylococcus, Corynebacteriaceae, Carnobacteriacea, Peptoniphilaceae, Staphylococcaceae, Corynebacterium, Alloiococcus, Haemophilus, Granulicatella,* and *Moraxella*	Teo et al. (2015); Stewart et al. (2017); Copeland et al. (2018); Wen et al. (2018); Kang and Kang (2021)
7.	Gastrointestinal tract	*Bacteroides, Clostridium, Lactobacillus, Escherichia, Shigella, Coprobacillus, Faecalibacterium prausnitzii, Bifidobacterium, Lysinibacillus, Pseudomonas, Bacillus, Arthrobacter, Solibacillus, Lactococcus, Helicobacter pylori, Stentrophomonas, Prevotella, Veillonella, Enterococcus, Streptococcus, Akkermansia muciniphila,* and *Eubacterium rectale*	Pei et al. (2004); Maldonado-Contreras et al (2011); Zoetendal et al. (2012); Guinane et al. (2013); Liu et al. (2013); Khan, van Dijl, and Harmsen (2014); Angelakis et al. (2015); Sundin et al. (2017); Gong et al. (2019); Fan et al. (2020); James et al. (2020)

4.3 GLOBAL INSIGHTS FROM THE HUMAN MICROBIOME PROJECT AND RELATED INITIATIVES

The NIH started the historic Human Microbiome Project (HMP) in 2007 with the goal of methodically describing the microbial populations found throughout the human body and their function in both health and illness. The project, which was carried out in two main stages, has produced foundational resources for microbiome research (2012). The HMP1 study analyzed 5,177 16S rRNA gene sequencing profiles and 681 shotgun metagenomes spanning up to 18 body sites and three time points each from 242 healthy adults and established baseline ranges of taxonomic diversity within and between body sites, as well as strain personalization and functional commonalities.

The primary goal of Phase I (HMP1) was to determine the baseline diversity of microbial communities in health. About 300 healthy individuals' samples from various body locations (oral cavity, skin, stomach, nasal passages, and vagina) were examined using 16S rRNA and shotgun metagenomics (Kong et al. 2012; Lewis et al. 2015; Frank et al. 2011). The iHMP examined three model conditions:

- Microbiomes linked to pregnancy and preterm birth (Fettweis et al. 2019).
- Inflammatory bowel disease, including ulcerative colitis, Crohn's disease, and inflammatory bowel disease (IBD) (Manichanh et al. 2006; Huang et al. 2017).
- Prediabetes and type 2 diabetes (Pickup 2004; Cowie et al. 2009; Qin et al. 2012).

These investigations used host immunological profiling, metagenomics, metatranscriptomics, metabolomics, and clinical information to clarify the mechanisms underlying the relationship between microbial changes and disease progression. By offering previously unheard-of insights into resilience, host responses, and temporal dynamics, this approach established a standard for future microbiome-based systems biology. The major events in metagenomics are highlighted in Figure 4.1.

4.3.1 Global Projects

In parallel with the HMP, a number of worldwide initiatives have improved the global microbial diversity inventory, highlighting environmental and population-specific factors:

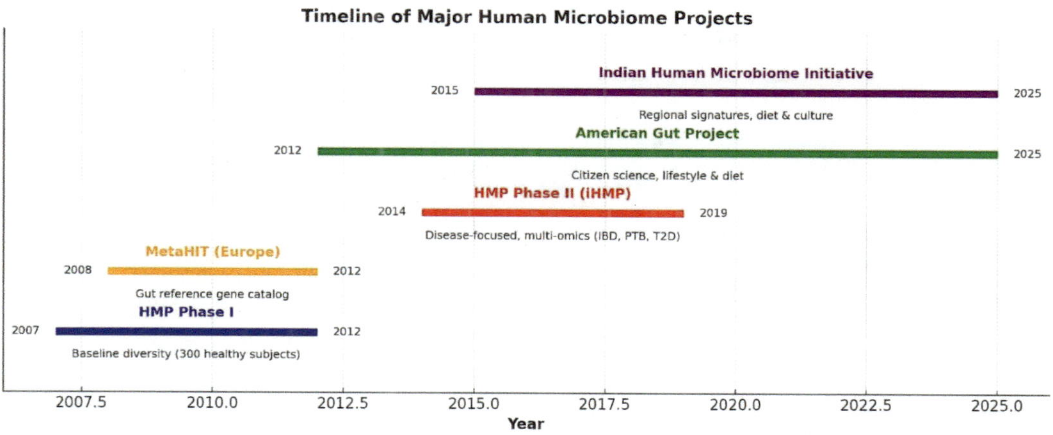

FIGURE 4.1 Timeline of major human microbiome research projects (2007–2025).

- *MetaHIT*

The European Commission–funded MetaHIT (METAgenomics of the Human Intestinal Tract) research created one of the largest gut microbial gene catalogs and identified enterotypes, linking microbial composition to metabolic health and disease (Qin et al. 2010).

- *American Gut Project*

The American Gut Project was started in November 2012 in collaboration with the Earth Microbiome Project and the Human Food Project, a citizen science initiative that integrated lifestyle and dietary data with gut microbiome profiles, highlighting the effects of diet, antibiotics, and lifestyle on microbial diversity (McDonald et al. 2018; Feng et al. 2019).

- *Indian Human Microbiome Initiative*

The Indian Human Microbiome Initiative (IHMI) aims to map the microbiome of diverse Indian populations, revealing links between microbial composition, diet, geography, and Ayurvedic Prakriti phenotypes. IHMI creates baseline data (https://ibdc.dbtindia.gov.in/indasecure/human-microbiome) to provide a foundation for disease modeling, personalized probiotics, and ethnicity-specific microbiome interventions (Dhotre et al. n.d.).

4.4 SEQUENCING AND BIOINFORMATICS: REVOLUTIONIZING MICROBIOME RESEARCH

Human microbiome research has changed significantly over the last 25 years due to the development of new sequencing technologies and bioinformatics. This advancement has made human microbiome research more advanced and reliable (Weinstock 2012). There are two major kinds of sequencing methods employed in microbiome research.

4.4.1 Amplicon Sequencing versus Shotgun Metagenomics Sequencing

In microbiome research, amplicon sequencing—often called targeted gene sequencing—is among the most common strategies used to study microbial diversity (Matsuo et al. 2021). The 16S rRNA gene particular region is chosen for amplification because different regions can influence the

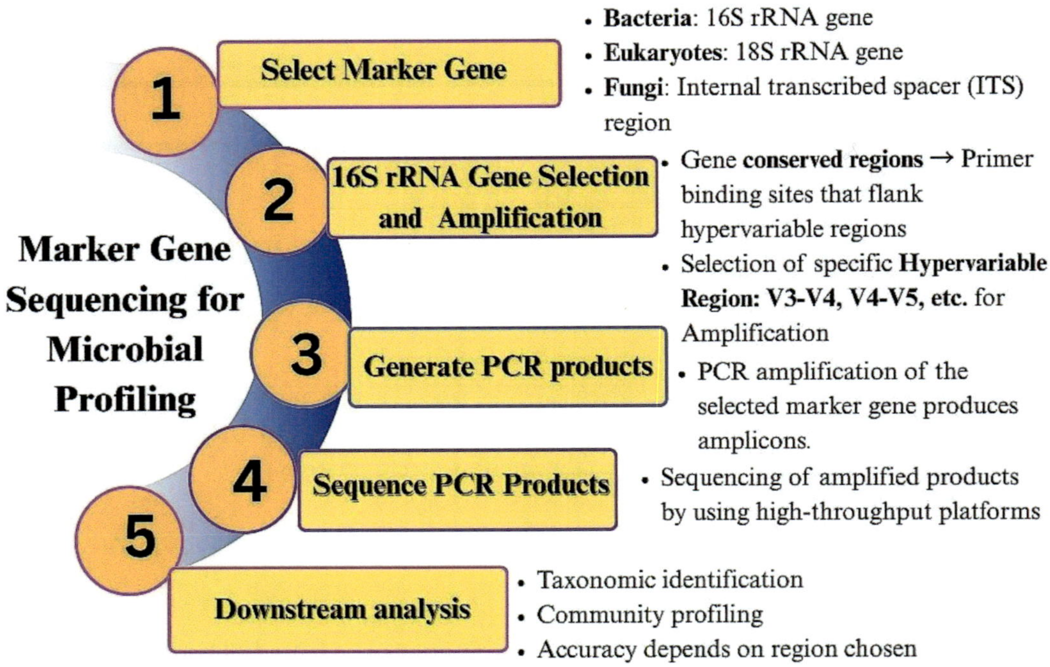

FIGURE 4.2 Marker gene or targeted gene sequencing for microbial profiling steps.

accuracy and depth of downstream analysis, as each offers a different degree of taxonomic resolution (Varliero et al. 2023). The procedure of the amplicon-based method is described in Figure 4.2. However, shotgun metagenomic sequencing represents a more inclusive and untargeted strategy for exploring microbial communities. Instead of focusing on a specific genetic marker, this method sequences all the DNA present in a sample (Rausch et al. 2019; Kumar et al. 2019). Because every fragment of DNA is captured, this approach is particularly helpful for investigating pathogenic variations or identifying probiotic traits (Fang et al. 2018). It also allows the generation of metagenome-assembled genomes (MAGs), which make it possible to rebuild nearly complete microbial genomes directly from environmental or host-associated samples (Mirete et al. 2025; Kumar, Gangani et al. 2024). The shotgun-based method is described in Figure 4.3.

Amplicon sequencing and shotgun metagenomic sequencing both have certain strengths and limitations because their approaches differ. A detailed comparison of the two sequencing methods is given in Table 4.2.

4.4.2 Bioinformatics: The Backbone of Microbiome Research

The progression in sequencing techniques, along with bioinformatics, has become the backbone of microbiome research (González, Fullaondo, and Odriozola 2024). In 16S rRNA gene sequencing, QIIME and mothur are two widely accepted pipelines used not only for taxonomic classification but also for preprocessing and post-processing steps. Amplicon data analysis depends on reference databases such as SILVA, Greengenes, NCBI, and Ribosomal Database Project (RDP), which serve as the foundation for taxonomic assignment (Almeida et al. 2018; Kumar et al. 2025). As each database has a different update frequency, taxonomic coverage, and computational efficiency, the choice of database is essential for downstream analysis and obtaining accurate results (Ceccarani and Severgnini 2023). The comparative analysis of major 16S rRNA and shotgun sequencing-based databases is discussed in Tables 4.3 and 4.4.

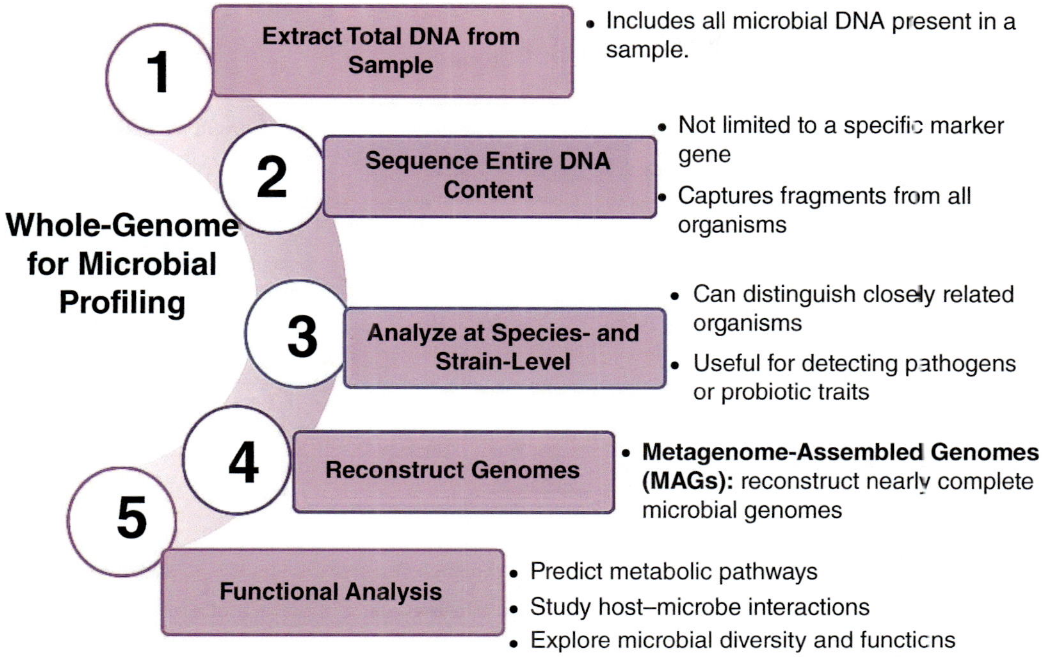

FIGURE 4.3 Whole genome or untargeted gene sequencing for microbial profiling steps.

TABLE 4.2
Comparison of Shotgun Metagenomics and Amplicon Sequencing

Aspect	Shotgun Metagenomics	Amplicon Sequencing	References
Strengths	• High taxonomic and functional resolution. • Detects novel genes and pathways. • Enable deep understanding of microbial interactions.	• Low cost and high scalability. • Standardized, reliable workflows. • Suitable for large sample sizes.	Jovel et al. (2016); Minich et al. (2018)
Limitations	• High cost and computational demand. • Requires deep sequencing coverage and advanced bioinformatics. • Host DNA contamination may obscure microbial signals.	• PCR biases and primer mismatches. • Limited resolution for closely related species. • Lacks direct functional information.	Tremblay, Schreiber, and Greer (2022); Liu et al. (2025); Sze and Schloss (2019)
Techniques/approach	Sequences all DNA fragments present in a sample for both taxonomic and functional profiling	Targets specific marker genes (e.g., 16SrRNA, ITS or 18SrRNA) for microbial identification.	Hernández Medina et al. (2022); Marcos-Zambrano et al. (2021)
Technological advances	Long-read sequencing improves assembly accuracy and species-level classification.	Long-read sequencing enables full length 16S rRNA gene analysis with better precision.	Buetas et al. (2024)

TABLE 4.3
Comparison of Commonly Used 16S rRNA Databases.

Database	Update frequency	Taxonomic Coverage	Computational Requirement	Current Status/Limitation
Greengenes	Discontinued	Moderate	Low	No longer updated; outdated taxonomy
RDP (Ribosomal Database Project)	Discontinued	Moderate	Low	Limited updates; older classification system
SILVA	Continuous	Broad (Bacteria, Archaea, Eukarya)	High	Comprehensive, but require significant resources
NCBI	Continuous	Very Broad	High	Updated regular; diverse but complex taxonomy

TABLE 4.4
Shotgun Metagenomic Taxonomic Profiling Tools

Tool	Core strategy	Strengths	References
MetaPhlAn/ MetaPhlAn 4	Clade-specific marker genes	Higher accuracy, low false positives, species- and strain-level profiling	(Beghini et al., 2021; Blanco-Míguez et al., 2023)
Kraken 2	k-mer based alignment	Rapid classification, scalable for large datasets	(Wood et al., 2019)

TABLE 4.5
Functional Profiling Tools in Metagenomics

Tool	Approach	Output Type	Key Features/Limitations
HUMAnN/ HUMAnN2	Direct translation-based functional reconstruction	Species-level pathway and gene abundances	High accuracy; requires annotated database
PICRUSt/ PICRUSt2	Functional prediction from 16S sequences	Predict gene families and pathways	Fast; limited to Bacteria and Archaea only

As currently understood, amplicon sequencing mainly focuses on taxonomic composition, whereas shotgun sequencing extends this to functional analysis, allowing even strain-level diversity to be observed. For taxonomic profiling, shotgun-based workflows mainly use classifiers that apply distinct algorithms (Beghini et al. 2021).

In addition to direct classification, shotgun metagenomic workflows also employ assembly-based strategies. De novo assemblers such as MEGAHIT and metaSPAdes reconstruct longer contiguous sequences (*contigs*) from short reads. MEGAHIT provides ultrafast and memory-efficient assembly for massive datasets, while metaSPAdes better manages genetic complexity and polymorphism typical of diverse microbial communities. Combined, these assemblers enable the recovery of high-quality MAGs and provide deeper insights into microbial genomic structures (Mirete et al. 2025). Functional profiling of shotgun metagenomic data commonly uses tools such as HUMAnN, HUMAnN2, PICRUSt, and PICRUSt2, which are discussed in Table 4.5 (Franzosa et al. 2018; Sulit et al. 2023).

These are advantages of shotgun metagenomics workflows; however, these workflows also have certain limitations, including the need for high computational power and sufficient storage to process large volumes of data (Tremblay, Schreiber, and Greer 2022). Short-read assemblies may also suffer from fragmentation, especially in low-abundance species, repetitive regions, or highly diverse

TABLE 4.6
Overview of Multi-Omics Approaches in Microbiome Studies

Approach	Focus	Function	Key Insights	References
Meta-trascriptomics	Analyze community-wide RNA	To identify actively *expressed genes* and infer functional profiles	Reveals microbial composition and genetic potential, showing actual microbial activity	Aguiar-Pulido et al. (2016)
Meta-proteomics	All proteins produced by a microbial community	To link *gene expression* to protein	Provides direct evidence of microbial activity and identifies which species perform specific metabolic roles	Van Den Bossche et al. (2021)
Metabolomics	Small molecules (metabolites) in a biological sample	To capture *biochemical activity* and metabolic state	Reflects the functional outcome of microbial metabolism, showing pathway changes and community homeostasis	Aguiar-Pulido et al. (2016)

microbial populations. These challenges can reduce assembly continuity and strain-level resolution (Quince et al. 2021).

4.4.3 Beyond DNA: Multi-Omics Approaches

As researchers have learned, sequencing methods have their own limitations. To resolve this issue, multi-omics approaches such as meta-transcriptomics, metagenomics, and metabolomics provide a real-time view of microbial functions and interactions within their environment (Franzosa et al. 2015). The multi-omics approaches associated with metagenomics are depicted in Table 4.6.

Together, these multi-omics approaches provide a clearer picture of community functions and interactions by linking what microbes are genetically capable of with what they are actually doing in their environment.

4.5 CORE FUNCTIONS OF THE HUMAN MICROBIOME

The human microbiome, composed of trillions of microorganisms, plays a central role in maintaining health through nutrient synthesis, metabolism, and signaling functions, supporting proper physiological functions across multiple body systems (Evans, Morris, and Marchesi 2013). The functionally relevant microbes are depicted in Figure 4.4.

I. Gut microbiota functions:

Gut microbes are responsible for several roles in humans, including the synthesis of essential vitamins (Gholami et al. 2023), fermentation of undigested carbohydrates (Hossain, Amarasena, and Mayengbam 2022), production of neuroactive compounds (Wu et al. 2021), and involvement in bile acid metabolism (Rowland et al. 2018).

II. Microbiota functions beyond the gut:

Gut microbiome research is more commonly preferred, but microbial communities at other body sites also play crucial roles in host health. Some of the microbes present on the skin, vagina, and oral routes are equally important and play a crucial role in several illnesses. The major functions of microbiota involved in different body parts are described in Table 4.7.

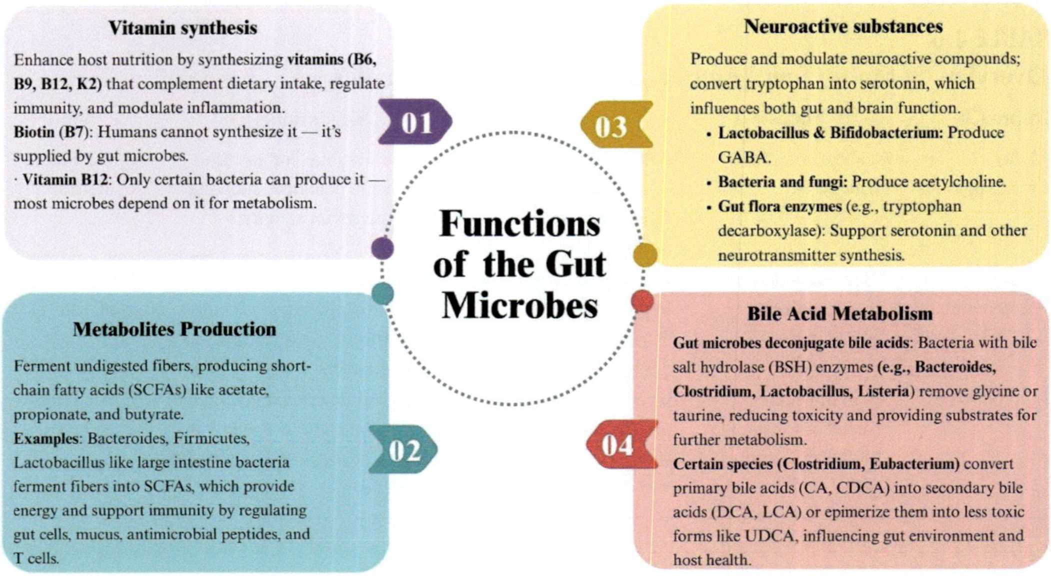

FIGURE 4.4 Major functional contributions of the gut microbiota to host health.

TABLE 4.7
Microbiota Functions Beyond the Gut

Body Site	Main Functions	Examples	Role of Metabolites/ Molecules	References
Skin microbiota	The skin hosts diverse microbes that protect against pathogens through immune surveillance, antimicrobial release, and competition.	*Cutibacterium acnes* maintains acidic conditions in hair follicles and prevent invasion by rivals. *Staphylococcus epidermidis* provides both antimicrobial protection and anti-inflammatory signals, supporting a balanced state of homeostatic immunity.	Antimicrobial peptides, anti-inflammatory signals	Yang et al. (2024)
Oral microbiota	Convert dietary nitrate →nitrite →nitric oxide (NO) for cardiovascular health.	*Veillonella*, *Actinomyces*, *Haemophilus*, *Neisseria*, and *Streptococcus* convert dietary nitrate from leafy vegetables into nitrite and nitric oxide (NO).	Nitrite, NO, H_2S	Li et al. (2022)
	High oral H_2S can cause bad breath and inflammation; at physiological levels it has antioxidant and vascular benefits.	*Prevotella* and *Porphyromonas* metabolize sulfur-containing amino acids to release hydrogen sulfide (H_2S).		

TABLE 4.7 (Continued)
Microbiota Functions Beyond the Gut

Body Site	Main Functions	Examples	Role of Metabolites/ Molecules	References
Respiratory microbiota	Upper airway microbes support immunity and block pathogens.	The upper airway contains microbes, such as *Cornebacterium*, *Dolosigranulum*, *Moraxella*, and *Streptococcus*, which support immunity and block pathogens.		Zhao et al. (2023)
	Lower airway microbes maintain balance between immigration and clearance.	The lower airway, once thought sterile, contains fewer microbes due to limited nutrients and high oxygen but still harbors species such as *Prevotella*, *Veillonella*, *Coprococcus*, and *Dorea*.		
Cervicoveginal microbiota	Maintain vaginal acidity (lactic acid), regulate immunity, and protect tissue (H_2O_2, bacteriocins, and peptides) by suppressing pathogens.	Lactobacillus species: *L. crispatus* dominance is linked to stability, while *L. iners* often associated with dominance is with dysbiosis and higher infection risk.	Lactic acid (particularly D-lactate), hydrogen peroxide, bacteriocins, and peptides	Avitabile et al. (2024)

4.6 DYSBIOSIS AND ITS ROLE IN HUMAN DISEASES

The human gastrointestinal system contains ~10^{14} microbes, mostly from the *Firmicutes* and *Bacteroidetes* phyla. These microorganisms regulate immunity, metabolism, and barrier integrity as part of a dynamic ecosystem. In a balanced state (eubiosis), microbial populations and host immune responses interact to maintain homeostasis, supporting metabolism, pathogen defense, and digestion.

Dysbiosis refers to an imbalanced microbial community, typically characterized by (1) reduced diversity, (2) increased pathogenic taxa, and/or (3) loss of beneficial commensals. This imbalance alters microbial functions, such as immune regulation, nutrient metabolism, and barrier maintenance, and is linked to local and systemic disorders. Dysbiosis contributes to increased gut permeability, abnormal immune responses, chronic inflammation, and altered metabolite profiles. Diseases associated with dysbiosis include IBD, obesity, type 2 diabetes, allergic disorders, colorectal cancer, and neurodevelopmental conditions such as autism.

- ***Dysbiosis in IBD (Crohn's Disease and Ulcerative Colitis):*** For many years, researchers searched for a single microbial cause of IBD. However, the focus shifted to dysbiosis as a consequence of microbial imbalance. An imbalance in gut microbiota, including increased *Enterobacteriaceae* and reduced beneficial *Firmicutes* and *Bacteroides*, has been observed (Li et al. 2015). Dysbiosis is also a contributor to Crohn's disease, characterized by an increase in *Ruminococcus gnavus* and a reduction in butyrate-producing bacteria, which damages epithelial barrier function (Joossens et al. 2011). In addition, reduced bile acid metabolism is caused by deconjugating bacteria, which disrupt anti-inflammatory

TABLE 4.8
List of Bacterial Based Cancer Therapies in Clinical Trials

S.No.	Bacterial Strain	Condition	Study Phase	Reference
1.	Live BCG bacteria	Non-muscle invasive bladder cancer	Phase III	Jiang and Redelman-Sidi (2022)
2.	Clostridium novyi-NT spores	Solid tumor malignancies	Phase I	Janku et al. (2021)
3.	LBP (Live bacteria product)	Non-small cell lung cancer/melanoma/ renal cell carcinoma	Phase I	Ijaz et al. (2024)
4.	BacTRL-IL-12	Cancer-solid tumors	Phase I	Dong et al. (2024)

pathways and cause inflammation. Collectively, these findings indicate that IBD arises from complex host–microbe interactions rather than a single pathogen.

- ***Obesity and Type 2 Diabetes:*** Obesity is a complex metabolic disorder and is increasingly linked to intestinal dysbiosis in addition to diet and lifestyle factors. Obese individuals typically show reduced microbial diversity, with an increased *Firmicutes-to-Bacteroidetes* ratio that enhances energy harvest and fat accumulation (Turnbaugh et al. 2006). In type 1 and type 2 diabetes, microbial imbalance occurs, causing a loss of diversity and a reduction in butyrate production. These changes link dysbiosis to chronic inflammation and metabolic dysfunction (Cani et al. 2007).
- ***Neurodegenerative Disorders (Alzheimer's and Parkinson's):*** Intestinal dysbiosis in Parkinson's disease (PD) and Alzheimer's disease (AD) is linked to increased gut permeability, allowing endotoxins such as LPS to enter blood circulation, triggering systemic inflammation and neuroinflammation that influence AD and PD disease progression. In AD, dysbiosis drives immune activation, oxidative stress, and metabolic alterations that promote amyloid-beta and tau pathology, while in PD, it disrupts short-chain fatty acid (SCFA) synthesis, impairs gut barrier function, and increases alpha-synuclein aggregation (Ullah et al. 2023).
- ***Colorectal Cancer (CRC):*** Intestinal dysbiosis in colorectal cancer is characterized by reduced SCFA- and butyrate-producing bacteria (e.g., *Bifidobacteria*, *Prevotella*) and increased pathogenic bacteria (e.g., *Firmicutes*, *Bacteroidetes*, *Enterobacteriaceae*, *Fusobacteria*), which contribute to tumorigenesis. Dysbiosis has been shown to induce cancer in preclinical research, such as germ-free (GF) mouse models and antibiotic therapies, indicating a causative connection. Functionally, dysbiosis is linked to epithelial proliferation, chronic inflammation, and neoplastic transformation through the release of toxic microbial metabolites, the overproduction of immune modulators such as alpha-defensins, and the lack of protective SCFAs such as butyrate (Elinav et al. 2013). Therapeutically important studies are depicted in Table 4.8.

4.7 MACHINE LEARNING APPROACHES

Human microbiome research is important due to its applications in biotechnology and health science. Scientists have mainly used human microbiome studies for pathogen identification and to understand their effects on human health, so that they can find cures for particular diseases. However, recently, they have become more interested in studying the effects of pathogens on genetics, the influence of environmental factors on the human microbiome, and the role of diet in shaping the microbiome, in order to better understand its role in clinical applications.

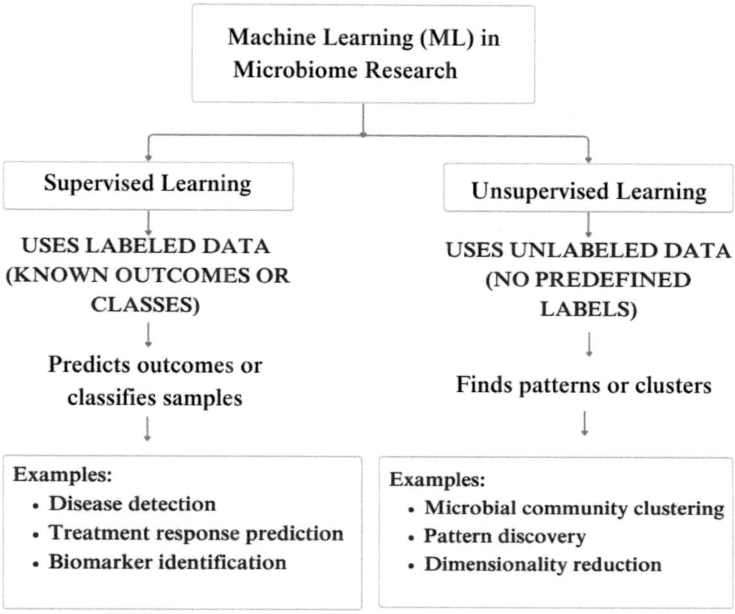

FIGURE 4.5 Workflow of a machine learning-based classification pipeline.

Advances in high-throughput sequencing have greatly expanded human microbiome research, but the resulting large and complex datasets pose analytical challenges (Moreno-Indias et al. 2021). Machine learning (ML) is important due to its applications in microbiome research and health science. Scientists have mainly used ML for disease detection, prognosis, and prediction of treatment responses, so that they can improve diagnosis and therapeutic outcomes. However, recently, they have become more interested in applying ML to forecast microbiome changes influenced by drugs, diets, and environmental factors, in order to understand complex host–microbiome interactions (Papoutsoglou et al. 2023). The machine learning classification flow chart is depicted in Figure 4.5.

4.7.1 Functional Roles of Machine Learning in Human Microbiome Research

In human microbiome research, ML models are applied for diagnosis, phenotype prediction, and biomarker discovery. Supervised models such as LR, SVM, Random Forest (RF), and GB enable disease classification and early detection. Other advanced machine learning methods are used for different tasks. Table 4.9 explains each task in detail, along with the models used and corresponding examples.

4.8 CHALLENGES IN THE FIELD

The efficient translation of microbiome science into clinical and therapeutic applications is still hampered by a number of fundamental challenges, despite notable advancements in the field.

1. ***High inter-individual variability:*** It is difficult to define a "universal healthy microbiome" because of the enormous variation in microbiome composition among individuals, even among those who are considered healthy. Interventions that are effective for one

TABLE 4.9
Applications for Machine Learning in Microbiome Data Analysis

Task	ML Model	Strength	Weakness	Examples
Diagnosis	Logistic Regression (LR)	Simple, interpretable	Limited for non-linear or complex data	LR applied for diagnosis, e.g., Classification of women as bacterial vaginosis positive vs bacterial vaginosis negative (Ojo et al. 2025).
	Support Vector Machines (SVMs)	Effective for small datasets, handles non-linear boundaries	Sensitive to parameter tuning, less interpretable	SVM applied for early diagnosis of disease, e.g., Type 2 diabetes mellitus (T2DM) risk prediction (Cui and Zhang 2013).
	Random Forest (RF)	Robust, handles high-dimensional data, good accuracy	Less interpretable, can overfit if not tuned	RF applied for disease prediction, e.g., colorectal cancer classification from stool microbiome (Young et al. 2021).
	Gradient Boosting (GB)	High predictive accuracy, captures complex patterns	Computationally intensive, sensitive to overfitting	GB applied for diagnosis of different types of diseases, e.g., GB used in early prediction of breast cancer (Islam et al. 2024) and heart disease prediction (Alsabhan and Alfadhly 2025).
Phenotype Prediction	RF/GB	Accurate predictions, robust, handles complex features	May overfit small datasets	RF and GB both applied for predicting host traits, e.g., Prediction of Body Mass Index (BMI) (Delnevo et al. 2021).
	DeepMicro (Autoencoder)	Condenses features, improves prediction and clustering	Performance depends on dataset size, autoencoder choice	DeepMicro applied for phenotype prediction, e.g., drug response prediction, food allergy, forensic human identification (Oh and Zhang 2020).
	phyLoSTM (CNN-LSTM)	Captures temporal changes, improves disease risk prediction	Requires large, well-sampled time-series data	phyLoSTM applied for Prediction of host phenotypes from longitudinal microbiome data, e.g., allergy from peanut (Sharma and Xu 2021).
	MDeep (CNN-based)	Uses phylogenetic information; robust to even with imperfect phylogenetic tree; strong predictive power	Limited interpretability, requires phylogenetic tree	MDeep applied for predicting host traits or health outcomes using microbiome data, e.g., prediction of new-onset rheumatoid arthritis (NORA) vs healthy controls using intestinal microbiome (Wang et al. 2020).
Biomarker/ Feature Selection	Supervised Models (RF, Deep Forest, SVM, KNN, LASSO, Elastic Net)	Improves model interpretability, identify key microbial features, handles high-dimensional data (RF, Deep Forest), automatically perform feature selection (LASSO, Elastic Net)	May overfit small datasets, Performance depends on model choice, Less interpretable (RF, Deep Forest)	RF applied for selecting important microbial features (biomarkers) and disease classification, e.g., colorectal cancer classification from stool microbiome data (Kolisnik et al. 2023). SVM applied for biomarker identification, e.g., SVM model is used to identify individuals at higher risk of developing type 2 diabetes mellitus (T2DM) (Abbas et al. 2019).
	Unsupervised (Clustering, NMF, DMM)	Explores hidden patterns in data	Indirect association between features and outcomes, Complex interpretation, post-processing needed	Hierarchical clustering applied for exploring microbiota-host relationships, e.g., transcriptome changes in colonocytes (Richards et al. 2019). NMF applied to identifying key features of microbial communities, e.g., NMF is used in identifying differences in metabolic pathways in Inflammatory Bowel Disease patients and healthy control (Cai, Gu, and Kenney 2017)

population or individual may not be equally effective for another, highlighting the need for population-specific or even personalized microbiome-based approaches.

2. ***Limited strain-level resolution:*** Current metagenomic approaches provide taxonomic classification at the genus or species level, but they are not able to accurately capture strain-level variations (Pust and Tümmler 2021). These differences can affect the results of important functional analyses between strains. Recent developments such as single-cell sequencing (Gross et al. 2015), strain-resolved metagenomics, and long-read sequencing methods can help overcome this limitation.
3. ***Reproducibility and methodological inconsistencies:*** The results obtained from different studies are difficult to compare due to variations in laboratory and computational approaches. Comparisons can be challenging because results depend on DNA extraction methods, sequencing platforms (e.g., Illumina, Nanopore, and PacBio), and bioinformatics pipelines. To address this challenge, standardized metadata frameworks, such as MIxS: Minimum Information about any (x) Sequence proposed by the Genomic Standards Consortium, are being adopted to improve reproducibility and comparability (Glass et al. 2014).
4. ***Causality versus correlation:*** Most microbiome studies remain observational, describing associations rather than causal relationships. It is often unclear whether microbiome changes cause disease, result from disease, or merely correlate with another factor. Cause-and-effect relationships are starting to be demonstrated using controlled interventions, microbial culture studies, animal models, and longitudinal investigations.

4.9 FUTURE DIRECTIONS AND CLINICAL APPLICATIONS

4.9.1 Multi-Omics Integration

It is impossible to fully understand the complexity of the human microbiome using a single omics method. Multi-omics integration offers a detailed system for connecting host interactions, metabolic outputs, and microbial activity to taxonomic makeup. This integration provides functional insights into health and illness and allows for accurate functional profiling of the microbiome (Priya et al. 2022). Multi-omics methods—metagenomics (Sharpton 2014; Bai et al. 2022), metatranscriptomics (Shakya, Lo, and Chain 2019), metabolomics (Ankrah et al. 2021; Magnúsdóttir et al. 2017; Heinken, Hertel, and Thiele 2021), and proteomics (Yizhak et al. 2010)—are discussed here using the flow chart in Figure 4.6.

4.9.2 Microbiome-Based Therapies

4.9.2.1 Next-Gen Probiotics

Conventional probiotics such as *Lactobacillus* and *Bifidobacterium* offer broad advantages but frequently lack strain-specific effectiveness or colonization potential. Next-generation probiotics use microbial strains that have been genetically modified or carefully selected (Cuello-Garcia et al. 2015; Depommier et al. 2019). These bacteria can be engineered to emit antimicrobial peptides against infections, alter immunological pathways, or create therapeutic metabolites (such as butyrate and neurotransmitter precursors). Engineered *Escherichia coli* Nissle derivatives, for instance, are being investigated as live biotherapeutics to treat metabolic dysfunctions and inflammatory bowel disease (Coqueiro et al. 2019).

4.9.2.2 FMT (Fecal Microbiota Transplantation)

The goal of FMT is to restore microbial balance by transferring the complete gut microbial population from a healthy donor to a recipient (Gagliardi et al. 2018). In addition, FMT is being tested in current studies for neurological disorders, including Parkinson's disease (Finegold et al. 2002) and autism spectrum disorder, Alzheimer's disease, obesity, ulcerative colitis, and type 2 diabetes. The

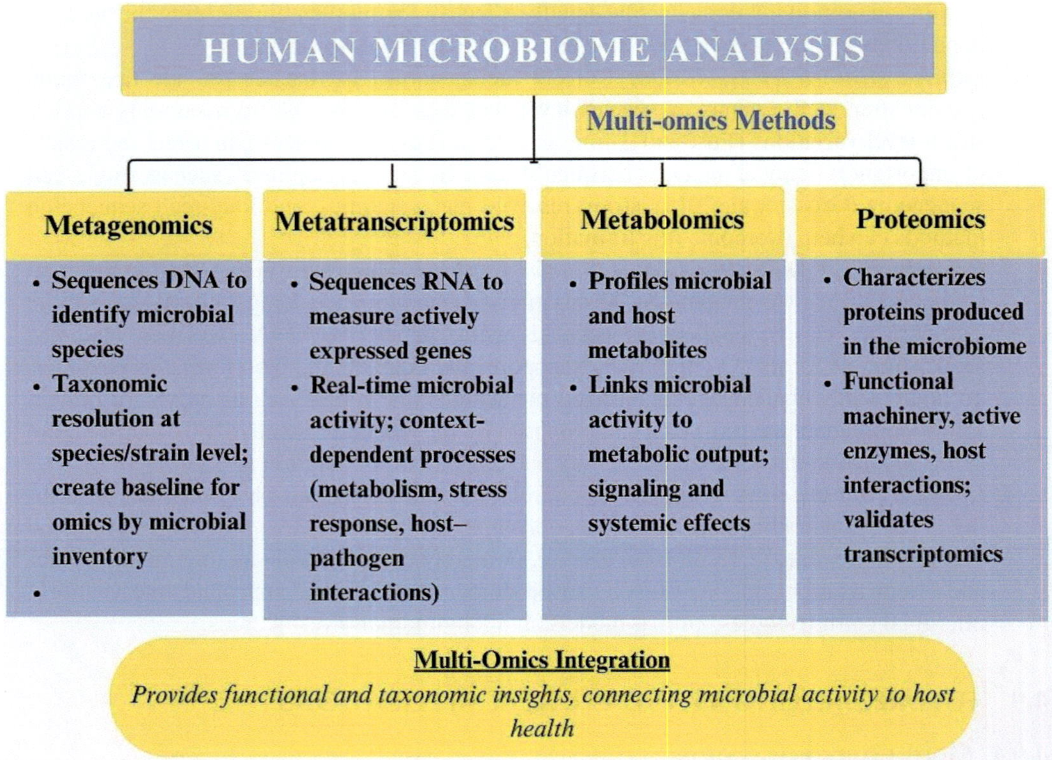

FIGURE 4.6 Multi-omics approaches in microbiome research.

TABLE 4.10
Engineered Microbes and Their Therapeutic Effects on Host Health

S.No.	Engineered/Naive Strain	Cloned Gene	Desired Target	Effect
1.	*Lactococcus lactis*	Human Trefoil Factor 1	Oral mucosa	Reduced severity of oral mucositis (Caluwaerts et al. 2010)
2.	*Escherichia coli*	CsgA-TFF þ trefoil factor	Gut epithelium during colitis	Treated colitis with mucosal healing and immunomodulation (Praveschotinunt et al. 2019)
3.	*Lactic-acid bacteria*	Elafin	Intestinal inflammation during inflammatory bowel disease	Improved treatment against intestinal dysfunction (Motta et al. 2012)
4.	*Bacteroides ovatus*	Transforming growth factor-b1	Intestinal inflammation during colitis	Improvement of colitis treatment by production of transforming growth factor-b (Hamady et al. 2011)
5.	*Lactobacillus acidophilus*		Urease activity	Reduced hyperammonemia (Noonan et al. 2020)
6.	*Bifidobacterium, Streptococcusthermophilus, and Lactobacillus*			Ulcerative colitis (Bibiloni et al. 2005; Tursi et al. 2010)

treatment is effective in restoring colonization resistance, normalizing metabolite synthesis, and restoring microbial diversity. However, it remains difficult to standardize donor screening, preparation, and delivery techniques (Petrof and Khoruts 2014). Genetically engineered microbes involved in therapy are illustrated in Table 4.10.

4.9.2.3 Postbiotics

Postbiotics use microbial metabolites, cell components, or released compounds with bioactive properties instead of live organisms such as probiotics or FMT. Peptidoglycans, bacteriocins, extracellular polysaccharides, and short-chain fatty acids (butyrate, acetate, and propionate) are a few examples. These substances prevent pathogen colonization, control host immunological responses, and strengthen the integrity of the epithelial barrier. Postbiotics are safer, more stable, and simpler to standardize since they do not contain live bacteria. This makes them a desirable treatment option for susceptible patient groups, such as those with impaired immune systems.

REFERENCES

Abbas, H. T., L. Alic, M. Erraguntla, J. X. Ji, M. Abdul-Ghani, Q. H. Abbasi, and M. K. Qaraqe. 2019. "Predicting Long-Term Type 2 Diabetes with Support Vector Machine Using Oral Glucose Tolerance Test." *PLoS One* 14 (12).

Aguiar-Pulido, Vanessa, Wenrui Huang, Victoria Suarez-Ulloa, Trevor Cickovski, Kalai Mathee, and Giri Narasimhan. 2016. "Metagenomics, Metatranscriptomics, and Metabolomics Approaches for Microbiome Analysis: Supplementary Issue: Bioinformatics Methods and Applications for Big Metagenomics Data." *Evol Bioinform* 12 (S1):5–16. https://doi.org/10.4137/ebo.s36436.

Almeida, Alexandre, Alex L. Mitchell, Aleksandra Tarkowska, and Robert D. Finn. 2018. "Benchmarking Taxonomic Assignments Based on 16S rRNA Gene Profiling of the Microbiota from Commonly Sampled Environments." *GigaScience* 7 (5). https://doi.org/10.1093/gigascience/giy054.

Alsabhan, Waleed, and Abdullah Alfadhly. 2025. "Effectiveness of Machine Learning Models in Diagnosis of Heart Disease: A Comparative Study." *Sci Rep* 15 (1):24568. https://doi.org/10.1038/s41598-025-09423-y.

Angelakis, E., F. Armougom, F. Carrière, D. Bachar, R. Laugier, J. C. Lagier, C. Robert, C. Michelle, B. Henrissat, and D. Raoult. 2015. "A Metagenomic Investigation of the Duodenal Microbiota Reveals Links with Obesity." *PLoS One* 10 (9).

Ankrah, Nana Y. D., David B. Bernstein, Matthew Biggs, Maureen Carey, Melinda Engevik, Beatriz García-Jiménez, Meiyappan Lakshmanan, Alan R. Pacheco, Snorre Sulheim, and Gregory L. Medlock. 2021. "Enhancing Microbiome Research Through Genome-Scale Metabolic Modeling." *mSystems* 6 (6):e00599.21. https://doi.org/10.1128/mSystems.00599-21.

Asao, Kazunobu, Noriyasu Hashida, Satoru Ando, Daisuke Motooka, Hiroyuki Kurakami, Shota Nakamura, Daisuke Yamashita, Kazuichi Maruyama, Satoshi Kawasaki, Tomomi Yamada, Tetsuya Iida, and Kohji Nishida. 2019. "Conjunctival Dysbiosis in Mucosa-Associated Lymphoid Tissue Lymphoma." *Sci Rep* 9 (1):8424. https://doi.org/10.1038/s41598-019-44861-5.

Asnicar, F., A. M. Thomas, A. Passerini, L. Waldron, and N. Segata. 2024. "Machine Learning for Microbiologists." *Nat Rev Microbiol* 22 (4):191–205.

Avitabile, Elisa, Laura Menotti, Vanessa Croatti, Barbara Giordani, Carola Parolin, and Beatrice Vitali. 2024. "Protective Mechanisms of Vaginal Lactobacilli Against Sexually Transmitted Viral Infections." *Int J Mol Sci* 25 (17):9168.

Bai, G. H., S. C. Lin, Y. H. Hsu, and S. Y. Chen. 2022. "The Human Virome: Viral Metagenomics, Relations with Human Diseases, and Therapeutic Applications." *Viruses* 14 (2).

Bay, L., C. J. Barnes, B. G. Fritz, J. Thorsen, M. E. M. Restrup, L. Rasmussen, J. K. Sørensen, A. B. Hesselvig, A. Odgaard, A. J. Hansen, and T. Bjarnsholt. 2020. "Universal Dermal Microbiome in Human Skin." *mBio* 11 (1):e2945.19.

Beghini, F., L. J. McIver, A. Blanco-Míguez, L. Dubois, F. Asnicar, S. Maharjan, A. Mailyan, P. Manghi, M. Scholz, A. M. Thomas, M. Valles-Colomer, G. Weingart, Y. Zhang, M. Zolfo, C. Huttenhower, E. A. Franzosa, and N. Segata. 2021. "Integrating Taxonomic, Functional, and Strain-Level Profiling of Diverse Microbial Communities with bioBakery 3." *Elife* 4 (10):65088.

Belkaid, Yasmine, and Timothy W. Hand. 2014. "Role of the Microbiota in Immunity and Inflammation." *Cell* 157 (1):121–41. https://doi.org/10.1016/j.cell.2014.03.011.

Bibiloni, R., R. N. Fedorak, G. W. Tannock, K. L. Madsen, P. Gionchetti, M. Campieri, C. De Simone, and R. B. Sartor. 2005. "VSL#3 Probiotic-Mixture Induces Remission in Patients with Active Ulerative Colitis." *Am J Gastroenterol* 100 (7):1539–46.

Bisht, D., V. Kumar, S. Gupta, and M. Sachan. 2025. "Navigating Circulating Tumour Cells and Tumour Microemboli: Where We Stand in Bridging Their Molecular Dynamics with Clinical Translation." *Biochim Biophys Acta Rev Cancer* 1880 (4):189384. https://doi.org/10.1016/j.bbcan.2025.189384.

Blanco-Míguez, A., F. Beghini, F. Cumbo, L. J. McIver, and K. N. Thompson, M. Zolfo, . . . N. Segata. 2023. "Extending and Improving Metagenomic Taxonomic Profiling with Uncharacterized Species using MetaPhlAn 4." *Nat Biotechnol* 41 (11):1633–44.

Buetas, Elena, Marta Jordán-López, Andrés López-Roldán, Giuseppe D'Auria, Llucia Martínez-Priego, Griselda De Marco, Miguel Carda-Diéguez, and Alex Mira. 2024. "Full-Length 16S rRNA Gene Sequencing by PacBio Improves Taxonomic Resolution in Human Microbiome Samples." *BMC Genom* 25 (1):310. https://doi.org/10.1186/s12864-024-10213-5.

Bull, M. J., and N. T. Plummer. 2014. "Part 1: The Human Gut Microbiome in Health and Disease." *Integr Med* 13 (6):17–22.

Cai, Yun, Hong Gu, and Toby Kenney. 2017. "Learning Microbial Community Structures with Supervised and Unsupervised Non-Negative Matrix Factorization." *Microbiome* 5 (1):110. https://doi.org/10.1186/s40168-017-0323-1.

Caluwaerts, S., K. Vandenbroucke, L. Steidler, S. Neirynck, P. Vanhoenacker, S. Corveleyn, B. Watkins, S. Sonis, B. Coulie, and P. Rottiers. 2010. "AG013, a Mouth Rinse Formulation of Lactococcus Lactis Secreting Human Trefoil Factor 1, Provides a Safe and Efficacious Therapeutic Tool for Treating Oral Mucositis." *Oral Oncol* 46 (7):564–70.

Cani, P. D., J. Amar, M. A. Iglesias, M. Poggi, C. Knauf, D. Bastelica, A. M. Neyrinck, F. Fava, K. M. Tuohy, C. Chabo, A. Waget, E. Delmée, B. Cousin, T. Sulpice, B. Chamontin, J. Ferrières, J. F. Tanti, G. R. Gibson, L. Casteilla, N. M. Delzenne, M. C. Alessi, and R. Burcelin. 2007. "Metabolic Endotoxemia Initiates Obesity and Insulin Resistance." *Diabetes* 56 (7):1761–72.

Ceccarani, Camilla, and Marco Severgnini. 2023. "A Comparison Between Greengenes, SILVA, RDP, and NCBI Reference Databases in Four Published Microbiota Datasets." *bioRxiv*. https://doi.org/10.1101/2023.04.12.535864.

Chang, H. W., D. Yan, R. Singh, J. Liu, X. Lu, D. Ucmak, K. Lee, L. Afifi, D. Fadrosh, J. Leech, K. S. Vasquez, M. M. Lowe, M. D. Rosenblum, T. C. Scharschmidt, S. V. Lynch, and W. Liao. 2018. "Alteration of the Cutaneous Microbiome in Psoriasis and Potential Role in Th17 Polarization." *Microbiome* 6 (1):18–533.

Chaudhari, D. S., D. P. Dhotre, D. M. Agarwal, A. H. Gaike, D. Bhalerao, P. Jadhav, D. Mongad, H. Lubree, V. P. Sinkar, U. K. Patil, S. Salvi, A. Bavdekar, S. K. Juvekar, and Y. S. Shouche. 2020. "Gut, Oral and Skin Microbiome of Indian Patrilineal Families Reveal Perceptible Association with Age." *Sci Rep* 10 (1):20–62195.

Copeland, E., K. Leonard, R. Carney, J. Kong, M. Forer, Y. Naidoo, B. G. G. Oliver, J. R. Seymour, S. Woodcock, C. M. Burke, and N. W. Stow. 2018. "Chronic Rhinosinusitis: Potential Role of Microbial Dysbiosis and Recommendations for Sampling Sites." *Front Cell Infect Microbiol* 8 (57).

Coqueiro, A. Y., R. Raizel, A. Bonvini, J. Tirapegui, and M. M. Rogero. 2019. "Probiotics for Inflammatory Bowel Diseases: A Promising Adjuvant Treatment." *Int J Food Sci Nutr* 70 (1):20–9.

Costello, E. K., K. Stagaman, L. Dethlefsen, B. J. Bohannan, and D. A. Relman. 2012. "The Application of Ecological Theory Toward an Understanding of the Human Microbiome." *Science* 336 (6086):1255–62.

Cowie, C. C., K. F. Rust, E. S. Ford, M. S. Eberhardt, D. D. Byrd-Holt, C. Li, D. E. Williams, E. W. Gregg, K. E. Bainbridge, S. H. Saydah, and L. S. Geiss. 2009. "Full Accounting of Diabetes and Pre-Diabetes in the U.S. Population in 1988–1994 and 2005–2006." *Diabetes Care* 32 (2):287–94.

Crielaard, W., E. Zaura, A. A. Schuller, S. M. Huse, R. C. Montijn, and B. J. Keijser. 2011. "Exploring the Oral Microbiota of Children at Various Developmental Stages of Their Dentition in the Relation to Their Oral Health." *BMC Med Genom* 4 (22):1755–8794.

Cuello-Garcia, C. A., J. L. Brożek, A. Fiocchi, R. Pawankar, J. J. Yepes-Nuñez, L. Terracciano, S. Gandhi, A. Agarwal, Y. Zhang, and H. J. Schünemann. 2015. "Probiotics for the Prevention of Allergy: A Systematic Review and Meta-Analysis of Randomized Controlled Trials." *J Allergy Clin Immunol* 136 (4):952–61.

Cui, H., and X. Zhang. 2013. "Alignment-Free Supervised Classification of Metagenomes by Recursive SVM." *BMC Genom* 14 (641):1471–2164.

Delnevo, Giovanni, Giacomo Mancini, Marco Roccetti, Paola Salomoni, Elena Trombini, and Federica Andrei. 2021. "The Prediction of Body Mass Index from Negative Affectivity through Machine Learning: A Confirmatory Study." *Sensors* 21 (7):2361.

Depommier, C., A. Everard, C. Druart, H. Plovier, M. Van Hul, S. Vieira-Silva, G. Falony, J. Raes, D. Maiter, N. M. Delzenne, M. de Barsy, A. Loumaye, M. P. Hermans, J. P. Thissen, W. M. de Vos, and P. D. Cani. 2019. "Supplementation with Akkermansia Muciniphila in Overweight and Obese Human Volunteers: A Proof-of-Concept Exploratory Study." *Nat Med* 25 (7):1096–103.

Dewhirst, F. E., T. Chen, J. Izard, B. J. Paster, A. C. Tanner, W. H. Yu, A. Lakshmanan, and W. G. Wade. 2010. "The Human Oral Microbiome." *J Bacteriol* 192 (19):5002–17.

Dhotre, D., M. Inamdar, V. Shah, A. Gupta, S. Bhanushali, S. Bodkhe, N. Jiya, M. Gujare, A. Hagir, A. Deshpande, S. Gurav, S. Kumar, G. Tillu, S. Ozarkar, S. Indira, D. Agarwal, S. Juvekar, G. Makharia, and Y. S. Shouche. n.d. "Indian Human Microbiome Initiative (IHMI) to Study Interkingdom Microbiome Interactions with Diet of Healthy Adults from Select Tribal and Non-Tribal Communities." https://ibdc.dbtindia.gov.in/indasecure/humanmicrobiome.

Di Paola, M., C. Sani, A. M. Clemente, A. Iossa, E. Perissi, G. Castronovo, M. Tanturli, D. Rivero, F. Cozzolino, D. Cavalieri, F. Carozzi, C. De Filippo, and M. G. Torcia. 2017. "Characterization of Cervico-Vaginal Microbiota in Women Developing Persistent High-Risk Human Papillomavirus Infection." *Sci Rep* 7 (1):17–9842.

Dong, C., D. Tan, H. Sun, Z. Li, L. Zhang, Y. Zheng, S. Liu, Y. Zhang, and Q. He. 2024. "Interleukin-12 Delivery Strategies and Advances in Tumor Immunotherapy." *Curr Issues Mol Biol* 46 (10):11548–79.

Elinav, E., R. Nowarski, C. A. Thaiss, B. Hu, C. Jin, and R. A. Flavell. 2013. "Inflammation-Induced Cancer: Crosstalk Between Tumours, Immune Cells and Microorganisms." *Nat Rev Cancer* 13 (11):759–71.

Eloe-Fadrosh, E. A., and D. A. Rasko. 2013. "The Human Microbiome: From Symbiosis to Pathogenesis." *Annu Rev Med* 64:145–63.

Evans, J. M., L. S. Morris, and J. R. Marchesi. 2013. "The Gut Microbiome: The Role of a Virtual Organ in the Endocrinology of the Host." *J Endocrinol* 218 (3):13–131.

Fan, H. N., P. Zhu, Y. M. Lu, J. H. Guo, J. Zhang, G. Q. Qu, and J. S. Zhu. 2020. "Mild Changes in the Mucosal Microbiome During Terminal Ileum Inflammation." *Microb Pathog* 142 (104104):28.

Fang, X., J. M. Monk, S. Nurk, M. Akseshina, Q. Zhu, C. Gemmell, C. Gianetto-Hill, N. Leung, R. Szubin, J. Sanders, P. L. Beck, W. Li, W. J. Sandborn, S. D. Gray-Owen, R. Knight, E. Allen-Vercoe, B. O. Palsson, and L. Smarr. 2018. "Metagenomics-Based, Strain-Level Analysis of Escherichia Coli from a Time-Series of Microbiome Samples from a Crohn's Disease Patient." *Front Microbiol* 9 (2559).

Felten, A., C. Barreau, C. Bizet, P. H. Lagrange, and A. Philippon. 1999. "Lactobacillus Species Identification, H2O2 Production, and Antibiotic Resistance and Correlation with Human Clinical Status." *J Clin Microbiol* 37 (3):729–33.

Feng, Y., Y. Duan, Z. Xu, N. Lyu, F. Liu, S. Liang, and B. Zhu. 2019. "An Examination of Data from the American Gut Project Reveals That the Dominance of the Genus Bifidobacterium Is Associated with the Diversity and Robustness of the Gut Microbiota." *Microbiologyopen* 8 (12):30.

Fettweis, Jennifer M., Myrna G. Serrano, J. Paul Brooks, David J. Edwards, Philippe H. Girerd, Hardik I. Parikh, Bernice Huang, Tom J. Arodz, Laahirie Edupuganti, Abigail L. Glascock, Jie Xu, Nicole R. Jimenez, Stephany C. Vivadelli, Stephen S. Fong, Nihar U. Sheth, Sophonie Jean, Vladimir Lee, Yahya A. Bokhari, Ana M. Lara, Shreni D. Mistry, Robert A. Duckworth, Steven P. Bradley, Vishal N. Koparde, X. Valentine Orenda, Sarah H. Milton, Sarah K. Rozycki, Andrey V. Matveyev, Michelle L. Wright, Snehalata V. Huzurbazar, Eugenie M. Jackson, Ekaterina Smirnova, Jonas Korlach, Yu-Chih Tsai, Molly R. Dickinson, Jamie L. Brooks, Jennifer I. Drake, Donald O. Chaffin, Amber L. Sexton, Michael G. Gravett, Craig E. Rubens, N. Romesh Wijesooriya, Karen D. Hendricks-Muñoz, Kimberly K. Jefferson, Jerome F. Strauss, and Gregory A. Buck. 2019. "The Vaginal Microbiome and Preterm Birth." *Nat Med* 25 (6):1012–21. https://doi.org/10.1038/s41591-019-0450-2.

Finegold, S. M., D. Molitoris, Y. Song, C. Liu, M. L. Vaisanen, E. Bolte, M. McTeague, R. Sandler, H Wexler, E. M. Marlowe, M. D. Collins, P. A. Lawson, P. Summanen, M. Baysallar, T. J. Tomzynski, E. Read, E. Johnson, R. Rolfe, P. Nasir, H. Shah, D. A. Haake, P. Manning, and A. Kaul. 2002. "Gastrointestinal Microflora Studies in Late-Onset Autism." *Clin Infect Dis* 35 (S1):S6–16.

Fodor, Anthony A., Todd Z. DeSantis, Kristine M. Wylie, Jonathan H. Badger, Yuzhen Ye, Theresa Hepburn, Ping Hu, Erica Sodergren, Konstantinos Liolios, Heather Huot-Creasy, Bruce W. Birren, and Ashlee M. Earl. 2012. "The 'Most Wanted' Taxa from the Human Microbiome for Whole Genome Sequencing." *PLoS One* 7 (7):e41294. https://doi.org/10.1371/journal.pone.0041294.

Frank, D. N., C. E. Robertson, C. M. Hamm, Z. Kpadeh, T. Zhang, H. Chen, W. Zhu, R. B. Sartor, E. C. Boedeker, N. Harpaz, N. R. Pace, and E. Li. 2011. "Disease Phenotype and Genotype Are Associated with Shifts in Intestinal-Associated Microbiota in Inflammatory Bowel Diseases." *Inflamm Bowel Dis* 17 (1):179–84.

Franzosa, E. A., L. J. McIver, G. Rahnavard, L. R. Thompson, M. Schirmer, G. Weingart, K. S. Lipson, R. Knight, J. G. Caporaso, N. Segata, and C. Huttenhower. 2018. "Species-Level Functional Profiling of Metagenomes and Metatranscriptomes." *Nat Methods* 15 (11):962–8.

Franzosa, E. A., T. Hsu, A. Sirota-Madi, A. Shafquat, G. Abu-Ali, X. C. Morgan, and C. Huttenhower. 2015. "Sequencing and Beyond: Integrating Molecular 'Omics' for Microbial Community Profiling." *Nat Rev Microbiol* 13 (6):360–72.

Gagliardi, A., V. Totino, F. Cacciotti, V. Iebba, B. Neroni, G. Bonfiglio, M. Trancassini, C. Passariello, F. Pantanella, and S. Schippa. 2018. "Rebuilding the Gut Microbiota Ecosystem." *Int J Environ Res Public Health* 15 (8).

Gevers, D., R. Knight, J. F. Petrosino, K. Huang, A. L. McGuire, B. W. Birren, K. E. Nelson, O. White, B. A. Methé, and C. Huttenhower. 2012. "The Human Microbiome Project: A Community Resource for the Healthy Human Microbiome." *PLoS Biol* 10 (8):14.

Gholami, H., J. A. Chmiel, J. P. Burton, and S. Maleki Vareki. 2023. "The Role of Microbiota-Derived Vitamins in Immune Homeostasis and Enhancing Cancer Immunotherapy." *Cancers* 15 (4).

Glass, E. M., Y. Dribinsky, P. Yilmaz, H. Levin, R. Van Pelt, D. Wendel, A. Wilke, J. A. Eisen, S. Huse, A. Shipanova, M. Sogin, J. Stajich, R. Knight, F. Meyer, and L. M. Schriml. 2014. "MIxS-BE: A MIxS Extension Defining a Minimum Information Standard for Sequence Data from the Built Environment." *Isme J* 8 (1):1–3.

Gong, J., L. Li, X. Zuo, and Y. Li. 2019. "Change of the Duodenal Mucosa-Associated Microbiota Is Related to Intestinal Metaplasia." *BMC Microbiol* 19 (1):19–1666.

González, A., A. Fullaondo, and A. Odriozola. 2024. "Techniques, Procedures, and Applications in Microbiome Analysis." *Adv Genet* 111:81–115.

Gross, A., J. Schoendube, S. Zimmermann, M. Steeb, R. Zengerle, and P. Koltay. 2015. "Technologies for Single-Cell Isolation." *Int J Mol Sci* 16 (8):16897–919.

Guinane, C. M., A. Tadrous, F. Fouhy, C. A. Ryan, E. M. Dempsey, B. Murphy, E. Andrews, P. D. Cotter, C. Stanton, and R. P. Ross. 2013. "Microbial Composition of Human Appendices from Patients Following Appendectomy." *mBio* 4 (1):e366.12.

Gupta, A., S. Pahal, V. Kumar, and A. Chaudhary. 2025. "Role of Monoclonal Antibodies in Alzheimer's Disease: Successes, Failures, and Future Directions." *Neurol Sci.* https://doi.org/10.1007/s10072-025-08456-5.

Hamady, Z. Z., N. Scott, M. D. Farrar, M. Wadhwa, P. Dilger, T. R. Whitehead, R. Thorpe, K. T. Holland, J. P. Lodge, and S. R. Carding. 2011. "Treatment of Colitis with a Commensal Gut Bacterium Engineered to Secrete Human TGF-β1 Under the Control of Dietary Xylan 1." *Inflamm Bowel Dis* 17 (9):1925–35.

Heinken, A., J. Hertel, and I. Thiele. 2021. "Metabolic Modelling Reveals Broad Changes in Gut Microbial Metabolism in Inflammatory Bowel Disease Patients with Dysbiosis." *NPJ Syst Biol Appl* 7 (1):21–178.

Hernández Medina, R., S. Kutuzova, K. N. Nielsen, J. Johansen, L. H. Hansen, M. Nielsen, and S. Rasmussen. 2022. "Machine Learning and Deep Learning Applications in Microbiome Research." *ISME Commun* 2 (1):22–182.

Ho, B. S., E. X. P. Ho, C. W. Chu, S. Ramasamy, M. Bigliardi-Qi, P. F. de Sessions, and P. L. Bigliardi. 2019. "Microbiome in the Hair Follicle of Androgenetic Alopecia Patients." *PLoS One* 14 (5).

Hossain, K. S., S. Amarasena, and S. Mayengbam. 2022. "B Vitamins and Their Roles in Gut Health." *Microorganisms* 10 (6).

Huang, C., X. Yi, H. Long, G. Zhang, H. Wu, M. Zhao, and Q. Lu. 2020. "Disordered Cutaneous Microbiota in Systemic Lupus Erythematosus." *J Autoimmun* 108 (102391):26.

Huang, H., M. Fang, L. Jostins, M. Umićević Mirkov, G. Boucher, C. A. Anderson, V. Andersen, I. Cleynen, A. Cortes, F. Crins, M. D'Amato, V. Deffontaine, J. Dmitrieva, E. Docampo, M. Elansary, K. K. Farh, A. Franke, A. S. Gori, P. Goyette, J. Halfvarson, T. Haritunians, J. Knight, I. C. Lawrance, C. W. Lees, E. Louis, R. Mariman, T. Meuwissen, M. Mni, Y. Momozawa, M. Parkes, S. L. Spain, E. Théâtre, G. Trynka, J. Satsangi, S. van Sommeren, S. Vermeire, R. J. Xavier, R. K. Weersma, R. H. Duerr, C. G. Mathew, J. D. Rioux, D. P. B. McGovern, J. H. Cho, M. Georges, M. J. Daly, and J. C. Barrett. 2017. "Fine-Mapping Inflammatory Bowel Disease Loci to Single-Variant Resolution." *Nature* 547 (7662):173–8.

Huang, Shi, Fang Yang, Xiaowei Zeng, Jie Chen, Rui Li, Ting Wen, Chun Li, Wei Wei, Jiquan Liu, Lan Chen, Catherine Davis, and Jian Xu. 2011. "Preliminary Characterization of the Oral Microbiota of Chinese Adults with and Without Gingivitis." *BMC Oral Health* 11 (1):33. https://doi.org/10.1186/1472-6831-11-33.

Human Microbiome Project. 2012. "Structure, Function and Diversity of the Healthy Human Microbiome." *Nature* 486 (7402):207–14.

Ijaz, Muhammad, Mohsin Khurshid, Jingsi Gu, Ikram Hasan, Shubham Roy, Zia Ullah, Simin Liang, Jing Cheng, Yinghe Zhang, Chao Mi, and Bing Guo. 2024. "Breaking Barriers in Cancer Treatment:

Nanobiohybrids Empowered by Modified Bacteria and Vesicles." *Nanoscale* 16 (18):8759–77. https://doi.org/10.1039/d3nr06666e.

Islam, Taminul, Md Alif Sheakh, Mst Sazia Tahosin, Most Hasna Hena, Shopnil Akash, Yousef A. Bin Jardan, Gezahign FentahunWondmie, Hiba-Allah Nafidi, and Mohammed Bourhia. 2024. "Predictive Modeling for Breast Cancer Classification in the Context of Bangladeshi Patients by Use of Machine Learning Approach with Explainable AI." *Sci Rep* 14 (1):8487. https://doi.org/10.1038/s41598-024-57740-5.

James, K. R., T. Gomes, R. Elmentaite, N. Kumar, E. L. Gulliver, H. W. King, M. D. Stares, B. R. Bareham, J. R. Ferdinand, V. N. Petrova, K. Polański, S. C. Forster, L. B. Jarvis, O. Suchanek, S. Howlett, L. K. James, J. L. Jones, K. B. Meyer, M. R. Clatworthy, K. Saeb-Parsy, T. D. Lawley, and S. A. Teichmann. 2020. "Distinct Microbial and Immune Niches of the Human Colon." *Nat Immunol* 21 (3):343–53.

Janku, Filip, Halle Huihong Zhang, Abdulmohammad Pezeshki, Sanjay Goel, Ravi Murthy, Andrea Wang-Gillam, Dale R. Shepard, Thorunn Helgason, Tyler Masters, David S. Hong, Sarina A Piha-Paul, Daniel D. Karp, Mark Klang, Steven Y. Huang, Divya Sakamuri, Anjali Raina, Jean Torrisi, Stephen B. Solomon, Alice Weissfeld, Ernest Trevino, Gary DeCrescenzo, Amanda Collins, Maria Miller, Jennifer L. Salstrom, Ronald L. Korn, Linping Zhang, Saurabh Saha, Alexey A. Leontovich, David Tung, Brent Kreider, Mary Varterasian, Khashayarsha Khazaie, and Mrinal M. Gounder. 2021. "Intratumoral Injection of Clostridium novyi-NT Spores in Patients with Treatment-Refractory Advanced Solid Tumors." *Clin Cancer Res* 27 (1):96–106. https://doi.org/10.1158/1078-0432.ccr-20-2065.

Jervis-Bardy, J., L. E. X. Leong, L. E. Papanicolas, K. L. Ivey, S. Chawla, C. M. Woods, C. Frauenfelder, E. H. Ooi, and G. B. Rogers. 2019. "Examining the Evidence for an Adult Healthy Middle Ear Microbiome." *mSphere* 4 (5):e456.19.

Jiang, S., and G. Redelman-Sidi. 2022. "BCG in Bladder Cancer Immunotherapy." *Cancers* 14 (13).

Joossens, M., G. Huys, M. Cnockaert, V. De Preter, K. Verbeke, P. Rutgeerts, P. Vandamme, and S. Vermeire. 2011. "Dysbiosis of the Faecal Microbiota in Patients with Crohn's Disease and Their Unaffected Relatives." *Gut* 60 (5):631–7.

Jovel, Juan, Jordan Patterson, Weiwei Wang, Naomi Hotte, Sandra O'Keefe, Troy Mitchel, Troy Perry, Dina Kao, Andrew L. Mason, Karen L. Madsen, and Gane K.-S. Wong. 2016. "Characterization of the Gut Microbiome Using 16S or Shotgun Metagenomics." *Front Microbiol* 7. https://doi.org/10.3389/fmicb.2016.00459.

Kang, H. M., and J. H. Kang. 2021. "Effects of Nasopharyngeal Microbiota in Respiratory Infections and Allergies." *Clin Exp Pediatr* 64 (11):543–51.

Kennedy, Beatrice, Sari Peura, Ulf Hammar, Silvia Vicenzi, Anna Hedman, Catarina Almqvist, Ellika Andolf, Göran Pershagen, Johan Dicksved, Stefan Bertilsson, and Tove Fall. 2019. "Oral Microbiota Development in Early Childhood." *Sci Rep* 9 (1):19025. https://doi.org/10.1038/s41598-019-54702-0.

Khan, M. T., J. M. van Dijl, and H. J. Harmsen. 2014. "Antioxidants Keep the Potentially Probiotic but Highly Oxygen-Sensitive Human Gut Bacterium Faecalibacterium Prausnitzii Alive at Ambient Air." *PLoS One* 9 (5).

Kolbe, A. R., E. Castro-Nallar, D. Preciado, and M. Pérez-Losada. 2019. "Altered Middle Ear Microbiome in Children with Chronic Otitis Media with Effusion and Respiratory Illnesses." *Front Cell Infect Microbiol* 9 (339).

Kolisnik, Tyler, Arielle Kae Sulit, Sebastian Schmeier, Frank Frizelle, Rachel Purcell, Adam Smith, and Olin Silander. 2023. "Identifying Important Microbial and Genomic Biomarkers for Differentiating Right-Versus Left-Sided Colorectal Cancer Using Random Forest Models." *BMC Cancer* 23 (1):647. https://doi.org/10.1186/s12885-023-10848-9.

Kong, H. H., J. Oh, C. Deming, S. Conlan, E. A. Grice, M. A. Beatson, E. Nomicos, E. C. Polley, H. D. Komarow, P. R. Murray, M. L. Turner, and J. A. Segre. 2012. "Temporal Shifts in the Skin Microbiome Associated with Disease Flares and Treatment in Children with Atopic Dermatitis." *Genome Res* 22 (5):850–9.

Kumar, V., S. Gangani, R. Shukla, S. K. Prajapati, H. Shekhar, and V. Shukla. 2024. "Next-Generation Sequencing and Genomic Data Analysis." In A. Chaudhary, S. K. Sethi, and A. Verma (Eds.), *Unraveling New Frontiers and Advances in Bioinformatics* (pp. 49–76). Springer Nature. https://doi.org/10.1007/978-981-97-7123-3_3.

Kumar, V., A. Pandey, A. Arora, P. Gautam, D. Bisht, S. Gupta, A. Chaurasia, and M. Sachan. 2023. "Diagnostics and Therapeutic Potential of miR-205 and miR-34a in Ovarian Cancer Management: A miRNA-Target-Based Analysis." *DNA Cell Biol* 42 (3):151–62. https://doi.org/10.1089/dna.2022.0487.

Kumar, V., J.K. Sharma, S. Kumar, J. Gulia, N. P. Singh, and A. R. Santal. 2025. "Metagenomic and Proteomic Approach for Bioremediation of Environmental Pollutants." In M. P. Shah (Ed.), *Development in Waste Water Treatment Research and Processes*. Elsevier. https://doi.org/10.1016/B978-0-443-13615-3.00020-0.

Kumar, V., R. Shukla, S. Gangani, H. Shekhar, and V. Shukla. 2024. "Metagenomics and Microbiome Analysis." In A. Chaudhary, S. K. Sethi, and A. Verma (Eds.), *Unraveling New Frontiers and Advances in Bioinformatics* (pp. 77–92). Springer Nature. https://doi.org/10.1007/978-981-97-7123-3_4.

Kumar, V., A. Singh, P. Gautam, and M. Sachan. 2019. "Techniques/Tools to Study Epigenetic Biomarkers in Human Cancer Detection." In S. Paul (Ed.), *Biomedical Engineering and Its Applications in Healthcare* (pp. 327–51). Springer. https://doi.org/10.1007/978-981-13-3705-5_14.

Lappan, R., K. Imbrogno, C. Sikazwe, D. Anderson, D. Mok, H. Coates, S. Vijayasekaran, P. Bumbak, C. C. Blyth, S. E. Jamieson, and C. S. Peacock. 2018. "A Microbiome Case-Control Study of Recurrent Acute Otitis Media Identified Potentially Protective Bacterial Genera." *BMC Microbiol* 18 (1):18–1154.

Lewis, J. D., E. Z. Chen, R. N. Baldassano, A. R. Otley, A. M. Griffiths, D. Lee, K. Bittinger, A. Bailey, E. S. Friedman, C. Hoffmann, L. Albenberg, R. Sinha, C. Compher, E. Gilroy, L. Nessel, A. Grant, C. Chehoud, H. Li, G. D. Wu, and F. D. Bushman. 2015. "Inflammation, Antibiotics, and Diet as Environmental Stressors of the Gut Microbiome in Pediatric Crohn's Disease." *Cell Host Microbe* 18 (4):489–500.

Li, J., J. Butcher, D. Mack, and A. Stintzi. 2015. "Functional Impacts of the Intestinal Microbiome in the Pathogenesis of Inflammatory Bowel Disease." *Inflamm Bowel Dis* 21 (1):139–53.

Li, S., G. Yi, H. Peng, Z. Li, S. Chen, H. Zhong, Y. Chen, Z. Wang, Q. Deng, and M. Fu. 2019. "How Ocular Surface Microbiota Debuts in Type 2 Diabetes Mellitus." *Front Cell Infect Microbiol* 9 (202).

Li, Y., M. Zhu, Y. Liu, B. Luo, J. Cui, L. Huang, and K. Chen. 2022. "The Oral Microbiota and Cardiometabolic Health: A Comprehensive Review and Emerging Insights." *Front Immunol* 13 (1010368).

Liu, N., T. Ando, K. Ishiguro, O. Maeda, O. Watanabe, K. Funasaka, M. Nakamura, R. Miyahara, N. Ohmiya, and H. Goto. 2013. "Characterization of Bacterial Biota in the Distal Esophagus of Japanese Patients with Reflux Esophagitis and Barrett's Esophagus." *BMC Infect Dis* 13 (130):1471–2334.

Liu, S., J. S. Rodriguez, V. Munteanu, C. Ronkowski, N. K. Sharma, M. Alser, F. Andreace, R. Blekhman, D. Błaszczyk, R. Chikhi, K. A. Crandall, K. Della Libera, D. Francis, A. Frolova, A. S. Gancz, N. E. Huntley, P. Jaiswal, T. Kosciolek, P. P. Łabaj, W. Łabaj, T. Luan, C. Mason, A. M. Moustafa, H. S. Muralidharan, O. Mutlu, N. Mansouri Ghiasi, A. Rahnavard, F. Sun, S. Tian, B. T. Tierney, E. Van Syoc, R. Vicedomini, J. P. Zackular, A. Zelikovsky, K. Zielińska, E. Ganda, E. R. Davenport, M. Pop, D. Koslicki, and S. Mangul. 2025. "Analysis of Metagenomic Data." *Nat Rev Methods Primers* 5 (5):24–376.

Ma, Ziqi, Tao Zuo, Norbert Frey, and Ashraf Yusuf Rangrez. 2024. "A Systematic Framework for Understanding the Microbiome in Human Health and Disease: From Basic Principles to Clinical Translation." *Signal Transduct Target Ther* 9 (1):237. https://doi.org/10.1038/s41392-024-01946-6.

Magnúsdóttir, S., A. Heinken, L. Kutt, D. A. Ravcheev, E. Bauer, A. Noronha, K. Greenhalgh, C. Jäger, J. Baginska, P. Wilmes, R. M. Fleming, and I. Thiele. 2017. "Generation of Genome-Scale Metabolic Reconstructions for 773 Members of the Human Gut Microbiota." *Nat Biotechnol* 35 (1):81–9.

Maldonado-Contreras, A., K. C. Goldfarb, F. Godoy-Vitorino, U. Karaoz, M. Contreras, M. J. Blaser, E. L. Brodie, and M. G. Dominguez-Bello. 2011. "Structure of the Human Gastric Bacterial Community in Relation to Helicobacter Pylori Status." *Isme J* 5 (4):574–9.

Manichanh, C., L. Rigottier-Gois, E. Bonnaud, K. Gloux, E. Pelletier, L. Frangeul, R. Nalin, C. Jarrin, P. Chardon, P. Marteau, J. Roca, and J. Dore. 2006. "Reduced Diversity of Faecal Microbiota in Crohn's Disease Revealed by a Metagenomic Approach." *Gut* 55 (2):205–11.

Marcos-Zambrano, L. J., K. Karaduzovic-Hadziabdic, T. Loncar Turukalo, P. Przymus, V. Trajkovik, O. Aasmets, M. Berland, A. Gruca, J. Hasic, K. Hron, T. Klammsteiner, M. Kolev, L. Lahti, M. B. Lopes, V. Moreno, I. Naskinova, E. Org, I. Paciência, G. Papoutsoglou, R. Shigdel, B. Stres, B. Vilne, M. Yousef, E. Zdravevski, I. Tsamardinos, E. Carrillo de Santa Pau, M. J. Claesson, I. Moreno-Indias, and J. Truu. 2021. "Applications of Machine Learning in Human Microbiome Studies: A Review on Feature Selection, Biomarker Identification, Disease Prediction and Treatment." *Front Microbiol* 12 (634511).

Matsuo, Y., S. Komiya, Y. Yasumizu, Y. Yasuoka, K. Mizushima, T. Takagi, K. Kryukov, A. Fukuda, Y. Morimoto, Y. Naito, H. Okada, H. Bono, S. Nakagawa, and K. Hirota. 2021. "Full-Length 16S rRNA Gene Amplicon Analysis of Human Gut Microbiota Using MinION™ Nanopore Sequencing Confers Species-Level Resolution." *BMC Microbiol* 21 (1):21–2094.

McDonald, D., E. Hyde, J. W. Debelius, J. T. Morton, A. Gonzalez, G. Ackermann, A. A. Aksenov, B. Behsaz, C. Brennan, Y. Chen, L. DeRight Goldasich, P. C. Dorrestein, R. R. Dunn, A. K. Fahimipour, J. Gaffney, J. A. Gilbert, G. Gogul, J. L. Green, P. Hugenholtz, G. Humphrey, C. Huttenhower, M. A. Jackson, S. Janssen, D. V. Jeste, L. Jiang, S. T. Kelley, D. Knights, T. Kosciolek, J. Ladau, J. Leach, C. Marotz, D. Meleshko, A. V. Melnik, J. L. Metcalf, H. Mohimani, E. Montassier, J. Navas-Molina, T. T. Nguyen, S. Peddada, P. Pevzner, K. S. Pollard, G. Rahnavard, A. Robbins-Pianka, N. Sangwan, J. Shorenstein, L. Smarr, S. J. Song, T. Spector, A. D. Swafford, V. G. Thackray, L. R. Thompson, A. Tripathi, Y.

Vázquez-Baeza, A. Vrbanac, P. Wischmeyer, E. Wolfe, Q. Zhu, and R. Knight. 2018. "American Gut: An Open Platform for Citizen Science Microbiome Research." *mSystems* 3 (3):e31.18.

Minich, J. J., G. Humphrey, R. A. S. Benitez, J. Sanders, A. Swafford, E. E. Allen, and R. Knight. 2018. "High-Throughput Miniaturized 16S rRNA Amplicon Library Preparation Reduces Costs While Preserving Microbiome Integrity." *mSystems* 3 (6):e166.18.

Mirete, Salvador, Mercedes Sánchez-Costa, Jorge Díaz-Rullo, Carolina González de Figueras, Pablo Martínez-Rodríguez, and José Eduardo González-Pastor. 2025. "Metagenome-Assembled Genomes (MAGs): Advances, Challenges, and Ecological Insights." *Microorganisms* 13 (5):985.

Moreno-Indias, I., L. Lahti, M. Nedyalkova, I. Elbere, G. Roshchupkin, M. Adilovic, O. Aydemir, B. Bakir-Gungor, E. C. Santa Pau, D. D'Elia, M. S. Desai, L. Falquet, A. Gundogdu, K. Hron, T. Klammsteiner, M. B. Lopes, L. J. Marcos-Zambrano, C. Marques, M. Mason, P. May, L. Pašić, G. Pio, S. Pongor, V. J. Promponas, P. Przymus, J. Saez-Rodriguez, A. Sampri, R. Shigdel, B. Stres, R. Suharoschi, J. Truu, C. O. Truică, B. Vilne, D. Vlachakis, E. Yilmaz, G. Zeller, A. L. Zomer, D. Gómez-Cabrero, and M. J. Claesson. 2021. "Statistical and Machine Learning Techniques in Human Microbiome Studies: Contemporary Challenges and Solutions." *Front Microbiol* 12 (635781).

Motta, J. P., L. G. Bermúdez-Humarán, C. Deraison, L. Martin, C. Rolland, P. Rousset, J. Boue, G. Dietrich, K. Chapman, P. Kharrat, J. P. Vinel, L. Alric, E. Mas, J. M. Sallenave, P. Langella, and N. Vergnolle. 2012. "Food-Grade Bacteria Expressing Elafin Protect Against Inflammation and Restore Colon Homeostasis." *Sci Transl Med* 4 (158):3004212.

Nair, S. U., Y. Lin, S. Jain, D. S. Chaudhari, R. Shukla, V. Kumar, R. T. Mankowski, H. Meng, T. W. Buford, G. A. Kuchel, H. Yadav, and MiaGB Consortium. 2024. "Unique Gut Microbiome Signatures Associated with Physical and Cognitive Function." *Alzheimer's Dementia* 20 (S2):e092044. https://doi.org/10.1002/alz.092044.

Nelson, K. E., G. M. Weinstock, S. K. Highlander, K. C. Worley, H. H. Creasy, J. R. Wortman, D. B. Rusch, M. Mitreva, E. Sodergren, A. T. Chinwalla, M. Feldgarden, D. Gevers, B. J. Haas, R. Madupu, D. V. Ward, B. W. Birren, R. A. Gibbs, B. Methe, J. F. Petrosino, R. L. Strausberg, G. G. Sutton, O. R. White, R. K. Wilson, S. Durkin, M. G. Giglio, S. Gujja, C. Howarth, C. D. Kodira, N. Kyrpides, T. Mehta, D. M. Muzny, M. Pearson, K. Pepin, A. Pati, X. Qin, C. Yandava, Q. Zeng, L. Zhang, A. M. Berlin, L. Chen, T. A. Hepburn, J. Johnson, J. McCorrison, J. Miller, P. Minx, C. Nusbaum, C. Russ, S. M. Sykes, C. M. Tomlinson, S. Young, W. C. Warren, J. Badger, J. Crabtree, V. M. Markowitz, J. Orvis, A. Cree, S. Ferriera, L. L. Fulton, R. S. Fulton, M. Gillis, L. D. Hemphill, V. Joshi, C. Kovar, M. Torralba, K. A. Wetterstrand, A. Abouellleil, A. M. Wollam, C. J. Buhay, Y. Ding, S. Dugan, M. G. FitzGerald, M. Holder, J. Hostetler, S. W. Clifton, E. Allen-Vercoe, A. M. Earl, C. N. Farmer, K. Liolios, M. G. Surette, Q. Xu, C. Pohl, K. Wilczek-Boney, and D. Zhu. 2010. "A Catalog of Reference Genomes from the Human Microbiome." *Science* 328 (5981):994–9.

Noonan, S., M. Zaveri, E. Macaninch, and K. Martyn. 2020. "Food & Mood: A Review of Supplementary Prebiotic and Probiotic Interventions in the Treatment of Anxiety and Depression in Adults." *BMJ Nutr Prev Health* 3 (2):351–62.

Oh, Min, and Liqing Zhang. 2020. "DeepMicro: Deep Representation Learning for Disease Prediction Based on Microbiome Data." *Sci Rep* 10 (1):6026. https://doi.org/10.1038/s41598-020-63159-5.

Ojo, Diandra P., Cameron Celeste, Dion Ming, Ruogu Fang, and Ivana K. Parker. 2025. "Population-Level Predictive Variation in Machine Learning Diagnosis of Symptomatic Bacterial Vaginosis." *npj Women's Health* 3 (1):45. https://doi.org/10.1038/s44294-025-00092-w.

Paller, A. S., H. H. Kong, P. Seed, S. Naik, T. C. Scharschmidt, R. L. Gallo, T. Luger, and A. D. Irvine. 2019. "The Microbiome in Patients with Atopic Dermatitis." *J Allergy Clin Immunol* 143 (1):26–35.

Papoutsoglou, G., S. Tarazona, M. B. Lopes, T. Klammsteiner, E. Ibrahimi, J. Eckenberger, P. Novielli, A. Tonda, A. Simeon, R. Shigdel, S. Béreux, G. Vitali, S. Tangaro, L. Lahti, A. Temko, M. J. Claesson, and M. Berland. 2023. "Machine Learning Approaches in Microbiome Research: Challenges and Best Practices." *Front Microbiol* 14 (1261889).

Pei, Z., E. J. Bini, L. Yang, M. Zhou, F. Francois, and M. J. Blaser. 2004. "Bacterial Biota in the Human Distal Esophagus." *Proc Natl Acad Sci U S A* 101 (12):4250–5.

Petrof, E. O., and A. Khoruts. 2014. "From Stool Transplants to Next-Generation Microbiota Therapeutics." *Gastroenterol* 146 (6):1573–82.

Pickup, J. C. 2004. "Inflammation and Activated Innate Immunity in the Pathogenesis of Type 2 Diabetes." *Diabetes Care* 27 (3):813–23.

Prajapati, S. K., V. Kumar, R. Shukla, C. Szekeres, Y. Ma, S. Jain, and H. Yadav. 2024. "Microbiome-Immune Interplay in Aging Brains: A Pilot Investigation from the MiaGB Cohort." *Alzheimer's Dementia* 20 (S2):e090734. https://doi.org/10.1002/alz.090734.

Praveschotinunt, Pichet, Anna M. Duraj-Thatte, Ilia Gelfat, Franziska Bahl, David B. Chou, and Neel S. Joshi. 2019. "Engineered E. Coli Nissle 1917 for the Delivery of Matrix-Tethered Therapeutic Domains to the Gut." *Nat Commun* 10 (1):5580. https://doi.org/10.1038/s41467-019-13336-6.

Priya, Sambhawa, Michael B. Burns, Tonya Ward, Ruben A. T. Mars, Beth Adamowicz, Eric F. Lock, Purna C. Kashyap, Dan Knights, and Ran Blekhman. 2022. "Identification of Shared and Disease-Specific Host Gene-Microbiome Associations Across Human Diseases Using Multi-Omic Integration." *Nat Microbiol* 7 (6):780–95. https://doi.org/10.1038/s41564-022-01121-z.

Pust, Marie-Madlen, and Burkhard Tümmler. 2021. "Identification of Core and Rare Species in Metagenome Samples Based on Shotgun Metagenomic Sequencing, Fourier Transforms and Spectral Comparisons." *ISME Commun* 1 (1):2. https://doi.org/10.1038/s43705-021-00010-6.

Qin, Junjie, Yingrui Li, Zhiming Cai, Shenghui Li, Jianfeng Zhu, Fan Zhang, Suisha Liang, Wenwei Zhang, Yuanlin Guan, Dongqian Shen, Yangqing Peng, Dongya Zhang, Zhuye Jie, Wenxian Wu, Youwen Qin, Wenbin Xue, Junhua Li, Lingchuan Han, Donghui Lu, Peixian Wu, Yali Dai, Xiaojuan Sun, Zesong Li, Aifa Tang, Shilong Zhong, Xiaoping Li, Weineng Chen, Ran Xu, Mingbang Wang, Qiang Feng, Meihua Gong, Jing Yu, Yanyan Zhang, Ming Zhang, Torben Hansen, Gaston Sanchez, Jeroen Raes, Gwen Falony, Shujiro Okuda, Mathieu Almeida, Emmanuelle LeChatelier, Pierre Renault, Nicolas Pons, Jean-Michel Batto, Zhaoxi Zhang, Hua Chen, Ruifu Yang, Weimou Zheng, Songgang Li, Huanming Yang, Jian Wang, S. Dusko Ehrlich, Rasmus Nielsen, Oluf Pedersen, Karsten Kristiansen, and Jun Wang. 2012. "A Metagenome-Wide Association Study of Gut Microbiota in Type 2 Diabetes." *Nature* 490 (7418):55–60. https://doi.org/10.1038/nature11450.

Qin, Junjie, Ruiqiang Li, Jeroen Raes, Manimozhiyan Arumugam, Kristoffer Solvsten Burgdorf, Chaysavanh Manichanh, Trine Nielsen, Nicolas Pons, Florence Levenez, Takuji Yamada, Daniel R. Mende, Junhua Li, Junming Xu, Shaochuan Li, Dongfang Li, Jianjun Cao, Bo Wang, Huiqing Liang, Huisong Zheng, Yinlong Xie, Julien Tap, Patricia Lepage, Marcelo Bertalan, Jean-Michel Batto, Torben Hansen, Denis Le Paslier, Allan Linneberg, H. Bjørn Nielsen, Eric Pelletier, Pierre Renault, Thomas Sicheritz-Ponten, Keith Turner, Hongmei Zhu, Chang Yu, Shengting Li, Min Jian, Yan Zhou, Yingrui Li, Xiuqing Zhang, Songgang Li, Nan Qin, Huanming Yang, Jian Wang, Søren Brunak, Joel Doré, Francisco Guarner, Karsten Kristiansen, Oluf Pedersen, Julian Parkhill, Jean Weissenbach, Maria Antolin, François Artiguenave, Hervé Blottiere, Natalia Borruel, Thomas Bruls, Francesc Casellas, Christian Chervaux, Antonella Cultrone, Christine Delorme, Gérard Denariaz, Rozenn Dervyn, Miguel Forte, Carsten Friss, Maarten van de Guchte, Eric Guedon, Florence Haimet, Alexandre Jamet, Catherine Juste, Ghalia Kaci, Michiel Kleerebezem, Jan Knol, Michel Kristensen, Severine Layec, Karine Le Roux, Marion Leclerc, Emmanuelle Maguin, Raquel Melo Minardi, Raish Oozeer, Maria Rescigno, Nicolas Sanchez, Sebastian Tims, Toni Torrejon, Encarna Varela, Willem de Vos, Yohanan Winogradsky, Erwin Zoetendal, Peer Bork, S. Dusko Ehrlich, Jun Wang, and H. I. T. Consortium Meta. 2010. "A Human Gut Microbial Gene Catalogue Established by Metagenomic Sequencing." *Nature* 464 (7285):59–65. https://doi.org/10.1038/nature08821.

Quince, C., S. Nurk, S. Raguideau, R. James, O. S. Soyer, J. K. Summers, A. Limasset, A. M. Eren, R. Chikhi, and A. E. Darling. 2021. "STRONG: Metagenomics Strain Resolution on Assembly Graphs." *Genome Biol* 22 (1):21–2419.

Rausch, P., M. Rühlemann, B. M. Hermes, S. Doms, T. Dagan, K. Dierking, H. Domin, S. Fraune, J. von Frieling, U. Hentschel, F. A. Heinsen, M. Höppner, M. T. Jahn, C. Jaspers, K. A. B. Kissoyan, D. Langfeldt, A. Rehman, T. B. H. Reusch, T. Roeder, R. A. Schmitz, H. Schulenburg, R. Soluch, F. Sommer, E. Stukenbrock, N. Weiland-Bräuer, P. Rosenstiel, A. Franke, T. Bosch, and J. F. Baines. 2019. "Comparative Analysis of Amplicon and Metagenomic Sequencing Methods Reveals Key Features in the Evolution of Animal Metaorganisms." *Microbiome* 7 (1):19–743.

Richards, Allison L., Amanda L. Muehlbauer, Adnan Alazizi, Michael B. Burns, Anthony Findley, Francesco Messina, Trevor J. Gould, Camilla Cascardo, Roger Pique-Regi, Ran Blekhman, and Francesca Luca. 2019. "Gut Microbiota Has a Widespread and Modifiable Effect on Host Gene Regulation." *mSystems* 4 (5). https://doi.org/10.1128/msystems.00323-18.

Rinninella, E., P. Raoul, M. Cintoni, F. Franceschi, G. A. D. Miggiano, A. Gasbarrini, and M. C. Mele. 2019. "What Is the Healthy Gut Microbiota Composition? A Changing Ecosystem Across Age, Environment, Diet, and Diseases." *Microorganisms* 7 (1).

Rowland, Ian, Glenn Gibson, Almut Heinken, Karen Scott, Jonathan Swann, Ines Thiele, and Kieran Tuohy. 2018. "Gut Microbiota Functions: Metabolism of Nutrients and Other Food Components." *Eur J Nutr* 57 (1):1–24. https://doi.org/10.1007/s00394-017-1445-8.

Ruuskanen, Matti O., Deepti Vats, Renuka Potbhare, Ameeta RaviKumar, Eveliina Munukka, Richa Ashma, and Leo Lahti. 2022. "Towards Standardized and Reproducible Research in Skin Microbiomes." *Environ Microbiol* 24 (9):3840–60. https://doi.org/10.1111/1462-2920.15945.

Sarfraz, H., B. J. Czerniecki, K. Stankiewicz, M. V. Jaglal, S. Jain, H. Yadav, M. Liu, R. Shukla, V. Kumar, M. Alhomsi, M. Hathout, F. Tuz Zahra, A. Darwin, U. Joshi, A. Sen, J. Muzaffar, and S. Bari. 2024. "Evaluating Gut Microbiome as a Biomarker of Pathological Complete Response in Patients with Early Locally Advanced Triple Negative Breast Cancer (LA TNBC): A Pilot Study." *J Clin Oncol* 42(S16):593. https://doi.org/10.1200/JCO.2024.42.16_suppl.593.

Schnorr, S. L., M. Candela, S. Rampelli, M. Centanni, C. Consolandi, G. Basaglia, S. Turroni, E. Biagi, C. Peano, M. Severgnini, J. Fiori, R. Gotti, G. De Bellis, D. Luiselli, P. Brigidi, A. Mabulla, F. Marlowe, A. G. Henry, and A. N. Crittenden. 2014. "Gut Microbiome of the Hadza Hunter-Gatherers." *Nat Commun* 5 (3654).

Shakya, M., C. C. Lo, and P. S. G. Chain. 2019. "Advances and Challenges in Metatranscriptomic Analysis." *Front Genet* 10 (904).

Sharma, Divya, and Wei Xu. 2021. "phyLoSTM: A Novel Deep Learning Model on Disease Prediction from Longitudinal Microbiome Data." *Bioinformatics* 37 (21):3707–14. https://doi.org/10.1093/bioinformatics/btab482.

Sharpton, Thomas J. 2014. "An Introduction to the Analysis of Shotgun Metagenomic Data." *Front Plant Sci* 5 (209). https://doi.org/10.3389/fpls.2014.00209.

Stewart, C. J., J. M. Mansbach, M. C. Wong, N. J. Ajami, J. F. Petrosino, C. A. Camargo, Jr., and K. Hasegawa. 2017. "Associations of Nasopharyngeal Metabolome and Microbiome with Severity Among Infants with Bronchiolitis: A Multiomic Analysis." *Am J Respir Crit Care Med* 196 (7):882–91.

Sulit, A. K., T. Kolisnik, F. A. Frizelle, R. Purcell, and S. Schmeier. 2023. "MetaFunc: Taxonomic and Functional Analyses of High Throughput Sequencing for Microbiomes." *Gut Microbiome* 12 (4).

Sulyanto, Rosalyn M., Zachary A. Thompson, Clifford J. Beall, Eugene J. Leys, and Ann L. Griffen. 2019. "The Predominant Oral Microbiota Is Acquired Early in an Organized Pattern." *Sci Rep* 9 (1):10550. https://doi.org/10.1038/s41598-019-46923-0.

Sundin, O. H., A. Mendoza-Ladd, M. Zeng, D. Diaz-Arévalo, E. Morales, B. M. Fagan, J. Ordoñez, P. Velez, N. Antony, and R. W. McCallum. 2017. "The Human Jejunum Has an Endogenous Microbiota That Differs from Those in the Oral Cavity and Colon." *BMC Microbiol* 17 (1):17–1059.

Suzuki, T., T. Sutani, H. Nakai, K. Shirahige, and S. Kinoshita. 2020. "The Microbiome of the Meibum and Ocular Surface in Healthy Subjects." *Invest Ophthalmol Vis Sci* 61 (2):18.

Sze, Marc A., and Patrick D. Schloss. 2019. "The Impact of DNA Polymerase and Number of Rounds of Amplification in PCR on 16S rRNA Gene Sequence Data." *mSphere* 4 (3). https://doi.org/10.1128/msphere.00163-19.

Teo, S. M., D. Mok, K. Pham, M. Kusel, M. Serralha, N. Troy, B. J. Holt, B. J. Hales, M. L. Walker, E. Hollams, Y. A. Bochkov, K. Grindle, S. L. Johnston, J. E. Gern, P. D. Sly, P. G. Holt, K. E. Holt, and M. Inouye. 2015. "The Infant Nasopharyngeal Microbiome Impacts Severity of Lower Respiratory Infection and Risk of Asthma Development." *Cell Host Microbe* 17 (5):704–15.

Tremblay, Julien, Lars Schreiber, and Charles W Greer. 2022. "High-Resolution Shotgun Metagenomics: The More Data, the Better?" *Brief Bioinform* 23 (6). https://doi.org/10.1093/bib/bbac443.

Turnbaugh, P. J., R. E. Ley, M. A. Mahowald, V. Magrini, E. R. Mardis, and J. I. Gordon. 2006. "An Obesity-Associated Gut Microbiome with Increased Capacity for Energy Harvest." *Nature* 444 (7122):1027–31.

Tursi, A., G. Brandimarte, A. Papa, A. Giglio, W. Elisei, G. M. Giorgetti, G. Forti, S. Morini, C. Hassan, M. A. Pistoia, M. E. Modeo, S. Rodino, T. D'Amico, L. Sebkova, N. Sacca, E. Di Giulio, F. Luzza, M. Imeneo, T. Larussa, S. Di Rosa, V. Annese, S. Danese, and A. Gasbarrini. 2010. "Treatment of Relapsing Mild-to-Moderate Ulcerative Colitis with the Probiotic VSL#3 as Adjunctive to a Standard Pharmaceutical Treatment: A Double-Blind, Randomized, Placebo-Controlled Study." *Am J Gastroenterol* 105 (10):2218–27.

Ullah, H., S. Arbab, Y. Tian, C. Q. Liu, Y. Chen, L. Qijie, M. I. U. Khan, I. U. Hassan, and K. Li. 2023. "The Gut Microbiota-Brain Axis in Neurological Disorder." *Front Neurosci* 17 (1225875).

Valenti, P., L. Rosa, D. Capobianco, M. S. Lepanto, E. Schiavi, A. Cutone, R. Paesano, and P. Mastromarino. 2018. "Role of Lactobacilli and Lactoferrin in the Mucosal Cervicovaginal Defense." *Front Immunol* 9 (376).

Van Den Bossche, T., M. Ø Arntzen, D. Becher, D. Benndorf, V. G. H. Eijsink, C. Henry, P. D. Jagtap, N. Jehmlich, C. Juste, B. J. Kunath, B. Mesuere, T. Muth, P. B. Pope, J. Seifert, A. Tanca, S. Uzzau, P. Wilmes, R. L. Hettich, and J. Armengaud. 2021. "The Metaproteomics Initiative: A Coordinated Approach for Propelling the Functional Characterization of Microbiomes." *Microbiome* 9(1):243, December 20. https://doi.org/10.1186/s40168-021-01176-w.

Varliero, G., P. H. Lebre, M. I. Stevens, P. Czechowski, T. Makhalanyane, and D. A. Cowan. 2023. "The Use of Different 16S rRNA Gene Variable Regions in Biogeographical Studies." *Environ Microbiol Rep* 15 (3):216–28.

Vellend, M. 2010. "Conceptual Synthesis in Community Ecology." *Q Rev Biol* 85 (2):183–206.

Wang, Ye, Tathagata Bhattacharya, Yuchao Jiang, Xiao Qin, Yue Wang, Yunlong Liu, Andrew J Saykin, and Li Chen. 2020. "A Novel Deep Learning Method for Predictive Modeling of Microbiome Data." *Brief Bioinform* 22 (3). https://doi.org/10.1093/bib/bbaa073.

Weinstock, G. M. 2012. "Genomic Approaches to Studying the Human Microbiota." *Nature* 489 (7415):250–6.

Wen, Z., G. Xie, Q. Zhou, C. Qiu, J. Li, Q. Hu, W. Dai, D. Li, Y. Zheng, and F. Wen. 2018. "Distinct Nasopharyngeal and Oropharyngeal Microbiota of Children with Influenza A Virus Compared with Healthy Children." *Biomed Res Int* 19 (6362716).

Wood, D. E., J. Lu, and B. Langmead. 2019. "Improved Metagenomic Analysis with Kraken 2." *Genome Biol* 20 (1):019–1891.

Wu, L., Z. Tang, H. Chen, Z. Ren, Q. Ding, K. Liang, and Z. Sun. 2021. "Mutual Interaction Between Gut Microbiota and Protein/Amino Acid Metabolism for Host Mucosal Immunity and Health." *Anim Nutr* 7 (1):11–16.

Wylie, K. M., R. M. Truty, T. J. Sharpton, K. A. Mihindukulasuriya, Y. Zhou, H. Gao, E. Sodergren, G. M. Weinstock, and K. S. Pollard. 2012. "Novel Bacterial Taxa in the Human Microbiome." *PLoS One* 7 (6):13.

Yang, Y., J. Huang, A. Zeng, X. Long, N. Yu, and X. Wang. 2024. "The Role of the Skin Microbiome in Wound Healing." *Burns Trauma* 4 (12).

Yizhak, K., T. Benyamini, W. Liebermeister, E. Ruppin, and T. Shlomi. 2010. "Integrating Quantitative Proteomics and Metabolomics with a Genome-Scale Metabolic Network Model." *Bioinformatics* 26 (12).

Young, Caroline, Henry M. Wood, Ramakrishnan Ayloor Seshadri, Pham Van Nang, Carlos Vaccaro, Luis Contreras Melendez, Mayilvahanan Bose, Mai Van Doi, Tamara Alejandra Piñero, Camilo Tapia Valladares, Julieta Arguero, Alba Fuentes Balaguer, Kelsey N. Thompson, Yan Yan, Curtis Huttenhower, and Philip Quirke. 2021. "The Colorectal Cancer-Associated Faecal Microbiome of Developing Countries Resembles That of Developed Countries." *Genome Med* 13 (1):27. https://doi.org/10.1186/s13073-021-00844-8.

Zhao, L., J. L. Luo, M. K. Ali, E. Spiekerkoetter, and M. R. Nicolls. 2023. "The Human Respiratory Microbiome: Current Understandings and Future Directions." *Am J Respir Cell Mol Biol* 68 (3):245–55.

Zhou, X., C. J. Brown, Z. Abdo, C. C. Davis, M. A. Hansmann, P. Joyce, J. A. Foster, and L. J. Forney. 2007. "Differences in the Composition of Vaginal Microbial Communities Found in Healthy Caucasian and Black Women." *Isme J* 1 (2):121–33.

Zoetendal, E. G., J. Raes, B. van den Bogert, M. Arumugam, C. C. Booijink, F. J. Troost, P. Bork, M. Wels, W. M. de Vos, and M. Kleerebezem. 2012. "The Human Small Intestinal Microbiota Is Driven by Rapid Uptake and conversion of Simple Carbohydrates." *Isme J* 6 (7):1415–26.

5 Metagenomics in Diagnostics and Disease Detection

Lovely Sinha[1] *and Jitendra Kumar Sharma*[2*]
[1]Department of Pulmonary Medicine, AIIMS Patna, Bihar, India
[2]Centre for Biotechnology, M. D. University, Rohtak, Haryana, India
*Corresponding authors

5.1 INTRODUCTION

The influence of microbiota on human health is evidently present in the public domain. Microbiota present in gut or on skin, such as bacteria, viruses, archaea, protozoa, and fungi, have been evaluated for their role in disease progression. Microbial infectious disease is a serious global health concern, resulting in 13 million cases of morbidity per year. The use of antimicrobial medicines and changes in the environment result in microbial dysbiosis, further leading to major health problems. The recent COVID-19 pandemic is the worst example of such an incidence, urging technological advancement in fast, accurate, and broad diagnostic tools for early detection. Even with modern diagnostic tools, the chance of wrong diagnosis and long waiting times is still a matter of concern. Together, these can lead to high mortality and long-term illness. This highlights the importance of developing diagnostic systems that can detect unknown infections quickly and accurately (Pindi et al., 2009). Conventional diagnostic methods, including culture-based assays and targeted PCR, frequently fail to detect pathogenic organisms. Many studies indicated that around 40% of cases of gastroenteritis and 60% of cases of encephalitis are undiagnosed or wrongly diagnosed with other diseases. Some new pathogens, such as Middle East Respiratory Syndrome Coronavirus (MERS-CoV) and the H7N9 influenza virus, need upgradation in existing diagnostics.

Metagenomics has emerged as a robust solution to address the limitations of culture-based methodologies. It makes it possible to analyze clinical or environmental specimens directly without culturing them in the laboratory. Sequence-based functional metagenomics helps in the identification of both existing and novel pathogens (Kumar, Shukla et al., 2024). It is also capable of revealing antimicrobial resistance and functional activities in complex systems. Among available tools, metagenomic next-generation sequencing (mNGS) stands out for its ability to detect rare or unexpected pathogens from very small samples with high sensitivity and accuracy (Kumar, Gangani et al., 2024). Besides identifying pathogens, mNGS can also predict antimicrobial resistance, allowing more personalized and effective treatment plans (De Abreu et al., 2021). Unlike traditional methods that depend on specific tests or observable traits, metagenomics offers a complete and unbiased approach. It reduces diagnostic time and increases the chance of finding hidden or rare pathogens. Its growing use in hospitals and public health systems shows its potential to transform how we detect infectious diseases, manage outbreaks, and monitor antimicrobial resistance. Recent studies have shown the clinical applicability of metagenomics-based diagnostics in diseases such as dementia and cancers, including breast cancer, colorectal cancer, and ovarian cancer (Nair et al., 2024; Prajapati et al., 2024; Sarfraz et al., 2024; Mishra et al., 2024; James et al., 2024).

DOI: 10.1201/9781003670155-5

5.2 MICROBIAL MECHANISMS IN DISEASE PROGRESSION

Recent literature shows that microorganisms influence disease progression across metabolic, inflammatory, infectious, neurological, and neoplastic disorders by shaping host immunity, metabolism, barrier function, and local tissue microenvironments. Microbial communities act both locally, for example, at mucosal surfaces and within tumors, and systemically via translocated components and small metabolites. Because microbes can change rapidly in response to diet, drugs, and infection, they act as dynamic drivers of disease trajectory and as actionable biomarkers and therapeutic targets (Di Vincenzo et al., 2024; Lu et al., 2022; Yang et al., 2023).

Moreover, recent studies also reported that short-chain fatty acids, secondary bile acids, and tryptophan metabolites act as ligands for host receptors and epigenetic modifiers, altering immune cell differentiation, epithelial repair, and systemic metabolism. These metabolite pathways explain how distant organs respond to gut microbial changes and how microbiota composition affects disease course (Lu et al., 2022; Xie et al., 2024). Mechanistic work links increased intestinal permeability to the passage of microbial products, such as lipopolysaccharide, into the circulation, which activates innate immune receptors and sustains low-grade systemic inflammation that worsens metabolic, autoimmune, and infectious outcomes. Interventional studies that restore barrier integrity reverse some downstream inflammatory signals (Dmytriv et al., 2024; Escalante et al., 2025).

Sequencing and spatial studies have found bacterial, fungal, and viral sequences inside multiple tumor types. Mechanisms reported include microbial modulation of local immune cells, bacterial production of metabolites or enzymes that drive genotoxic stress, and recruitment of pro-tumor inflammatory pathways, all of which can accelerate tumor growth and metastasis (Shi et al., 2024; Yang et al., 2023). Clinical and preclinical studies show that specific gut community structures and microbial metabolites correlate with response or resistance to immune checkpoint inhibitors. Mechanisms include priming of antigen-presenting cells, modulation of T cell phenotypes, and effects on immune-related adverse events; fecal microbiota transfer and metabolite modulation are being trialed to change outcomes (Kiousi et al., 2023; Xie et al., 2024). Work examining biofilm formation on tissues and devices shows that multispecies biofilms protect bacteria from immune clearance and antibiotics, promote local immune dysregulation, and create niches that sustain chronic infection and tissue damage. Horizontal gene transfer within biofilms accelerates the spread of virulence and resistance traits (Shi et al., 2024; M. Wang et al., 2023).

Genomic surveillance and mechanistic biofilm studies show that conjugation, transduction, and transformation spread resistance and virulence factors across commensal and pathogenic taxa, changing infection severity and limiting therapeutic options (Shi et al., 2024; M. Wang et al., 2023). Studies link dysbiosis to microglial activation, altered blood–brain barrier integrity, and metabolite-driven neuroinflammation. Animal models provide mechanistic chains from specific microbial taxa or metabolites to synaptic dysfunction and progressive neuronal loss, and human cohort work finds early microbiome shifts in prodromal disease states (C. Chen et al., 2025; Loh et al., 2024).

In addition, investigations show that localized microbial enzymes can convert dietary or environmental compounds into genotoxic metabolites, and they can biotransform drugs in ways that alter efficacy or toxicity. These mechanisms link microbiome composition to both cancer risk and variable drug responses (Xie et al., 2024; Yang et al., 2023). Multiple studies report microbial DNA or metabolite fingerprints in blood and tissue that correlate with disease stage, progression, and prognosis, supporting the idea that microbial signals can serve as minimally invasive biomarkers (Macura et al., 2024; Zong et al., 2024). Similarly, recent publications describe how viral infections reshape bacterial communities and vice versa, altering host susceptibility and clinical severity. Mechanistic models show that viruses can change mucosal immunity and nutrient niches, enabling bacterial overgrowth or translocation that drives worse outcomes (Xia et al., 2024; Xie et al., 2024). Key microbial mechanisms linking specific pathogens to cancer progression are summarized in Table 5.1.

TABLE 5.1
Microbial Mechanisms in Disease Progression Selected Examples from Recent Literature

Sr. No.	Disease	Microorganisms	Mechanistic Insight	Implication for Progression	References
1	Colorectal cancer	*Fusobacterium nucleatum*	*F. nucleatum* produces formate which activates the aryl hydrocarbon receptor pathway in colorectal epithelial cells and promotes cancer cell stemness and proliferation	Microbe-derived formate acts as an oncometabolite that directly promotes tumor growth and may increase aggressiveness and recurrence	Ternes et al. (2022)
2	Colorectal cancer	*Fusobacterium nucleatum*	A distinct Fn clade found in CRC and oral cavity suggests oral origin and circulating seeding; strain concordance supports translocation and tumor tropism	Strain transmission from oral site to colon can seed tumors and sustain intratumoral microbial populations that modify the tumor microenvironment	Zepeda-Rivera et al. (2024)
3	Colorectal cancer	*Fusobacterium nucleatum*	Fn modulates immune infiltration, increases pro-inflammatory cytokines, and promotes chemoresistance via multiple host signaling pathways described in CRC models	Microbial influence on immune contexture can accelerate progression and reduce therapy efficacy	Ou et al. (2022)
4	Gastric cancer	*Helicobacter pylori*	Chronic infection with CagA positive strains dysregulates NF-κB, β-catenin and inflammasome pathways leading to genomic instability and stepwise gastric carcinogenesis	Persistent H. pylori signaling creates a pro-tumor inflammatory and genotoxic environment that promotes progression from gastritis to cancer	Duan et al. (2025)
5	Gastric pathology/ cancer	*Helicobacter pylori*	CagA activates the NLRP3 inflammasome in gastric epithelial cells, enhancing inflammatory cytokine release and epithelial pathology	Inflammasome activation links H. pylori virulence directly to inflammation-driven tissue damage and progression risk	Wan et al. (2025)
6	Atherosclerosis/ cardiovascular disease	*Porphyromonas gingivalis*	Pg influences endothelial function, promotes foam cell formation and skews T cell balance, partly via gingipain proteases and LPS-mediated signaling	Oral pathogen activity can accelerate plaque formation and instability, linking periodontitis to faster CVD progression	J. Zhang et al. (2021)
7	Atherosclerosis	*Porphyromonas gingivalis*	Gingipains and other virulence factors degrade host junctional proteins and promote systemic inflammation and macrophage dysfunction in vascular tissues	Mechanisms explain how oral pathogens contribute to endothelial breach, lipid uptake and necrotic core expansion in plaques	Ruan et al. (2023)

(Continued)

TABLE 5.1 (*Continued*)
Microbial Mechanisms in Disease Progression Selected Examples from Recent Literature

Sr. No.	Disease	Microorganisms	Mechanistic Insight	Implication for Progression	References
8	Metabolic disease/ multi-system	Diverse gut microbes	Microbial metabolites including SCFAs, secondary bile acids and TMAO modulate host receptors and metabolic signaling leading to insulin resistance, inflammation and organ fat deposition	Functional metabolite shifts link microbiome changes to progression of obesity, NAFLD and cardiometabolic disease	Liu et al. (2025)
9	Neurodegeneration	Gut microbiota, SCFA producers	Microbial metabolites and LPS modulate microglial activation, blood–brain barrier integrity and amyloid processing, thereby influencing Alzheimer and Parkinson phenotypes in models	Microbiome-driven neuroinflammation contributes to neurodegenerative disease progression and modifies pathology severity	Loh et al. (2024)
10	Non-alcoholic fatty liver disease	Gut microbial consortia and metabolites	Dysbiosis correlates with altered metabolite profiles that promote hepatic inflammation, lipid accumulation and fibrosis via gut-liver immune signaling	Microbial signatures and metabolites are mechanistically implicated in NAFLD progression and may predict transition to NASH and fibrosis	Long et al. (2024)

CRC: Colorectal cancer; SCFAs: Short-chain fatty acids; TMAO: Trimethylamine N-oxide; NAFLD: Non-alcoholic fatty liver disease; NASH: Non-alcoholic steatohepatitis; LPS: Lipopolysaccharide

5.3 MICROBIAL DRIVERS OF DISEASE PROGRESSION: INSIGHTS FROM METAGENOMIC SEQUENCING

Building on these broader insights, it becomes evident that microorganisms are not merely contributors but active drivers in the onset and trajectory of diverse human diseases. The integration of metagenomic approaches has transformed this understanding from a conceptual framework into a practical diagnostic and therapeutic tool. By enabling unbiased detection and characterization of microbial communities within diseased tissues, metagenomics allows researchers to trace the molecular fingerprints of pathogens, uncover their metabolic and immunological interactions, and identify microbial markers predictive of disease progression. The following sections outline how this approach is reshaping our understanding of specific disease classes, ranging from cancer and neurological disorders to respiratory, gastrointestinal, bloodstream, and osteoarticular infections, where microbial activity, community imbalance, or horizontal gene transfer directly influence pathogenesis, host response, and treatment outcomes. Each case illustrates how metagenomic evidence has deepened mechanistic clarity, improved diagnostic precision, and opened new paths for targeted intervention.

5.3.1 Microbial Oncogenesis

Recent advances in metagenomic sequencing have opened promising avenues for the noninvasive detection and monitoring of microbe-associated oncogenesis. For example, a multinational shotgun fecal and salivary metagenomic study by Nagata et al. (2022) demonstrated that specific microbial signatures could discriminate patients with colorectal cancer from controls, with stool-based taxa such as *Fusobacterium nucleatum* showing elevated abundance in cases. In pancreatic ductal adenocarcinoma (PDAC), Khan et al. used metagenomic 16S rRNA analyses of tumor versus adjacent normal tissue and found differential abundances of genera such as *Delftia*, *Streptomyces*, and *Cutibacterium*, along with correlations with PD-L1 expression, suggesting that tumor-resident microbial profiles may hold diagnostic or prognostic value (Khan et al., 2023). In a recent study, Chen et al. combined metagenome-derived microbial signatures with the serum tumor marker CA19–9 to detect PDAC, reporting improved discrimination compared to CA19–9 alone (Y. Chen et al., 2025). Parallel reviews have noted that genome-resolved metagenomics reconstruction of metagenome-assembled genomes (MAGs) enables strain-level resolution and functional annotation of microbes, thereby enhancing biomarker discovery in oncology and precision microbiome medicine (Kim et al., 2024). Importantly, methodological reviews caution that while metagenomics holds diagnostic potential in oncology, issues such as host DNA depletion, low microbial biomass in tissue samples, contamination risk, and the need for rigorous bioinformatic pipelines remain major barriers to clinical implementation (Nam et al., 2023). Considering all information from the illustrations, the emerging body of evidence supports the feasibility of microbiome-based metagenomic diagnostics in cancer settings but underscores that further validation, standardization of workflows, and large-scale prospective studies are required before widespread clinical adoption.

5.3.2 Neurological and Autoimmune Conditions

Recent publications between 2018 and 2025 report that metagenomic sequencing approaches have been applied as diagnostic tools in both neurological and autoimmune conditions, with authors describing case series, cohort studies, and methodological evaluations that document diagnostic yields, candidate microbial biomarkers, and practical limitations. Multiple clinical studies and systematic evaluations find that unbiased mNGS of cerebrospinal fluid increases pathogen detection in suspected central nervous system infections compared with conventional testing and that mNGS results altered clinical management in a subset of cases; one multiyear clinical series reported higher sensitivity for mNGS versus standard assays and quantified diagnostic performance metrics for CNS infection cohorts (Benoit et al., 2024; Y.-Y. Chen et al., 2022).

In parallel, studies of plasma microbial cell-free DNA (mcfDNA) sequencing describe high positivity rates for microbial detection in selected clinical populations and propose that circulating microbial signatures can identify tissue-associated or systemic infectious signals that are not captured by routine culture or targeted PCR, with applications suggested for noninvasive surveillance and adjunctive diagnosis (Park et al., 2023; H. Zhang et al., 2024). In the neurodegenerative domain, large shotgun fecal metagenomic studies and subsequent machine-learning meta-analyses report reproducible taxonomic and functional differences between Parkinson's disease patients and controls and document the classification performance of models trained on metagenomic features, with pooled analyses reporting area under the curve values in the range of approximately 0.70 to 0.72 across aggregated cohorts (Romano et al., 2025; Wallen et al., 2022). For autoimmune disorders, reviews and functional metagenomic analyses identify microbial gene and pathway signatures that correlate with disease status and with host immune or genetic risk loci, and several retrospective studies have constructed predictive classifiers based on metagenomic features that discriminate autoimmune cases from controls in published datasets (Berryman et al., 2024; Zehrh et al., 2023). Methodological literature published during this period emphasizes critical laboratory and analytic steps required for diagnostic reliability, including host DNA depletion, rigorous negative controls, contamination filtering, standardized pipelines, and validation across independent cohorts; reanalysis papers specifically document that failure to control for low biomass and reagent contamination can produce spurious disease associations (Y.-Y. Chen et al., 2022; Wallen et al., 2022). Finally, several recent studies propose integrated, multimodal diagnostic frameworks that combine metagenomic microbial features with host omics or clinical biomarkers and report improved retrospective classification performance for selected neurologic and autoimmune phenotypes, while concurrently calling for prospective validation before clinical implementation (Villette et al., 2025; Ying et al., 2024).

Metagenomics increases the detection capability of infectious diseases and helps in the study of interactions between the body, bacteria, and disease development. Using DNA sequencing and computer analysis allows fast diagnosis and pathogen identification. Its role is significant in disease classification and in providing a deeper understanding of bacterial effects on human health.

5.3.3 Respiratory Infections

mNGS has demonstrated superior diagnostic performance compared to conventional microbiological tests in respiratory tract infections, with recent clinical studies revealing significantly enhanced pathogen detection capabilities across diverse patient populations and clinical settings. Chen et al. conducted a prospective study of 188 patients with suspected pulmonary infection, demonstrating that mNGS achieved an 86.17% positive detection rate versus 67.55% for conventional microbiological tests ($P < 0.01$), while identifying 96 distinct pathogens comprising 59 bacteria, 18 fungi, 15 viruses, and 4 special pathogens compared to only 28 species detected through conventional methods (S. Chen et al., 2025). A comprehensive meta-analysis of 24 studies further validated these findings, revealing mNGS pathogen detection positive rates of 80.48% compared to 45.78% for conventional methods, with particular advantages in detecting fungi, viruses, and rare pathogens, while demonstrating reduced susceptibility to antibiotic pretreatment effects (Lv et al., 2023). The technology has proven especially valuable in immunocompromised populations, where Zhao et al. found that mNGS detected mixed infections in 58.82% of cases versus 17.96% for conventional tests ($P < 0.05$), while significantly reducing undetected pathogen rates from 38.73% to 4.90% in this high-risk population (Zhao et al., 2025).

Bronchoalveolar lavage fluid (BALF) has emerged as the optimal specimen type for mNGS respiratory diagnostics, with multiple studies demonstrating superior performance compared to other sample types. Jiang et al. reported that BALF samples achieved 82.14% positivity rates compared to 48.21% for blood samples in 56 pneumonia patients, with frequent detection of clinically relevant pathogens, including *Acinetobacter baumannii*, *Klebsiella pneumoniae*, and *Pseudomonas aeruginosa* (L. Jiang et al., 2024). Han et al. demonstrated that BALF mNGS independently diagnosed

65.6% of coinfections and identified resistance genes matching phenotypic resistance patterns in 69.4% of multidrug-resistant pathogens, highlighting the technology's capacity for both pathogen detection and antimicrobial resistance characterization (Han et al., 2024). In intensive care unit settings, the diagnostic advantages become even more pronounced, with DNA and RNA sequencing achieving positive pathogen detection rates of 96.69% and 96.43%, respectively, compared to culture test positive rates of only 39.34% (Zhou et al., 2022).

The clinical impact of mNGS extends beyond enhanced detection rates to include significant improvements in treatment optimization and patient outcomes. Studies consistently demonstrate rapid diagnostic turnaround times, with median intervals from sample submission to result feedback ranging from 18–24 hours for mNGS versus 56–83 hours for conventional methods, enabling faster therapeutic decision-making in critically ill patients. Treatment effectiveness rates based on mNGS results consistently exceed 90% across diverse patient populations, with particularly notable improvements observed in immunocompromised patients, where treatment effectiveness reached 65.69% versus 56.40% for conventional approaches ($P < 0.05$) (Zhao et al., 2025). These findings establish mNGS as a transformative diagnostic approach that enables rapid, comprehensive pathogen identification crucial for optimizing antimicrobial therapy, supporting enhanced antimicrobial stewardship initiatives, and improving overall patient outcomes in respiratory infections.

5.3.4 Gastrointestinal Disorders

mNGS is a competent diagnostic approach for gastrointestinal disorders, showing exceptional effectiveness in inflammatory bowel disease (IBD) diagnosis through comprehensive microbiome-based models. Zheng et al. developed advanced diagnostic tests by analyzing metagenomic data from 5,979 fecal samples, which included diverse geographic and ethnic populations. Among those samples, 10 bacterial species were identified that are known pathogens causing ulcerative colitis, and 9 were related to Crohn's disease. The test accuracy was statistically significant, with area under the curve values >0.90 for differentiating IBD from controls, while maintaining satisfactory performance in transethnic validation cohorts (Zheng et al., 2024). The metagenomics-based models demonstrated area under the curve values of 0.81 (95% CI: 0.80–0.82) for IBD patient differentiation, exceeding clinical utility thresholds, with multiplex droplet digital PCR tests targeting selected IBD-associated bacterial species showing superior performance compared to fecal calprotectin. Comprehensive analysis of 4,406 samples from 13 IBD cohorts worldwide identified 1,175 taxa, revealing decreased microbial diversity and richness in UC and CD patients compared to controls, with IBD associated with 2.2% of microbial alterations. Beyond IBD, metagenomics has proven valuable for *Clostridioides difficile* detection, where studies demonstrated 86% detection rates in known positive samples while identifying additional pathogens beyond conventional methods. Advanced shotgun sequencing approaches achieved 98% *C. difficile* detection using stringent read count criteria, providing species-level bacterial resolution. Meta-analysis studies by Jiang et al. (2021) identified 87 species and 65 pathway markers consistently changed across multiple datasets for Crohn's disease, ulcerative colitis, and colorectal cancer, constructing high-performance multidisease models with AUROC values of 0.87 and true-positive rates up to 90%. These findings establish metagenomics as a cornerstone technology for precision medicine in gastrointestinal disorders, enabling comprehensive characterization of microbial communities, functional pathway analysis, and identification of disease-specific biomarkers for improved diagnostic accuracy and clinical management (P. Jiang et al., 2021).

5.3.5 Sepsis and Bloodstream Infections

mNGS has transformed the diagnostic approach for sepsis and bloodstream infections, demonstrating significant enhancements in diagnostic precision and clinical applicability compared with conventional culture-based methods (Yu et al., 2024). Recent technological advancements, notably

Probe-Capture Metagenomics devised by researchers from multiple medical institutions, have mitigated the primary challenge related to patient-derived human DNA contamination, which generally constitutes over 99% of the nucleotide sequence in blood samples (R. Wang et al., 2025). This novel methodology integrates hybrid capture technology with mNGS, yielding detection rates of 51.6% in contrast to 17.4% achieved via blood culturing alone ($p < 0.001$). Sun et al. (2025) reported that a prospective study involving 184 patients with sepsis demonstrated significant analytical performance, achieving 100% sensitivity and 87.1% specificity. The integration of pathogen identification with the host's transcriptional profile using RNA sequencing of plasma samples signifies a paradigm change toward comprehensive diagnostic platforms (Hogan et al., 2021). This comprehensive methodology, utilizing biomarkers such as CD177 elevation and HLA-DRA suppression, attained 99% sensitivity in microbiologically confirmed sepsis cases and accurately assessed sepsis in 74% of suspected cases and 89% of unclear cases (Kalantar et al., 2022). Clinical implementation studies have consistently shown the superiority of mNGS across diverse medical care settings. The diagnostic sensitivity of mNGS varies from 84.4% to 85.25%, whereas standard blood culture methods exhibit a range of 36.09% to 85.25%. Moreover, response times have been clearly reduced, decreasing from 111.0 to 19.1 hours (Li et al., 2024; Sun et al., 2025; Wu et al., 2023). Clinical implementation have demonstrated diverse therapeutic effects in practical applications. Retrospective investigations performed at the University of Maryland School of Medicine indicate that the Karius test, utilizing cell-free DNA analysis, has demonstrated a beneficial therapeutic effect in 16–43% of patients. Optimal efficacy has been observed in organ transplant recipients (71.4% positive impact) and sepsis patients (71.4% positive impact) when provided within 7 days after initiating antimicrobial therapy (Noe et al., 2021; Olthoff et al., 2024).Comparative analyses indicate that plasma samples outperform whole blood samples. In this context, mNGS analysis of plasma exhibits less interference from patient-derived DNA, at approximately 97%, compared with 99.9% in whole blood. The improved capacity to identify pathogens in plasma is crucial, especially for gram-negative bacteria, which exhibited detection rates of 78.26% compared with 17% for gram-positive cocci (Yu et al., 2024). The diagnostic advantages extend beyond bacterial identification, encompassing enhanced recognition of physiological responses and superior viral detection via patient transcriptional mechanisms. This evidence supports the use of mNGS as an essential instrument for comprehensive diagnosis of infectious diseases in critically ill patients. Clinical advantages have been evidenced by modifications in antimicrobial treatment in 34.8% of patients and a two-point elevation in SOFA scores in 22.3% of sepsis patients following drug alterations informed by mNGS (Sun et al., 2025).

5.3.6 Osteoarticular Infections

mNGS has the potential to revolutionize the diagnosis of osteoarticular infections, with significant advances in pathogen detection rates and clinical outcomes compared with traditional cultivation-based techniques. Recent prospective studies show that pathogen detection with mNGS is 75.33%, compared with 36.67% obtained via traditional testing methods, representing more than double the diagnostic sensitivity (Q. Zhang et al., 2025). Improvements in performance are particularly evident in prosthetic joint infections (PJI), where systematic studies show combined diagnostic sensitivity and specificity of 93% and 95%, respectively, with an area under the curve of 0.96 (95% CI: 0.93-0.97) (Tan et al., 2022). Diagnostic capacity includes several types of samples, with particular emphasis on syringe fluid extracted from removed prosthetic devices, which performs exceptionally well. Tan et al. (2024) and C. Zhang et al. (2019) identified pathogens in 100% of cultivation-negative PJI cases by removing bacterial biofilms and increasing bacterial concentrations through centrifugation. Analyzing free microbial DNA in plasma is a noninvasive approach that has a sensitivity of 93.5% for viral infections and 83.6% for systemic infections, compared with traditional techniques showing sensitivities of 37.7% and 14.3%, respectively (Pang et al., 2024). The clinical significance of mNGS extends beyond improved detection rates to include comprehensive pathogen identification. Studies identified 47 distinct organisms, compared with only 9 organisms detected using traditional methods. This research improved detection of specific pathogens, including *Staphylococcus lugdunensis*, *Mycobacterium marinum*, and *Neisseria subflava* (Q. Zhang et al., 2025). Clinical benefits are substantial, with 63.72% of mNGS-positive patients receiving targeted antimicrobial therapy and 27.43% validating empirical treatments. This is particularly important in settings involving prior antibiotic exposure, as failure rates of conventional cultures increase by a factor of 4.8 for each additional risk factor (Ding et al., 2024; Q. Zhang et al., 2025). Despite technical limitations in identifying organisms with thick cell walls, such as *Mycobacterium tuberculosis*, T-SPOT techniques achieve a sensitivity of 90.91% compared with mNGS, which has a sensitivity of 69.69%. The higher specificity of mNGS (100% versus 68.32%) improves overall diagnostic accuracy in osteoarticular infections (Q. Zhang et al., 2025).

5.4 CASE STUDIES: METAGENOMICS IN OUTBREAK INVESTIGATION

mNGS is transforming the approach scientists use to investigate outbreaks. Conventional whole-genome sequencing depends on culturing pathogens from infected patients, which limits its ability to identify asymptomatic carriers or unculturable infections. Metagenomics enables direct sequencing of complex clinical samples, such as fecal or nasal swabs. This approach obviates the need for culture and enables detection of a much broader array of microorganisms.

Using bioinformatic tools, scientists can extract, assemble, and compare pathogen genomes from these samples, making it possible to map disease transmission with a level of precision that was previously unachievable.

A clear example of this shift is observed in studies of *Clostridioides difficile*. Earlier, it was believed that most infections originated in hospitals. However, whole-genome analyses showed that many patients were already carrying the bacteria before hospital admission, often acquiring it from community sources. Since treating a single case of *C. difficile* infection can cost approximately $22,000, metagenomic screening to identify community carriers early can save both money and lives. This finding highlights the value of metagenomics not only for diagnosis but also for prevention.

The COVID-19 pandemic demonstrated the potential of metagenomics to improve public health. Metagenomics made it possible to identify the SARS-CoV-2 virus within one month of its initial reported case. Rapid identification enabled monitoring of the virus and design of preventive tests. Metagenomics was also used for detection, transmission tracking, and evolutionary analysis of other prominent viruses, such as Zika and Ebola. Metagenomic data assist in monitoring viral mutations and geographic dispersal. These attributes render metagenomics a crucial instrument for preventing epidemics, facilitating early detection, and averting future pandemics. The use of metagenomic sequencing for outbreak investigation has proven to be both effective and economical, supporting rapid identification, improved understanding of transmission patterns, and increased preparedness for future infectious threats (Trivett et al., 2025; Buddle et al., 2024).

5.5 CHALLENGES IN METAGENOMICS FOR DIAGNOSTICS

Metagenomics possesses significant potential for clinical diagnosis; however, several considerable challenges must be addressed prior to its routine implementation. These include technological, computational, and ethical challenges that must be resolved before adoption in diagnostic laboratories. A major technical challenge involves obtaining high-quality DNA from clinical or environmental samples. Sample contaminants, such as humic acids and other inhibitors, interfere with sequencing. Removing these contaminants frequently results in DNA loss, thereby decreasing result accuracy. It is also difficult to assemble millions of short DNA fragments into complete genomes, particularly when analyzing complex microbial communities containing repetitive genetic sequences. This process requires robust computational infrastructure and substantial processing time. The large volume of sequencing data generated creates additional challenges for analysis and storage. Existing reference databases are biased toward extensively studied and culturable species, leading to gaps in identification of novel or rare microorganisms. Moreover, the absence of universal standards for data processing, combined with a shortage of trained bioinformatics professionals, complicates accurate interpretation of results. Addressing these issues is essential for effective clinical use of metagenomics.

5.5.1 Technical Limitations

mNGS still has technical constraints that can affect diagnostic accuracy. It cannot always distinguish between DNA from viable pathogens and DNA from dead or nonpathogenic microbes, which can lead to false-positive results. The method also provides only relative abundance data, and because host DNA often predominates, low-level pathogens may remain undetected.

Variation in DNA extraction efficiency among microorganisms is another challenge. Pathogens such as *Mycobacterium tuberculosis* or fungi with thick cell walls often yield low DNA quantities, requiring specialized extraction methods. Furthermore, short DNA fragments make it difficult to assemble full genes, including those related to antimicrobial resistance, and to assign them confidently to specific organisms.

5.5.2 Data Interpretation and Standardization

Interpretation of mNGS results remains a major challenge. The lack of standardized procedures across all stages—from sample collection to data analysis—leads to variability among laboratories. Positive controls cannot encompass all pathogens, while standard negative controls often fail to capture background noise present in clinical samples. This complicates differentiation of true signals from contamination and establishment of universal cutoff values for identifying genuine infections. Developing validated reference databases and consistent analytical workflows is necessary to improve consistency and reliability of mNGS results across institutions.

5.5.3 Ethical and Privacy Concerns

Metagenomic sequencing raises ethical and privacy concerns. Because the approach is untargeted, it may reveal unexpected information, such as genetic disease risks or the presence of unrelated pathogens in a patient's sample. This raises ethical questions regarding informed consent, patient rights, and clinician responsibilities for reporting incidental findings.

In addition, sequencing generates large volumes of sensitive human and microbial data. Without appropriate management, this information poses risks to privacy and data security. Establishing strong policies for data protection, ownership, and accountability is essential to maintain patient trust and uphold ethical standards.

5.6 FUTURE PERSPECTIVES

In the coming years, metagenomics can transform the diagnostic methodology of infectious diseases. As sequencing technologies continue to advance, they are becoming faster, more accurate, and less expensive. This progress will help move mNGS from a specialized research tool to a routine part of clinical diagnostics.

The main advantage of mNGS is its ability for early detection of complex or unknown pathogens, especially in critically ill patients who are undiagnosed after traditional testing. It can also be used to monitor changes in microbial communities in the body. Regular metagenomic profiling may predict disease risk before symptoms appear.

In the future, metagenomics could also play an important role in personalized medicine. By combining sequencing data with large clinical databases, it will be possible to not only identify

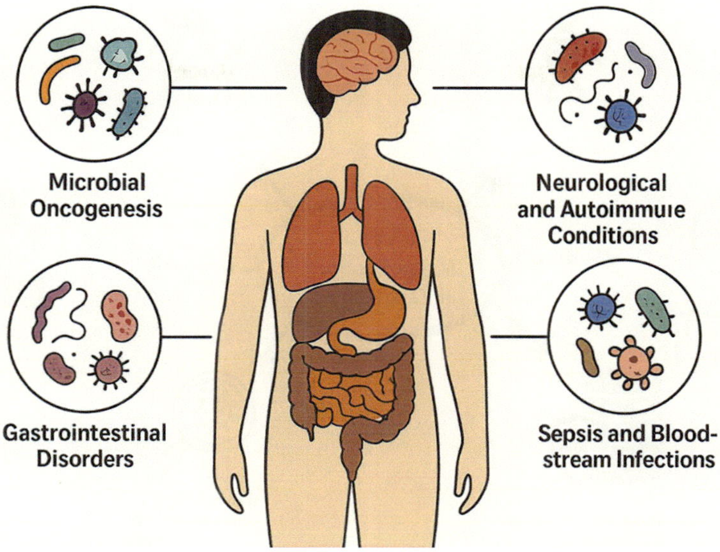

FIGURE 5.1 Role of microorganisms in disease progression.

pathogens but also predict their resistance patterns and virulence factors. This will allow doctors to choose the most effective treatment more quickly, improving outcomes in serious infections.

Combining metagenomics with other technologies, such as artificial intelligence and machine learning, will make data interpretation faster and more precise. These systems can recognize complex patterns in large datasets that would be difficult for humans to detect. The development of portable sequencing platforms may also make metagenomic testing available at the bedside or even at home.

The future of metagenomics in diagnostics looks very promising. With continued research, better data analysis tools, and standardized protocols, mNGS has the potential to become an essential part of modern healthcare, supporting faster diagnosis, improved treatment, and stronger disease prevention strategies. Key clinical applications of metagenomics are depicted in Figures 5.1 and 5.2.

5.7 CONCLUSION

The mNGS has brought changes in pathogen detection and diagnostic methodologies, as it allows identification of microbes directly from clinical samples. This technique uses culture-independent methods to identify pathogens. This approach makes it valuable while diagnosing uncultivable pathogens. The mNGS offers a comprehensive assessment of infection-associated microbial communities. Regardless of its enormous clinical efficiency, mNGS cannot yet displace established diagnostic methods. Widespread acceptance of the technique faces several challenges, which include standardized procedures, data interpretation, sequencing cost, and technological expertise. Presently, its applications exhibit an utmost ability in complicated cases, especially in critically ill patients. Besides its role in diagnosing pathogenic microorganisms, mNGS also demonstrates great potential for oncological research and revealing microbiome–disease interactions. Integration of traditional methods with metagenomic analysis can improve the accuracy, speed, and cost-efficiency of diagnosis. In order to achieve this, work must be done toward standardization and validation methods, and its reliable clinical application. Through systematic development, mNGS may bridge translational research and clinical application, becoming a cornerstone of precision diagnostics.

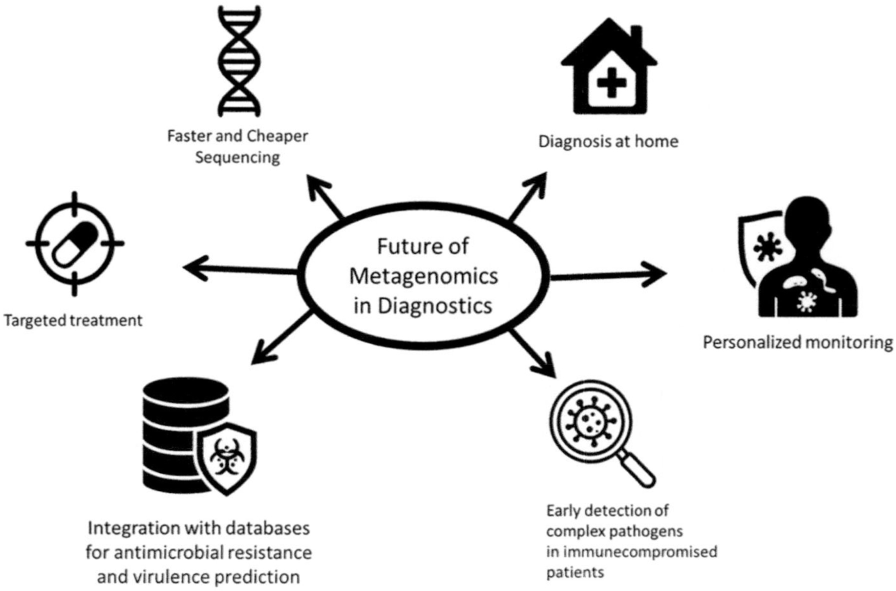

FIGURE 5.2 Clinical applications of Metagenomics.

ACKNOWLEDGMENT

I acknowledge the use of ChatGPT [https://chatgpt.com/] in helping me to review my writing at the final stage of preparing the chapter and drawing the figures.

REFERENCES

Benoit, Patrick, Noah Brazer, Mikael De Lorenzi-Tognon, Emily Kelly, Venice Servellita, Miriam Oseguera, . . . Charles Y. Chiu. (2024). Seven-year performance of a clinical metagenomic next-generation sequencing test for diagnosis of central nervous system infections. *Nature Medicine*, 30(12), 3522–3533. https://doi.org/10.1038/s41591-024-03275-1

Berryman, Meghan A., Jorma Ilonen, Eric W. Triplett, and Johnny Ludvigsson. (2024). Functional metagenomic analysis reveals potential inflammatory triggers associated with genetic risk for autoimmune disease. *Journal of Autoimmunity*, 148, 103290. https://doi.org/10.1016/j.jaut.2024.103290

Buddle, S., L. Forrest, N. Akinsuyi, L. M. Martin Bernal, T. Brooks, C. Venturini, . . . Edwards, D. J. (2024). Evaluating metagenomics and targeted approaches for diagnosis and surveillance of viruses. *Genome Medicine*, 16(1), Article 111. https://doi.org/10.1186/s13073-024-01378-5

Chen, Ce, Guo-Qing Wang, Dai-di Li, and Feng Zhang. (2025). Microbiota-gut-brain axis in neurodegenerative diseases: Molecular mechanisms and therapeutic targets. *Molecular Biomedicine*, 6(1), 64. https://doi.org/10.1186/s43556-025-00307-1

Chen, H. B., J. Liu, Y. Zhang, H. Huang, and L. N. Wang. (2025). Application of metagenomic next-generation sequencing in the diagnosis of pathogens in patients with diabetes complicated by community-acquired pneumonia. *Open Life Sciences*, 20(1), Article 20221048. https://doi.org/10.1515/biol-2022-1048

Chen, Song, Tanglin Ouyang, Kaiyang Wang, Xuan Hou, Rong Zhang, Meiyong Li, . . . Bo Huang. (2025). Application of metagenomic next-generation sequencing in pathogen detection of lung infections. *Frontiers in Cellular and Infection Microbiology*, 15, 1513603. https://doi.org/10.3389/fcimb.2025.1513603

Chen, Y., L. C. Fan, Y. H. Chai, and J. F. Xu. (2022). Advantages and challenges of metagenomic sequencing for the diagnosis of pulmonary infectious diseases. *The Clinical Respiratory Journal*, 16(10), 646–656. https://doi.org/10.1111/crj.13538

Chen, Ying-Ying, Yan Guo, Xin-Hong Xue, and Feng Pang. (2022). Application of metagenomic next-generation sequencing in the diagnosis of infectious diseases of the central nervous system after empirical treatment. *World Journal of Clinical Cases*, 10(22), 7760–7771. https://doi.org/10.12998/wjcc.v10.i22.7760

Chen, Yueying, Fulin Nian, Jia Chen, Qiuyu Jiang, Tianli Yuan, Haokang Feng, . . . Ling Dong. (2025). Metagenomic microbial signatures for noninvasive detection of pancreatic cancer. *Biomedicines*, 13(4), 1000. https://doi.org/10.3390/biomedicines13041000

De Abreu, V. A., J. Perdigão, and S. Almeida. (2021). Metagenomic approaches to analyze antimicrobial resistance: An overview. *Frontiers in Genetics*, 11, Article 575592. https://doi.org/10.3389/fgene.2020.575592

Di Vincenzo, Federica, Angelo Del Gaudio, Valentina Petito, Loris Riccardo Lopetuso, and Franco Scaldaferri. (2024). Gut microbiota, intestinal permeability, and systemic inflammation: A narrative review. *Internal and Emergency Medicine*, 19(2), 275–293. https://doi.org/10.1007/s11739-023-03374-w

Ding, Haiqi, Jiexin Huang, Lan Lin, Yang Chen, Qijin Wang, Wenbo Li, . . . Wenming Zhang. (2024). Shedding light on negative cultures in osteoarticular infections: leveraging mNGS to unravel risk factors and microbial profiles. *Frontiers in Cellular and Infection Microbiology*, 14, 1457639. https://doi.org/10.3389/fcimb.2024.1457639

Dmytriv, Tetiana R., Kenneth B. Storey, and Volodymyr I. Lushchak. (2024). Intestinal barrier permeability: The influence of gut microbiota, nutrition, and exercise. *Frontiers in Physiology*, 15, 1380713. https://doi.org/10.3389/fphys.2024.1380713

Duan, Yantao, Yonghu Xu, Yi Dou, and Dazhi Xu. (2025). Helicobacter pylori and gastric cancer: Mechanisms and new perspectives. *Journal of Hematology & Oncology*, 18(1), 10. https://doi.org/10.1186/s13045-024-01654-2

Escalante, Jonathan, Olivia Artaiz, Shanti Diwakarla, and Rachel M. McQuade. (2025). Leaky gut in systemic inflammation: Exploring the link between gastrointestinal disorders and age-related diseases. *GeroScience*, 47(1), 1–22. https://doi.org/10.1007/s11357-024-01451-2

Han, Dongsheng, Fei Yu, Dan Zhang, Qing Yang, Ruting Shen, Shufa Zheng, and Yu Chen. (2024). Applicability of bronchoalveolar lavage fluid and plasma metagenomic next-generation sequencing assays in the diagnosis of pneumonia. *Open Forum Infectious Diseases*, 11(1), ofad631. https://doi.org/10.1093/ofid/ofad631

Handelsman, J., M. R. Rondon, S. F. Brady, J. Clardy, and R. M. Goodman. (1998). Molecular biological access to the chemistry of unknown soil microbes: A new frontier for natural products. *Chemistry & Biology*, 5(10), R245–R249. https://doi.org/10.1016/S1074-5521(98)90108-9

Hogan, Catherine A., Shangxin Yang, Omai B. Garner, Daniel A. Green, Carlos A. Gomez, Jennifer Dien Bard, . . . Niaz Banaei. (2021). Clinical impact of metagenomic next-generation sequencing of plasma cell-free DNA for the diagnosis of infectious diseases: A multicenter retrospective cohort study. *Clinical Infectious Diseases: An Official Publication of the Infectious Diseases Society of America*, 72(2), 239–245. https://doi.org/10.1093/cid/ciaa035

James, A. S., N. A. Adil, D. Goltz, D. Tangudu, D. S. Chaudhari, R. Shukla, . . . H. Yadav. (2024). Abnormalities in gut virome signatures linked with cognitive impairment in older adults. *Gut Microbes*, 16(1), 2431648. https://doi.org/10.1080/19490976.2024.2431648

Jiang, Lingyu, Lin Han, Yonglong Zhong, Meng Zhang, Jianliang Li, Guanhua Rao, and Shulin Xiang. (2024). High utility of bronchoalveolar lavage fluid metagenomic next-generation sequencing approach for etiological diagnosis of pneumonia. *BMC Infectious Diseases*, 24(1), 1232. https://doi.org/10.1186/s12879-024-10108-6

Jiang, Puzi, Sicheng Wu, Qibin Luo, Xing-Ming Zhao, and Wei-Hua Chen. (2021). Metagenomic analysis of common intestinal diseases reveals relationships among microbial signatures and powers multidisease diagnostic models. *mSystems*, 6(3), e00112.21. https://doi.org/10.1128/mSystems.00112-21

Kalantar, Katrina L., Lucile Neyton, Mazin Abdelghany, Eran Mick, Alejandra Jauregui, Saharai Caldera, . . . Charles R. Langelier. (2022). Integrated host-microbe plasma metagenomics for sepsis diagnosis in a prospective cohort of critically ill adults. *Nature Microbiology*, 7(11), 1805–1816. https://doi.org/10.1038/s41564-022-01237-2

Khan, Sheema, Goutam Banerjee, Saini Setua, Daleniece Higgins Jones, Bhavin V. Chauhan, Anupam Dhasmana, . . . Subhash C. Chauhan. (2023). Metagenomic analysis unveils the microbial landscape of pancreatic tumors. *Frontiers in Microbiology*, 14, 1275374. https://doi.org/10.3389/fmicb.2023.1275374

Kim, Nayeon, Junyeong Ma, Wonjong Kim, Jungyeon Kim, Peter Belenky, and Insuk Lee. (2024). Genome-resolved metagenomics: A game changer for microbiome medicine. *Experimental & Molecular Medicine*, 56(7), 1501–1512. https://doi.org/10.1038/s12276-024-01262-7

Kiousi, Despoina E., Antonia Z. Kouroutzidou, Konstantinos Neanidis, Emmanuel Karavanis, Dimitrios Matthaios, Aglaia Pappa, and Alex Galanis. (2023). The role of the gut microbiome in cancer immunotherapy: Current knowledge and future directions. *Cancers*, 15(7), 2101. https://doi.org/10.3390/cancers15072101

Kumar, V., R. Shukla, S. Gangani, H. Shekhar, and V. Shukla. (2024). Metagenomics and Microbiome analysis. In A. Chaudhary, S. K. Sethi, and A. Verma (Eds.), *Unraveling New Frontiers and Advances in Bioinformatics* (pp. 77–92). Springer Nature. https://doi.org/10.1007/978-981-97-7123-3_4

Kumar, V., S. Gangani, R. Shukla, S. K. Prajapati, H. Shekhar, and V. Shukla. (2024). Next-generation sequencing and genomic data analysis. In A. Chaudhary, S. K. Sethi, and A. Verma (Eds.), *Unraveling New Frontiers and Advances in Bioinformatics* (pp. 49–76). Springer Nature. https://doi.org/10.1007/978-981-97-7123-3_3

Li, Chaozhong, Fang Wang, Wanrong Li, Guihu Sun, Dexing Yang, Ting Yang, . . . Zhaohui Meng. (2024). The diagnostic value of metagenomic next-generation sequencing in critically ill patients with sepsis: A retrospective cohort study. *Medicine*, 103(41), e39987. https://doi.org/10.1097/MD.0000000000039987

Liu, Wenwen, Lei Wang, Jinmei Ou, Daiyin Peng, Yue Zhang, Weidong Chen, and Yanyan Wang. (2025). Gut microbiota metabolites and chronic diseases: Interactions, mechanisms, and therapeutic strategies. *International Journal of Molecular Sciences*, 26(8), 3752. https://doi.org/10.3390/ijms26083752

Loh, Jian Sheng, Wen Qi Mak, Li Kar Stella Tan, Chu Xin Ng, Hong Hao Chan, Shiau Hueh Yeow, . . . Kooi Yeong Khaw. (2024). Microbiota-gut-brain axis and its therapeutic applications in neurodegenerative diseases. *Signal Transduction and Targeted Therapy*, 9(1), 37. https://doi.org/10.1038/s41392-024-01743-1

Long, Qichen, Fengming Luo, Binghui Li, Ziyang Li, Zhe Guo, Zhiyang Chen, . . . Min Hu. (2024). Gut microbiota and metabolic biomarkers in metabolic dysfunction–associated steatotic liver disease. *Hepatology Communications*, 8(3). https://doi.org/10.1097/HC9.0000000000000310

Lu, Yuting, Xiangliang Yuan, Miao Wang, Zhihao He, Hongzhong Li, Ji Wang, and Qin Li. (2022). Gut microbiota influence immunotherapy responses: Mechanisms and therapeutic strategies. *Journal of Hematology & Oncology*, 15(1), 47. https://doi.org/10.1186/s13045-022-01273-9

Lv, Minjie, Changjun Zhu, Chenghua Zhu, Jing Yao, Lixu Xie, Changwen Zhang, . . . Ganzhu Feng. (2023). Clinical values of metagenomic next-generation sequencing in patients with severe pneumonia: A systematic review and meta-analysis. *Frontiers in Cellular and Infection Microbiology*, 13, 1106859. https://doi.org/10.3389/fcimb.2023.1106859

Macura, Barbara, Aneta Kiecka, and Marian Szczepanik. (2024). Intestinal permeability disturbances: Causes, diseases and therapy. *Clinical and Experimental Medicine*, 24(1), 232. https://doi.org/10.1007/s10238-024-01496-9

Mishra, S., L. Buddendorff, J. Hoover, R. Shukla, V. Kumar, S. Jain, and H. Yadav. (2024). Microbiome role in cognitive impairment: MiaGB cohort insights. *Innovation in Aging*, 8(S1), 811. https://doi.org/10.1093/geroni/igae098.2629

Nagata, Naoyoshi, Suguru Nishijima, Yasushi Kojima, Yuya Hisada, Koh Imbe, Tohru Miyoshi-Akiyama, . . . Takashi Kawai. (2022). Metagenomic identification of microbial signatures predicting pancreatic cancer from a multinational study. *Gastroenterology*, 163(1), 222–238. https://doi.org/10.1053/j.gastro.2022.03.054

Nair, S. U., Y. Lin, S. Jain, D. S. Chaudhari, R. Shukla, V. Kumar, . . . MiaGB Consortium. (2024). Unique gut microbiome signatures associated with physical and cognitive function. *Alzheimer's & Dementia*, 20(S2), e092044. https://doi.org/10.1002/alz.092044

Nam, Nguyen, Hoang Do, Kieu Loan Trinh, and Nae Lee. (2023). Metagenomics: An effective approach for exploring microbial diversity and functions. *Foods*, 12(11), 2140. https://doi.org/10.3390/foods12112140

Noe, Myint M., Akira A. Shishido, Kapil Saharia, and Paul Luethy. (2021). 661. Clinical utility and impact of the metagenomic microbial plasma cell-free DNA next-generation sequencing assay on treatment decision: A single-center retrospective study. *Open Forum Infectious Diseases*, 8(S1), S432–S433. https://doi.org/10.1093/ofid/ofab466.858

Olthoff, Michael, Takaaki Kobayashi, Meredith G. Parsons, Bradley Ford, Kunatum Prasidthrathsint, Lemuel Non, . . . Dilek Ince. (2024). Impact of metagenomic next-generation sequencing on clinical decision-making at an academic medical center, a retrospective study, Iowa, 2020–2022. *Antimicrobial Stewardship & Healthcare Epidemiology*, 4(1), e39. https://doi.org/10.1017/ash.2024.31

Ou, Suwen, Hufei Wang, Yangbao Tao, Kangjia Luo, Jinhua Ye, Songlin Ran, . . . Rui Huang. (2022). Fusobacterium nucleatum and colorectal cancer: From phenomenon to mechanism. *Frontiers in Cellular and Infection Microbiology*, 12, 1020583. https://doi.org/10.3389/fcimb.2022.1020583

Pang, Feng, Wenbin Xu, Hui Zhao, Shuai Chen, Yaxian Tian, Juanjuan Fu, . . . Xiuqin Jia. (2024). Comprehensive evaluation of plasma microbial cell-free DNA sequencing for predicting bloodstream and local infections in clinical practice: A multicenter retrospective study. *Frontiers in Cellular and Infection Microbiology*, 13, 1256099. https://doi.org/10.3389/fcimb.2023.1256099

Park, Sarah Y., Eliza J. Chang, Nathan Ledeboer, Kevin Messacar, Martin S. Lindner, Shivkumar Venkatasubrahmanyam, . . . Frederick S. Nolte. (2023). Plasma microbial cell-free DNA sequencing from over 15,000 patients identified a broad spectrum of pathogens. *Journal of Clinical Microbiology*, e01855.22. https://doi.org/10.1128/jcm.01855-22

Pindi, P. K., N. Kishore, and S. Reddy. (2009). Metagenomics: Principles, methodology and biotechnological potentials. In A. S. Verma and A. Das (Eds.), *A Textbook of Molecular Biotechnology* (p. 343). I.K. International Publishing House.

Prajapati, S. K., V. Kumar, R. Shukla, C. Szekeres, Y. Ma, S. Jain, and H. Yadav. (2024). Microbiome-immune interplay in aging brains: A pilot investigation from the MiaGB cohort. *Alzheimer's & Dementia*, 20(S2), e090734. https://doi.org/10.1002/alz.090734

Romano, Stefano, Jakob Wirbel, Rebecca Ansorge, Christian Schudoma, Quinten Raymond Ducarmon, Arjan Narbad, and Georg Zeller. (2025). Machine learning-based meta-analysis reveals gut microbiome alterations associated with Parkinson's disease. *Nature Communications*, 16(1), 4227. https://doi.org/10.1038/s41467-025-56829-3

Ruan, Qijun, Peng Guan, Weijuan Qi, Jiatong Li, Mengying Xi, Limin Xiao, . . . Jia Ni. (2023). Porphyromonas gingivalis regulates atherosclerosis through an immune pathway. *Frontiers in Immunology*, 14, 1103592. https://doi.org/10.3389/fimmu.2023.1103592

Sarfraz, H., B. J. Czerniecki, K. Stankiewicz, M. V. Jaglal, S. Jain, H. Yadav, . . . Bari, S. (2024). Evaluating gut microbiome as a biomarker of pathological complete response in patients with early locally advanced triple negative breast cancer (LA TNBC): A pilot study. *Journal of Clinical Oncology*, 42(S16), 593. https://doi.org/10.1200/JCO.2024.42.16_suppl.593

Shi, Zhuangzhuang, Zhaoming Li, and Mingzhi Zhang. (2024). Emerging roles of intratumor microbiota in cancer: Tumorigenesis and management strategies. *Journal of Translational Medicine*, 22(1), 837. https://doi.org/10.1186/s12967-024-05640-7

Sun, Qin, Ran Teng, Qiankun Shi, Yun Liu, Xing Cai, Bin Yang, . . . Ling Liu. (2025). Clinical implement of probe-capture metagenomics in sepsis patients: A multicentre and prospective study. *Clinical and Translational Medicine*, 15(4), e70297. https://doi.org/10.1002/ctm2.70297

Tan, Jun, Lingxiao Wu, Lijuan Zhan, Minkui Sheng, Zhongxin Tang, Jianzhong Xu, and Haijun Ma. (2024). Optimal selection of specimens for metagenomic next-generation sequencing in diagnosing periprosthetic joint infections. *Frontiers in Cellular and Infection Microbiology*, 14, 1356804. https://doi.org/10.3389/fcimb.2024.1356804

Tan, Jun, Yang Liu, Sabrina Ehnert, Andreas K. Nüssler, Yang Yu, Jianzhong Xu, and Tao Chen. (2022). The effectiveness of metagenomic next-generation sequencing in the diagnosis of prosthetic joint infection: A systematic review and meta-analysis. *Frontiers in Cellular and Infection Microbiology*, 12, 875822. https://doi.org/10.3389/fcimb.2022.875822

Ternes, Dominik, Mina Tsenkova, Vitaly Igorevich Pozdeev, Marianne Meyers, Eric Koncina, Sura Atatri, . . . Elisabeth Letellier. (2022). The gut microbial metabolite formate exacerbates colorectal cancer progression. *Nature Metabolism*, 4(4), 458–475. https://doi.org/10.1038/s42255-022-00558-0

Trivett, H., A. C. Darby, and O. Oyebode. (2025). Academic and clinical perspectives of metagenome sequencing as a diagnostic tool for infectious disease: An interpretive phenomenological study. *BMC Infectious Diseases*, 25(1), Article 448. https://doi.org/10.1186/s12879-025-13825-0

Villette, Rémy, Júlia Ortís Sunyer, Polina V. Novikova, Velma T. E. Aho, Viacheslav A. Petrov, Oskar Hickl, . . . Paul Wilmes. (2025). Integrated multi-omics highlights alterations of gut microbiome functions in prodromal and idiopathic Parkinson's disease. *Microbiome*, 13(1), 200. https://doi.org/10.1186/s40168-025-02227-2

Wallen, Zachary D., Ayse Demirkan, Guy Twa, Gwendolyn Cohen, Marissa N. Dean, David G. Standaert, . . . Haydeh Payami. (2022). Metagenomics of Parkinson's disease implicates the gut microbiome in multiple disease mechanisms. *Nature Communications*, 13(1), 6958. https://doi.org/10.1038/s41467-022-34667-x

Wan, Chuandan, Ping Wang, Yeqiong Xu, Yanping Zhu, Huanhuan Chen, Xuexian Cao, and Yulan Gu. (2025). Mechanism and role of H. pylori CagA-induced NLRP3 inflammasome in gastric cancer immune cell infiltration. *Scientific Reports*, 15(1), 14335. https://doi.org/10.1038/s41598-025-98301-8

Wang, J., Y. Han, and J. Feng. (2023). Metagenomic next-generation sequencing for the diagnosis of urinary tract infections. *Frontiers in Cellular and Infection Microbiology*, 13, Article 1127166. https://doi.org/10.3389/fcimb.2023.1127166

Wang, Man, Fei Yu, and Peifeng Li. (2023). Intratumor microbiota in cancer pathogenesis and immunity: From mechanisms of action to therapeutic opportunities. *Frontiers in Immunology*, 14, 1269054. https://doi.org/10.3389/fimmu.2023.1269054

Wang, Ruixuan, Shitong Cheng, Wei Du, Fengshuo Sun, Yang Gao, Zheming Zhu, . . . Xiaoxu Han. (2025). An innovative fungal-specific targeted next-generation sequencing method: Analytical performance and a single-center prospective clinical study. *Journal of Translational Medicine*, 23(1), 820. https://doi.org/10.1186/s12967-025-06784-w

Wu, Chen, Xin Yu, Wei Gai, Yanxia Liu, Ying Qi, Yafeng Zheng, . . . Bing Chen. (2023). Diagnostic value of plasma and blood cells metagenomic next-generation sequencing in patients with sepsis. *Biochemical and Biophysical Research Communications*, 683, 149079. https://doi.org/10.1016/j.bbrc.2023.10.011

Xia, Liliang, Xiaokuan Zhu, Ying Wang, and Shun Lu. (2024). The gut microbiota improves the efficacy of immune-checkpoint inhibitor immunotherapy against tumors: From association to cause and effect. *Cancer Letters*, 598, 217123. https://doi.org/10.1016/j.canlet.2024.217123

Xie, Jindong, Manqing Liu, Xinpei Deng, Yuhui Tang, Shaoquan Zheng, Xueqi Ou, . . . Yutian Zou. (2024). Gut microbiota reshapes cancer immunotherapy efficacy: Mechanisms and therapeutic strategies. *iMeta*, 3(1), e156. https://doi.org/10.1002/imt2.156

Yang, Li, Aitian Li, Ying Wang, and Yi Zhang. (2023). Intratumoral microbiota: Roles in cancer initiation, development and therapeutic efficacy. *Signal Transduction and Targeted Therapy*, 8(1), 35. https://doi.org/10.1038/s41392-022-01304-4

Ying, Chao, Yuan Li, Hui Zhang, Shimin Pang, Shuwen Hao, Songnian Hu, and Lifang Zhao. (2024). Probing the diagnostic values of plasma cf-nDNA and cf-mtDNA for Parkinson's disease and multiple system atrophy. *Frontiers in Neuroscience*, 18, 1488820. https://doi.org/10.3389/fnins.2024.1488820

Yu, Juan, Li Zhang, Deyu Gao, Jie Wang, Yi Li, and Ning Sun. (2024). Comparison of metagenomic next-generation sequencing and blood culture for diagnosis of bloodstream infections. *Frontiers in Cellular and Infection Microbiology*, 14, 1338861. https://doi.org/10.3389/fcimb.2024.1338861

Zehrh, Itrat, Umme Habiba, Matilde Rosalia Picco, Sidra Hafiza Basheer, Umer Abdul Rehman, Owaisa Haider, and Shahzaib Khoso. (2023). Metagenomics and machine learning-based precision medicine approaches for autoimmune diseases. *Preprint*, April 11. https://doi.org/10.20944/preprints202304.0209.v1

Zepeda-Rivera, Martha, Samuel S. Minot, Heather Bouzek, Hanrui Wu, Aitor Blanco-Míguez, Paolo Manghi, . . . Christopher D. Johnston. (2024). A distinct *Fusobacterium nucleatum* clade dominates the colorectal cancer niche. *Nature*, 628(8007), 424–432. https://doi.org/10.1038/s41586-024-07182-w

Zhang, Chongjing, Xinyu Fang, Zida Huang, Wenbo Li, Chao-Fan Zhang, Bin Yang, and Wenming Zhang. (2019). Value of mNGS in sonication fluid for the diagnosis of periprosthetic joint infection. *Arthroplasty (London, England)*, 1(1), 9. https://doi.org/10.1186/s42836-019-0006-4

Zhang, Hui, Ruobing Liang, Yunzhu Zhu, Lifen Hu, Han Xia, Jiabin Li, and Ying Ye. (2024). Metagenomic next-generation sequencing of plasma cell-free DNA improves the early diagnosis of suspected infections. *BMC Infectious Diseases*, 24(1), 187. https://doi.org/10.1186/s12879-024-09043-3

Zhang, L., J. Wang, and X. Zhang. (2021). Detection of Cyclospora cayetanensis in human fecal specimens by next-generation sequencing. *Journal of Clinical Microbiology*, 59(3), e02434.20. https://doi.org/10.1128/JCM.02434-20

Zhang, Qiang, Qiushi Wang, Feng Zhang, Xiuping Li, Yuhong Sun, Li Wang, and Zhongfa Zhang. (2025). Diagnostic value of metagenomic next-generation sequencing in patients with osteoarticular infections: A prospective study. *Microbiology Spectrum*, 13(5), e01064.24. https://doi.org/10.1128/spectrum.01064-24

Zhao, Junjie, Runxi Zhuge, Bangchuan Hu, Yesong Wang, Xingxing Wang, Yi Zhang, . . . Honglong Fang. (2025). Clinical impact of bronchoalveolar lavage fluid metagenomic next-generation sequencing in immunocompromised patients with severe community-acquired pneumonia in ICU: A multicenter retrospective study. *Infection*, 53(5), 1911–1927. https://doi.org/10.1007/s15010-025-02520-0

Zheng, Jiaying, Qianru Sun, Mengjing Zhang, Chengyu Liu, Qi Su, Lin Zhang, . . . Siew C. Ng. (2024). Noninvasive, microbiome-based diagnosis of inflammatory bowel disease. *Nature Medicine*, 30(12), 3555–3567. https://doi.org/10.1038/s41591-024-03280-4

Zhou, Jing-Jiang, Wei-Chao Ding, Yan-Cun Liu, Yu-Lei Gao, Lei Xu, Run-Lu Geng, . . . Yan-Fen Chai. (2022). Diagnostic value of metagenomic next-generation sequencing for pulmonary infection in intensive care unit and non-intensive care unit patients. *Frontiers in Cellular and Infection Microbiology*, 12, 929856. https://doi.org/10.3389/fcimb.2022.929856

Zong, Zhen, Wenjuan Zeng, Yin Li, Menghui Wang, Yuke Cao, Xifu Cheng, . . . Xingen Zhu. (2024). Intratumor microbiota and colorectal cancer: Comprehensive and lucid review. *Chinese Journal of Cancer Research = Chung-Kuo Yen Cheng Yen Chiu*, 36(6), 683–699. https://doi.org/10.21147/j.issn.1000-9604.2024.06.07

6 Viromics
The Study of Viral Communities

Brandi C. Miller[1], *Rohit Shukla*[1], *Mansi Gautam*[2], *Vivek Kumar*[1], *and Rajan Kumar Singh*[3]*
[1]USF Center for Microbiome Research, Microbiomes Institute, University of South Florida, Tampa, FL, USA.
[2]Department of Biomedical Sciences, Galgotia University, Greater Noida, Uttar Pradesh, India
[3]Department of Biotechnology, Parul Institute of Technology, Faculty of Engineering and Technology, Parul University, Vadodara, India.
*Corresponding author

6.1 INTRODUCTION

The collection of viruses that reside inside a microbiome is collectively referred to as the virome. Virome components are differentiated based on the hosts they infect and are categorized into eukaryotic viruses and prokaryotic viruses. These viruses can be further classified as DNA or RNA viruses based on the genetic material they carry. Inside the human gut, these viruses coexist with bacteria, fungi, and other microbes, forming a complex hidden ecosystem. These viruses affect microbial growth patterns, exchange genes, and interact with the human body. Moreover, they maintain microbial balance in the gut, which further impacts metabolism, immunity, and even brain function. Viromics is a term that involves metagenomics-based and culture-free approaches to study the virome. In viromics, DNA from microbiome samples such as stool, saliva, and skin is used for sequencing and further annotation. Although viromics is an extension of microbiome research, it is technically more challenging. In contrast to bacteria, viruses lack a common gene, such as 16S rRNA, which makes identification difficult. Additionally, many viral genomes are unique, meaning that there are few or no related sequences in current databases. This unexplored part of viral diversity is often called "viral dark matter" (Garmaeva et al., 2019). These challenges make virome research exciting but also complex.

For many years, microbiome research was almost exclusively focused on bacteria. However, the significance of viruses has been explored more recently and has become a rapidly emerging area of microbiome research. Recent studies have revealed that viruses can affect bacterial growth, influence the immune system, and transmit genes that aid bacterial adaptation. This approach reveals several indirect ways in which viruses affect human health. Dysregulation of gut virus composition and function (dysbiosis) has been correlated with several diseases, such as inflammatory bowel disease (IBD), obesity, and neurological disorders (Gregory et al., 2020). A recent study by James et al. (2024) reported that altered gut virome patterns are linked to cognitive decline in older adults, suggesting that gut viruses might even influence brain health (James et al., 2024). Recent improvements in next-generation sequencing have revolutionized virome research. As a result of these innovations, several studies have reported thousands of new viral species in the human gut. The findings of these studies serve as large viral catalogs that act as maps for studying interactions between viruses, bacteria, and human hosts. For instance, Zeng et al. (2024) reported early-life impacts of gut viruses on newborns that regulate immunological

DOI: 10.1201/9781003670155-6

and metabolic systems (Zeng et al., 2024). These results demonstrate that the virome is an integral component of microbial ecology that maintains human health from birth and is not merely a reservoir of potential disease.

Viromics studies are now used in combination with other scientific approaches, including immunological profiling, transcriptomics, and metabolomics. These integrative multi-omics approaches allow multidimensional study of viruses and other microbes and their communication with the human body. Understanding these relationships is indispensable for improving disease diagnosis, prevention, and treatment. In this chapter, we discuss methodologies such as sample collection, sequencing methods, and analysis that are fundamental to virome research. Moreover, we discuss the impacts of the gut virome on health, their association with human disease pathology, and their therapeutic potential. Finally, we address current challenges and future directions in this rapidly growing field, emphasizing how virome research is reshaping understanding of the microbiome and human health.

6.2 VIROMICS: TOOLS AND METHODOLOGIES

Development of a well-defined experimental workflow and optimization of computational pipelines are fundamental for virome research. Because viruses are less abundant in samples, careful handling is required at each step, from sequencing to annotation, to avoid contamination and bias. Viromics consists of three essential components, including (1) sample preparation and enrichment, (2) sequencing, and (3) computational analysis; each component is important to uncover the true viral landscape.

6.2.1 Sample Preparation and Enrichment

Any virome study is based on the precise and careful enrichment of virus-like particles (VLPs) from complex biological materials such as stool, saliva, or soil. The enrichment process generally begins with filtration using sequential 0.45 μm and 0.22 μm membranes to remove other kinds of cells, including bacterial or eukaryotic cells. After that, the filtrate is concentrated through ultracentrifugation via a simple high-speed pellet or through a density-gradient separation technique using sucrose or iodixanol solutions. These methods have been standardized and adopted in the NetoVIR (Netherlands Virome) workflow, which can confirm reproducibility across sample types (Conceição-Neto et al., 2015). After enrichment, the samples are treated with DNase I and RNase A enzymes to digest free nucleic acids that come from lysed bacterial or host cells; this step ensures that only genome-protected, intact viral nucleic acids are present (Thurber et al., 2009). When enveloped and non-enveloped viruses are targeted for sequencing, it is important to ensure that buffers remain isotonic and free of detergents so that fragile viral membranes remain preserved. In the case of non-enveloped viruses only, detergent treatment or mild chloroform can be used to remove bacterial outer-membrane vesicles and lipids (Conceição-Neto et al., 2015). Despite these efforts, cellular contaminants may still be present. Thus, ViromeQC should be used regularly to quantify nonviral DNA within virome datasets and confirm the purity of the viral fraction (Zolfo et al., 2019). Proper handling of all these procedures at this stage is very important because errors introduced during sample handling can propagate through sequencing and downstream analyses.

6.2.2 Sequencing Approaches

Nucleic acid extracted using the above-described protocol can be used as input for sequencing to identify the composition and structure of viral communities. Illumina platforms are still considered the most optimal with regard to viral sequencing, as they produce short reads with high accuracy. These platforms perform well for smaller and abundant viral genomes but fail to resolve larger or highly mosaic genomes, which limits assembly continuity. In a previous study, the authors faced such challenges using VirSorter-assisted assemblies; thus, they established best practices for short-read virome analysis (Roux et al., 2015). To overcome these limitations, technologies such as Oxford Nanopore and PacBio HiFi (which produce long reads) have emerged and have played an integral role in viromics and metagenomics research. They can generate kilobase-scale reads that capture complete genomes, produce high-quality contigs, and preserve structural variants that short reads often miss (Saroj et al., 2023). One study reported that long-read sequencing can profile marine viral populations more accurately than short-read approaches (Warwick-Dugdale et al., 2019). However, both of these technologies are often applied to gut virome studies because they perform well in hybrid assembly pipelines to produce complete viral contigs. The portability of the long-read Nanopore sequencer has demonstrated its potential in Ebola outbreaks, where a handheld sequencer was implemented for real-time viral surveillance using an on-site genome-monitoring procedure (Quick et al., 2016). Together, these sequencing technologies have transformed viromics from static observation to dynamic ecological monitoring.

6.2.3 Computational Pipelines and Tools

Sequencing produces raw reads, but an advanced computational workflow is required to obtain meaningful information from these reads. These workflows contextualize viral sequences and profile them for downstream analysis. Because viruses lack a universal housekeeping gene, such as bacterial 16S rRNA, viral classification depends on machine-learning methods, sequence similarity, and compositional signatures rather than traditional alignment-based methods. Among these tools, VirFinder is widely used and applies a k-mer frequency-based machine-learning model to identify viral versus nonviral sequences without requiring homology to known genomes (Ren et al., 2017). Based on the VirFinder concept, DeepVirFinder uses deep-learning architectures and shows enhanced sensitivity for short or fragmented contigs that share little similarity with reference data (J. Ren et al., 2020). To identify known bacteriophages, virMine uses a BLAST-based approach by comparing assembled contigs against a curated database of phage genomes to characterize known viruses with high precision (Garretto et al., 2019). VirSorter detects viral signals by searching for hallmark genes and genome-organization patterns, enabling discovery of both integrated prophages and free viruses (Roux et al., 2015). After viral contigs are identified, their quality and completeness must be assessed. CheckV is a widely used tool that provides precise metrics for evaluating metagenome-assembled viral genomes (vMAGs). It also estimates genome completeness and identifies host contamination by comparison against a comprehensive database of curated viral sequences (Nayfach et al., 2021). ViromeQC, originally designed for presequencing quality control, can also assess host DNA proportions after assembly to ensure dataset integrity (Zolfo et al., 2019).

After viral contigs are identified, annotation is required to assign putative gene functions and identify auxiliary metabolic genes (AMGs) that may modulate host pathways. Researchers also use the VIBRANT tool, which merges protein-similarity and machine-learning signatures for viral recovery and annotates AMGs from metagenomic assemblies (Kieft et al., 2020). Many researchers also use Prokka to identify genes and proteins (Seemann, 2014) and then refine their functions with HMMER searches against curated domain databases such as Pfam (Finn et al., 2016).

For taxonomic classification, current best practices follow the genome-based framework of the International Committee on Taxonomy of Viruses (ICTV) (gene—sharing networks, sequence similarity, and protein-cluster relatedness) to define species-level and higher-rank taxa. Recent ICTV/JGV updates codify classifications beyond morphology-only criteria (Zerbini et al., 2025). In parallel, reference resources such as Integrated Microbial Genomes/Virus (IMG/VR), which contains cultured and uncultured viral genomes from metagenomes (Paez-Espino et al., 2016), and the Gut

Virome Database (GVD), a large human gut–specific catalog (Gregory et al., 2020), can be used to classify viral sequences.

Collectively, these annotation and taxonomy pipelines have transformed viral metagenomics data into a structured biological framework. By applying ICTV classification principles and integrating genome-based host inference, modern viromics provides a comprehensive understanding of how viral diversity, function, and microbial interactions shape ecosystem stability and host health.

6.3 GUT VIROME IN HUMAN DISEASES

Emerging studies have described the role of the gut virome, including both eukaryotic viruses and bacteriophages, in human disease pathology. Although a core gut virome remains relatively stable, the composition of viral components, especially phages, is highly variable from among individuals and is modifiable by several factors (Cao et al., 2022). Age, diet, geography, and antibiotic use strongly influence the gut virome composition, which has been linked with the development of many human conditions and may increase susceptibility to disease onset (Cao et al., 2022). Other mechanisms include modulation of the bacteriome and mucosal immune system by phages, exchange of genomic material, metabolic changes or production of bioactive molecules, and direct effects of eukaryotic viruses on mammalian cells. The ratio of phages to bacterial cells exceeds 10-fold, indicating a prominent role for bacterial targeting by viruses in human health and disease pathogenesis (Minot et al., 2013). However, current knowledge of the gut virome, particularly its role in disease progression, remains preliminary, and substantial work is required to fill these knowledge gaps. Emerging studies have linked gut virome structure and function with inflammation and disease pathologies, including gastrointestinal conditions, neurological disorders, metabolic syndromes, cardiovascular diseases (CVDs), infectious diseases, and cancer. Additional studies are necessary to elucidate the potential of gut virome composition and function as biomarkers or diagnostic tools for human disease.

6.3.1 Chronic Inflammatory and Gastrointestinal Disorders

Inflammation is a hallmark of many human diseases and often originates from perturbations in the gut. Changes in gut virome composition have been linked with chronic inflammatory bowel diseases (IBDs), including Chron's disease (CD) and ulcerative colitis (UC) (Wu, Cheng, et al., 2025). Elevations in phages targeting *Escherichia*, *Klebsiella*, *Enterococcus_B*, *Streptococcus*, and *Veillonella*, and depletion of phages targeting *Prevotella*, *Ruminococcus_E*, *Bifidobacterium*, and *Blautia* were observed in a large cross-population cohort study and were confirmed to accelerate colitis progression using fecal virome transplantation (FVT) (Tian et al., 2024). Similarly, the gut of individuals with CD or UC was characterized by phage enrichment, a concurrent decline in bacterial diversity, distinct phageomes, and increased abundance of *Caudovirales*, irrespective of age (Majzoub et al., 2024).These unique virome signatures were consistent across multiple cohorts, further confirming that the phageome, as well as the viral family *Caudovirales*, plays prominent roles in IBD development (Norman et al., 2015). Increased virulence functions of viral and bacterial components and higher abundances of *Escherichia* and *Enterobacteria* phages have also been reported (Zuo et al., 2019). Eukaryotic viruses can directly impair host cell function and contribute to IBD pathology. *Hepadnaviridae* abundances were significantly enriched in the gut of patients with UC, whereas *Herpesviridae* was increased in CD (Ungaro et al., 2019). Although different signatures were observed in UC and CD, both were linked with inflammation, suggesting a role for these eukaryotic viruses in IBDs. Colorectal cancer is characterized by similar physiological changes—including inflammation and disruptions in barrier function and bowel habits—as IBDs. Studies have reported *Fusobacterium* as a major bacterial taxon associated with oncogenesis in colorectal cancer, while *Siphoviridae* and *Myoviridae* phages showed the highest virome associations (Hannigan et al., 2018). Collectively, these data provide strong evidence that gut virome dysbiosis is present in IBDs and can lead to immune dysregulation, heightened inflammation, and epithelial barrier disruption.

The major mechanism underlying these changes is metabolite production through bacterial lysis or transcytosis—both of which activate intestinal immune cells and modulate immune responses (Metzger et al., 2018). Treatment with phages in germ-free mice activated immune cell expansion, and *Lactobacillus*, *Escherichia*, and *Bacteroides* phages worsened colitis progression through toll-like receptor and interferon-γ (IFN-γ) signaling (Gogokhia et al., 2019). Furthermore, in UC, reduced phage abundance was linked with higher IFN-γ levels in the gut mucosa (Gogokhia et al., 2019). *Caudovirales* phage expansion and diversity have been associated with inflammatory markers such as albumin, IBD diagnosis, and increased barrier dysfunction. Although still emerging, these data underscore the role of bacteria–phage interactions in IBDs. Colitis symptoms induced by dextran sodium sulfate were significantly attenuated by a phage cocktail targeting adherent invasive *Escherichia coli* (Galtier et al., 2017).

This therapeutic study strengthens understanding that viruses, particularly phages, have a causal role in IBD development and gut inflammation; however, additional preclinical studies with similar designs are warranted. Viruses can also interact with genes or integrate into genomes to impact intestinal cells and exacerbate IBD progression. Norovirus infection has been linked with intestinal immune dysregulation—such as altered Paneth cell biology and increased tumor necrosis factor-α (TNF-α) or IFN-γ—and increased susceptibility to CD in *atg16L1* mutant mice (Cadwell et al., 2010). The CD virome has demonstrated fewer phage–bacteria interactions, particularly involving bacterial pathogens, and these signatures showed a causal role in worsening intestinal inflammation when transplanted into mice (Cao et al., 2024). Together, these proof-of-concept and preliminary human studies have begun to characterize gut virome involvement in IBDs and virus–bacteria–host interactions; however, further studies are required to fully elucidate the therapeutic potential of viromics in IBDs and other gastrointestinal conditions.

6.3.2 Neurological Disorders

Neurological diseases impact the brain, spinal cord, or central nervous system. Common neurological diseases include mild cognitive impairment (MCI), Alzheimer's disease (AD), and Parkinson's disease. Recent studies report that older adults with MCI or dementia have a distinct microbiome signature, including differences in both composition and function, compared to cognitively healthy, age-matched controls (James et al., 2024). Our recent pilot observational analysis from the Microbiome in Aging-Gut and Brain (MiaGB) consortium revealed a decline in multiple *Escherichia* viruses and phages in the gut of older adults with MCI or dementia, with concurrent increases in *Lactobacillus* and *Lactococcus* phages (i.e., *Lactobacillus* phage Sha1 and *Lactococcus* phage P335 *sensu lato*), providing insight into phage–bacteria relationships (James et al., 2024). In another study, metagenomics analysis revealed that individuals with AD + amyloid-β (Aβ, the major neurotoxic protein aggregate in AD) demonstrated significantly lower viral diversity, as well as marked reductions in *Siphoviridae* and several *Lactococcus* phages (i.e., *Lactococcus* phage bIL286, *Lactococcus* phage bIL309, and *Lactococcus* phage BK5 T), which together produced a strong predictive model (Ghorbani et al., 2023). Thus, phages may have strong potential to discriminate healthy controls from individuals with AD + Aβ. Mouse studies have shown that there are unique transkingdom signatures in AD, some of which are specific to AD (Ghorbani et al., 2023). Phage therapy beneficially modulated the gut microbiome and was detected in the brain in a sex-dependent manner; however, the roles of gut viruses and phages in males versus females warrant further study (Grabowski et al., 2023). Collectively, these studies are very preliminary and merely show associations between gut virome signatures and neurodegenerative conditions. There is a paucity of studies that profile the gut virome and assess gut–brain functions (including barrier functions and inflammation) concurrently. It is important to design and conduct both observational and mechanistic studies so that we can better characterize the gut virome in neurodegenerative diseases and elucidate gut–brain functions in the context of phages.

Unique virome structures and functions have been implicated in mental and psychiatric health conditions in a few recent studies. Higher diversity of viruses was linked with worse schizophrenia outcomes, as assessed by the Positive and Negative Syndrome Scale for schizophrenia (Wu, Tao, et al., 2025). Furthermore, the gut of patients with schizophrenia harbored 171 unique viral operational taxonomic units, with enrichments in *Siphoviridae* and *Flandersviridae* (Y. Ren et al., 2025). There were also significant distinctions in virus–bacteria interactions, specifically among *Akkermansia* and *Clostridia*, compared to controls, with strong diagnostic potential (Y. Ren et al., 2025). Additionally, in a mixed cohort of psychiatric diseases, sequences for HIV-1, Torque Teno, Pegi, Herpes, and Papilloma viruses were detected in blood and stool (Yolken et al., 2021). However, it remains elusive whether these viral signatures were linked with psychiatric disease outcomes in this cohort, and no associative studies were conducted. Further studies are necessary to fully delineate the role of the gut virome in psychiatric conditions and their mechanisms.

Gut virome signatures may impact brain function via the gut–brain axis, which is the bidirectional connection between the gut and brain and includes immune cells, neurons, neurotransmitters, hormones, and bioactive metabolites such as short-chain fatty acids. Gut–brain axis signaling can impact gut functions, leading to leaky gut, and also regulates many behaviors, including mood, anxiety, and eating behaviors (Cryan et al., 2019; Miller et al., 2023). Metabolites derived from the gut microbiota can modulate peripheral organs, including the brain—this modulation is demonstrated by changes in inflammatory tone and blood–brain barrier integrity (Liu et al., 2024). Specific microbiota signatures, such as those acquired during natural aging or in AD, are linked with cognitive impairment and increased neuroinflammation (Mishra et al., 2024). The phage class *Caudoviricetes* has been linked with impairments in circulating immune cells (such as monocytes), increased cytokine production (i.e., interleukin [IL]-6 and TNF-α), changes in bacterial diversity and abundance, and altered brain gene expression in a mouse model of stress (Ritz et al., 2024). Furthermore, transplantation of *Caudovirales* phages from cognitively healthy donors to mice resulted in marked improvements in executive functions, such as learning and memory, and beneficially modulated gene expression in the prefrontal cortex (Mayneris-Perxachs et al., 2022). Although limited, these studies provide mechanistic insights into how the gut virome (particularly phages) modulates the gut–brain axis in the context of CI and other neurodegenerative diseases.

6.3.3 Metabolic and Cardiovascular Diseases

Classical studies have suggested that the gut microbiome contributes to the pathogenesis of obesity, metabolic syndrome, nonalcoholic fatty liver disease (NAFLD), type 2 diabetes mellitus (T2D), atherosclerosis, and CVD, and chronic, low-grade inflammation is a hallmark of these conditions (Fan & Pedersen, 2021). However, the role of the gut virome is less well understood. FVT studies have demonstrated that virome-mediated changes in the gut microbiome contribute to obesity phenotypes, such as changes in weight gain during high-fat diet (HFD) administration (Borin et al., 2023). However, this study did not characterize the virome signatures linked with these phenotypic changes. Feeding with a Western diet (commonly linked with obesity and metabolic conditions) was associated with increased abundances of temperate *Caudovirales* phages, which target *Bacilli*, *Negativicutes*, and *Bacteroidia* and are advantageous for bacterial host adaptation (M. Kim & Bae, 2016). In HFD-induced obesity, enrichment of *Microviridae* phages was linked with a functional decline in integrase genes (Schulfer et al., 2020). Virome structures in metabolic diseases have also been linked with functional changes and alterations in outcome measures. Increases in viral RNA and DNA were positively correlated with body weight, fat mass, and fasting blood glucose—common metabolic parameters (Yadav et al., 2016). Mice on an HFD administered lean FVT demonstrated a significant reduction in weight gain as well as preserved glucose tolerance;

these changes were associated with favorable shifts in the gut microbiota structure, plasma metabolome, and gene expression of obesity- and T2D-related genes (Rasmussen et al., 2020).

Cohort studies have also characterized the gut virome in metabolic disease. Individuals with metabolic syndrome had decreased virome richness and diversity, altered phage diversity (increases in *Streptococcaceae*- and *Bacteroidaceae*-infecting phages and decreases in *Bifidobacteriaceae*-infecting phages), and specific enrichment of the *Candidatus Heliusviridae* phage family (De Jonge et al., 2022). There was a similar decline in viral diversity and abundance in obesity and T2D compared to metabolic syndrome, with prominent enrichments in *Escherichia* phage, *Geobacillus* phage, and *Lactobacillus* phage (Yang et al., 2021). Obese children or those with metabolic syndrome had significant expansion of phage diversity (Bikel et al., 2021); however, the opposite trend was observed in NAFLD and advanced NAFLD (characterized by higher liver fibrosis) (Kumar et al., 2019; Lang et al., 2020). The gut virome of patients with T2D was enriched with *Micromonas pusilla* virus, *Cellulophaga* phage, *Bacteroides* phage, and *Halovirus*, with concurrent decreases in 13 viral families (including *Hokovirus*, *Klosneuvirus*, and *Catovirus*) (Yang et al., 2021). In a mouse model of T2D, we reported that enrichment of *Enterobacteriaceae* was linked with impairments in glucose tolerance and insulin sensitivity, as well as accelerated body weight loss—largely through microbiome changes (Kumar, Shukla, et al., 2024; Miller et al., 2021). Interestingly, the gut of patients with T2D is also rich in *Enterobacteriaceae* as well as phages targeting *Enterobacteriaceae* (Chen et al., 2021). These signatures were linked with increased systemic circulation of inflammatory components such as lipopolysaccharide, which has high deposition in *Enterobacteriaceae* and other Gram-negative bacteria (Chen et al., 2021; Kumar, Gangani, et al., 2024). As presented here, gut virome signatures have been characterized in the context of obesity and T2D, two very common metabolic conditions with high prevalence rates globally. Further exploration is necessary to confirm virome structure and function in these metabolic diseases, as well as liver diseases such as NAFLD, so that we can deepen our understanding of virus–host and/or phage–bacteria interactions in these conditions.

CVDs, which are the leading cause of death worldwide, have been less studied in the context of the virome. In individuals with coronary heart disease (CHD), the abundance of *Virgaviridae* was significantly higher, while that of *Microviridae* was lower, with 14 divergent *Microviridae* viruses present (Guo et al., 2017; Gupta et al., 2025).The gut of patients with atherosclerotic CVD was enriched with phages targeting *Enterobacteriaceae* and *Streptococcus*, which comprise several opportunistic pathogens (Jie et al., 2017). Interestingly, the functional landscape of the microbiome in atherosclerotic CVD, characterized by an increased inflammatory tone and decreased fermentative capacity, was similar to that reported in obesity and/or T2D and liver cirrhosis. Although the taxonomy may vary, these data suggest that the functional changes in these cardiometabolic diseases are similar. Hypertension, which is a major risk factor for all cardiometabolic diseases, also is linked with a unique gut virome with biomarker potential. There were two major viral classes in patients with hypertension, with *Cnaphalocrocis medinalis* granulovirus, *Cronobacter* phage CR3, and *Streptococcus* virus phiAbc2 demonstrating the highest biomarker potential (Fujimoto et al., 2022).These integrative studies have reported taxonomic changes in cardiometabolic diseases, with some highlighting how compositional changes impact gut functions. However, conducting more studies will be monumental in developing our understanding of how virome changes pathology in metabolic diseases and CVDs, which currently are the leading causes of mortality worldwide and continue to place health, social, and economic burdens on society.

6.3.4 Infectious Diseases and Cancer

Viruses represent an important fraction of the human virome, influencing disease susceptibility well beyond acute infections, and can be involved in cancer development. A select group of oncogenic viruses has been shown to initiate or promote cancer development through mechanisms such as genome integration, chronic inflammation, and interference with host immune or tumor-suppressor pathways. The most researched examples are hepatitis B virus (HBV), Epstein–Barr virus (EBV), and human papillomavirus (HPV). These viruses have unique oncogenic transformation mechanisms.

- **Human papillomavirus (HPV)** is a primary cause of anogenital and oral cancer. Moreover, it is the etiologic agent of all forms of cervical cancer. HPV produces two key oncoproteins, E6 and E7, which inhibit cellular tumor suppressors. The molecular mechanistic study by Moody and Laimins (2010) reported that the E6 gene is involved in ubiquitin-mediated degradation of the tumor suppressor *p53.* Similarly, the E7 gene inactivates the *retinoblastoma (Rb)* protein, leading to unregulated cell division (Moody & Laimins, 2010)
- Epstein–Barr virus (EBV) is a gammaherpesvirus infecting over 90% of adults. It establishes lifelong latency in B lymphocytes and epithelial cells and is a causative agent of several cancers, including Burkitt lymphoma, Hodgkin lymphoma, nasopharyngeal carcinoma, and gastric carcinoma. Its oncogenic potential stems primarily from latent membrane protein 1 (LMP1), which mimics a constitutively active CD40 receptor, engaging TRAF6 and IRAK1 to trigger nuclear factor kappa B (NF-κB), JAK/STAT, and mitogen-activated protein kinase (MAPK) signaling, which promotes proliferation and inhibits apoptosis. Similarly, Gires et al. demonstrated the dependence of LMP1's receptor-like activity and NF-κB in the pathogenesis of EBV (Cahir-McFarland et al., 2000; Kieser, 1997; Kumar et al., 2023; Luftig et al., 2003). Through these pathways and by modulating host epigenetic and microRNA networks, EBV ensures latent persistence, immune evasion, and conditions conducive to malignant transformation.
- Hepatitis B virus (HBV) and hepatitis C virus (HCV) are principal causes of hepatocellular carcinoma (HCC). HBV integrates segments of its DNA, including the regulatory gene HBx, into hepatocyte genomes, where HBx interferes with p53 signaling, DNA repair, and cell-cycle control (Feitelson et al., 2009; Chen et al., 2009). HCV, while nonintegrating, drives chronic inflammation and oxidative stress and rewires lipid metabolism, in large part via the core and NS5A proteins: core targets lipid droplets to couple assembly and steatosis, and NS5A bridges replication sites to lipid droplets and perturbs Ca^{2+}/cytokine signaling, together promoting metabolic stress and inducing a proinflammatory tone (Bisht et al., 2025; Dionisio et al., 2009; Lindenbach & Rice, 2013; Miyanari et al., 2007; Salloum et al., 2013).

Beyond classical oncogenesis, chronic viral infections can reshape systemic immunity and profoundly influence the gut ecosystem. Viral persistence modifies cytokine signaling, weakens epithelial barriers, and disturbs microbial balance. Enteric and systemic viruses, such as noroviruses, adenoviruses, and bacteriophages, can act as ecological modulators that amplify intestinal permeability and metabolic stress. A pivotal study by Norman et al. (2015) analyzed the fecal virome of patients with IBD and demonstrated that intestinal virome dysbiosis is strongly associated with mucosal inflammation, with expansion of *Caudovirales* bacteriophages correlating with disease activity (Norman et al., 2015). Subsequent clinical metagenomic profiling by Zuo et al. (2019) further showed that disruption of gut viral communities accompanies epithelial injury and altered immune tone in gastrointestinal disorders, indicating that virome imbalance may exacerbate host inflammatory responses (Hildebrand et al., 2019).

Together, these findings establish that perturbations of the enteric virome—whether by chronic infection or host immune dysfunction—can extend beyond the intestine, shaping systemic inflammation and metabolic health. Integrating virome dynamics into models of chronic disease thus provides a more complete picture of how viral, microbial, and immune networks collectively influence human physiology.

6.4 ENVIRONMENTAL AND PUBLIC HEALTH PERSPECTIVES

Studies on the virome extend beyond the gut, and these investigations are integral for (1) characterizing environmental viromes (i.e., in water, soil, and marine ecosystems) and (2) understanding how external factors or viromes impact human health and disease, especially from a public health standpoint. Environmental reservoirs, including water and soil, are important for shaping microbial and ecological functions (i.e., carbon and nutrient cycling) but are also key sources of viral exposure in humans (Wilhelm & Suttle, 1999). Viruses can facilitate microbial evolution through horizontal gene transfer, which is pivotal in evolutionary change and in enhancing antibiotic resistance (Abbasi & Alam, 2025). As a result, expanding our knowledge of virome profiles in natural reservoirs is indispensable for understanding how they influence the environment as well as human disease. This section presents some critical roles of studying viromics from both environmental and public health perspectives.

6.4.1 Environmental Viromics

Advanced techniques such as next-generation sequencing allow us to better assess entire viral communities not only in the human body but also in the surrounding environment, such as oceans, soil, and water. The past few decades of research have revealed high viral diversity in many complex ecosystems, but the basis or source of this profound viral richness remains elusive. In lake and ocean environments, viruses are highly abundant, outnumbering bacterial and archaeal species, with the majority being phages (Weynberg, 2018). pH, salinity, seasonal variations, and the abundances of other microbes, such as bacteria and archaea, shape viral communities in these ecosystems (Zheng et al., 2024). Geography is also a key factor driving these differences in viral abundance and diversity—open ocean water has a distinct viral ecosystem compared to deep-sea (benthic) niches. The lifestyle of viruses—either lytic or lysogenic—is also niche-specific and is important for determining virus–environment interactions.

Recycling of organic matter is a major function carried out by viruses found in ocean and other aquatic ecosystems. Carbon, nitrogen, and phosphorus compounds are recycled during biogeochemical cycles. Viral shunt is a specific process impacting deep-sea ecosystems whereby viruses lyse bacterial (prokaryotic) cells, releasing carbon and other nutrients that serve as food sources for other organisms (Danovaro et al., 2008). Another process, called viral decay, is important for rechanneling viral components into the food web and preventing viral overload in marine ecosystems. It is estimated that benthic viruses (located at the bottom of the ocean) release up to 50 megatons of carbon annually, underscoring the importance of viruses for the production of organic compounds in these environments (Dell'Anno et al., 2015).

Viruses are also critical regulators of genetic diversity and evolutionary change and can instigate disease outbreaks. They can transfer genes or genomes and host fragments and integrate them into other hosts through horizontal gene transfer. One common route of horizontal gene transfer is from phages to bacteria but can also occur between viruses and eukaryotes. For example, viral glycosyltransferases have shown impacts in algae, protozoan mitochondria, and mammalian tissues (Irwin et al., 2021). Because of the potential to transfer genetic components from one reservoir to another, environmental viromics is critical for maintaining public health in drinking water and soils, which can spread infections if not properly managed (Hegarty et al., 2022).

6.4.2 Public Health Applications

Surveillance for and identification of viral pathogens is a common public health application for viromics. A recent example is the identification of SARS-CoV-2, the causative agent of the COVID-19 pandemic. COVID-19 was considered a global public health crisis, resulting in over 1 million deaths in the United States alone and an estimated near 10 million deaths worldwide

(Gebru et al., 2021). Thus, extensive research has been conducted to sequence the virus and understand the mechanisms it uses to enter cells and cause infection. Although the mechanisms of viral entry, hijacking of host machinery, and viral replication remain conserved among different virus species, viral genomes are highly variable (Jackson et al., 2022). As a result, it is important to employ viromics to understand how these viruses cause infection and mutate and to identify virulence factors that contribute to disease pathogenesis. These studies are also important from an epidemiological standpoint so that vaccines can be developed to mitigate severe disease outcomes. Such is the case with the influenza virus—continual surveillance is performed to identify the strains that are most likely to be circulating, and based on these data, a new vaccine is developed each year (Gomez Lorenzo & Fenton, 2013). The COVID-19 vaccine, which implemented novel mRNA technology, was also developed based on extensive viromics studies that used integrative approaches to understand viral pathogenesis and virulence factors. Genomics studies are necessary for determining which viral components contribute to disease pathogenesis (i.e., the spike protein of SARS-CoV-2) and how viral mutants in these genes affect virulence (Li, 2016). Collectively, these approaches are indispensable for keeping the public safe and mitigating high disease burden from viruses found in animals as well as natural reservoirs such as soil and water.

6.5 THERAPEUTIC AND DIAGNOSTIC FRONTIERS

As demonstrated herein, the gut virome has prominent roles in the development of many diseases, including inflammatory and infectious diseases, gastrointestinal conditions, neurological conditions, metabolic dysfunctions, and cancer. The gut virome and phageome both have significant therapeutic potential for combating these ailments and reducing disease burden. Current studies are investigating the safety, feasibility, and overall efficacy of multiple modalities for targeting the human gut virome and phageome, including (1) phage therapy and virome modulation, (2) FVT, and (3) machine learning and artificial intelligence (AI). Although very new, these technologies have reported potential benefits for diagnosing and treating gut virome–mediated diseases.

6.5.1 Phage Therapy and Virome Modulation

Studies determining the efficacy of phage therapy have gained tremendous momentum in recent decades. Phage therapy involves the use of phages to target, infect, and kill specific bacteria, most prominently pathogens or infectious species. Most studies have reported the safety of phage therapy for bacteria that are the causative agents of urinary tract infections (UTIs), cystic fibrosis, joint and orthopedic infections, and gastrointestinal infections. A few studies have even reported the potential of phage therapy for combating multidrug-resistant pathogens. A multistrain phage cocktail targeting uropathogenic *Escherichia coli* significantly reduced *E. coli* in the urine within four hours in uncomplicated UTI and alleviated clinical symptoms (P. Kim et al., 2024). A similar study will implement phage therapy in cystic fibrosis, with the major outcome being a reduction in pathogen (*Pseudomonas aeruginosa*) load (Tamma et al., 2022). Phage therapies have also been implemented in *Clostridioides difficile* infection (CDI) (Quraishi et al., 2017), *Klebsiella pneumoniae*-associated IBD (Federici et al., 2022), and *Enterococcus faecalis*-associated alcoholic liver disease and have attenuated disease progression (Duan et al., 2019). Many of these studies have also implemented FVT, which we will discuss in the next section, to provide proof of concept that viral components are driving these infectious and/or inflammatory phenotypes. Instead of completely depleting microbial communities, as with antibiotics, phage therapy can target pathogens or specific disease-causing microbes and preserve beneficial bacteria (i.e., *Lactobacillus* and *Bifidobacterium*). Phage therapy is also advantageous in that it can target biofilms (which contribute to the persistence of bacterial infections) and even antimicrobial resistance, which is becoming a major global public health concern. However, more studies are warranted to better demonstrate its efficacy in different disease pathologies and to understand how it modulates not only the microbiome but also other key players, such as the metabolome.

6.5.2 Fecal Virome Transplantation

Another approach for modulating the gut microbiome is fecal microbiota transplantation (FMT). During FMT, stool from a healthy donor is transferred to a patient; FMT is a common, safe, and effective treatment for recurrent infections such as CDI or others that are antibiotic resistant (Paaske et al., 2025). FMT has also shown improvement in clinical outcomes and remission in UC (Green et al., 2020). FVT (fecal virome transplantation) has emerged as a plausible therapeutic option based on the principles of FMT; however, it involves the transfer of only viral components from a healthy donor to a patient. FVT has demonstrated promising effects in preclinical models. In the context of CDI, modified FVT has been effective in eliminating pathogen load, with solvent/detergent treatment and chemostatic propagation demonstrating the best effects (Rasmussen et al., 2024). It is important to note that modifying the FVT is imperative for eliminating eukaryotic components or other harmful viral fragments. Administration of eukaryotic virus-depleted FVT in obese mice rescued blood glucose clearance and improved metabolic syndrome, suggesting a role for bacteriophages in metabolic disease progression (Mao et al., 2024). In humans, FVT studies are very limited and include small sample sizes. Sterile fecal filtrate reduced CDI burden in five subjects via bacterial modulation by phages, which persisted for at least 6 months (Ott et al., 2017). Although well tolerated, fecal filtrate demonstrated no beneficial effects in patients with metabolic syndrome compared to those not given FVT (Wortelboer et al., 2023).

Although safety, feasibility, and efficacy studies in humans are extremely limited and remain controversial, clinical trials continue to be designed to provide further evidence of FVT (NCT04960306). One limitation of FVT is the risk of eukaryotic viral infections—thus, FVT applications are focused on depleting eukaryotic viruses from virome therapies or inactivating them prior to administration in the host. Additional studies not only need to focus on the efficacy of FVT but also need to continue validating their long-term safety profiles.

6.5.3 Machine Learning and AI Applications

Although wet-lab approaches such as phage therapy, virome modulation, and FVT are emerging and have significant therapeutic potential, there are limitations. One major limitation is that much of the human gut virome is unculturable, meaning that it cannot be studied using preclinical models such as mice and cell culture. Viruses require host cells to replicate, which can be difficult to recapitulate in the laboratory, and some have unique nutritional needs that cannot be met in a laboratory setting. Therefore, machine learning approaches have emerged, allowing us to combine metagenomic sequencing with bioinformatics to expand viromics studies. Machine learning and AI together allow us to analyze complex metagenomic datasets and perform integrative analyses to link virome signatures with their functions (i.e., through gene ontology or KEGG analysis) or the HUMAnN (HMP Unified Metabolic Analysis Network) database (Beghini et al., 2021). These tools can also be used to predict disease outbreaks linked to viruses, predict the diagnostic capacity of viral biomarkers, and identify certain features, such as antimicrobial genes, that can affect viral functions. Deep machine learning applications such as DeepVirFinder can also enhance the accuracy of identifying many viral contigs (J. Ren et al., 2020). From a public health perspective, machine learning and AI have demonstrated high-throughput and promising efficacy in diagnosing and determining the prognosis of viral infections, such as COVID-19. Machine learning and AI approaches significantly aided in the diagnosis of infection based on imaging and blood tests, as well as prognosis, in several cohorts (Cheng et al., 2025). Random forest, convolutional/deep neural networks, and Bayesian modeling are some emerging algorithms for estimating viral prevalence, disease occurrence, and even host–viral interactions based on sequencing of multiple sample types, including human and environmental samples (Varughese et al., 2018). RNA viruses can be especially difficult to classify using metagenomics approaches due to their high divergence. However, algorithms such as LucaProt have improved viral discovery from several sample types, including air and hot springs (Varughese et al., 2018). The rapid emergence of machine learning and AI has greatly enhanced our ability to identify, classify, and predict viral composition and disease outbreaks, but continued studies are monumental for improving our understanding of viromes in humans, animals, and the environment and for uncovering their interconnections.

6.6 CONCLUSION AND FUTURE PERSPECTIVES

The study of gut viruses living in the digestive system has opened a new window into how microbes and viruses work together to affect human health. Viruses, which were previously believed to be merely background noise in microbiome studies, are now understood to have important functions in regulating bacterial populations, transmitting genes, and affecting how the immune system reacts to environmental stimuli. As sequencing and bioinformatics tools continue to improve, we are beginning to understand how changes in the virome can contribute to diseases such as inflammatory bowel disease, cancer, obesity, and neurological disorders. Future research should focus on *long-term and multi-omics studies*, in which viral data are analyzed together with bacterial, metabolic, and immune data. Such combined and integrative approaches will help move the field from simply observing viral patterns to understanding cause-and-effect relationships in human disease. Technological advances will also make *viral discovery and classification* more accurate. Building better viral databases, improving host-prediction models, and creating standardized protocols for virus isolation will allow scientists to study viruses that are currently unknown or unclassified—the so-called "viral dark matter."

From an applied point of view, the future of virome research is full of possibilities. *Phage therapy,* for example, could offer new ways to treat antibiotic-resistant infections. *FVT* and *virome-based probiotics* might help restore healthy microbial balance in patients with gut or metabolic disorders. AI and machine-learning tools could also make it possible to track viral changes in real time, supporting early diagnosis and personalized treatments.

In summary, integrating virome science into mainstream microbiome and medical research will transform how we understand the connections among viruses, microbes, and humans. The coming years are likely to see the virome move from a hidden component of the microbiome to a central player in precision medicine and human health.

REFERENCES

Abbasi, M., & Alam, M. (2025). Understanding the roles of viruses as key players in environmental dynamics and ecosystem functioning. *Discover Viruses*, *2*(1), 10. https://doi.org/10.1007/s44370-025-00015-y

Beghini, F., McIver, L. J., Blanco-Míguez, A., Dubois, L., Asnicar, F., Maharjan, S., Mailyan, A., Manghi, P., Scholz, M., Thomas, A. M., Valles-Colomer, M., Weingart, G., Zhang, Y., Zolfo, M., Huttenhower, C., Franzosa, E. A., & Segata, N. (2021). Integrating taxonomic, functional, and strain-level profiling of diverse microbial communities with bioBakery 3. *eLife*, *10*, e65088. https://doi.org/10.7554/eLife.65088

Bikel, S., López-Leal, G., Cornejo-Granados, F., Gallardo-Becerra, L., García-López, R., Sánchez, F., Equihua-Medina, E., Ochoa-Romo, J. P., López-Contreras, B. E., Canizales-Quinteros, S., Hernández-Reyna, A., Mendoza-Vargas, A., & Ochoa-Leyva, A. (2021). Gut dsDNA virome shows diversity and richness alterations associated with childhood obesity and metabolic syndrome. *iScience*, *24*(8), 102900. https://doi.org/10.1016/j.isci.2021.102900

Bisht, D., Kumar, V., Gupta, S., & Sachan, M. (2025). Navigating circulating tumour cells and tumour microemboli: Where we stand in bridging their molecular dynamics with clinical translation. *Biochimica et Biophysica Acta (BBA)—Reviews on Cancer*, *1880*(4), 189384. https://doi.org/10.1016/j.bbcan.2025.189384

Borin, J. M., Liu, R., Wang, Y., Wu, T.-C., Chopyk, J., Huang, L., Kuo, P., Ghose, C., Meyer, J. R., Tu, X. M., Schnabl, B., & Pride, D. T. (2023). Fecal virome transplantation is sufficient to alter fecal microbiota and drive lean and obese body phenotypes in mice. *Gut Microbes*, *15*(1), 2236750. https://doi.org/10.1080/19490976.2023.2236750

Cadwell, K., Patel, K. K., Maloney, N. S., Liu, T.-C., Ng, A. C. Y., Storer, C. E., Head, R. D., Xavier, R., Stappenbeck, T. S., & Virgin, H. W. (2010). Virus-plus-susceptibility gene interaction determines Crohn's disease gene Atg16L1 phenotypes in intestine. *Cell*, *141*(7), 1135–1145. https://doi.org/10.1016/j.cell.2010.05.009

Cahir-McFarland, E. D., Davidson, D. M., Schauer, S. L., Duong, J., & Kieff, E. (2000). NF-κB inhibition causes spontaneous apoptosis in Epstein-Barr virus-transformed lymphoblastoid cells. *Proceedings of the National Academy of Sciences*, *97*(11), 6055–6060. https://doi.org/10.1073/pnas.100119497

Cao, Z., Fan, D., Sun, Y., Huang, Z., Li, Y., Su, R., Zhang, F., Li, Q., Yang, H., Zhang, F., Miao, Y., Lan, P., Wu, X., & Zuo, T. (2024). The gut ileal mucosal virome is disturbed in patients with Crohn's disease and exacerbates intestinal inflammation in mice. *Nature Communications*, *15*(1), 1638. https://doi.org/10.1038/s41467-024-45794-y

Cao, Z., Sugimura, N., Burgermeister, E., Ebert, M. P., Zuo, T., & Lan, P. (2022). The gut virome: A new microbiome component in health and disease. *EBioMedicine*, *81*, 104113. https://doi.org/10.1016/j.ebiom.2022.104113

Chen, C. J., Yang, H. I., Iloeje, U. H., & Reveal-HBV Study Group. (2009). Hepatitis B virus DNA levels and outcomes in chronic hepatitis B. *Hepatology*, *49*(S5), S72–S84. https://doi.org/10.1002/hep.22884

Chen, Q., Ma, X., Li, C., Shen, Y., Zhu, W., Zhang, Y., Guo, X., Zhou, J., & Liu, C. (2021). Enteric phageome alterations in patients with type 2 diabetes. *Frontiers in Cellular and Infection Microbiology*, *10*, 575084. https://doi.org/10.3389/fcimb.2020.575084

Cheng, Y., Cheng, R., Xu, T., Tan, X., & Bai, Y. (2025). Machine learning techniques applied to COVID-19 prediction: A systematic literature review. *Bioengineering (Basel, Switzerland)*, *12*(5), 514. https://doi.org/10.3390/bioengineering12050514

Conceição-Neto, N., Zeller, M., Lefrère, H., De Bruyn, P., Beller, L., Deboutte, W., Yinda, C. K., Lavigne, R., Maes, P., Ranst, M. V., Heylen, E., & Matthijnssens, J. (2015). Modular approach to customise sample preparation procedures for viral metagenomics: A reproducible protocol for virome analysis. *Scientific Reports*, *5*(1), 16532. https://doi.org/10.1038/srep16532

Cryan, J. F., O'Riordan, K. J., Cowan, C. S. M., Sandhu, K. V., Bastiaanssen, T. F. S., Boehme, M., Codagnone, M. G., Cussotto, S., Fulling, C., Golubeva, A. V., Guzzetta, K. E., Jaggar, M., Long-Smith, C. M., Lyte, J. M., Martin, J. A., Molinero-Perez, A., Moloney, G., Morelli, E., Morillas, E., . . . Dinan, T. G. (2019). The microbiota-gut-brain axis. *Physiological Reviews*, *99*(4), 1877–2013. https://doi.org/10.1152/physrev.00018.2018

Danovaro, R., Dell'Anno, A., Corinaldesi, C., Magagnini, M., Noble, R., Tamburini, C., & Weinbauer, M. (2008). Major viral impact on the functioning of benthic deep-sea ecosystems. *Nature*, *454*(7208), 1084–1087. https://doi.org/10.1038/nature07268

De Jonge, P. A., Wortelboer, K., Scheithauer, T. P. M., Van Den Born, B.-J. H., Zwinderman, A. H., Nobrega, F. L., Dutilh, B. E., Nieuwdorp, M., & Herrema, H. (2022). Gut virome profiling identifies a widespread bacteriophage family associated with metabolic syndrome. *Nature Communications*, *13*(1), 3594. https://doi.org/10.1038/s41467-022-31390-5

Dell'Anno, A., Corinaldesi, C., & Danovaro, R. (2015). Virus decomposition provides an important contribution to benthic deep-sea ecosystem functioning. *Proceedings of the National Academy of Sciences*, *112*(16). https://doi.org/10.1073/pnas.1422234112

Dionisio, N., Garcia-Mediavilla, M. V., Sanchez-Campos, S., Majano, P. L., Benedicto, I., Rosado, J. A., Salido, G. M., & Gonzalez-Gallego, J. (2009). Hepatitis C virus NS5A and core proteins induce oxidative stress-mediated calcium signalling alterations in hepatocytes. *Journal of Hepatology*, *50*(5), 872–882. https://doi.org/10.1016/j.jhep.2008.12.026

Duan, Y., Llorente, C., Lang, S., Brandl, K., Chu, H., Jiang, L., White, R. C., Clarke, T. H., Nguyen, K., Torralba, M., Shao, Y., Liu, J., Hernandez-Morales, A., Lessor, L., Rahman, I. R., Miyamoto, Y., Ly, M., Gao, B., Sun, W., . . . Schnabl, B. (2019). Bacteriophage targeting of gut bacterium attenuates alcoholic liver disease. *Nature*, *575*(7783), 505–511. https://doi.org/10.1038/s41586-019-1742-x

Fan, Y., & Pedersen, O. (2021). Gut microbiota in human metabolic health and disease. *Nature Reviews Microbiology*, *19*(1), 55–71. https://doi.org/10.1038/s41579-020-0433-9

Federici, S., Kredo-Russo, S., Valdés-Mas, R., Kviatcovsky, D., Weinstock, E., Matiuhin, Y., Silberberg, Y., Atarashi, K., Furuichi, M., Oka, A., Liu, B., Fibelman, M., Weiner, I. N., Khabra, E., Cullin, N., Ben-Yishai, N., Inbar, D., Ben-David, H., Nicenboim, J., . . . Elinav, E. (2022). Targeted suppression of human IBD-associated gut microbiota commensals by phage consortia for treatment of intestinal inflammation. *Cell*, *185*(16), 2879–2898.e24. https://doi.org/10.1016/j.cell.2022.07.003

Feitelson, M. A., Reis, H. M. G. P. V., Lale Tufan, N., Sun, B., Pan, J., & Lian, Z. (2009). Putative roles of hepatitis B x antigen in the pathogenesis of chronic liver disease. *Cancer Letters*, *286*(1), 69–79. https://doi.org/10.1016/j.canlet.2008.12.010

Finn, R. D., Coggill, P., Eberhardt, R. Y., Eddy, S. R., Mistry, J., Mitchell, A. L., Potter, S. C., Punta, M., Qureshi, M., Sangrador-Vegas, A., Salazar, G. A., Tate, J., & Bateman, A. (2016). The Pfam protein families database: Towards a more sustainable future. *Nucleic Acids Research*, *44*(D1), D279–D285. https://doi.org/10.1093/nar/gkv1344

Fujimoto, K., Miyaoka, D., & Uematsu, S. (2022). Characterization of the human gut virome in metabolic and autoimmune diseases. *Inflammation and Regeneration*, *42*(1), 32. https://doi.org/10.1186/s41232-022-00218-6

Galtier, M., De Sordi, L., Sivignon, A., De Vallée, A., Maura, D., Neut, C., Rahmouni, O., Wannerberger, K., Darfeuille-Michaud, A., Desreumaux, P., Barnich, N., & Debarbieux, L. (2017). Bacteriophages targeting adherent invasive *Escherichia coli* strains as a promising new treatment for Crohn's disease. *Journal of Crohn's and Colitis*, jjw224. https://doi.org/10.1093/ecco-jcc/jjw224

Garmaeva, S., Sinha, T., Kurilshikov, A., Fu, J., Wijmenga, C., & Zhernakova, A. (2019). Studying the gut virome in the metagenomic era: Challenges and perspectives. *BMC Biology*, *17*(1), 84. https://doi.org/10.1186/s12915-019-0704-y

Garretto, A., Hatzopoulos, T., & Putonti, C. (2019). virMine: Automated detection of viral sequences from complex metagenomic samples. *PeerJ*, *7*, e6695. https://doi.org/10.7717/peerj.6695

Gebru, A. A., Birhanu, T., Wendimu, E., Ayalew, A. F., Mulat, S., Abasimel, H. Z., Kazemi, A., Tadesse, B. A., Gebru, B. A., Deriba, B. S., Zeleke, N. S., Girma, A. G., Munkhbat, B., Yusuf, Q. K., Luke, A. O., & Hailu, D. (2021). Global burden of COVID-19: Situational analysis and review. *Human Antibodies*, *29*(2), 139–148. https://doi.org/10.3233/HAB-200420

Ghorbani, M., Ferreira, D., & Maioli, S. (2023). A metagenomic study of gut viral markers in amyloid-positive Alzheimer's disease patients. *Alzheimer's Research & Therapy*, *15*(1), 141. https://doi.org/10.1186/s13195-023-01285-8

Gogokhia, L., Buhrke, K., Bell, R., Hoffman, B., Brown, D. G., Hanke-Gogokhia, C., Ajami, N. J., Wong, M. C., Ghazaryan, A., Valentine, J. F., Porter, N., Martens, E., O'Connell, R., Jacob, V., Scherl, E., Crawford, C., Stephens, W. Z., Casjens, S. R., Longman, R. S., & Round, J. L. (2019). Expansion of bacteriophages is linked to aggravated intestinal inflammation and colitis. *Cell Host & Microbe*, *25*(2), 285–299.e8. https://doi.org/10.1016/j.chom.2019.01.008

Gomez Lorenzo, M. M., & Fenton, M. J. (2013). Immunobiology of influenza vaccines. *Chest*, *143*(2), 502–510. https://doi.org/10.1378/chest.12-1711

Grabowski, Ł., Pierzynowska, K., Kosznik-Kwaśnicka, K., Stasiłojć, M., Jerzemowska, G., Węgrzyn, A., Węgrzyn, G., & Podlacha, M. (2023). Sex-dependent differences in behavioral and immunological responses to antibiotic and bacteriophage administration in mice. *Frontiers in Immunology*, *14*, 1133358. https://doi.org/10.3389/fimmu.2023.1133358

Green, J. E., Davis, J. A., Berk, M., Hair, C., Loughman, A., Castle, D., Athan, E., Nierenberg, A. A., Cryan, J. F., Jacka, F., & Marx, W. (2020). Efficacy and safety of fecal microbiota transplantation for the treatment of diseases other than *Clostridium difficile* infection: A systematic review and meta-analysis. *Gut Microbes*, *12*(1), 1854640. https://doi.org/10.1080/19490976.2020.1854640

Gregory, A. C., Zablocki, O., Zayed, A. A., Howell, A., Bolduc, B., & Sullivan, M. B. (2020). The gut virome database reveals age-dependent patterns of virome diversity in the human gut. *Cell Host & Microbe*, *28*(5), 724–740.e8. https://doi.org/10.1016/j.chom.2020.08.003

Guo, L., Hua, X., Zhang, W., Yang, S., Shen, Q., Hu, H., Li, J., Liu, Z., Wang, X., Wang, H., Zhou, C., & Cui, L. (2017). Viral metagenomics analysis of feces from coronary heart disease patients reveals the genetic diversity of the Microviridae. *Virologica Sinica*, *32*(2), 130–138. https://doi.org/10.1007/s12250-016-3896-0

Gupta, A., Pahal, S., Kumar, V., & Chaudhary, A. (2025). Role of monoclonal antibodies in Alzheimer's disease: Successes, failures, and future directions. *Neurological Sciences*. https://doi.org/10.1007/s10072-025-08456-5

Hannigan, G. D., Duhaime, M. B., Ruffin, M. T., Koumpouras, C. C., & Schloss, P. D. (2018). Diagnostic potential and interactive dynamics of the colorectal cancer virome. *mBio*, *9*(6), e02248.18. https://doi.org/10.1128/mBio.02248-18

Hegarty, B., Dai, Z., Raskin, L., Pinto, A., Wigginton, K., & Duhaime, M. (2022). A snapshot of the global drinking water virome: Diversity and metabolic potential vary with residual disinfectant use. *Water Research*, *218*, 118484. https://doi.org/10.1016/j.watres.2022.118484

Hildebrand, F., Moitinho-Silva, L., Blasche, S., Jahn, M. T., Gossmann, T. I., Huerta-Cepas, J., Hercog, R., Luetge, M., Bahram, M., Pryszlak, A., Alves, R. J., Waszak, S. M., Zhu, A., Ye, L., Costea, P. I., Aalvink, S., Belzer, C., Forslund, S. K., Sunagawa, S., . . . Bork, P. (2019). Antibiotics-induced monodominance of a novel gut bacterial order. *Gut*, *68*(10), 1781–1790. https://doi.org/10.1136/gutjnl-2018-317715

Irwin, N. A. T., Pittis, A. A., Richards, T. A., & Keeling, P. J. (2021). Systematic evaluation of horizontal gene transfer between eukaryotes and viruses. *Nature Microbiology*, *7*(2), 327–336. https://doi.org/10.1038/s41564-021-01026-3

Jackson, C. B., Farzan, M., Chen, B., & Choe, H. (2022). Mechanisms of SARS-CoV-2 entry into cells. *Nature Reviews Molecular Cell Biology*, *23*(1), 3–20. https://doi.org/10.1038/s41580-021-00418-x

James, A. S., Adil, N. A., Goltz, D., Tangudu, D., Chaudhari, D. S., Shukla, R., Kumar, V., Kumar, A., Masternak, M. M., Holland, P., Labyak, C., Golden, A., Dangiolo, M., Arikawa, A. Y., Kociolek, J., Fraser, A., Williams, C., Agronin, M., Aymat, M., . . . Yadav, H. (2024). Abnormalities in gut virome signatures linked with cognitive impairment in older adults. *Gut Microbes*, *16*(1), 2431648. https://doi.org/10.1080/19490976.2024.2431648

Jie, Z., Xia, H., Zhong, S.-L., Feng, Q., Li, S., Liang, S., Zhong, H., Liu, Z., Gao, Y., Zhao, H., Zhang, D., Su, Z., Fang, Z., Lan, Z., Li, J., Xiao, L., Li, J., Li, R., Li, X., . . . Kristiansen, K. (2017). The gut microbiome in atherosclerotic cardiovascular disease. *Nature Communications*, *8*(1), 845. https://doi.org/10.1038/s41467-017-00900-1

Kieft, K., Zhou, Z., & Anantharaman, K. (2020). VIBRANT: Automated recovery, annotation and curation of microbial viruses, and evaluation of viral community function from genomic sequences. *Microbiome*, *8*(1), 90. https://doi.org/10.1186/s40168-020-00867-0

Kieser, A. (1997). Epstein-Barr virus latent membrane protein-1 triggers AP-1 activity via the c-Jun N-terminal kinase cascade. *The EMBO Journal*, *16*(21), 6478–6485. https://doi.org/10.1093/emboj/16.21.6478

Kim, M., & Bae, J. (2016). Spatial disturbances in altered mucosal and luminal gut viromes of diet-induced obese mice. *Environmental Microbiology*, *18*(5), 1498–1510. https://doi.org/10.1111/1462-2920.13182

Kim, P., Sanchez, A. M., Penke, T. J. R., Tuson, H. H., Kime, J. C., McKee, R. W., Slone, W. L., Conley, N. R., McMillan, L. J., Prybol, C. J., & Garofolo, P. M. (2024). Safety, pharmacokinetics, and pharmacodynamics of LBP-EC01, a CRISPR-Cas3-enhanced bacteriophage cocktail, in uncomplicated urinary tract infections due to Escherichia coli (ELIMINATE): The randomised, open-label, first part of a two-part phase 2 trial. *The Lancet Infectious Diseases*, *24*(12), 1319–1332. https://doi.org/10.1016/S1473-3099(24)00424-9

Kumar, V., Gangani, S., Shukla, R., Prajapati, S. K., Shekhar, H., & Shukla, V. (2024). Next-generation sequencing and genomic data analysis. In A. Chaudhary, S. K. Sethi, & A. Verma (Eds.), *Unraveling New Frontiers and Advances in Bioinformatics* (pp. 49–76). Springer Nature Singapore. https://doi.org/10.1007/978-981-97-7123-3_3

Kumar, V., Pandey, A., Arora, A., Gautam, P., Bisht, D., Gupta, S., Chaurasia, A., & Sachan, M. (2023). Diagnostics and therapeutic potential of miR-205 and miR-34a in ovarian cancer management: A miRNA-target-based analysis. *DNA and Cell Biology*, *42*(3), 151–162. https://doi.org/10.1089/dna.2022.0487

Kumar, V., Shukla, R., Gangani, S., Shekhar, H., & Shukla, V. (2024). Metagenomics and microbiome analysis. In A. Chaudhary, S. K. Sethi, & A. Verma (Eds.), *Unraveling New Frontiers and Advances in Bioinformatics* (pp. 77–92). Springer Nature Singapore. https://doi.org/10.1007/978-981-97-7123-3_4

Kumar, V., Singh, A., Gautam, P., & Sachan, M. (2019). Techniques/tools to study epigenetic biomarkers in human cancer detection. In S. Paul (Ed.), *Biomedical Engineering and Its Applications in Healthcare* (pp. 327–351). Springer Singapore. https://doi.org/10.1007/978-981-13-3705-5_14

Lang, S., Demir, M., Martin, A., Jiang, L., Zhang, X., Duan, Y., Gao, B., Wisplinghoff, H., Kasper, P., Roderburg, C., Tacke, F., Steffen, H.-M., Goeser, T., Abraldes, J. G., Tu, X. M., Loomba, R., Stärkel, P., Pride, D., Fouts, D. E., & Schnabl, B. (2020). Intestinal virome signature associated with severity of nonalcoholic fatty liver disease. *Gastroenterology*, *159*(5), 1839–1852. https://doi.org/10.1053/j.gastro.2020.07.005

Li, F. (2016). Structure, function, and evolution of coronavirus spike proteins. *Annual Review of Virology*, *3*(1), 237–261. https://doi.org/10.1146/annurev-virology-110615-042301

Lindenbach, B. D., & Rice, C. M. (2013). The ins and outs of hepatitis C virus entry and assembly. *Nature Reviews Microbiology*, *11*(10), 688–700. https://doi.org/10.1038/nrmicro3098

Liu, X., Liu, Y., Liu, J., Zhang, H., Shan, C., Guo, Y., Gong, X., Cui, M., Li, X., & Tang, M. (2024). Correlation between the gut microbiome and neurodegenerative diseases: A review of metagenomics evidence. *Neural Regeneration Research*, *19*(4), 833–845. https://doi.org/10.4103/1673-5374.382223

Luftig, M., Prinarakis, E., Yasui, T., Tsichritzis, T., Cahir-McFarland, E., Inoue, J.-I., Nakano, H., Mak, T. W., Yeh, W.-C., Li, X., Akira, S., Suzuki, N., Suzuki, S., Mosialos, G., & Kieff, E. (2003). Epstein–Barr virus latent membrane protein 1 activation of NF-κB through IRAK1 and TRAF6. *Proceedings of the National Academy of Sciences*, *100*(26), 15595–15600. https://doi.org/10.1073/pnas.2136756100

Majzoub, M. E., Paramsothy, S., Haifer, C., Parthasarathy, R., Borody, T. J., Leong, R. W., Kamm, M. A., & Kaakoush, N. O. (2024). The phageome of patients with ulcerative colitis treated with donor fecal microbiota reveals markers associated with disease remission. *Nature Communications*, *15*(1), 8979. https://doi.org/10.1038/s41467-024-53454-4

Mao, X., Larsen, S. B., Zachariassen, L. S. F., Brunse, A., Adamberg, S., Mejia, J. L. C., Larsen, F., Adamberg, K., Nielsen, D. S., Hansen, A. K., Hansen, C. H. F., & Rasmussen, T. S. (2024). Transfer of modified gut viromes improves symptoms associated with metabolic syndrome in obese male mice. *Nature Communications*, *15*(1), 4704. https://doi.org/10.1038/s41467-024-49152-w

Mayneris-Perxachs, J., Castells-Nobau, A., Arnoriaga-Rodríguez, M., Garre-Olmo, J., Puig, J., Ramos, R., Martínez-Hernández, F., Burokas, A., Coll, C., Moreno-Navarrete, J. M., Zapata-Tona, C., Pedraza, S., Pérez-Brocal, V., Ramió-Torrentà, L., Ricart, W., Moya, A., Martínez-García, M., Maldonado, R., & Fernández-Real, J.-M. (2022). Caudovirales bacteriophages are associated with improved executive function and memory in flies, mice, and humans. *Cell Host & Microbe*, *30*(3), 340–356.e8. https://doi.org/10.1016/j.chom.2022.01.013

Metzger, R., Krug, A., & Eisenächer, K. (2018). Enteric virome sensing—its role in intestinal homeostasis and immunity. *Viruses*, *10*(4), 146. https://doi.org/10.3390/v10040146

Miller, B., Mainali, R., Nagpal, R., & Yadav, H. (2021). A newly developed synbiotic yogurt prevents diabetes by improving the microbiome–intestine–pancreas axis. *International Journal of Molecular Sciences*, *22*(4), 1647. https://doi.org/10.3390/ijms22041647

Miller, B., Mishra, S. P., Jain, S., Marotta, F., & Yadav, H. (2023). Gut microbiota and eating disorders on the extremes of aging. In *Gut Microbiota in Aging and Chronic Diseases* (pp. 99–127). Springer. https://doi.org/10.1007/978-3-031-14023-5_6

Minot, S., Bryson, A., Chehoud, C., Wu, G. D., Lewis, J. D., & Bushman, F. D. (2013). Rapid evolution of the human gut virome. *Proceedings of the National Academy of Sciences*, *110*(30), 12450–12455. https://doi.org/10.1073/pnas.1300833110

Mishra, S. P., Jain, S., Wang, B., Wang, S., Miller, B. C., Lee, J. Y., Borlongan, C. V., Jiang, L., Pollak, J., Taraphder, S., Layden, B. T., Rane, S. G., & Yadav, H. (2024). Abnormalities in microbiota/butyrate/FFAR3 signaling in aging gut impair brain function. *JCI Insight*, *9*(3), e168443. https://doi.org/10.1172/jci.insight.168443

Miyanari, Y., Atsuzawa, K., Usuda, N., Watashi, K., Hishiki, T., Zayas, M., Bartenschlager, R., Wakita, T., Hijikata, M., & Shimotohno, K. (2007). The lipid droplet is an important organelle for hepatitis C virus production. *Nature Cell Biology*, *9*(9), 1089–1097. https://doi.org/10.1038/ncb1631

Moody, C. A., & Laimins, L. A. (2010). Human papillomavirus oncoproteins: Pathways to transformation. *Nature Reviews Cancer*, *10*(8), 550–560. https://doi.org/10.1038/nrc2886

Nayfach, S., Camargo, A. P., Schulz, F., Eloe-Fadrosh, E., Roux, S., & Kyrpides, N. C. (2021). CheckV assesses the quality and completeness of metagenome-assembled viral genomes. *Nature Biotechnology*, *39*(5), 578–585. https://doi.org/10.1038/s41587-020-00774-7

Norman, J. M., Handley, S. A., Baldridge, M. T., Droit, L., Liu, C. Y., Keller, B. C., Kambal, A., Monaco, C. L., Zhao, G., Fleshner, P., Stappenbeck, T. S., McGovern, D. P. B., Keshavarzian, A., Mutlu, E. A., Sauk, J., Gevers, D., Xavier, R. J., Wang, D., Parkes, M., & Virgin, H. W. (2015). Disease-specific alterations in the enteric virome in inflammatory bowel disease. *Cell*, *160*(3), 447–460. https://doi.org/10.1016/j.cell.2015.01.002

Ott, S. J., Waetzig, G. H., Rehman, A., Moltzau-Anderson, J., Bharti, R., Grasis, J. A., Cassidy, L., Tholey, A., Fickenscher, H., Seegert, D., Rosenstiel, P., & Schreiber, S. (2017). Efficacy of sterile fecal filtrate transfer for treating patients with clostridium difficile infection. *Gastroenterology*, *152*(4), 799–811.e7. https://doi.org/10.1053/j.gastro.2016.11.010

Paaske, S. E., Baumwall, S. M. D., Rubak, T., Birn, F. H., Rågård, N., Kelsen, J., Hansen, M. M., Svenningsen, L., Krarup, A. L., Fernis, C. M. C., Neumann, A., Lødrup, A. B., Glerup, H., Vinter-Jensen, L., Helms, M., Erikstrup, L. T., Grosen, A. K., Mikkelsen, S., Erikstrup, C., . . . Hvas, C. L. (2025). Real-world effectiveness of fecal microbiota transplantation for first or second clostridioides difficile infection. *Clinical Gastroenterology and Hepatology*, *23*(4), 602–611.e8. https://doi.org/10.1016/j.cgh.2024.05.038

Paez-Espino, D., Chen, I.-M. A., Palaniappan, K., Ratner, A., Chu, K., Szeto, E., Pillay, M., Huang, J., Markowitz, V. M., Nielsen, T., Huntemann, M., K. Reddy, T. B., Pavlopoulos, G. A., Sullivan, M. B., Campbell, B. J., Chen, F., McMahon, K., Hallam, S. J., Denef, V., . . . Kyrpides, N. C. (2016). IMG/VR: A database of cultured and uncultured DNA viruses and retroviruses. *Nucleic Acids Research*, *45*(D1), gkw1030. https://doi.org/10.1093/nar/gkw1030

Quick, J., Loman, N. J., Duraffour, S., Simpson, J. T., Severi, E., Cowley, L., Bore, J. A., Koundouno, R., Dudas, G., Mikhail, A., Ouédraogo, N., Afrough, B., Bah, A., Baum, J. H. J., Becker-Ziaja, B., Boettcher, J. P., Cabeza-Cabrerizo, M., Camino-Sánchez, Á., Carter, L. L., . . . Carroll, M. W. (2016). Real-time, portable genome sequencing for Ebola surveillance. *Nature*, *530*(7589), 228–232. https://doi.org/10.1038/nature16996

Quraishi, M. N., Widlak, M., Bhala, N., Moore, D., Price, M., Sharma, N., & Iqbal, T. H. (2017). Systematic review with meta-analysis: The efficacy of faecal microbiota transplantation for the treatment of recurrent and refractory *Clostridium difficile* infection. *Alimentary Pharmacology & Therapeutics*, *46*(5), 479–493. https://doi.org/10.1111/apt.14201

Rasmussen, T. S., Mao, X., Forster, S., Larsen, S. B., Von Münchow, A., Tranæs, K. D., Brunse, A., Larsen, F., Mejia, J. L. C., Adamberg, S., Hansen, A. K., Adamberg, K., Hansen, C. H. F., & Nielsen, D. S. (2024). Overcoming donor variability and risks associated with fecal microbiota transplants through bacteriophage-mediated treatments. *Microbiome*, *12*(1), 119. https://doi.org/10.1186/s40168-024-01820-1

Rasmussen, T. S., Mentzel, C. M. J., Kot, W., Castro-Mejía, J. L., Zuffa, S., Swann, J. R., Hansen, L. H., Vogensen, F. K., Hansen, A. K., & Nielsen, D. S. (2020). Faecal virome transplantation decreases symptoms of type 2 diabetes and obesity in a murine model. *Gut*, *69*(12), 2122–2130. https://doi.org/10.1136/gutjnl-2019-320005

Ren, J., Ahlgren, N. A., Lu, Y. Y., Fuhrman, J. A., & Sun, F. (2017). VirFinder: A novel k-mer based tool for identifying viral sequences from assembled metagenomic data. *Microbiome*, *5*(1), 69. https://doi.org/10.1186/s40168-017-0283-5

Ren, J., Song, K., Deng, C., Ahlgren, N. A., Fuhrman, J. A., Li, Y., Xie, X., Poplin, R., & Sun, F. (2020). Identifying viruses from metagenomic data using deep learning. *Quantitative Biology (Beijing, China)*, *8*(1), 64–77. https://doi.org/10.1007/s40484-019-0187-4

Ren, Y., Zhang, P., Yu, H., Li, S., & Jiang, H. (2025). Metagenome-based characterization of the gut virome in patients with schizophrenia. *Journal of Translational Medicine*, *23*(1), 895. https://doi.org/10.1186/s12967-025-06923-3

Ritz, N. L., Draper, L. A., Bastiaanssen, T. F. S., Turkington, C. J. R., Peterson, V. L., Van De Wouw, M., Vlckova, K., Fülling, C., Guzzetta, K. E., Burokas, A., Harris, H., Dalmasso, M., Crispie, F., Cotter, P. D., Shkoporov, A. N., Moloney, G. M., Dinan, T. G., Hill, C., & Cryan, J. F. (2024). The gut virome is associated with stress-induced changes in behaviour and immune responses in mice. *Nature Microbiology*, *9*(2), 359–376. https://doi.org/10.1038/s41564-023-01564-y

Roux, S., Enault, F., Hurwitz, B. L., & Sullivan, M. B. (2015). VirSorter: Mining viral signal from microbial genomic data. *PeerJ*, *3*, e985. https://doi.org/10.7717/peerj.985

Salloum, S., Wang, H., Ferguson, C., Parton, R. G., & Tai, A. W. (2013). Rab18 binds to hepatitis C virus NS5A and promotes interaction between sites of viral replication and lipid droplets. *PLoS Pathogens*, *9*(8), e1003513. https://doi.org/10.1371/journal.ppat.1003513

Saroj, D. B., Ahire, J. J., & Shukla, R. (2023). Genetic and phenotypic assessments for the safety of probiotic Bacillus clausii 088AE. *3 Biotech*, *13*(7), 238. https://doi.org/10.1007/s13205-023-03662-4

Schulfer, A., Santiago-Rodriguez, T. M., Ly, M., Borin, J. M., Chopyk, J., Blaser, M. J., & Pride, D. T. (2020). Fecal viral community responses to high-fat diet in mice. *mSphere*, *5*(1), e00833.19. https://doi.org/10.1128/mSphere.00833-19

Seemann, T. (2014). Prokka: Rapid prokaryotic genome annotation. *Bioinformatics*, *30*(14), 2068–2069. https://doi.org/10.1093/bioinformatics/btu153

Tamma, P. D., Souli, M., Billard, M., Campbell, J., Conrad, D., Ellison, D. W., Evans, B., Evans, S. R., Greenwood-Quaintance, K. E., Filippov, A. A., Geres, H. S., Hamasaki, T., Komarow, L., Nikolich, M. P., Lodise, T. P., Nayak, S. U., Norice-Tra, C., Patel, R., Pride, D., . . . For the Antibacterial Resistance Leadership Group. (2022). Safety and microbiological activity of phage therapy in persons with cystic fibrosis colonized with pseudomonas aeruginosa: Study protocol for a phase 1b/2, multicenter, randomized, double-blind, placebo-controlled trial. *Trials*, *23*(1), 1057. https://doi.org/10.1186/s13063-022-07047-5

Thurber, R. V., Haynes, M., Breitbart, M., Wegley, L., & Rohwer, F. (2009). Laboratory procedures to generate viral metagenomes. *Nature Protocols*, *4*(4), 470–483. https://doi.org/10.1038/nprot.2009.10

Tian, X., Li, S., Wang, C., Zhang, Y., Feng, X., Yan, Q., Guo, R., Wu, F., Wu, C., Wang, Y., Huo, X., & Ma, X. (2024). Gut virome-wide association analysis identifies cross-population viral signatures for inflammatory bowel disease. *Microbiome*, *12*(1), 130. https://doi.org/10.1186/s40168-024-01832-x

Ungaro, F., Massimino, L., Furfaro, F., Rimoldi, V., Peyrin-Biroulet, L., D'Alessio, S., & Danese, S. (2019). Metagenomic analysis of intestinal mucosa revealed a specific eukaryotic gut virome signature in early-diagnosed inflammatory bowel disease. *Gut Microbes*, *10*(2), 149–158. https://doi.org/10.1080/19490976.2018.1511664

Varughese, E. A., Brinkman, N. E., Anneken, E. M., Cashdollar, J. L., Fout, G. S., Furlong, E. T., Kolpin, D. W., Glassmeyer, S. T., & Keely, S. P. (2018). Estimating virus occurrence using Bayesian modeling in multiple drinking water systems of the United States. *Science of The Total Environment*, *619–620*, 1330–1339. https://doi.org/10.1016/j.scitotenv.2017.10.267

Warwick-Dugdale, J., Solonenko, N., Moore, K., Chittick, L., Gregory, A. C., Allen, M. J., Sullivan, M. B., & Temperton, B. (2019). Long-read viral metagenomics captures abundant and microdiverse viral populations and their niche-defining genomic islands. *PeerJ*, *7*, e6800. https://doi.org/10.7717/peerj.6800

Weynberg, K. D. (2018). Viruses in marine ecosystems: From open waters to coral reefs. In *Advances in Virus Research* (Vol. 101, pp. 1–38). Elsevier. https://doi.org/10.1016/bs.aivir.2018.02.001

Wilhelm, S. W., & Suttle, C. A. (1999). Viruses and nutrient cycles in the sea. *BioScience*, *49*(10), 781–788. https://doi.org/10.2307/1313569

Wortelboer, K., De Jonge, P. A., Scheithauer, T. P. M., Attaye, I., Kemper, E. M., Nieuwdorp, M., & Herrema, H. (2023). Phage-microbe dynamics after sterile faecal filtrate transplantation in individuals with metabolic syndrome: A double-blind, randomised, placebo-controlled clinical trial assessing efficacy and safety. *Nature Communications*, *14*(1), 5600. https://doi.org/10.1038/s41467-023-41329-z

Wu, Y., Cheng, R., Lin, H., Li, L., Jia, Y., Philips, A., Zuo, T., & Zhang, H. (2025). Gut virome and its implications in the pathogenesis and therapeutics of inflammatory bowel disease. *BMC Medicine*, *23*(1), 183. https://doi.org/10.1186/s12916-025-04016-y

Wu, Y., Tao, S., Xiao, L., Zhang, J., Tang, Y., Zhang, M., Liu, S., Huang, Y., Liu, Y., Xie, M., Zhao, Z., Lv, Q., Cai, J., Pei, K., Ma, Q., Yin, Y., Dai, M., Wei, M., Chen, Y., & Wang, Q. (2025). Gut virome alterations in schizophrenia: Identifying viral biomarkers associated with schizophrenia and treatment response. *Brain, Behavior, and Immunity*, *130*, 106080. https://doi.org/10.1016/j.bbi.2025.106080

Yadav, H., Jain, S., Nagpal, R., & Marotta, F. (2016). Increased fecal viral content associated with obesity in mice. *World Journal of Diabetes*, *7*(15), 316. https://doi.org/10.4239/wjd.v7.i15.316

Yang, K., Niu, J., Zuo, T., Sun, Y., Xu, Z., Tang, W., Liu, Q., Zhang, J., Ng, E. K. W., Wong, S. K. H., Yeoh, Y. K., Chan, P. K. S., Chan, F. K. L., Miao, Y., & Ng, S. C. (2021). Alterations in the gut virome in obesity and type 2 diabetes mellitus. *Gastroenterology*, *161*(4), 1257–1269.e13. https://doi.org/10.1053/j.gastro.2021.06.056

Yolken, R. H., Kinnunen, P. M., Vapalahti, O., Dickerson, F., Suvisaari, J., Chen, O., & Sabunciyan, S. (2021). Studying the virome in psychiatric disease. *Schizophrenia Research*, *234*, 78–86. https://doi.org/10.1016/j.schres.2021.04.006

Zeng, S., Almeida, A., Li, S., Ying, J., Wang, H., Qu, Y., Paul Ross, R., Stanton, C., Zhou, Z., Niu, X., Mu, D., & Wang, S. (2024). A metagenomic catalog of the early-life human gut virome. *Nature Communications*, *15*(1), 1864. https://doi.org/10.1038/s41467-024-45793-z

Zerbini, F. M., Crane, A., Kuhn, J. H., Simmonds, P., Lefkowitz, E. J., & ICTV Taxonomy Summary Consortium. (2025). Summary of taxonomy changes ratified by the international committee on taxonomy of viruses (ICTV)—general taxonomy proposals, 2025. *Journal of General Virology*, *106*(7). https://doi.org/10.1099/jgv.0.002116

Zheng, Y., Gao, Z., Wu, S., & Ruan, A. (2024). Community structure, drivers, and potential functions of different lifestyle viruses in Chaohu lake. *Viruses*, *16*(4), 590. https://doi.org/10.3390/v16040590

Zolfo, M., Pinto, F., Asnicar, F., Manghi, P., Tett, A., Bushman, F. D., & Segata, N. (2019). Detecting contamination in viromes using ViromeQC. *Nature Biotechnology*, *37*(12), 1408–1412. https://doi.org/10.1038/s41587-019-0334-5

Zuo, T., Lu, X.-J., Zhang, Y., Cheung, C. P., Lam, S., Zhang, F., Tang, W., Ching, J. Y. L., Zhao, R., Chan, P. K. S., Sung, J. J. Y., Yu, J., Chan, F. K. L., Cao, Q., Sheng, J.-Q., & Ng, S. C. (2019). Gut mucosal virome alterations in ulcerative colitis. *Gut*, *68*(7), 1169–1179. https://doi.org/10.1136/gutjnl-2018-318131

7 Environmental Metagenomics

Rajan Kumar Singh[1*], *Mistry Hetkumar Bhavesh*[1], *Mansi Gautam*[2], *Sirsat Hrudya Sonalkumar*[1], *and Aditya Sanjay Lakade*[1]

[1] Department of Biotechnology, Parul Institute of Technology, Faculty of Engineering and Technology, Parul University, India.

[2] Department of Biomedical Engineering, Galgotia University, Greater Noida, Uttar Pradesh, India.

*Corresponding author

7.1 INTRODUCTION

Metagenomics is the study of microbial genetic material taken directly from different environments, including ocean, soil, air, thermal vents, and extreme environment samples. This process does not involve in culturing samples in the lab. Metagenomics is frequently being employed in diagnosis of diseases such breast cancer, colorectal cancer, ovarian cancer, and age-related diseases (Madigan et al., 2025; Nair et al., 2024; Prajapati et al., 2024; Mishra et al., 2024; Sarfraz et al., 2024; James et al., 2024). Now, metagenomics is being employed in different kind of environmental samples. The environmental metagenomics method allows understanding of the diversity, structure, and function role of the microbiome present in that habitat (Lema et al., 2023; Mandal et al., 2021). Moreover, it also helps in taxonomic profiling and metabolic pathway analysis, linking microbial communities to significant ecological processes. Recent advancements in hybrid assembly, binning algorithms, and AI-assisted genome refinement can recover high-quality metagenome-assembled genomes (MAGs). That allows greater completeness and very low contamination (Qayyum et al., 2025; Yazhini et al., 2025; Belmann et al., 2025).

The idea of culture-free identification, that is, metagenomics, started advancing in the late 1990s. Initial studies involved Sanger sequencing and simple marker-gene studies. During the initial phase, unseen microbes were identified and called "microbial dark matter." A few years later, the Global Ocean Sampling Expedition project implemented the shotgun sequencing method that revealed the variety of microbes in the ocean. However, at that time, short reads often left the genomes broken into many small pieces. A significant advancement occurred with the use of long-read platforms, such as PacBio HiFi and Oxford Nanopore, along with high-throughput Illumina short reads. The shortcoming of other method was overcome by hybrid sequencing, which highlighted complex areas such as repetitive elements and mobile genetic elements (Eisenhofer et al., 2024; Lang et al., 2020; L. Liu et al., 2022). Recent advancements have refined binning software and AI-based refined metagenomics tools. This refinement helps in rebuilding biosynthetic gene clusters, plasmids, and prophages with far greater accuracy. This has made it easier to understand how microbes exchange genes and adapt to the environments around them (Nayfach et al., 2021; S. Liu et al., 2022).

Over time, environmental metagenomics began to look far beyond just soil and surface water. Several other areas, such as deep-sea hydrothermal vents and polar ice fields to dry deserts and high-altitude cloud zones, are being discovered (Kuang et al., 2023; Cavicchioli et al., 2021; Kumar et al., 2025). This reveals the dynamics of environmental metagenomics that let us understand

DOI: 10.1201/9781003670155-7

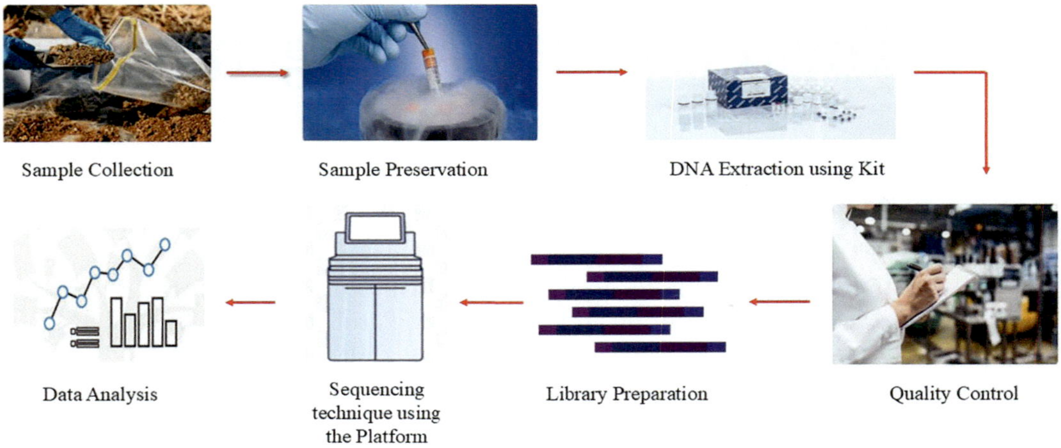

FIGURE 7.1 Environmental Metagenomics Workflow.

the key biogeochemical cycles such as carbon fixation, nitrogen and sulfur transformations, and phosphorus regulation. Moreover, it also allows comparative analysis of microbial communities from different ecosystems, revealing functional and taxonomic patterns across these diverse environments (Nayfach et al., 2021). The complete process involved in metagenomics is depicted in Figure 7.1.

Metagenomics has an emerging role in industrial and scientific research. Researchers are understanding the impact of various microorganisms on greenhouse gas emissions, letting us understand climate change and its management (Cavicchioli et al., 2021). Real-time monitoring of pathogens or microbes is also possible using portable sequencing. This process expedites monitoring and intervention (De Brauwer et al., 2022; Nwachukwu & Babalola, 2022; S. Liu et al., 2022; Belmann et al., 2025). The upcoming subsections of this chapter will deal with different aspects of environmental metagenomics and let us understand the process and hurdles associated with it.

7.2 TYPES OF ENVIRONMENTAL SAMPLE COLLECTION AND PRESERVATION STRATEGIES

Environmental samples are varied to represent a broad range of environments (soil, sediment, freshwater, marine water, atmospheric aerosols, and extreme habitats such as permafrost, hypersaline lakes, deep-sea hydrothermal vents, and acidic hot springs), depending upon potential zones that have microbial diversity (Cavicchioli et al., 2021; Kuang et al., 2023). Selection of sites is based on environmental gradients (temperature, salinity, pH, and redox potential). In addition, regulatory paperwork, such as logistical access, ethical clearances, and permits for sensitive or protected sites, must be obtained (De Brauwer et al., 2022; Nayfach et al., 2021).

Collection methods must be adapted to limit contamination and bias. Different environmental samples have different methods of collection. For soil, the topsoil (0–10 cm) is carefully drilled using sterile corers and collected in sterile containers. Freshwater and marine samples should be filtered on-site through a 0.22 μm membrane filter, and concentrated cells are preserved in a buffer (Sunagawa et al., 2015). In the case of atmospheric aerosols, samples are collected using high-volume air samplers fitted with quartz filters and subsequently placed in aseptic Petri dishes. During collection, all tools and containers should be UV-sterilized or autoclaved to prevent cross-sample contamination (Brill & Stillman, 1989). To prevent decay and community alteration, it is essential to conserve the integrity of nucleic acids at the point of collection. Common protocols include freezing on ice or dry ice, followed by cryopreservation at −80 °C, or immersion in liquid nitrogen and stabilization using DNA preservative buffers for field-based holding (De Brauwer et al., 2022). Most importantly, gentle handling of samples, avoiding aggressive bead beating, and reducing the number of freeze–thaw cycles are basic requirements to maintain DNA quality for further steps (Liu et al., 2025; Klingström et al., 2018).

Before DNA extraction, samples should remain at 4 °C; otherwise, they should be transported on dry ice to ensure sample integrity. Similarly, filtered marine samples should be sealed in sterile cryovials and placed directly into dry ice storage until shipping (Sunagawa et al., 2020). Cryopreservation is used for storing long-term or when extraction cannot be done instantly, ultra-low freezing maintains DNA quality for months to years. Samples should be preserved at −80 °C or in liquid nitrogen (−196 °C) immediately to prevent enzymatic degradation and shifts in community compositions (Eisenhofer et al., 2024; L. Liu et al., 2022). Additionally, samples in 95% EtOH, or 20–25% DMSO-EDTA that had been stored at 4°C for up to six hours displayed little shearing when subjected to gentle lysis upon thawing (Dahn et al., 2022).

In cases where samples cannot be maintained at low temperatures, samples should be preserved with nuclease-inhibiting media (e.g., DNA Shield or alcohol-based fixatives), which also stabilize the DNA and prevent its denaturation (De Brauwer et al., 2022). These reagents allow storage at room temperature for days, facilitating sampling campaigns in remote or difficult-to-access regions. For example, airborne microbial samples collected into liquid media can be directly processed with preservation buffers to maintain the integrity of nucleic acids and the viability of microbes, making downstream molecular analysis much more dependable (Hong et al., 2021).

Field sampling for metagenomics encounters contamination and heterogeneity problems that require rigorous quality control. Field and extraction blanks and reagent controls help identify carryover, whereas bleach or UV decontamination with single-use consumables limits background noise (De Brauwer et al., 2022). Rich metadata, including GPS, time, pH, and weather, concatenated with harmonized SOPs ensure that data are comparable across studies and can be integrated at a global scale (De Brauwer et al., 2022; Nayfach et al., 2021). Recently developed field-deployable workflows, such as portable extraction and sequencing kits, also support immediate investigation on-site, leading to adaptive sampling and shortened turnaround times in the event of outbreaks or ecological transitions (Russell et al., 2025).

7.3 DNA EXTRACTION FROM ENVIRONMENTAL SAMPLES

Extraction of quality nucleic acids from environmental samples is a multi-step process. Some of the most important considerations when carrying out DNA extraction are strong lysis methods, effective removal of inhibitors, maintenance of nucleic acid integrity, and application of strict contamination controls over (Lema et al., 2023). These are key variables that directly influence yield, purity, and representativeness of metagenomic datasets (Kuang et al., 2023; Eisenhofer et al., 2024; Kumar, Gangani et al., 2024). Several commercial kits are available for DNA isolation. These kits are based on two different cell lysis methods: chemical lysis and mechanical shearing or bead beating. The choice of kit is based on individual discretion (Lever et al., 2015).

In brief, bead-based lysis involves small beads; approximately 400 μL of sample is mixed with lysis buffer and vortexed at 3000 rpm for 15 minutes. This process ensures effective rupture of cells and release of DNA into solution. Further, samples are transferred to columns and spun at 6000 rpm for 1 minute, after which samples are washed and eluted. Several bead-based kits are available from companies such as Qiagen, Zymo Research, and Thermo Fisher (Eisenhofer et al., 2024).

Chemical lysis methods involve the use of detergents such as sodium dodecyl sulfate (SDS) for disrupting membranes and denaturing proteins, polysaccharide solubilizers such as CTAB for polysaccharide-rich samples like plant-associated soils, and chaotropic salts. Guanidinium thiocyanate is used for denaturing proteins and inactivating nucleases, which enables simultaneous DNA extraction (Lema et al., 2023). Combining these chemical agents with mechanical lysis further enhances cell disruption, thus improving nucleic acid recovery from diverse microbial consortia. In some protocols, lysozyme (10–20 mg/mL) treatment for 30–60 minutes at 37 °C is also employed (Yuan et al., 2012). In parallel, proteinase K, a broad-spectrum protease, is commonly used at 0.1–0.4 mg/mL and incubated at 55–65 °C for 30–60 minutes to promote cellular lysis (Lema et al., 2023). These enzymatic treatments, alongside mechanical disruption, improve nucleic acid recovery from otherwise resistant microbial taxa (Mandal et al., 2021).

7.3.1 Quality Control and Library Preparation

After DNA isolation, quality control is the foundation for achieving successful metagenomic sequencing. Several quality control strategies are employed, including NanoDrop, Qubit, and TapeStation, for prepared libraries (Mandal et al., 2021). Once DNA concentration is obtained, commercially available library preparation kits are used. Several library preparation methods are available from companies such as Illumina, Swift Bioscience, Takara, and Zymo. The choice of kit depends on the research objective, whether short-read or long-read sequencing, as well as DNA yield and sample type (Goller & Sjogren, 2023).

The key process of library preparation involves fragmentation of genomic DNA into appropriately sized fragments (e.g., 300–600 bp), followed by end repair and A-tailing. Synthetic adapters containing platform-specific sequences are ligated to DNA ends, enabling recognition and amplification during sequencing. Size selection and cleanup processes are then performed to enrich fragments of the desired length. Depending on the protocol, amplification may be performed to increase library yield, although PCR-free methods are preferred in metagenomics (Eisenhofer et al., 2024). Finally, prepared libraries undergo quality assessment, including quantification and fragment size distribution analysis, to ensure optimal sequencing performance. The choice of library preparation strategy, whether for shotgun metagenomics or amplicon sequencing, is dictated by research objectives and sample complexity. After library preparation, QC is performed, and libraries are pooled in equimolar concentrations. They are then loaded onto sequencing platforms for data generation. The process involved in library preparation is depicted in Figure 7.2.

7.4 METAGENOMIC SEQUENCING PLATFORM AND BIOINFORMATICS ANALYSIS

7.4.1 Sequencing Platforms

The choice of sequencing platform is dependent upon the research goal, read length, accuracy,

cost, and computational resources. Illumina platforms such as MiSeq and NovaSeq are the most widely used platforms. Similarly, Oxford Nanopore technology is used for long reads and real-time analysis. This technology offers genome assembly, strain-level resolution, and structural variation studies. However, the error rate in Oxford Nanopore technology is high. Another platform is PacBio HiFi sequencing; it also generates long reads and provides higher accuracy. Moreover, it enables identification of complex microbial communities and complete genome reconstruction. Another system is Ion Torrent; however, it is widely used for targeted genomics. Overall, the choice of platform is dependent on sequencing depth, resolution, accuracy, read length, and, most importantly, cost (L. Liu et al., 2022; Eisenhofer et al., 2024; Lang et al., 2020). The comparison of different platforms is depicted in Table 7.1.

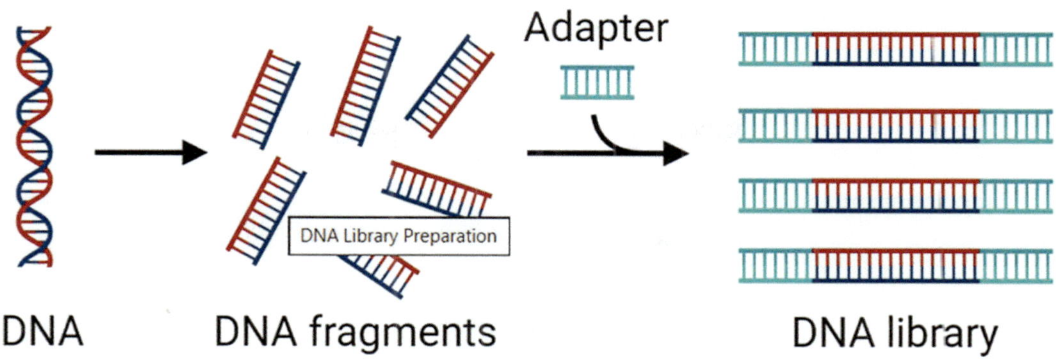

FIGURE 7.2 Workflow illustrating the process of library preparation.

TABLE 7.1
Various Metagenomics Approaches: Advantages, Limitations, and Uses

Approach	Advantages	Limitations	Uses	References
Shotgun Metagenomics	Comprehensive taxonomic + functional profiling; detects novel genes; supports MAG recovery; captures viruses & eukaryotes	High cost; computationally intensive; incomplete recovery of rare taxa	Characterize microbial diversity, Recover functional potential, Detect antimicrobial resistance (AMR) genes	Lema et al. (2023); Qayyum et al. (2025)
Amplicon Sequencing	Low cost; simple analysis; effective for comparing community composition	Primer bias; no functional resolution; limited taxonomic depth	Identification of microbial taxa, Phylogenetic analysis	Mandal et al. (2021)
Illumina	High accuracy; cost-efficient for large sets; strong for variant	Short reads fragment assemblies	Whole-genome sequencing, Epigenomics	Lema et al. (2023)
PacBio HiFi	Long, accurate reads; resolves repeats; enables complete	Higher cost; lower throughput	High-quality de novo genome assembly, Resolving complex genomic regions	Eisenhofer et al. (2024)
Oxford Nanopore	Ultra-long reads; portable; real-time sequencing	Higher raw error rates; requires high-quality DNA	Real-time sequencing, Transcriptomics	L. Liu et al. (2022)

7.4.2 Bioinformatics Pipeline

Bioinformatic workflows must be reproducible, modular, and well documented. A typical pipeline of metagenomic analysis involves retaining high-quality reads and removing adapters and low-quality reads using FastQC, MultiQC, and fastp (Lema et al., 2023). Next, Bowtie2 is used to filter host or contaminant sequences, which are aligned to reference genomes (Eisenhofer et al., 2024).

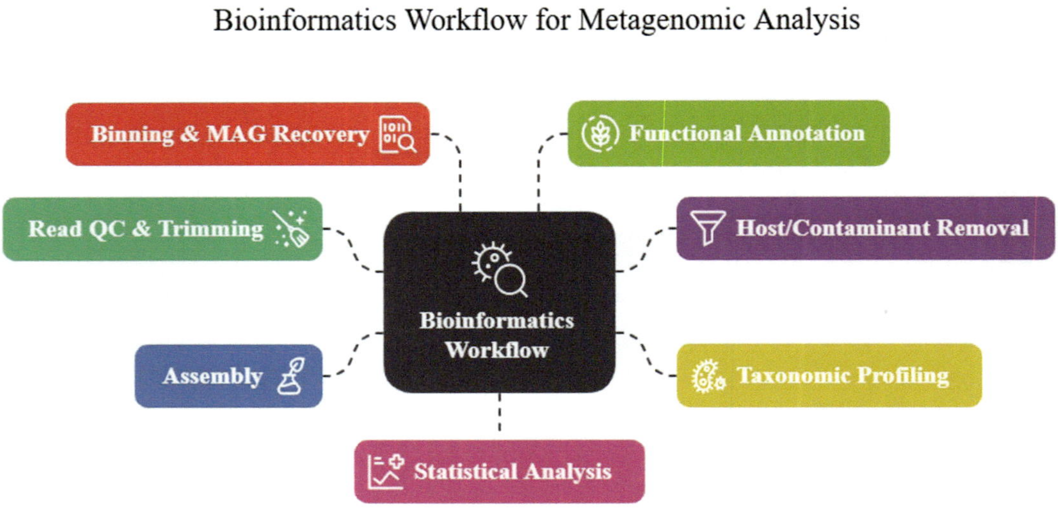

FIGURE 7.3 Metagenomics data analysis framework integrated with various steps.

Further, taxonomic profiling is performed using Kraken2 + Bracken or MetaPhlAn3. Both tools use curated databases for accurate classification (Nayfach et al., 2021). For deeper genomic insight, MEGAHIT or metaSPAdes are used for assembly preparation of short reads, and Flye or Canu are used for long reads (Lang et al., 2020; Eisenhofer et al., 2024).

Binning and MAG recovery follow, using MetaBAT2, MaxBin2, and CONCOCT to group contigs, while CheckM assesses genome quality, and GTDB-Tk assigns taxonomy (Eisenhofer et al., 2024). In the case of functional annotation, Prodigal is used for gene prediction, and EggNOG-mapper or DRAM is used for annotation. Similarly, antiSMASH is used for biosynthetic clusters, while CARD is used for resistance genes (Lema et al., 2023). Finally, phyloseq and vegan in R are used to assess diversity, while ALDEx2 and ANCOM detect differential abundance using compositionality-aware methods. The advanced bioinformatics analysis steps are depicted in Figure 7.3.

7.5 APPLICATIONS OF ENVIRONMENTAL METAGENOMICS

Environmental metagenomics is an area of science that explores microbial communities from different environmental sources without culturing in the lab. This approach helps in identifying functional metabolic pathways, microbial functional connections to ecosystem-scale processes, and discovering new species and environmental indicators (Lema et al., 2023; Nwachukwu & Babalola, 2022; Kuang et al., 2023). Several recent studies employed this approach for detection of pollution-degrading microbes and genes such as monooxygenase (alkB), CYP153, todC1/C2 for hydrocarbons; mer, ars, and czc for metals; and opd and atzA/B/C for pesticides, and identified taxa such as *Alcanivorax*, *Cycloclasticus*, *Pseudomonas*, *Shewanella*, and *Sphingomonas* (Lema et al., 2023; S. Liu et al., 2022). Similarly, in climate-related studies, methane-cycling methanotrophs (*Methylocystis*, *Methylomonas*) were identified using the *mcrA* gene. Carbon fixation pathways involving *acsB*, *fhs*, *dmdA*, and *dddP* genes have also been connected to microbial activity influencing global carbon flow (Cavicchioli et al., 2021; Sunagawa et al., 2020).

From an agricultural perspective, genes such as *amoA*, *nifH*, *phoD*, and PKS/NRPS clusters were found to be involved in maintaining soil health, improving plant pliability, resisting stress, and promoting growth (Lema et al., 2023; Nayfach et al., 2021). Metagenomics has facilitated the discovery of enzymes from extremophiles such as thermophiles, halophiles, and acidophiles,

TABLE 7.2
Applications of Environmental Metagenomics

Application	Environment	Key Marker Genes/Pathways	Example Taxa	Representative References
Bioremediation: Oil spills, heavy metals, pesticides	Marine spills, contaminated sediments, agricultural soils	*alkB, CYP153, xylE* (hydrocarbon degradation); *mer, ars, czc* (metal resistance); *opd, mpd* (pesticide degradation)	*Alcanivorax, Cycloclasticus, Sphingomonas, Pseudomonas*	Khan et al. (2022); Lema et al. (2023); Nwachukwu & Babalola (2022)
Climate change: Methane cycling, carbon sequestration	Permafrost, wetlands, marine sediments	*mcrA* (methanogenesis); *pmoA, mmoX* (methane oxidation); Wood–Ljungdahl pathway (carbon fixation)	*Methanobacteriales, Methanosarcinales, Methylocystis*	Cavicchioli et al. (2021); Kuang et al. (2023); Sunagawa et al. (2020); Eisenhofer et al. (2024)
Agriculture: Soil health, plant–microbe interactions	Rhizosphere soils, bulk agricultural soils	*nifH, amoA, nirK/S* (N cycling); phosphorus solubilization genes; PKS/NRPS biosynthetic clusters	*Bacillus, Pseudomonas, Streptomyces,* rhizobia	Nwachukwu & Babalola (2022); Lema et al. (2023); Nayfach et al. (2021)
Water quality: Pathogen detection, wastewater monitoring	Drinking water systems, recreational waters, urban wastewater	Pathogen marker genes (e.g., *ctxA* in *V. cholerae*); AMR genes (*blaCTX-M, blaNDM*); viral genomes (SARS-CoV-2, polio)	*Vibrio cholerae, Salmonella, Cryptosporidium*	Demirci et al. (2025); Seabra et al. (2024); Brumfield et al. (2022)
Extreme environments: Thermophiles, halophiles, acidophiles	Hot springs, hypersaline lakes, acidic rivers	Thermostable polymerases; halo-tolerant proteases; acid-resistant glycosidases	*Thermus, Thermococcus, Halobacterium,* acidophilic fungi	Lema et al. (2023); Mukhtar (2022); Sunagawa et al. (2020)

supporting industrial applications that require stable cellulases, proteases, and other biocatalysts (Mukhtar, 2022). Recent studies using metagenomics approach offers real-time monitoring of water quality and associated pathogens. Authors have identified pathogens such as *Vibrio*, *Salmonella*, *Cryptosporidium*, Rotavirus A, human beta herpesvirus, and SARS-CoV-2, even when present in low abundance (Brumfield et al., 2022; Bisht et al., 2025; Seabra et al., 2024). The applicability of environmental metagenomics is depicted in Table 7.2.

7.5.1 Case Study

Global programs demonstrate the strength of metagenomics. The Tara Oceans Project investigated microbial diversity and functional genes in oceans, uncovering biogeographic signatures, nitrogen fixation hotspots, and virus–host networks (Sunagawa et al., 2020; Cavicchioli et al., 2021; Lema et al., 2023; Eisenhofer et al., 2024). The Earth Microbiome Project (EMP) sampled soil microbiomes globally, representing core taxa, nutrient-cycle genes, and cross-biome functional patterns (Nayfach et al., 2021; Nwachukwu & Babalola, 2022; Yazhini et al., 2025). Recent research in extreme habitats such as hydrothermal vents and polar ice assisted in discovering new kinds of organisms and genes involved in nutrient cycle (Kuang et al., 2023; S. Liu et al., 2022; Lema et al., 2023; L. Liu et al., 2022; Lang et al., 2020). Another study comparing polluted versus unpolluted sites reported a higher abundance of AMR and stress-response genes in microbes (Belmann et al., 2025; Qayyum et al., 2025). Collectively, applications of environmental metagenomics provide insight into microbial life and its interaction with ecosystems. Changes in climate, or vice versa, can impact ecosystem processes (Lema et al., 2023; Kuang et al., 2023; Sunagawa et al., 2020).

7.6 ETHICAL, LEGAL, AND SOCIAL IMPLICATIONS (ELSI)

There are ethical, legal, and social implications when handling microbes and genetic material. Laboratories must comply with Biosafety Levels (BSLs), while digital safeguards such as genetic data encryption and cyber-biosecurity protect sensitive sequences (Harikrishnan et al., 2021). For environmental release, whether for bioremediation or advanced safeguard risk assessments, genetic devices (e.g., kill switches) and post-release metagenomic monitoring help lessen ecological risks (Cavicchioli et al., 2021). Meanwhile, AI-driven metagenomic pipelines yield robust data, effective analytical capabilities, open-source models, and regionally compliant cloud workflows, bringing transparency, strong frameworks, and sovereign data governance (Qayyum et al., 2025; Belmann et al., 2025). Public trust in metagenomics depends on transparency and preventing misinformation about its influence on ecosystems and health. Transparent reporting, such as plain-language summaries and involving stakeholders in advisory boards, fosters understanding. Additionally, integrating social science helps tailor outreach efforts to build long-lasting trust across diverse communities and disciplines.

7.7 FUTURE PERSPECTIVES

Understanding environmental metagenomics is rooted in multi-omics, integrating metagenomics with metabolomics and metatranscriptomics, to move beyond static directories toward dynamic functional profiling. This integrated layers approach enables researchers to observe gene abundance, expression, responses, and metabolic outputs in real-time environmental contexts such as nutrient influx or pollutant influx (Ballard et al., 2024). For example, *nifH* gene abundance can be integrated with transcript levels and ammonium outcomes, revealing active nitrogen turnover. Computational integration relies on ontologies such as KEGG and MetaCyc, structural knowledge network platforms such as Cytoscape and GNPS, and statistical methods such as sparse canonical correlation analysis (sCCA) to map gene–metabolite relationships.

The application of artificial intelligence in metagenomic workflows could enhance understanding from assembly to ecological forecasting. Graph neural networks and transformer-based models improve contig scaffolding and binning of complex data (Belmann et al., 2025; Qayyum et al., 2025). Generative AI predicts enzyme functions for unannotated genes, accelerating discovery in biofuel production and pollutant degradation (Yazhini et al., 2025). Machine-learning models help connect environmental metadata to microbial shifts under climate scenarios, while cloud-based elastic computing drastically decreases processing time, allowing global-scale monitoring.

Genomics is based on four technological pillars. First, portable nanopore sequencing, enabled by improved pore chemistry and edge-AI, supports real-time genetic profiling and assembly in remote or resource-limited settings. Drone-assisted sampling and rapid detection enhance its utility. Second, multi-omics integrates different layers of the microbial community. Third, AI analyzes databases to improve speed and accuracy for real-time monitoring and predictive modeling. Fourth, enriching databases through MAG integration, single-cell genomics, and global initiatives such as the Earth Microbiome Project and Tara Oceans improves annotation and taxonomic resolution for uncultured microbes (Nayfach et al., 2021; Gupta et al., 2025; Sunagawa et al., 2020). Together, these four pillars of advancement, AI, portable sequencing, multi-omics integration, and enriched reference libraries, will enable real-time ecological monitoring, drive biotechnological innovation, enhance predictive models, and democratize access to metagenomic science.

7.8 CONCLUSION

Environmental metagenomics has provided a fundamentally new view of the invisible microbial world that controls the health of our planet. Innate microbial diversity, long hidden in the form of "microbial dark matter," is now being illuminated using powerful sequencing technologies and computational developments. It has been demonstrated that microbes are not isolated actors but integral players in ecosystems that influence the flow of energy and matter across diverse environments. From hydrothermal vents to agricultural soils, metagenomics is revealing the extraordinary adaptability and diversity of life while simultaneously offering practical solutions ranging from bioremediation to water quality monitoring and the discovery of novel biomolecules with industrial potential. However, with scientific progress, challenges also increase. Technical aspects are important, but so are data reliability, ethical governance, and equitable access. Responsible sharing of genomic data, protection of biodiversity knowledge, and community involvement will determine the sustainability of applying this knowledge. Looking to the future, the integration of metagenomics with other omics disciplines, aided by artificial intelligence and portable sequencing technologies, promises to transform the field into a real-time environmental monitoring framework. "As more discoveries are made about microbial life, we'll be able to gain a more complete picture of organisms, which will not only advance science but will also serve as a basis for finding ways to meet the environmental challenges facing us as a global species." In essence, environmental metagenomics is not only a scientific approach but also an access point to understanding the hidden networks that sustain life on Earth, demonstrating that the smallest forms of life can have the greatest impact on the future of the planet.

REFERENCES

Ballard, J. L., Wang, Z., Li, W., Shen, L., & Long, Q. (2024). Deep learning-based approaches for multi-omics data integration and analysis. *BioData Mining*, *17*, 38. https://doi.org/10.1186/s13040-024-00352-3.

Belmann, P., Osterholz, B., Kleinbölting, N., Sczyrba, A., & Goesmann, A. (2025). Metagenomics-toolkit: A flexible, efficient cloud-based metagenomics workflow with ML-enabled resource allocation. *NAR Genomics and Bioinformatics*, *7*(3), lqaf093. https://doi.org/10.1093/nargab/lqaf093.

Bisht, D., Kumar, V., Gupta, S., & Sachan, M. (2025). Navigating circulating tumour cells and tumour microemboli: Where we stand in bridging their molecular dynamics with clinical translation. *Biochimica et Biophysica Acta (BBA)—Reviews on Cancer*, *1880*(4), 189384. https://doi.org/10.1016/j.bbcan.2025.189384

Brill, S. J., & Stillman, B. (1989). Yeast replication factor-A functions in the unwinding of the SV40 origin of DNA replication. *Nature*, *342*(6245), 92–95. https://doi.org/10.1038/342092a0.

Brumfield, K. D., Leddy, M., Usmani, M., Cotruvo, J. A., Tien, C.-T., Dorsey, S., . . . Colwell, R. R. (2022). Microbiome analysis for wastewater surveillance during COVID-19. *mBio*, *13*(4), e00591.22. https://doi.org/10.1128/mbio.00591-22.

Cavicchioli, R., Ripple, W. J., Timmis, K. N., Azam, F., Bakken, L. R., Baylis, M., . . . Webster, N. S. (2021). Scientists' warning to humanity: Microorganisms and climate change. *Nature Reviews Microbiology*, *19*(5), 287–302. https://doi.org/10.1038/s41579-020-00460-0.

Dahn, H. A., Mountcastle, J., Balacco, J., Winkler, S., Bista, I., Schmitt, A. D., . . . Fedrigo, O. (2022). Benchmarking ultra-high molecular weight DNA preservation methods for long-read and long-range sequencing. *GigaScience*, *11*, giac068. https://doi.org/10.1093/gigascience/giac068.

De Brauwer, M., Chariton, A., Clarke, L. J., Cooper, M. K., Di Battista, J., Furlan, E., . . . Noble, L. M. (2022). Environmental DNA protocol development guide for biomonitoring.

Demirci, M., Çubuk, C., Dasdemir, F., Saribas, A. S., Balcıoglu, E. B., Ozbey, D., . . . Kocazeybek, B. S. (2025). Comparative microbial metagenomic analysis of drinking water plants and wastewater treatment plants in Istanbul. *Frontiers in Microbiology*, *16*. https://doi.org/10.3389/fmicb.2025.1488268.

Eisenhofer, R., Nesme, J., Santos-Bay, L., Olsson-Francis, K., Tetu, S. G., & Tyson, G. W. (2024). A comparison of short-read, HiFi long-read, and hybrid strategies for genome-resolved metagenomics. *Microbiology Spectrum*, *12*(4), e03590.23. https://doi.org/10.1128/spectrum.03590-23.

Goller, C., & Sjogren, C. (2023). *Quantifying and Checking Genomic DNA*. https://www.protocols.io/view/quantifying-and-checking-genomic-dna-ch2at8ae.

Gupta, A., Pahal, S., Kumar, V., & Chaudhary, A. (2025). Role of monoclonal antibodies in Alzheimer's disease: Successes, failures, and future directions. *Neurological Sciences*. https://doi.org/10.1007/s10072-025-08456-5.

Harikrishnan, R., Jang, I. S., Balasundaram, C., & Heo, M. S. (2021). Current scenarios of biocontainment and biosafety considerations in microbial applications. *Aquaculture Reports*, *20*, 100718. https://doi.org/10.1016/j.aqrep.2021.100718.

Hong, S., Kim, M.-W., & Jang, J. (2021). Physical collection and viability of airborne bacteria collected under electrostatic field with different sampling media and protocols towards rapid detection. *Scientific Reports*, *11*(1), 14598. https://doi.org/10.1038/s41598-021-94033-7.

James, A. S., Adil, N. A., Goltz, D., Tangudu, D., Chaudhari, D. S., Shukla, R., . . . Yadav, H. (2024). Abnormalities in gut virome signatures linked with cognitive impairment in older adults. *Gut Microbes*, *16*(1), 2431648. https://doi.org/10.1080/19490976.2024.2431648.

Khan, D., Kabiraj, A., Roy, R. K., Let, M., Majhi, K., & Bandopadhyay, R. (2022). Bioremediation of heavy metals by metagenomic approaches. In V. Kumar & I. S. Thakur (Eds.), *Omics Insights in Environmental Bioremediation* (pp. 393–413). Springer Nature. https://doi.org/10.1007/978-981-19-4320-1_17.

Klingström, T., Bongcam-Rudloff, E., & Pettersson, O. V. (2018). A comprehensive model of DNA fragmentation for the preservation of High Molecular Weight DNA. *bioRxiv*, 254276. https://doi.org/10.1101/254276.

Kuang, B., Xiao, R., Hu, Y., Wang, Y., Zhang, L., Wei, Z., . . . Pan, W. (2023). Metagenomics reveals biogeochemical processes carried out by sediment microbial communities in a shallow eutrophic freshwater lake. *Frontiers in Microbiology*, *13*. https://doi.org/10.3389/fmicb.2022.1112669.

Kumar, V., Gangani, S., Shukla, R., Prajapati, S. K., Shekhar, H., & Shukla, V. (2024). Next-generation sequencing and genomic data analysis. In A. Chaudhary, S. K. Sethi, & A. Verma (Eds.), *Unraveling New Frontiers and Advances in Bioinformatics* (pp. 49–76). Springer Nature. https://doi.org/10.1007/978-981-97-7123-3_3

Kumar, V., Sharma, J. K., Kumar, S., Gulia, J., Singh, N. P., & Santal, A. R. (2025). Metagenomic and proteomic approach for bioremediation of environmental pollutants. In M. P. Shah (Ed.), *Development in Waste Water Treatment Research and Processes*. Elsevier. https://doi.org/10.1016/B978-0-443-13615-3.00020-0.

Kumar, V., Shukla, R., Gangani, S., Shekhar, H., & Shukla, V. (2024). Metagenomics and microbiome analysis. In A. Chaudhary, S. K. Sethi, & A. Verma (Eds.), *Unraveling New Frontiers and Advances in Bioinformatics* (pp. 77–92). Springer Nature Singapore. https://doi.org/10.1007/978-981-97-7123-3_4.

Lang, D., Zhang, S., Ren, P., Liang, F., Sun, Z., Meng, G., . . . Liu, S. (2020). Comparison of the two up-to-date sequencing technologies for genome assembly: HiFi reads of Pacific biosciences sequel II system and ultralong reads of Oxford nanopore. *GigaScience*, *9*(12), giaa123. https://doi.org/10.1093/gigascience/giaa123.

Lema, N. K., Gemeda, M. T., & Woldesemayat, A. A. (2023). Recent advances in metagenomic approaches, applications, and challenges. *Current Microbiology*, *80*(1), 347–360. https://doi.org/10.1007/s00284-022-03165-w.

Lever, M. A., Torti, A., Eickenbusch, P., Michaud, A. B., Šantl-Temkiv, T., & Jørgensen, B. B. (2015). A modular method for the extraction of DNA and RNA, and the separation of DNA pools from diverse environmental sample types. *Frontiers in Microbiology*, *6*. https://doi.org/10.3389/fmicb.2015.00476.

Liu, L., Yang, Y., Deng, Y., & Zhang, T. (2022). Nanopore long-read-only metagenomics enables complete and high-quality genome reconstruction from mock and complex metagenomes. *Microbiome*, *10*, 209. https://doi.org/10.1186/s40168-022-01407-4.

Liu, S., Moon, C. D., Zheng, N., Leahy, S. C., Wang, L., & Sun, W. (2022). Opportunities and challenges of using metagenomic data to bring uncultured microbes into cultivation. *Microbiome*, *10*, 76. https://doi.org/10.1186/s40168-022-01228-5.

Liu, S., Rodriguez, J. S., Munteanu, V., Ronkowski, C., Sharma, N. K., Alser, M., . . . Mangul, S. (2025). Analysis of metagenomic data. *Nature Reviews Methods Primers*, *5*(1), 5. https://doi.org/10.1038/s43586-024-00376-6.

Mandal, S. D., Panda, A. K., Kumar, N. S., Bisht, S. S., & Jin, F. (Eds.). (2021). *Metagenomics and Microbial Ecology: Techniques and Applications*. CRC Press. https://doi.org/10.1201/9781003042570.

Mishra, S., Buddendorff, L., Hoover, J., Shukla, R., Kumar, V., Jain, S., & Yadav, H. (2024). Microbiome role in cognitive impairment: MiaGB cohort insights. *Innovation in Aging*, *8*(S1), 811. https://doi.org/10.1093/geroni/igae098.2629

Mukhtar, S. (2022). Discovery of novel enzymes from extreme environments through metagenomic approaches: A mini-review. *International Journal of Frontiers in Life Science Research*, *2*(2), 37–45. https://doi.org/10.53294/ijflsr.2022.2.2.0042.

Nair, S. U., Lin, Y., Jain, S., Chaudhari, D. S., Shukla, R., Kumar, V., . . . MiaGB Consortium. (2024). Unique gut microbiome signatures associated with physical and cognitive function. *Alzheimer's & Dementia*, *20*(S2), e092044. https://doi.org/10.1002/alz.092044

Nayfach, S., Eloe-Fadrosh, E. A., Parks, D. H., Roux, S., Chen, I. M. A., Huson, D. H., . . . Kyrpides, N. C. (2021). A genomic catalog of Earth's microbiomes. *Nature Biotechnology*, *39*(4), 499–509. https://doi.org/10.1038/s41587-020-0718-6.

Nwachukwu, B. C., & Babalola, O. O. (2022). Metagenomics: A tool for exploring key microbiome with the potentials for improving sustainable agriculture. *Frontiers in Sustainable Food Systems*, *6*, 886987. https://doi.org/10.3389/fsufs.2022.886987.

Prajapati, S. K., Kumar, V., Shukla, R., Szekeres, C., Ma, Y., Jain, S., & Yadav, H. (2024). Microbiome-immune interplay in aging brains: A pilot investigation from the MiaGB cohort. *Alzheimer's & Dementia*, *20*(S2), e090734. https://doi.org/10.1002/alz.090734.

Qayyum, H., Ishaq, Z., Ali, A., Khan, S., Ashraf, M. F., & Tariq, M. (2025). Genome-resolved metagenomics from short-read sequencing data in the era of artificial intelligence. *Functional & Integrative Genomics*, *25*, 124. https://doi.org/10.1007/s10142-025-00789-1.

Qayyum, H., Raziq, M. F., Manzoor, H., Zaidi, S. S. A., Ali, A., & Kayani, M. U. R. (2025). Efficient De Novo assembly and recovery of microbial genomes from complex metagenomes using a reduced set of k-mers. *Interdisciplinary Sciences: Computational Life Sciences*. https://doi.org/10.1007/s12539-025-00722-6.

Russell, J. N., Kos, D., Yacoub, E., Sies, A. N., Warr, B., Jelinski, M., Ruzzini, A., & Cameron, A. D. S. (2025). Enhanced metagenomic surveillance for bovine respiratory disease pathogens and antimicrobial resistance by hybridization capture sequencing. *Applied and Environmental Microbiology*, e00977.25. https://doi.org/10.1128/aem.00977-25.

Sarfraz, H., Czerniecki, B. J., Stankiewicz, K., Jaglal, M. V., Jain, S., Yadav, H., . . . Bari, S. (2024). Evaluating gut microbiome as a biomarker of pathological complete response in patients with early locally advanced triple negative breast cancer (LA TNBC): A pilot study. *Journal of Clinical Oncology*, *42*(S16), 593. https://doi.org/10.1200/JCO.2024.42.16_suppl.593.

Seabra, S. G., Santos, A. F., Nunes, M., Manjate, F., Silva, A., Crespo, T., . . . Parreira, R. (2024). Viral pathogen surveillance in wastewater using metagenomic approaches: Obstacles and prospects. *European Journal of Public Health*, *34*(S3), ckae144.2136. https://doi.org/10.1093/eurpub/ckae144.2136.

Sunagawa, S., Acinas, S. G., Bork, P., Bowler, C., Eveillard, D., Gorsky, G., . . . Wincker, P. (2020). Tara oceans: Towards global ocean ecosystems biology. *Nature Reviews Microbiology*, *18*(8), 428–445. https://doi.org/10.1038/s41579-020-0364-5.

Sunagawa, S., Coelho, L. P., Chaffron, S., Kultima, J. R., Labadie, K., Salazar, G., . . . Bork, P. (2015). Structure and function of the global ocean microbiome. *Science*, *348*(6237), 1261359. https://doi.org/10.1126/science.1261359.

Yazhini, A., Morice, E., Jochheim, A., Hernandez, J., & Roux, S. (2025). Evaluation of metagenome binning: Advances and challenges. *bioRxiv*. https://doi.org/10.1101/2025.02.21.639465.Yuan, S., Cohen, D. B., Ravel, J., Abdo, Z., & Forney, L. J. (2012). Evaluation of methods for the extraction and purification of DNA from the human microbiome. *PLoS One*, *7*(3), e33865. https://doi.org/10.1371/journal.pone.0033865.

8 Metagenomics in Agriculture and Food Science

Sorabh Chaudhary[1#], *Divyanshi Srivastava*[2#], *Meenu Singh*[2], *Sushma Sagar*[3], *Monu*[2*], *and Pankaj Verma*[2*]

[1] Department of Biology, Faculty of Science and Humanities, SRM Institute of Science and Technology, Delhi-NCR Campus, Modinagar, Ghaziabad, Uttar Pradesh, India

[2] Department of Biotechnology, School of Engineering, IILM University, Greater Noida, Uttar Pradesh

[3] Division of Vegetable Sciences, ICAR-Indian Agricultural Research Institute (IARI), Pusa, New Delhi, India

*Corresponding author(s): Dr. Pankaj Verma and Dr. Monu

Sorabh Chaudhary and Divyanshi Srivastava shared first author

8.1 INTRODUCTION

Metagenomics is defined as the culture-independent analysis of genetic material retrieved directly from environmental or host-associated samples. This technique involves shotgun whole-genome sequencing, marker-gene (e.g., 16S rRNA/ITS) approaches, and genome-resolved workflows (metagenome-assembled genes) that describe the taxonomic composition, functional potential, and genomic diversity of microbial populations (Sadurski et al. 2024). Genome assembly has been technically challenging for many years for uncultivated microbial taxa. However, breakthroughs in genome-resolved metagenomics have brought significant changes in research, improving trait prediction, metabolic reconstruction, and the ability to link genes to ecological roles. Metagenomics has been employed in several domains, such as environmental and human disease detection. The applicability of metagenomics has been evaluated in diseases such as breast cancer, colorectal cancer, and Alzheimer's disease (Nair et al. 2024; James et al. 2024; Prajapati et al. 2024; Mishra et al. 2024; Sarfraz et al. 2024; Madigan et al. 2025). The process of metagenomics produces a high volume of data, and the accurate analysis of these is important for downstream application. Therefore, multiple computational methods have been developed for de novo genome assembly from metagenome shotgun sequencing data, leading to the rapid accumulation of draft genomes in the form of metagenome-assembled genomes (MAGs).

In agricultural and food systems, metagenomics is applied across multiple domains, including soil and rhizosphere ecology, crop phyllosphere studies, endophyte profiling, livestock gut microbiomes, and food microbiomes (Kim et al. 2024; Kumar et al. 2025). Microbial ecology began with the cultivation and morphological characterization of microorganisms. Although marker-gene surveys provided limited functional insights, they expanded observable diversity and allowed comparative analysis of soil and plant microbiomes. Foundational syntheses during the 2000s–2010s established how cultivation bias masked the majority of microbial diversity, marking the beginning of culture-independent sequencing approaches (Simon and Daniel 2011). Historical approaches to studying the microbiome relied entirely on *in vitro* culture to isolate practicable samples. However, these approaches cultivated only a small fraction of microbes, leading to a massive underestimation of microbiome biodiversity. This type of error is known as the "great plate count anomaly."

DOI: 10.1201/9781003670155-8

Studying microbiomes through metagenomic sequencing, which eliminates the requirement for laboratory culture, is now progressing as a promising method. As sequencing costs declined, shotgun metagenomics became viable for environmental samples. This transition enabled the discovery of novel enzymes, biosynthetic pathways, and resistance determinants pertinent to soil fertility, plant health, and food safety. Methodological studies through the early 2010s refined library preparation, assembly, binning, and functional annotation (Simon and Daniel 2009). A mechanistic approach to hypothesis-driven research emerged around 2020, leading to the development of translational outputs such as biomarkers for soil health and metagenomics-informed food surveillance (Inglis and Edwards 2022).

Soil is the residence of numerous agroecosystems, where microbiomes mediate crucial ecosystem services such as organic matter decomposition, nutrient cycling (N and P), disease suppression, and soil structure formation. It is only through metagenomics that the taxonomic and functional signatures of healthy versus degraded soils can be evaluated. These results provide statistical measures as bioindicators for soil health assessment, enabling targeted interventions to improve fertility and resilience (Galanova et al. 2025). Plant-associated microbiomes, such as those of the rhizosphere, endosphere, and phyllosphere, affect nutrient uptake, growth promotion, and resistance to pathogens and abiotic stress. Metagenomic profiling investigates associations between microbial genes and pathways involved in nutrient mobilization, phytohormone synthesis, and osmoprotectant production and beneficial phenotypes, thereby supporting breeding strategies and microbiome engineering (Y. Wang et al. 2024). Metagenomics also enables the discovery of functional consortia and metabolic pathways that can be targeted through diet, probiotics, or herd management to improve animal welfare, reduce emissions, and ensure product quality (Das et al. 2025). This approach also detects and characterizes complex microbial communities in fermented products and traces contamination or spoilage organisms, thereby providing culture-independent surveillance for foodborne pathogens and antimicrobial-resistance genes (Sadurski et al. 2024).

Metagenomics—the culture-independent analysis of collective microbial genomes from environmental samples—has revolutionized our understanding of complex microbial ecosystems in agriculture and food science. Figure 8.1 shows the implementation of metagenomics across various aspects. In agricultural systems, metagenomics provides insight into the exploration of soil- and plant-associated microbiomes, which helps advance sustainable farming methods by offering insights into nutrient cycling, plant growth promotion, and disease suppression (Yousuf et al. 2025). In food science, metagenomics helps characterize microbial communities involved in food fermentation, spoilage, and safety evaluation. By employing sophisticated metagenomic tools and approaches, including shotgun metagenomics, next-generation sequencing (NGS), and bioinformatics pipelines, scientists can precisely identify, quantify, and functionally annotate microbial populations (Nwachukwu and Babalola 2022). Applications and related case studies demonstrate its importance in increasing crop production, identifying food-associated pathogens, supporting fermentation processes, and developing probiotics and related functional foods. Despite its immense potential, metagenomics also faces challenges, including computational data requirements and difficulties in connecting functional outputs to microbial diversity. To overcome these obstacles and fully utilize metagenomics for agricultural and food innovation, further developments in sequencing technologies and bioinformatics are necessary (Lian et al. 2023).

8.2 METAGENOMICS IN AGRICULTURAL SYSTEMS

The dynamic interactions among the soil microbiome, flora, bacteria, and fungi are essential for achieving sustainable agriculture and fostering ecological restoration. Microbes play a crucial role in nutrient cycling and have a significant impact on the emission and reduction of greenhouse gases, which ultimately affect the global climate. Metagenomics has emerged as a powerful approach to strengthen synergistic interactions between plants and microbes, resulting in enhanced plant health

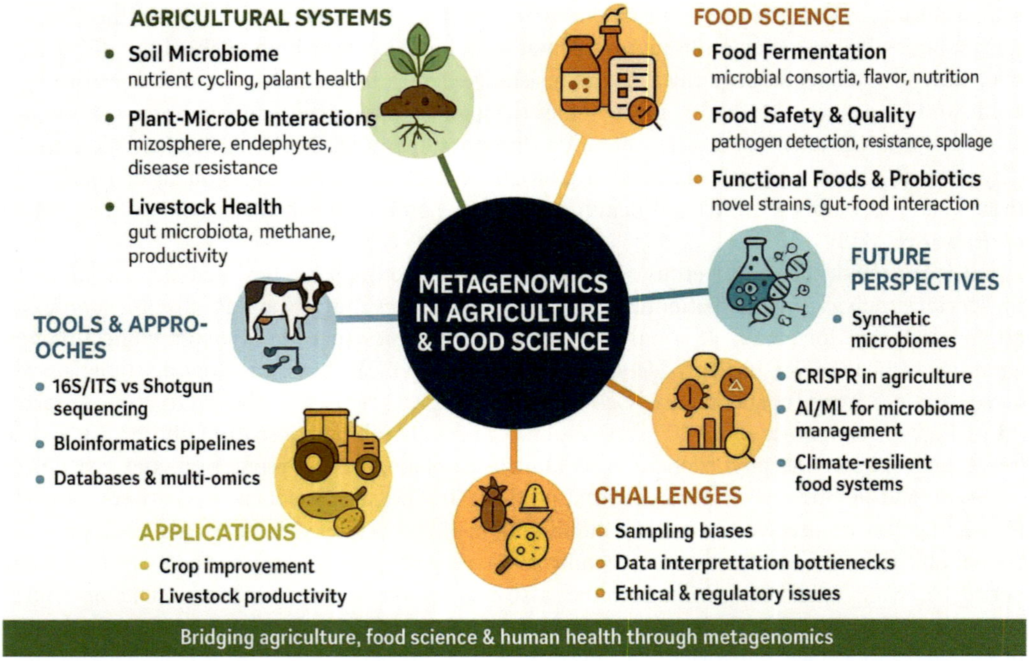

FIGURE 8.1 Metagenomics in agriculture and food sciences.

and resilience. Investigating the metagenome of soil microorganisms serves as a crucial tool for designing and evaluating ecosystem restoration strategies that align with sustainable agricultural practices in the context of climate change. Gaining insights into the specific functionalities and actions of soil microbes paves the way for developing restoration initiatives that fortify vital ecological components while generating tangible economic benefits through ecosystem services.

8.2.1 Soil Microbiome

According to Lederberg and McCray (2001), the microbiome refers to the collective group of microbes that inhabit a particular host or environment. Soil microorganisms are essential for managing important elements, including soil fertility, nutrient cycling (mobilization and absorption of essential nutrients), biodiversity, and plant health (Lederberg and McCray 2001; Kumar et al. 2022a; Chauhan et al. 2023). In an agricultural system, soil microorganisms serve as antagonists for controlling pests and diseases, as well as biostimulants and biofertilizers (Santos and Olivares 2021; Chaudhary et al. 2020; Lal et al. 2022; Verma et al. 2025). The composition and functionality of the soil microbiome evolve, thereby impacting soil health and nutrient accessibility. The intricacy and variability of soil microorganisms arise from different environments (Fierer 2017). Soil characteristics are influenced by both biotic factors, including plant species, microbial antagonists, and competitors, and abiotic factors, such as soil type, structure, moisture content, pH, nutrient availability, and other geographical aspects. Nitrogen fixation, nutrient solubilization, and pathogen suppression are crucial microbial functions that play important roles in plant growth promotion and enhancing ecosystem resilience of ecosystems (Iqbal et al. 2023; Pradhan et al. 2025). These activities respond rapidly to environmental stress, making soil enzyme activity a rapid and sensitive indicator of overall soil quality (Kracmarova et al. 2020). A diverse range of Plant Growth-Promoting Microorganisms (PGPMs) carry out various essential activities. This category includes endophytes, phosphorus solubilizing bacteria, and vesicular-arbuscular mycorrhiza (Mahmud et al. 2021; Kumar et al. 2022b). Self-replicating and self-sustaining organisms typically do not require ongoing inoculation. Their efficacy in plant growth promotion and soil resilience improvement is influenced by biotic and abiotic factors, plant species, and microbial community composition (Trivedi et al. 2016).

Endophytes refer to bacteria or fungi that reside within plant tissues. Phyllosphere microbes are present in aerial plant parts, encompassing both surface-dwelling epiphytes and internal endophytes, and are essential for maintaining plant health. Endophytes contribute to enhanced plant growth and stress tolerance, whereas phyllosphere microbes assist in nutrient uptake and improve resistance to various diseases (Lindow and Brandl 2003). In addition, phyllosphere microorganisms are responsible for enhancing crop productivity and plant fitness by altering physiological functions, and they play a significant role in contaminant removal (Stone et al. 2018; Chaudhary et al. 2024). Microbial diversity plays a crucial role in effective disease suppression. Biocontrol agents reduce pathogen populations through mechanisms such as competitive exclusion, antibiotic production, and the induction of systemic resistance. Metagenomics is considered an innovative approach for direct genetic examination of environmental samples. This technology has significantly improved our understanding of microbial communities by facilitating the study of uncultivated organisms within their natural habitats (Jagadesh et al. 2024). Metagenomics methodologies have unveiled critical insights into microbial diversity and functionality, and they provide essential tools for addressing challenges related to agriculture, ecosystem restoration, and climate change resilience. By analyzing genetic material, metagenomics annotates the roles of soil microbes in nutrient cycling, including nitrogen fixation, phosphorus solubilization, and carbon cycling. These analytical processes clarify microbial functions in plant–microbe interactions (Du et al. 2023). Metagenomic methods enhance our knowledge of microbial diversity, population structure, metabolic capabilities, and functional genes by effectively addressing cultivation limitations (H. Wang et al. 2024).

8.2.2 Plant–Microbe Interactions

Plant–microbe interactions are crucial for regulating plant physiology and significantly influence primary metabolism and the production of various secondary metabolites. These interactions are important because they determine key traits, including crop yield, nutritional quality, and resilience to stress (H. Wang et al. 2024; Vishwakarma, Upadhyay, et al. 2017). Complex microbial communities contribute to these processes by modifying plant-derived metabolites and influencing host metabolic pathways. The plant microbiome can be grouped into specific regions—the rhizosphere, the endosphere, and the phyllosphere (Vishwakarma, Mishra, et al. 2017; Shi et al. 2024). The development of large genomic databases has substantially increased our understanding of plant–microbe interactions. These resources have illuminated the molecular mechanisms through which plant genotypes and environmental factors shape microbiome composition, structure, and evolution. Moreover, the concordant heritability of diverse microbial taxa across studies suggests that beneficial plant–microbiome interactions have been conserved throughout evolution (Henry et al. 2021). Advanced techniques, such as targeted and shotgun high-throughput sequencing (HTS), have been essential in enhancing understanding of the plant microbiome. These strategies are crucial for differentiating beneficial microorganisms from pathogenic ones within microbial populations (Hu et al. 2020).

Genome-wide association studies (GWAS) have successfully identified quantitative trait loci (QTLs) in host plants that influence microbial diversity, morphology, and population structure in both the phyllosphere and rhizosphere (Demirjian et al. 2023). Researchers have combined GWAS with metagenomic profiling to accurately identify specific microbial taxa and potential plant genes that determine rhizosphere structure and functionality. A comprehensive investigation integrating metagenomics, metabolomics, and transcriptomics in barley identified specific loci that directly influence rhizosphere population composition, contrasting domesticated and wild genotypes (Escudero-Martinez et al. 2022). Expanding on this work, Oyserman et al. (2022) utilized quantitative genetics and microbiomics to identify heritable quantitative traits within rhizobacteria and molecular attributes of the rhizosphere microbiome. Collectively, these investigations demonstrate that essential components of microbiome assembly are influenced by host genetic factors, providing a framework for leveraging plant genetics to cultivate beneficial microbial communities.

8.2.3 Livestock and Animal Health

Livestock metagenomics involves the direct sequencing and analysis of combined genetic material (DNA/RNA) obtained from microbial communities inhabiting animals, particularly in regions such as the gut and oral cavity (Pawar et al. 2021). Using modern techniques, including 16S rRNA gene sequencing and whole-genome sequencing (WGS), this approach eliminates the need for culturing individual microbes. The complex nature of microbial communities often makes it difficult to obtain high-purity DNA, even with advances in computational technologies. Increasingly, veterinarians, animal nutritionists, and microbiologists are focusing on the examination of microorganisms in both domesticated and wildlife species (Oakley et al. 2013). Results from metagenomic studies of gut microbiota can improve genetic selection strategies aimed at enhancing animal health and productivity. Microbial profiles derived from fecal samples or oral swabs can act as "microbial fingerprints" associated with animals exhibiting high genetic quality, thereby serving as valuable markers for animal breeding programs. In a 2025 study, Legrand et al. utilized a genome-centric metagenomics approach to assemble a comprehensive set of MAGs from the microbiomes of Australian Angus cattle across various herds and life stages. Fecal MAGs were predominantly associated with *Bacillota* and *Bacteroidota,* whereas those derived from saliva and nasal mucus were primarily related to *Pseudomonadota, Actinobacterota,* and *Bacteroidota* (Legrand et al. 2025). Additionally, Opitz-Ríos et al. (2024) employed shotgun metagenomics to investigate microorganisms relevant to veterinary medicine, focusing on their potential for antimicrobial resistance (AMR) and essential virulence factors. Their findings revealed a wide diversity of microbiota, including known animal and human pathogens such as *Pasteurella multocida*, *Pseudomonas aeruginosa*, *Staphylococcus aureus*, and *Escherichia coli*. Metagenomic research has also demonstrated that the bovine milk microbiome is influenced by factors including health status, physiological conditions, and dietary composition (Couvillion et al. 2023). By analyzing these microbial communities, scientists can identify unique microbial fingerprints indicative of a healthy mammary gland, offering a potential strategy for the genetic selection of healthier dairy cattle (Andrews et al. 2019). In this manner, metagenomics introduces a novel, noninvasive approach to improving animal selection and breeding programs.

8.3 METAGENOMICS IN FOOD SCIENCE

8.3.1 Food Fermentation: Microbial Consortia in Traditional and Industrial Fermentation

Fermentation is one of the earliest and most significant biotechnological processes, involving diverse microbial communities that transform raw materials into foods that have longer shelf lives, enhanced flavors, and greater safety. Unlike single-strain fermentations, both traditional and industrial methods typically utilize a mix of bacteria, yeasts, and fungi. These microorganisms engage in intricate interactions through collaboration, competition, and ecological succession, which ultimately influence the nutritional, sensory, and functional attributes of the final product (Tamang et al. 2020).

8.3.1.1 Metagenomics in Identifying Key Functional Microbes and Pathways

Many microorganisms involved in fermentation cannot be cultivated under laboratory conditions; therefore, traditional methods provide only a limited view of microbial diversity. Metagenomics, which involves sequencing and analyzing the combined DNA of microbial communities, has become a crucial method for evaluating both taxonomic diversity and functional potential. Techniques such as 16S rRNA sequencing, ITS analysis, and shotgun metagenomics help reveal important microorganisms and pathways responsible for lactic acid production, proteolysis, and the synthesis of bioactive compounds. These methods also demonstrate variations in microbial populations during different fermentation phases and identify pathways linked to carbohydrate breakdown, vitamin production, and aroma compound synthesis (Pimentel et al. 2021). Such discoveries aid in the development of tailored starter cultures and enhance the management of industrial fermentation processes.

8.3.1.2 Optimization of Flavors, Texture, and Nutritional Quality

Microbial metabolism underlies the sensory and nutritional qualities of fermented foods. Fungi affect texture and drive biochemical changes through enzyme activity, yeasts provide aroma through volatile chemicals, and lactic acid bacteria contribute acidity and distinctive flavor. Optimization of microbial composition, raw materials, and process parameters improves these qualities. By encouraging bacteria that produce organic acids and beneficial esters and decreasing those associated with off-flavors, a desirable flavor profile can be achieved. Microbial production of exopolysaccharides and enzymatic modifications to structural polymers improve texture, whereas microbial synthesis of vitamins, amino acids, and peptides boosts nutritional value and reduces antinutritional factors. Modern studies integrate metagenomics with computational models and other omics techniques, including transcriptomics and metabolomics. This systems-level approach makes it possible to precisely control microbial interactions, guaranteeing consistent product quality in both large-scale food enterprises and small-scale artisanal activities (Marco et al. 2021).

8.3.2 Food Safety and Quality: Pathogen Detection and Outbreak Tracing

Maintaining food safety is extremely important for both agriculture and food-related industries because contamination with pathogens such as *Salmonella*, *Listeria monocytogenes*, and *E. coli* can cause widespread illness and lead to significant economic losses. Conventional culture-based and biochemical methods are effective but time-consuming and may fail to detect low-abundance or stressed microorganisms (Bintsis and Department of International Trade, TEI of West Macedonia, Kastoria, Greece 2017).

8.3.2.1 Antimicrobial-Resistance Genes in Food Chains

The rise of antimicrobial resistance (AMR) is a major global health concern, and food chains serve as significant reservoirs and transmission routes for resistance genes. Metagenomic sequencing allows the identification and monitoring of AMR genes within microbial communities across the farm-to-fork continuum. These genes include resistance determinants against clinically important antibiotics, which may spread through horizontal gene transfer among bacteria in food environments. By mapping resistomes, researchers can assess risks, track transmission routes, and design interventions to minimize the dissemination of resistant strains in food production systems (Elbehiry et al. 2025; Yaqub et al. 2024).

8.3.2.2 Monitoring Spoilage and Shelf-Life Prediction

Microbial-driven food spoilage leads to significant economic and nutritional losses. Conventional methods for predicting spoilage are based on visual inspection or microbial culturing, which are limited in sensitivity and scope. Metagenomics provides a detailed view of microbial succession during storage, enabling the identification of spoilage-associated taxa and their metabolic functions. By integrating metagenomic data with metabolomics and predictive modeling, it is possible to predict spoilage dynamics, optimize storage conditions, and improve shelf-life prediction. This systems-based approach enhances quality management and ensures consumer safety while reducing waste (Ercolini 2013).

8.3.3 Functional Foods and Probiotics

Functional foods and probiotics are designed to promote health beyond basic nutrition by delivering beneficial microbes or bioactive compounds. Metagenomic approaches have enabled the discovery of novel probiotic strains and provided insights into gut food microbiome interactions, revealing how diet shapes microbial communities and influences host health. These techniques also guide the development of functional foods and nutraceuticals enriched with metabolites such as short-chain fatty acids, vitamins, and anti-inflammatory compounds, ensuring safety, efficacy, and stability in modern dietary interventions (Damián et al. 2022).

8.3.3.1 Novel Probiotic Discovery through Metagenomic Screening

Recent metagenomic analyses have identified numerous previously uncharacterized probiotic candidates from food and gut microbiomes. Studies show that foodborne microbes, such as *Lacticaseibacillus paracasei*, can integrate into the human gut, contributing to microbial diversity and function. Tools such as meta-Probiotics now allow the detection of health-associated strains directly from gut metagenomes, offering insights into host–diet–microbe interactions (Zhang et al. 2025). Short-term supplementation studies also confirm that probiotics can rapidly alter microbiome composition and metabolic activity (Duysburgh et al. 2025; Limasset et al. 2020).

8.3.3.2 Metagenomics in Functional Food and Nutraceutical Design

Multi-omics approaches are increasingly applied to identify probiotic strains with targeted health benefits, such as immune modulation and pathogen inhibition (PubMed 2024). In aquaculture and human studies, metagenomics has informed the development of probiotic and synbiotic formulations that promote microbial balance and improve nutrient utilization. Emerging probiotics, including novel fiber-degrading and butyrate-producing species, hold potential for future functional food applications (Kwoji et al. 2023; Duncan and Sabater 2025).

8.4 METAGENOMIC TOOLS AND APPROACHES

Metagenomics offers extensive pipelines to assess shared genetic variation within microbial populations and to overcome the limitations of culture-dependent approaches. To understand how microorganisms function in agricultural and food systems, this section highlights the most prominent tools and strategies commonly used. These range from simple sequencing approaches to advanced techniques combined with different omics methodologies. The two most common methods are shotgun metagenomics and 16S/ITS amplicon sequencing, which depend on the investigation tools, the study's purpose, cost considerations, and specific data requirements for microbiome analysis (Usyk et al. 2023).

8.4.1 S rRNA/ITS Amplicon Sequencing

This method, known as 16S rRNA/ITS amplicon sequencing, focuses on obtaining and sequencing specific hypervariable regions of conserved marker genes. The 16S rRNA gene is commonly used for archaea and bacteria; however, the internal transcribed spacer (ITS) region is used for fungi. A key advantage of this approach is that it enables the processing of hundreds or even thousands of samples simultaneously. This strategy is also used to analyze large datasets and ecological surveys to observe how microbial communities evolve and expand over time (Hamilton et al. 2025).

A major limitation of this approach is its inability to distinguish between closely related species or strains. Functional predictions rely on estimates from databases such as PICRUSt2, which infer gene content based on known taxa. Recent studies have shown that 16S-based classifications can create inconsistencies in databases (such as *Methylobacterium* versus *Methylorubrum*), making cross-study comparisons difficult and highlighting the importance of genome-scale data for accurate taxonomy (X. Wang et al. 2024; Leducq et al. 2025).

8.4.2 Shotgun Metagenomics

Shotgun metagenomics sequences all DNA present in a sample to reveal gene diversity within microbial communities. It enables strain-level resolution, identification of novel genes, and reconstruction of near-complete microbial genomes from environmental samples, referred to as MAGs. Studies on wastewater-contaminated soils have demonstrated that shotgun sequencing is the only technique that reliably detects and predicts resistance to third-generation cephalosporins, as supported by culture-based investigations (Suh et al. 2022). For example, shotgun data in rhizosphere research identified two- to fivefold more functional gene orthologs (KEGG) compared with 16S-based predictions, indicating improved performance of nutrient cycle models (X. Wang et al. 2024; Rajguru et al. 2024; Kumar, Gangani, et al. 2024; Rajguru et al. 2024).

8.4.3 The Bioinformatics Pipeline: How to Learn About Biology from Raw Data

Regardless of the sequencing method used, raw data must undergo similar bioinformatics stages to extract meaningful information. Even as new tools emerge, the core stages remain consistent. Taxonomic profiling involves organizing 16S/ITS data into operational taxonomic units (OTUs) or amplicon sequence variants (ASVs). This process can be performed using tools such as QIIME 2 and mothur, along with databases including SILVA and Greengenes (Padhiary et al. 2024; Kumar et al., 2019; Faqir et al. 2024). For shotgun data, tools such as Bracken or MetaPhlAn 4 combined with Kraken2 enable rapid and accurate mapping of reads to reference genomes (H. Wang et al. 2024). Long-read sequencing and multi-omics integration are emerging technologies that allow deeper investigation of microbial communities. The field is rapidly evolving, moving beyond earlier approaches that relied solely on DNA profiling. Functional annotation links genes to databases such as KEGG, COG, and CAZy to evaluate metabolic potential. HUMAnN 3 is widely used to identify metabolic pathways and quantify gene abundance (Rajguru et al. 2024). Assemblers such as MetaSPAdes and MEGAHIT construct contigs from shotgun data, which can then be used to generate draft genomes (MAGs) using MetaBAT 2 (X. Wang et al. 2024; Kumar et al., 2024). In addition, statistical analyses and visualization tools, including PCoA and heat maps, help identify gene-level variation associated with environmental conditions, thereby improving interpretation of microbial community dynamics.

8.5 APPLICATIONS AND CASE STUDIES

8.5.1 Case Studies in Crop Improvement and Disease Management

Metagenomics has evolved from a basic technique to an advanced methodology. It is currently used extensively in food development and agricultural advancement. In this chapter, several case studies have shown multiple approaches that aid in crop yield and the association of metagenomics in food development. For instance, crop disease resistance, food safety, quality, and the effects of climate change on production have increased compared to earlier times. Metagenomics has established that the microbial communities, or microbiomes, in soil and plants play a critical role in improving nutrient availability, reducing environmental stress, and preventing disease. In conclusion, metagenomics can be successfully applied to improve crop quality and modern agricultural methods while developing sustainable plans for environmental preservation.

Case study 1: In this case study it has been demonstrated that natural soil microbial communities (natural soil microbiota) have been shown in a number of studies, including a multi-year field study carried out in China, to be able to be utilized in crops like Brassica napus (oilseed rape) to increase the uptake of beneficial selenium (Se) and decrease the absorption of toxic heavy metal cadmium (Cd).Additionally, shotgun metagenomics have been conducted to identify a core microbiome present across six field sites and various stages of plant development, highlighting the critical role of microbial diversity in overall crop development. The study identified many important bacterial species that help make up a synthetic community (SynCom). After being put back in the soil, the SynCom cut the amount of cadmium by about 52% and helped the roots take in 2.1 times more selenium. A metagenomic analysis showed that the number of microbial genes related to the glutathione metabolism pathway and the plant phenotype were strongly related ($R^2 = 0.71$). This suggests that there may be a mechanistic link between the two (Lei et al. 2025).

Case Study 2: This study, conducted in Shaanxi, China, investigated the effects of evolving from maize monocropping to a maize-soybean intercropping system on the rhizosphere microbiome and nutrient utilization efficiency over three years. The results demonstrated that intercropping significantly improves soybean yields by 17%. A metagenomic study also demonstrated that the intercropped system had significant genes related to the nitrogen cycle. The nxrA nitrifier gene increased by 31%, while the genes that help plants take up nitrate rose by 28%. In conclusion, they contend that alterations in the microbiome's function are the primary factor influencing yield enhancements, accounting for nearly 50% of the variations in crop performance (PERMANOVA $R^2 = 0.499$) (Raza et al. 2022).

Case Study 3: Biocontrol of Rice Blast Disease Using Nanoparticles: Researchers in Zhejiang, China, are using bio-formulated nanoparticles derived from Moringa and chitosan to combat *Magnaporthe oryzae,* the fungus responsible for rice blast. Nanoparticles were administered to rice plants in a greenhouse environment. We employed metagenomics to examine the microbiomes in the rhizosphere and phyllosphere, and RNA-seq to assess the effectiveness of the plant's defense system. The drug helped 77.7% of the time. A metagenomic investigation found that the plants that were treated had 93.5% fewer *M. oryzae* findings. The plant's photosynthetic rate went up by 18%, and the number of helpful microorganisms, such *Thiobacillus* and *Nitrospira,* was up too. This work illustrates the utilization of metagenomics in corroborating the processes of innovative biocontrol agents (Hafeez et al. 2024).

8.5.2 Case Studies in Livestock Productivity and Methane Reduction

The rumen microbiome plays a crucial role in methane production and feed digestion in ruminant animals. Metagenomics is facilitating environmentally sustainable cattle production. Numerous case studies have illustrated the importance of metagenomics in this context.

Case Study 1: Probiotics from Kombucha for Reducing Methane in the Lab. This case study included an *in vitro* examination at Seoul National University to evaluate the effectiveness of a Komagataeibacter-based symbiotic culture, similar to that found in kombucha, in reducing methane emissions inside an artificial rumen system. This study involved culturing rumen fluid from Holstein cows in batch fermenters, with the kombucha culture evaluated against a control group and the pharmaceutical ingredient 3-NOP. Every 12 hours, whole-metagenome sequencing was done. This demonstrated that the kombucha culture resulted in a 34.7% reduction in methane production over a 60-hour period. This did not affect the amount of volatile fatty acids (VFAs) produced. VFAs are a significant source of energy for animals. A metagenomic analysis revealed a 53% reduction in methanogenic archaea and a 44% increase in propionate, a more energy-efficient volatile fatty acid (Cannazza et al. 2021).

Case Study 2: In an in vitro study, it has been shown that only the red seaweed *Asparagopsis taxiformis* significantly altered the rumen prokaryotic community. Conversely, *Palmaria mollis* and *Mazzella japonica* exhibited no significant effect. Furthermore, 16S rRNA sequencing results showed that A. taxiformis caused a significant reduction in methanogenic archaea. Moreover, it also stopped important fibrolytic bacteria, like *Fibrobacter* and

Ruminococcus, as well as bacteria that make propionate. A. taxiformis, on the other hand, made other bacteria, like *Prevotella* and *Succinivibrio*, more common. This indicates that the microbes have the ability to adapt to the first change (O'Hara et al. 2023).

8.5.3 Case Studies in Fermented Foods and Pathogen Tracking

Metagenomics provides unique insights into the microbial populations that influence the taste, texture, and safety of fermented foods.

Case Study 1: This study examined phage dynamics during the industrial production of Gouda-type cheese at a rate of 10 tons per day, utilizing virome metagenomics. Monitoring viral predators in industrial cheese production Bacteriophages, viruses that infect bacteria, are vital to the dairy industry because they can disrupt fermentation. In this study, the samples from eight different locations have been collected and processed to curd. The qPCR techniques have been used to identify the microbiota's DNA, and results suggested that the primary lactococcal phage population has significantly changed during the initial six hours of fermentation. The study also showed that the phage population reached 10 plaque-forming units per milliliter, and after that, acidification (a pH lag of 0.25) was observed to be decreased. This approach (Yu et al. 2025) enables the identification and mitigation of phage outbreaks prior to their occurrence.

Case Study 2: This study highlights how Musalais wine gets its flavor: Musalais is known as a traditional fortified wine originating from Xinjiang, China. In the current study, it has been demonstrated that metagenomics is commonly used to elucidate the correlation between microbial succession and the emergence of its distinctive olfactory signature (Elbehiry et al. 2025). In this study, shotgun metagenomics and gas chromatography-mass spectrometry (GC-MS) were utilized to identify diverse bacterial types and the presence of 82 volatile fragrance substances in the products. Earlier, it is well known that Saccharomyces cerevisiae was the primary yeast used in the fermentation process, along with 24 other yeast species that improved the taste (Pei et al. 2025).

8.6 CHALLENGES AND LIMITATIONS

Metagenomics is one of the most advanced domains in agriculture and food science. However, its application is constrained by methodological, computational, and governance challenges, which may lead to misinterpretation of biological inferences. The primary domains posing challenges in agricultural and food metagenomics are described below.

8.6.1 Sampling and DNA Extraction Biases

Sampling design and DNA extraction methods are major sources of variation in metagenomic studies. Inconsistent sampling—such as nonrepresentative subsampling of heterogeneous soils or foods, low-biomass samples, storage and shipping effects, and variation in extraction-kit chemistry—introduces bias into taxonomic and functional profiles. These factors can mislead downstream ecological and food-safety interpretations. The use of different DNA extraction methods results in microbial profiles that differ from those present in the original samples (Galla et al. 2024). Spatially balanced sampling or composite sampling helps maintain inferential integrity; however, soil and food matrices are often patchy, leading to small-scale heterogeneity and high variance. To manage DNA extraction biases, the use of validated kits and standardized protocols is essential for maintaining consistency across samples (Masenya et al. 2024). Kit contamination, low-biomass samples, and storage or transport effects represent major sources of error. Low-biomass samples are particularly affected by environmental factors, such as adverse climatic conditions, in environmental metagenomics studies (Liu et al. 2025).

8.6.2 Bioinformatics Bottlenecks and Data Interpretation

Assembly and binning limitations: Genome reconstruction and accurate assignment of mobile genetic elements to host organisms are computationally intensive processes and represent major bottlenecks in metagenomic analysis. Antimicrobial-resistance genes (ARGs) are often located on mobile genetic elements (MGEs), such as plasmids or transposons. Linking ARGs to their host genomes within microbial communities is critical for understanding resistance dissemination. Reconstruction of metagenome-assembled resistance genes (mARGs) from diverse ecosystems poses significant challenges for *de novo* assembly. Many MGEs contain highly repetitive regions that exceed the average *k*-mer length used in de Bruijn graph construction. This complexity hampers accurate assembly of these regions and shared genes across replicons and genomes, making sequence reconstruction difficult (Kerkvliet et al. 2024). Another major limitation of metagenomic metabarcoding is constrained taxonomic resolution. Metabarcoding relies on amplification and sequencing of specific genomic regions; however, these regions often lack sufficient resolution to enable accurate species-level identification (Masenya et al. 2024; Kumar et al. 2023).

Large raw datasets, ranging from gigabytes to terabytes, require high-throughput computational infrastructure and cloud-based storage. Developing analytical frameworks capable of managing massive, complex datasets generated by agricultural systems remains a significant challenge. Advanced algorithms are required to integrate heterogeneous data sources, including genomic sequences, soil nutrient profiles, and climate variables. Limited data sharing and collaboration within the competitive agricultural sector further hinder reproducibility and integration. Additionally, farmers often lack technical training and are unable to effectively utilize advanced digital technologies (Sahu et al. 2023).

8.6.3 Ethical, Regulatory, and Intellectual-Property Concerns in Food/Agriculture Microbiome Data

Ethical, regulatory, and intellectual-property concerns in food and agricultural microbiome data involve issues of data ownership, privacy, and benefit sharing, particularly when samples originate from indigenous or local resources. Regulations must ensure responsible use of genetic information, while IP rights should balance innovation with equitable access to microbiome-derived technologies for sustainable agriculture and food production.

Access and benefit-sharing (ABS): New global legal frameworks have been developed to ensure the sharing of due benefits and to maintain integrity while eliminating biopiracy. Intergovernmental negotiations aim to protect and conserve biological resources and make them available under the presumption of benefit sharing. Screening programs for the selection of desirable traits in crop breeding require access to vast genetic resources from different sources and countries. Differing national and local ABS laws and regulations act as barriers to entry, and seed companies struggle to establish themselves in breeding practices. ABS procedures under the CBD/Nagoya Protocol differ from country to country, are often unclear and highly bureaucratic, and are still evolving. All of these factors intensify cross-border sampling, strain isolation, commercialization, and data sharing, as legal obligations vary according to the provider country (Ebert et al. 2023).

Digital sequence information (DSI): Discoveries that result from metagenomic datasets are subject to tracing and benefit distribution. Under ABS regimes, the sharing of sequence data and its ease of access raise policy questions that must be addressed and regulated according to established norms. Based on the gathering and exchange of physical material, both the Nagoya Protocol and the International Treaty on Plant Genetic Resources for Food and Agriculture (ITPGRFA) were established. The Nagoya Protocol states that digital assets (DSI) are now the subject of global discussion and controversy, as they are accessible, valued, managed, and employed in ways different from physical items. The principles of data integrity, accessibility, and ethical use are important in ensuring that these technological advancements benefit a broad spectrum of society, promoting food security and equitable resource distribution. Therefore, the link connecting DSI and farmers' rights is complex, and this complexity must be acknowledged (Krishnamoorthi et al. 2024).

Patent and ownership challenges: Complex legal and ethical formalities are involved in filing patents on microbes, consortia, or derived functions such as enzymes and biosynthetic pathways. Rapid technological advancements in the production of genetically modified organisms have led to increasing debate on the social, ethical, and legal implications of these technologies. For commercialization and successful business ventures, protection of these inventions through intellectual-property rights (IPR) regimes is crucial. It is important to ensure that innovations comply with regulatory frameworks while keeping pace with societal concerns. Ethical concerns are intertwined with genetically modified microorganisms, and questions such as whether scientists are "interfering with nature" or "playing god" continue to be debated globally (Meena and Mohanty 2020).

8.7 FUTURE PERSPECTIVES

Metagenomics will revolutionize agriculture and food science by enabling precision microbiome engineering, sustainable crop management, and enhanced food quality and safety. Data-driven, predictive methods for improving environmental sustainability, resilience, and productivity will be enabled through integration with AI, synthetic biology, and gene editing. Table 8.1 shows that

TABLE 8.1
List of Metagenomics Drives the Discovery of Novel CRISPR/Cas Systems and Genome-Editing Tools for Precision Crop Improvement and Functional Validation of Microbial Genes

Aspect	Concept	Applications	Role of Metagenomics	References
CRISPR and genome-editing techniques	Application of *CRISPR/Cas systems* for precision genome editing in plants and microbes	- Discovery of novel CRISPR systems and regulatory genes from environmental datasets. - Used for improving crop traits—stress resistance, nutritional value, and yield.	- Expands the catalog of natural Cas variants and anti-CRISPR elements. - Enables validation of gene functions derived from metagenomic studies.	Chen et al. (2024)
AI/ML integration for predictive microbiome management	Use of *Artificial Intelligence (AI)* and *Machine Learning (ML)* to predict and manage microbiome function	- Predict soil health, disease risk, and yield outcomes. - Optimize fertilizer timing and SynCom composition. - Deep learning enhances taxonomic binning, functional annotation, and phenotype prediction.	- Integrates metagenomic, metatranscriptomic, and soil metadata for robust predictive modeling. - Supports *Decision-Support Systems (DSS)* for microbiome management.	Przymus et al. (2025); Pace et al. (2025)
Metagenomics in climate-resilient food systems	Leveraging microbial communities to enhance climate resilience in crops	- Identifies microbial strategies for improving tolerance to drought, heat, and salinity. - Integrates microbiome knowledge into breeding and inoculation programs.	- Characterizes functional diversity and resilience of soil microbes. - Aids in designing adaptive and climate-resilient food systems.	Clagnan et al. (2024)

metagenomics drives the discovery of novel CRISPR/Cas systems and genome-editing tools for precision crop improvement and functional validation of microbial genes. Integration of AI/ML enhances predictive microbiome management by linking metagenomic data to soil health and yield outcomes. Together, these approaches enable climate-resilient, sustainable food systems through optimized microbial functions and targeted interventions.

8.8 SYNTHETIC MICROBIOMES FOR PRECISION AGRICULTURE

Synthetic microbial communities (SynComs) have been developed as a reductionist approach to harness microbial benefits and understand multispecies interactions. Metagenomics maps the ecological and functional routes used to select taxa and functional genes for incorporation into SynComs, eliminating dependence on single-strain inoculants. SynComs are defined, artificially designed microbial consortia that provide beneficial services to plants and soils, such as nutrient mobilization, disease suppression, drought tolerance, and customized metabolite production. Bottom-up and top-down approaches are employed, involving functional screening of strains and host selection, to develop SynComs with a precise focus. SynCom construction is being optimized by considering stability, resilience, and reproducibility (Xu et al. 2025). The application pathway of SynComs includes field metagenomic surveys (shotgun sequencing), reconstruction of MAGs, trait annotation, and *in vitro* assembly, followed by screening and field trials (Clagnan et al. 2024).

8.9 CRISPR AND GENOME-EDITING APPLICATIONS INFORMED BY METAGENOMICS

Novel CRISPR/Cas systems from environmental datasets, regulatory elements, and catalogs of natural genes are expanded through metagenomics. These discoveries are harnessed for precision editing, including crop trait improvement and the development of microbial inoculants with desired functions. Genome editing expands the validation of gene functions identified through metagenomics. CRISPR (Clustered Regularly Interspaced Short Palindromic Repeats), derived from the adaptive immune system of bacteria and archaea, deploys a guide RNA (gRNA) to direct Cas nucleases to specific DNA sequences, where precise double-strand breaks occur. CRISPR/Cas technology has emerged as a revolutionary genome-editing tool in agricultural biotechnology and is used for targeted genome editing, advancing both basic and applied agricultural research. Crop varieties developed using CRISPR/Cas systems exhibit enhanced traits, including improved resistance to biotic and abiotic stresses, increased nutritional value, and potential for higher yields. CRISPR/Cas systems enhance agricultural productivity and sustainability due to their simplicity, adaptability, cost-effectiveness, and public acceptance, as they enable precise genetic alterations without the integration of foreign DNA. New Cas variants and anti-CRISPR elements mined from metagenomes provide alternative toolsets with diverse PAM specificities and temperature optima, which are valuable for plants and soil microbes (Chen et al. 2024).

8.10 AI/ML INTEGRATION FOR PREDICTIVE MICROBIOME MANAGEMENT

AI and machine-learning models, ranging from classical random forests to deep learning approaches, are used to map complex, high-dimensional metagenomic structures to functional outcomes such as soil health indices, disease risk, and yield predictions. These models also suggest management strategies, including fertilizer timing and SynCom composition, enabling predictive microbiome science. Deep learning and ensemble models outperform simple statistical methods in tasks such as taxonomic binning, functional annotation, and prediction of plant phenotypes from microbiome data. These improvements result from robust integration of metagenomic, metatranscriptomic, and soil metadata. Microbe-based recommendations are generated through decision-support systems (DSS) that integrate ML predictions with agronomic constraints. Deep learning algorithms demonstrate strong capabilities in pattern recognition, feature extraction, and predictive modeling, enabling the discovery of hidden relationships within microbial ecosystems (Przymus et al. 2025; Pace et al. 2025).

8.11 METAGENOMICS IN CLIMATE-RESILIENT FOOD SYSTEMS

Metagenomics governs the mediation of crop responses to climatic stressors such as drought, heat, and salinity by microbial communities. It also facilitates the identification of microbiome-based strategies to enhance resilience through breeding and targeted inoculation. Integration of metagenomics into climate-resilience design can make food systems more adaptive. Metagenomics, culturomics, and metabolic fingerprinting characterize the diversity, function, and resilience of soil microbial communities, highlighting their functional potential and fundamental ecological roles in maintaining environmental balance (Clagnan et al. 2024).

8.12 CONCLUSION

Metagenomics is the study that provides valuable insight into microbial populations and their relationships with livestock, agriculture, ecosystems, and food security. Metagenomic studies help in the analysis of genetic and metabolic information of different microbes that play important roles in agriculture and sustainable field development. This technology is helpful in developing and exploring the use of different microbes as biofertilizers, biocontrol agents, and ecosystem and food quality maintenance tools. These techniques help in understanding the metabolic pathways of microbes and the genes responsible for making microbes effective biofertilizers, biocontrol agents, and contributors to abiotic stress resistance. This advanced information is helpful in reducing the dependency on chemical fertilizers and pesticides in agricultural systems. Microbiome studies also play an important role for livestock farmers, as they help to control disease resistance and improve feed quality and animal health. Metagenomics-based information is helpful in pathogen monitoring, food spoilage monitoring, and the identification of new microbial strains that can be used as probiotics for animals. It serves as a foundation for advanced technologies involving NGS technology and bioinformatics approaches, which are helpful in developing environmentally friendly strategies for sustainable food development and agricultural practices. Metagenomic studies are also helpful in developing regulatory frameworks to establish resilient agro-ecosystems, enhance food security, and preserve microbial information for future generations. Moreover, metagenomics provides valuable insights, but it also has several obstacles to its implementation at practical levels. Some of the major obstacles include computational analysis, data-based predictive analysis, sampling bias, and ethical considerations. Advanced metagenomics incorporates AI and CRISPR–Cas9 technology, which are helpful in predicting microbiome control and developing safe and clean global food systems.

REFERENCES

Andrews, Tucker, Deborah A. Neher, Thomas R. Weicht, and John W. Barlow. 2019. "Mammary Microbiome of Lactating Organic Dairy Cows Varies by Time, Tissue Site, and Infection Status." *PLoS One* 14 (11): e0225001. https://doi.org/10.1371/journal.pone.0225001.

Bintsis, Thomas and Department of International Trade, TEI of West Macedonia, Kastoria, Greece. 2017. "Foodborne Pathogens." *AIMS Microbiology* 3 (3): 529–63. https://doi.org/10.3934/microbiol.2017.3.529.

Cannazza, Pietro, Antti J. Rissanen, Dieval Guizelini, et al. 2021. "Characterization of Komagataeibacter Isolate Reveals New Prospects in Waste Stream Valorization for Bacterial Cellulose Production." *Microorganisms* 9 (11): 2230. https://doi.org/10.3390/microorganisms9112230.

Chaudhary, Sorabh, Sagar, Sushma, Lal Mehi, et al. 2020. "Biocontrol and Growth Enhancement Potential of *Trichoderma* spp. Against *Rhizoctonia Solani* Causing Sheath Blight Disease in Rice." *Journal of Environmental Biology* 15 (5): 1034. https://doi.org/10.22438/jeb/41/5/MRN-1303.

Chaudhary, Sorabh, Sunil Kumar Yadav, Pankaj Verma, Sushma Sagar, and Mehi Lal. 2024. "Botanical Insecticides for Crop Protection: Major Classes and Possible Mechanisms of Action." In *Agricultural Sciences*, edited by Sarita Kumar, vol. 16. IntechOpen. https://doi.org/10.5772/intechopen.1006743.

Chauhan, Poonam, Neha Sharma, Ashwani Tapwal, et al. 2023. "Soil Microbiome: Diversity, Benefits and Interactions with Plants." *Sustainability* 15 (19): 14643. https://doi.org/10.3390/su151914643.

Chen, Feng, Lu Chen, Zhao Yan, et al. 2024. "Recent Advances of CRISPR-Based Genome Editing for Enhancing Staple Crops." *Frontiers in Plant Science* 15: 1478398. https://doi.org/10.3389/fpls.2024.1478398.

Clagnan, Elisa, Manuela Costanzo, Andrea Visca, et al. 2024. "Culturomics- and Metagenomics-Based Insights into the Soil Microbiome Preservation and Application for Sustainable Agriculture." *Frontiers in Microbiology* 15: 1473666. https://doi.org/10.3389/fmicb.2024.1473666.

Couvillion, Sneha P., Katie E. Mostoller, Janet E. Williams, et al. 2023. "Interrogating the Role of the Milk Microbiome in Mastitis in the Multi-Omics Era." *Frontiers in Microbiology* 14: 1105675. https://doi.org/10.3389/fmicb.2023.1105675.

Damián, Morayma Ramírez, Naima G. Cortes-Perez, Erika T. Quintana, et al. 2022. "Functional Foods, Nutraceuticals and Probiotics: A Focus on Human Health." *Microorganisms* 10 (5): 1065. https://doi.org/10.3390/microorganisms10051065.

Das, Basanta Kumar, Ayushman Gadnayak, Hirak Jyoti Chakraborty, Smruti Priyambada Pradhan, Subhashree Subhasmita Raut, and Sanjoy Kumar Das. 2025. "Exploring Microbial Players for Metagenomic Profiling of Carbon Cycling Bacteria in Sundarban Mangrove Soils." *Scientific Reports* 15 (1): 4784. https://doi.org/10.1038/s41598-025-89418-x.

Demirjian, Choghag, Fabienne Vailleau, Richard Berthomé, and Fabrice Roux. 2023. "Genome-Wide Association Studies in Plant Pathosystems: Success or Failure?" *Trends in Plant Science* 28 (4): 471–85. https://doi.org/10.1016/j.tplants.2022.11.006.

Du, Tianyu, Qufei Hu, Wenjing Mao, et al. 2023. "Metagenomics Insights into the Functional Profiles of Soil Carbon, Nitrogen, and Phosphorus Cycles in a Walnut Orchard Under Various Regimes of Long-Term Fertilisation." *European Journal of Agronomy* 148: 126887. https://doi.org/10.1016/j.eja.2023.126887.

Duncan, Sylvia H., and Carlos Sabater. 2025. "Emerging Probiotics: Future Therapeutics for Human Gut Health." *FEMS Microbiology Ecology* 101 (8): fiaf077. https://doi.org/10.1093/femsec/fiaf077.

Duysburgh, Cindy, Walter Fiore, and Massimo Marzorati. 2025. "Survival and Impact on Microbial Diversity of Lacticaseibacillus Paracasei DG in a Simulation of Human Intestinal Microbial Ecosystem." *Nutrients* 17 (18): 2952. https://doi.org/10.3390/nu17182952.

Ebert, Andreas W., Johannes M. M. Engels, Roland Schafleitner, Theo Van Hintum, and Godfrey Mwila. 2023. "Critical Review of the Increasing Complexity of Access and Benefit-Sharing Policies of Genetic Resources for Genebank Curators and Plant Breeders: A Public and Private Sector Perspective." *Plants* 12 (16): 2992. https://doi.org/10.3390/plants12162992.

Elbehiry, Ayman, Eman Marzouk, Adil Abalkhail, et al. 2025. "Microbial Food Safety and Antimicrobial Resistance in Foods: A Dual Threat to Public Health." *Microorganisms* 13 (7): 1592. https://doi.org/10.3390/microorganisms13071592.

Ercolini, Danilo. 2013. "High-Throughput Sequencing and Metagenomics: Moving Forward in the Culture-Independent Analysis of Food Microbial Ecology." *Applied and Environmental Microbiology* 79 (10): 3148–55. https://doi.org/10.1128/AEM.00256-13.

Escudero-Martinez, Carmen, Max Coulter, Rodrigo Alegria Terrazas, et al. 2022. "Identifying Plant Genes Shaping Microbiota Composition in the Barley Rhizosphere." *Nature Communications* 13 (1): 3443. https://doi.org/10.1038/s41467-022-31022-y.

Faqir, Yahya, Abdul Qayoom, Elizabeth Erasmus, Marietjie Schutte-Smith, and Hendrik G. Visser. 2024. "A Review on the Application of Advanced Soil and Plant Sensors in the Agriculture Sector." *Computers and Electronics in Agriculture* 226: 109385. https://doi.org/10.1016/j.compag.2024.109385.

Fierer, Noah. 2017. "Embracing the Unknown: Disentangling the Complexities of the Soil Microbiome." *Nature Reviews Microbiology* 15 (10): 579–90. https://doi.org/10.1038/nrmicro.2017.87.

Galanova, Olesya O., Nikita A. Mitkin, Albina A. Danilova, et al. 2025. "Assessment of Soil Health Through Metagenomic Analysis of Bacterial Diversity in Russian Black Soil." *Microorganisms* 13 (4): 854. https://doi.org/10.3390/microorganisms13040854.

Galla, Giulio, Nadine Praeg, Theresa Rzehak, et al. 2024. "Comparison of DNA Extraction Methods on Different Sample Matrices within the Same Terrestrial Ecosystem." *Scientific Reports* 14 (1): 8715. https://doi.org/10.1038/s41598-024-59086-4.

Hafeez, Rahila, Junning Guo, Temoor Ahmed, et al. 2024. "Bio-Formulated Chitosan Nanoparticles Enhance Disease Resistance Against Rice Blast by Physiomorphic, Transcriptional, and Microbiome Modulation of Rice (*Oryza Sativa* L.)." *Carbohydrate Polymers* 334 (June): 122023. https://doi.org/10.1016/j.carbpol.2024.122023.

Hamilton, Allyson N., Sarah L. Jones, Christopher A. Baker, et al. 2025. "A Systematic Review and Meta-Analysis of Chemical Sanitizer Efficacy Against Biofilms of Listeria Monocytogenes, Salmonella Enterica, and STEC on Food Processing Surfaces." *Journal of Food Protection* 88 (5): 100495. https://doi.org/10.1016/j.jfp.2025.100495.

Henry, Lucas P., Marjolein Bruijning, Simon K. G. Forsberg, and Julien F. Ayroles. 2021. "The Microbiome Extends Host Evolutionary Potential." *Nature Communications* 12 (1): 5141. https://doi.org/10.1038/s41467-021-25315-x.

Hu, Jie, Zhong Wei, George A. Kowalchuk, Yangchun Xu, Qirong Shen, and Alexandre Jousset. 2020. "Rhizosphere Microbiome Functional Diversity and Pathogen Invasion Resistance Build Up During Plant Development." *Environmental Microbiology* 22 (12): 5005–18. https://doi.org/10.1111/1462-2920.15097.

Inglis, Laura K., and Robert A. Edwards. 2022. "How Metagenomics Has Transformed Our Understanding of Bacteriophages in Microbiome Research." *Microorganisms* 10 (8): 1671. https://doi.org/10.3390/microorganisms10081671.

Iqbal, Babar, Guanlin Li, Khulood Fahad Alabbosh, et al. 2023. "Advancing Environmental Sustainability through Microbial Reprogramming in Growth Improvement, Stress Alleviation, and Phytoremediation." *Plant Stress* 10: 100283. https://doi.org/10.1016/j.stress.2023.100283.

Jagadesh, M., Munmun Dash, Aradhna Kumari, et al. 2024. "Revealing the Hidden World of Soil Microbes: Metagenomic Insights into Plant, Bacteria, and Fungi Interactions for Sustainable Agriculture and Ecosystem Restoration." *Microbiological Research* 285: 127764. https://doi.org/10.1016/j.micres.2024.127764.

James, A. S., N. A. Adil, D. Goltz, et al. 2024. "Abnormalities in Gut Virome Signatures Linked with Cognitive Impairment in Older Adults." *Gut Microbes* 16 (1): 2431648. https://doi.org/10.1080/19490976.2024.2431648.

Kerkvliet, Jesse J., Alex Bossers, Jannigje G. Kers, Rodrigo Meneses, Rob Willems, and Anita C. Schürch. 2024. "Metagenomic Assembly Is the Main Bottleneck in the Identification of Mobile Genetic Elements." *PeerJ* 12: e16695. https://doi.org/10.7717/peerj.16695.

Kim, Nayeon, Junyeong Ma, Wonjong Kim, Jungyeon Kim, Peter Belenky, and Insuk Lee. 2024. "Genome-Resolved Metagenomics: A Game Changer for Microbiome Medicine." *Experimental & Molecular Medicine* 56 (7): 1501–12. https://doi.org/10.1038/s12276-024-01262-7.

Kracmarova, Martina, Hana Kratochvilova, Ondrej Uhlik, et al. 2020. "Response of Soil Microbes and Soil Enzymatic Activity to 20 Years of Fertilization." *Agronomy* 10 (10): 1542. https://doi.org/10.3390/agronomy10101542.

Krishnamoorthi, A, Nishant Kumar Salam, Dr. P. Kalaiselvi, et al. 2024. "Digital Sequencing Information in Agricultural Crops: A Comprehensive Review." *International Journal of Advanced Biochemistry Research* 8 (8): 1404–12. https://doi.org/10.33545/26174693.2024.v8.i8r.2019.

Kumar, V., S. Gangani, R. Shukla, S. K. Prajapati, H. Shekhar, and V. Shukla. 2024. "Next-Generation Sequencing and Genomic Data Analysis." In A. Chaudhary, S. K. Sethi, and A. Verma (Eds.), *Unraveling New Frontiers and Advances in Bioinformatics* (pp. 49–76). Springer Nature. https://doi.org/10.1007/978-981-97-7123-3_3

Kumar, V., A. Pandey, A. Arora, P. Gautam, D. Bisht, S. Gupta, A. Chaurasia, and M. Sachan. 2023. "Diagnostics and Therapeutic Potential of miR-205 and miR-34a in Ovarian Cancer Management: A miRNA-Target-Based Analysis." *DNA and Cell Biology* 42 (3): 151–62. https://doi.org/10.1089/dna.2022.0487.

Kumar, V., J. K. Sharma, S. Kumar, J. Gulia, N. P. Singh, and A. R. Santal. 2025. "Metagenomic and Proteomic Approach for Bioremediation of Environmental Pollutants." In Maulin P. Shah (Ed.), *Development in Waste Water Treatment Research and Processes*. Elsevier. https://doi.org/10.1016/B978-0-443-13615-3.00020-0.

Kumar, V., R. Shukla, S. Gangani, H. Shekhar, and V. Shukla. 2024. "Metagenomics and Microbiome Analysis." In A. Chaudhary, S. K. Sethi, and A. Verma (Eds.), *Unraveling New Frontiers and Advances in Bioinformatics* (pp. 77–92). Springer Nature Singapore. https://doi.org/10.1007/978-981-97-7123-3_4.

Kumar, V., A. Singh, P. Gautam, and M. Sachan. 2019. "Techniques/Tools to Study Epigenetic Biomarkers in Human Cancer Detection." In S. Paul (Ed.), *Biomedical Engineering and Its Applications in Healthcare*. Springer. https://doi.org/10.1007/978-981-13-3705-5_14.

Kumar, Vinay, Lata Jain, Pankaj Kaushal, and Ravindra Soni. 2021. "Fungal Endophytes and Their Applications as Growth Promoters and Biological Control Agents." In *Fungi Bio-Prospects in Sustainable Agriculture, Environment and Nano-Technology*. Elsevier. https://doi.org/10.1016/B978-0-12-821394-0.00012-3.

Kumar, Vinay, Anisha Srivastava, Lata Jain, Sorabh Chaudhary, Pankaj Kaushal, and Ravindra Soni. 2022. "Harnessing the Potential of Genetically Improved Bioinoculants for Sustainable Agriculture: Recent Advances and Perspectives." In *Trends of Applied Microbiology for Sustainable Economy*. Elsevier. https://doi.org/10.1016/B978-0-323-91595-3.00007-0.

Kwoji, Iliya Dauda, Olayinka Ayobami Aiyegoro, Moses Okpeku, and Matthew Adekunle Adeleke. 2023. "'Multi-Omics' Data Integration: Applications in Probiotics Studies." *npj Science of Food* 7 (1): 25. https://doi.org/10.1038/s41538-023-00199-x.

Lal, Mehi, Ashvani Kumar, Sorabh Chaudhary, R. K. Singh, Sanjeev Sharma, and Manoj Kumar 2022. "Antagonistic and Growth Enhancement Activities of Native *Pseudomonas* spp. Against Soil and Tuber-Borne Diseases of Potato (*Solanum tuberosum* L.)." *Egyptian Journal of Biological Pest Control* 32: 22. https://doi.org/10.1186/s41938-022-00522-w.

Lederberg, J., and A. T. McCray. 2001. "Ome SweetOmics—A Genealogical Treasury of Words." *Scientist* 15 (8). https://www.semanticscholar.org/paper/%60Ome-Sweet-%60Omics--A-Genealogical-Treasury-of-Words-Lederberg-McCray/c06ee544b5e87e82f7705c401e1eff5cc8e1f780.

Leducq, Jean-Baptiste, Louis-Philippe St-Amand, David Ross, and Steven W. Kembel. 2025. "A Phylogenomic and Metagenomic Meta-Analysis of Bacterial Diversity in the Phyllosphere Lifts a Veil on Hyphomicrobiales Dark Matter." *Preprint, Microbiology* 19. https://doi.org/10.1101/2025.08.19.671110.

Legrand, Thibault P. R. A., Pâmela A. Alexandre, Annaleise Wilson, Ryan J. Farr, Antonio Reverter, and Stuart E. Denman. 2025. "Genome-Centric Metagenomics Reveals Uncharacterised Microbiomes in Angus Cattle." *Scientific Data* 12 (1): 547. https://doi.org/10.1038/s41597-025-04919-8.

Lei, Zheng, Hua Zhang, Wenju Liu, et al. 2025. "Dynamic and Stable Core Microbiota Assist Plants in Enriching Selenium and Reducing Cadmium Absorption." *Advanced Science* 12 (25): e00862. https://doi.org/10.1002/advs.202500862.

Lian, Wen-Hui, Osama Abdalla Abdelshafy Mohamad, Lei Dong, et al. 2023. "Culturomics- and Metagenomics-Based Insights into the Microbial Community and Function of Rhizosphere Soils in Sinai Desert Farming Systems." *Environmental Microbiome* 18 (1): 4. https://doi.org/10.1186/s40793-023-00463-3.

Limasset, Antoine, Jean-François Flot, and Pierre Peterlongo. 2020. "Toward Perfect Reads: Self-Correction of Short Reads via Mapping on de Bruijn Graphs." *Bioinformatics* 36 (5): 1374–81. https://doi.org/10.1093/bioinformatics/btz102.

Lindow, Steven E., and Maria T. Brandl. 2003. "Microbiology of the Phyllosphere." *Applied and Environmental Microbiology* 69 (4): 1875–83. https://doi.org/10.1128/AEM.69.4.1875-1883.2003.

Liu, Shaopeng, Judith S. Rodriguez, Viorel Munteanu, et al. 2025. "Analysis of Metagenomic Data." *Nature Reviews Methods Primers* 5 (1): 5. https://doi.org/10.1038/s43586-024-00376-6.

Madigan, K. E., V. J. Maffei, R. Shukla, V. Kumar, H. Yadav, and R. B. Weinberg. 2025. "Fecal Microbiome Richness and Evenness Are Reduced and the Relative Abundance of Firmicutes Species Is Increased in Patients with Intestinal Methanogen Overgrowth." *Gastroenterology* 169 (S1): S1263. https://doi.org/10.1016/S0016-5085(25)03787-4.

Mahmud, Aliyu Ahmad, Sudhir K. Upadhyay, Abhishek K. Srivastava, and Ali Asger Bhojiya. 2021. "Biofertilizers: A Nexus Between Soil Fertility and Crop Productivity Under Abiotic Stress." *Current Research in Environmental Sustainability* 3: 100063. https://doi.org/10.1016/j.crsust.2021.100063.

Marco, Maria L., Mary Ellen Sanders, Michael Gänzle, et al. 2021. "The International Scientific Association for Probiotics and Prebiotics (ISAPP) Consensus Statement on Fermented Foods." *Nature Reviews Gastroenterology & Hepatology* 18 (3): 196–208. https://doi.org/10.1038/s41575-020-00390-5.

Masenya, Kedibone, Madira Coutlyne Manganyi, and Tshegofatso Bridget Dikobe. 2024. "Exploring Cereal Metagenomics: Unravelling Microbial Communities for Improved Food Security." *Microorganisms* 12 (3): 510. https://doi.org/10.3390/microorganisms12030510.

Meena, Sumer Singh, and Anee Mohanty. 2020. "Ethical, Patent, and Regulatory Issues in Microbial Engineering." In Vijai Singh, Ajay Kumar Singh, Poonam Bhargava, Madhvi Joshi, and Chaitanya G. Joshi (Eds.), *Engineering of Microbial Biosynthetic Pathways*. Springer. https://doi.org/10.1007/978-981-15-2604-6_8.

Mishra, S., L. Buddendorff, J. Hoover, R. Shukla, V. Kumar, S. Jain, and H. Yadav. 2024. "Microbiome Role in Cognitive Impairment: MiaGB Cohort Insights." *Innovation in Aging* 8 (S1), 811. https://doi.org/10.1093/geroni/igae098.2629.

Nair, S. U., Y. Lin, S. Jain, et al. 2024. "Unique Gut Microbiome Signatures Associated with Physical and Cognitive Function." *Alzheimer's & Dementia* 20 (S2): e092044. https://doi.org/10.1002/alz.092044.

Nwachukwu, Blessing Chidinma, and Olubukola Oluranti Babalola. 2022. "Metagenomics: A Tool for Exploring Key Microbiome with the Potentials for Improving Sustainable Agriculture." *Frontiers in Sustainable Food Systems* 6: 886987. https://doi.org/10.3389/fsufs.2022.886987.

O'Hara, Eóin, Stephanie A. Terry, Paul Moote, et al. 2023. "Comparative Analysis of Macroalgae Supplementation on the Rumen Microbial Community: Asparagopsis Taxiformis Inhibits Major Ruminal Methanogenic, Fibrolytic, and Volatile Fatty Acid-Producing Microbes in Vitro." *Frontiers in Microbiology* 14: 1104667. https://doi.org/10.3389/fmicb.2023.1104667.

Oakley, Brian B., Cesar A. Morales, J. Line, et al. 2013. "The Poultry-Associated Microbiome: Network Analysis and Farm-to-Fork Characterizations." *PLoS One* 8 (2): e57190. https://doi.org/10.1371/journal.pone.0057190.

Opitz-Ríos, Catherine, Alvaro Burgos-Pacheco, Francisca Paredes-Cárcamo, Javier Campanini-Salinas, and Daniel A. Medina. 2024. "Metagenomics Insight into Veterinary and Zoonotic Pathogens Identified in Urban Wetlands of Los Lagos, Chile." *Pathogens* 13 (9): 788. https://doi.org/10.3390/pathogens13090788.

Oyserman, Ben O., Stalin Sarango Flores, Thom Griffioen, et al. 2022. "Disentangling the Genetic Basis of Rhizosphere Microbiome Assembly in Tomato." *Nature Communications* 13 (1): 3228 https://doi.org/10.1038/s41467-022-30849-9.

Pace, Roberta, Vincenzo Schiano Di Cola, Maurilia Maria Monti, et al. 2025. "Artificial Intelligence in Soil Microbiome Analysis: A Potential Application in Predicting and Enhancing Soil Health—a Review." *Discover Applied Sciences* 7 (2): 85. https://doi.org/10.1007/s42452-024-06381-4.

Padhiary, Mrutyunjay, Debapam Saha, Raushan Kumar, Laxmi Narayan Sethi, and Avinash Kumar. 2024. "Enhancing Precision Agriculture: A Comprehensive Review of Machine Learning and AI Vision Applications in All-Terrain Vehicle for Farm Automation." *Smart Agricultural Technology* 8 (August): 100483. https://doi.org/10.1016/j.atech.2024.100483.

Pawar, Sachin S., Pallavi A. Mohanapure, Manoj P. Brahmane, Mukesh P. Bhendarkar, Avinash V. Nirmale, and Nitin P. Kurade. 2021. "Metagenomics: A Novel Tool for Livestock and Poultry Improvement: A Review." *Agricultural Reviews*. https://doi.org/10.18805/ag.R-2167.

Pei, Yongzeng, Mengrong Chen, and Qiling Chen. 2025. "Microbial Dynamics in a Musalais Wine Fermentation: A Metagenomic Study." *Foods* 14 (15): 2570. https://doi.org/10.3390/foods14152570.

Pimentel, Tatiana Colombo, Whyara Karoline Almeida Da Costa, Carlos Eduardo Barão, Michele Rosset, and Marciane Magnani. 2021. "Vegan Probiotic Products: A Modern Tendency or the Newest Challenge in Functional Foods." *Food Research International* 140: 110033. https://doi.org/10.1016/j.foodres.2020.110033.

Pradhan, Nikita, Shikha Singh, Garima Saxena, et al. 2025. "A Review on Microbe-Mineral Transformations and Their Impact on Plant Growth." *Frontiers in Microbiology* 16: 1549022. https://doi.org/10.3389/fmicb.2025.1549022.

Prajapati, S. K., V. Kumar, R. Shukla, C. Szekeres, Y. Ma, S. Jain, and H. Yadav. 2024. "Microbiome-Immune Interplay in Aging Brains: A Pilot Investigation from the MiaGB Cohort." *Alzheimer's & Dementia* 20 (S2): e090734. https://doi.org/10.1002/alz.090734.

Przymus, Piotr, Krzysztof Rykaczewski, Adrián Martín-Segura, et al. 2025. "Deep Learning in Microbiome Analysis: A Comprehensive Review of Neural Network Models." *Frontiers in Microbiology* 15 (January): 1516667. https://doi.org/10.3389/fmicb.2024.1516667.

Rajguru, Bhumi, Manju Shri, and Vaibhav D. Bhatt. 2024. "Exploring Microbial Diversity in the Rhizosphere: A Comprehensive Review of Metagenomic Approaches and Their Applications." *3 Biotech* 14 (10): 224. https://doi.org/10.1007/s13205-024-04065-9.

Raza, Muhammad Ali, Hassan Shehryar Yasin, Hina Gul, et al. 2022. "Maize/Soybean Strip Intercropping Produces Higher Crop Yields and Saves Water Under Semi-Arid Conditions." *Frontiers in Plant Science* 13: 1006720. https://doi.org/10.3389/fpls.2022.1006720.

Sadurski, Jan, Magdalena Polak-Berecka, Adam Staniszewski, and Adam Waśko. 2024. "Step-by-Step Metagenomics for Food Microbiome Analysis: A Detailed Review." *Foods* 13 (14): 2216. https://doi.org/10.3390/foods13142216.

Sahu, Sunil Kumar, Muhammad Waseem, and Mehtab Muhammad Aslam. 2023. "Editorial: Bioinformatics, Big Data and Agriculture: A Challenge for the Future." *Frontiers in Plant Science* 14: 1271305. https://doi.org/10.3389/fpls.2023.1271305.

Santos, Lidiane Figueiredo, and Fabio Lopes Olivares. 2021. "Plant Microbiome Structure and Benefits for Sustainable Agriculture." *Current Plant Biology* 26: 100198. https://doi.org/10.1016/j.cpb.2021.100198.

Sarfraz, H., B. J. Czerniecki, K. Stankiewicz, et al. 2024. "Evaluating Gut Microbiome as a Biomarker of Pathological Complete Response in Patients with Early Locally Advanced Triple Negative Breast Cancer (LA TNBC): A Pilot Study." *Journal of Clinical Oncology* 42 (S16): 593. https://doi.org/10.1200/JCO.2024.42.16_suppl.593.

Shi, Yixuan, Yuansheng He, Yuanxian Zheng, et al. 2024. "Characteristics of the Phyllosphere Microbial Community and Its Relationship with Major Aroma Precursors during the Tobacco Maturation Process." *Frontiers in Plant Science* 15: 1346154. https://doi.org/10.3389/fpls.2024.1346154.

Simon, Carola, and Rolf Daniel. 2009. "Achievements and New Knowledge Unraveled by Metagenomic Approaches." *Applied Microbiology and Biotechnology* 85 (2): 265–76. https://doi.org/10.1007/s00253-009-2233-z.

Simon, Carola, and Rolf Daniel. 2011. "Metagenomic Analyses: Past and Future Trends." *Applied and Environmental Microbiology* 77 (4): 1153–61. https://doi.org/10.1128/AEM.02345-10.

Stone, Bram W. G., Eric A. Weingarten, and Colin R. Jackson. 2018. "The Role of the Phyllosphere Microbiome in Plant Health and Function." In Jeremy A. Roberts (Ed.), *Annual Plant Reviews Online* (1st ed.). Wiley. https://doi.org/10.1002/9781119312994.apr0614.

Suh, Jungho, Cevat Tosun, Thomas Eck, and Soyoung An. 2022. "A Cross-Cultural Study of Value Priorities Between U.S. and Chinese Airbnb Guests: An Analysis of Social and Economic Benefits." *Sustainability* 15 (1): 223. https://doi.org/10.3390/su15010223.

Tamang, Jyoti Prakash, Paul D. Cotter, Akihito Endo, et al. 2020. "Fermented Foods in a Global Age: East Meets West." *Comprehensive Reviews in Food Science and Food Safety* 19 (1): 184–217. https://doi.org/10.1111/1541-4337.12520.

Trivedi, Pankaj, Manuel Delgado-Baquerizo, Chanda Trivedi, et al. 2016. "Microbial Regulation of the Soil Carbon Cycle: Evidence from Gene-Enzyme Relationships." *The ISME Journal* 10 (11): 2593–604. https://doi.org/10.1038/ismej.2016.65.

Usyk, Mykhaylo, Brandilyn A. Peters, Smruthi Karthikeyan, et al. 2023. "Comprehensive Evaluation of Shotgun Metagenomics, Amplicon Sequencing, and Harmonization of These Platforms for Epidemiological Studies." *Cell Reports Methods* 3 (1): 100391. https://doi.org/10.1016/j.crmeth.2022.100391.

Verma, Pankaj, Sorabh Chaudhary, Sushma Sagar, Avinash Singh, Meenu Singh, and Akash Tomar. 2025. "Morpho-Molecular Characterization of Native *Trichoderma* Spp. Isolates and Their Biocontrol Efficacy Against *Fusarium Oxysporum*." *Vegetos*, ahead of print (February 17). https://doi.org/10.1007/s42535-025-01168-3.

Vishwakarma, Kanchan, Mitali Mishra, Shruti Jain, et al. 2017. "Exploring the Role of Plant-Microbe Interactions in Improving Soil Structure and Function Through Root Exudation: A Key to Sustainable Agriculture." In Dhananjaya Pratap Singh, Harikesh Bahadur Singh, and Ratna Prabha (Eds.), *Plant-Microbe Interactions in Agro-Ecological Perspectives*. Springer. https://doi.org/10.1007/978-981-10-5813-4_23.

Vishwakarma, Kanchan, Neha Upadhyay, Nitin Kumar, et al. 2017. "Microbial Interactions in Litchi Rhizosphere." In Manoj Kumar, Vivek Kumar, Neera Bhalla-Sarin, and Ajit Varma (Eds.), *Lychee Disease Management*. Springer. https://doi.org/10.1007/978-981-10-4247-8_2.

Wang, Haocai, Hang Wang, Thomas W Crowther, et al. 2024. "Metagenomic Insights into Inhibition of Soil Microbial Carbon Metabolism by Phosphorus Limitation During Vegetation Succession." *ISME Communications* 4 (1): ycae128. https://doi.org/10.1093/ismeco/ycae128.

Wang, Xueling, Yongkuan Chi, and Shuzhen Song. 2024. "Important Soil Microbiota's Effects on Plants and Soils: A Comprehensive 30-Year Systematic Literature Review." *Frontiers in Microbiology* 15 (March): 1347745. https://doi.org/10.3389/fmicb.2024.1347745.

Wang, Yao, Peng Yu, and Yong-Xin Liu. 2024. "Metagenomic Analysis for Unveiling Agricultural Microbiome." *Agronomy* 14 (5): 981. https://doi.org/10.3390/agronomy14050981.

Xu, Xinming, Caja Dinesen, Adele Pioppi, Ákos T. Kovács, and Carlos N. Lozano-Andrade. 2025. "Composing a Microbial Symphony: Synthetic Communities for Promoting Plant Growth." *Trends in Microbiology* 33 (7): 738–51. https://doi.org/10.1016/j.tim.2025.01.006.

Yaqub, Muneer Oladipupo, Chinedu Eucharia Joseph, Aashika Jain, and Lekshmi K. Edison. 2024. "Resistome Mapping in Foodborne Pathogens: Understanding Role in the Transmission Dynamics of Resistance Genes." *Applied Microbiology* 4 (4): 1476–92. https://doi.org/10.3390/applmicrobiol4040102.

Yousuf, Salsabeel, Yao Wang, Peng Yu, and Yong-Xin Liu. 2025. "Metagenomic Analysis for Unveiling Agricultural Microbiome—2nd Edition." *Agronomy* 15 (6): 1419. https://doi.org/10.3390/agronomy15061419.

Yu, Jun-Hyeok, Gabriele Andrea Lugli, Marco Ventura, et al. 2025. "Distinct Shifts in Bacteriophage Diversity and Abundance During Various Stages of Gouda-Type Cheese Production." *Applied and Environmental Microbiology* 91 (9): e00651.25. https://doi.org/10.1128/aem.00651-25.

Zhang, Nina, Antti Karkman, Kaisa Hiippala, Reetta Satokari, Timo M. Takala, and Per E. J. Saris. 2025. "Comparative Genomics and Characterization of Highly Adhesive and Stress Resistant Lacticaseibacillus Paracasei Strain E." *Current Research in Food Science* 11: 101245. https://doi.org/10.1016/j.crfs.2025.101245.

9 Metagenomics in Bioremediation and Bioengineering

Jitendra Kumar Sharma[1*], *Lovely Sinha*[2], *and Amit Lath*[1]
[1]Centre for Biotechnology, M. D. University, Rohtak, Haryana, India
[2]Department of Pulmonary Medicine, AIIMS Patna, Bihar, India
*Corresponding authors

9.1 INTRODUCTION

Metagenomics is the study of genetic material extracted directly from environmental samples such as soil, water, or the human gut, rather than isolating or cultivating particular bacteria. It focuses on examining the entire genome, or metagenome, of all microbes in a community. Unlike the approach of traditional microbiology, which relies on growing single species in the lab, metagenomics examines the whole DNA in a habitat to show the diversity and functional potential of all microbes, including those that cannot be grown in laboratories (Nam et al., 2023; Kumar et al., 2025).

Metagenomics has a broad range of scientific and practical applications. It is critical to microbial diversity and ecology since it allows scientists to detect and classify microorganisms in varied settings, as well as comprehend how they interact within ecosystems. Metagenomics uses functional genomics to investigate the genes found in microbial communities in order to determine their metabolic and ecological roles. In environmental monitoring, it is used to detect variations in microbial populations induced by pollution, habitat disturbance, or climate change, making it a useful tool for assessing ecosystem health (Zhang et al., 2021).

In the study of human health and disease, metagenomics allows for the exploration of the human microbiome, which sheds insight on its relationships to digestion, immunity, metabolic balance, and other pathologies (Madigan et al., 2025; Nair et al., 2024; Prajapati et al., 2024; Mishra et al., 2024; Sarfraz et al., 2024; James et al., 2024). It also helps with biotechnology and drug development by enabling researchers to mine ambient DNA for new enzymes, antibiotics, and bioactive chemicals with medical or industrial applications. In agriculture, metagenomic research focuses on soil- and plant-associated bacteria to improve nutrient cycling, plant growth, and resistance to pests and stress conditions. Furthermore, bioremediation applications use metagenomics to find microbes that can degrade environmental toxins, providing long-term solutions for ecosystem restoration (Lema et al., 2023).

Microbial communities are the living architecture of all ecosystems. They control nutrient cycles, detoxify contaminants, and provide biogeochemical stability. Their involvement is essential in restoration and engineering. Ignoring them is like attempting to rebuild a city without understanding its plumbing or power grid. Microbes process what enters the system, control what leaves, and form the environment in which higher species can thrive (Rebello et al., 2021). In environmental restoration, microbial communities are the first to respond to disturbance. When ecosystems are harmed by pollution, overexploitation, or habitat loss, microorganisms rearrange faster than plants or animals, often determining the path of recovery. Metagenomic data show that resilience stems from functional redundancy, which occurs when numerous taxa can perform the same ecological task. This redundancy provides biological insurance: even if some species are eliminated, key services

DOI: 10.1201/9781003670155-9

like carbon fixation and the nitrogen cycle continue. The challenge is not simply determining which bacteria are present, but also understanding how their interactions restore system stability (Shade et al., 2012).

Microbial communities can be used as dynamic biocatalysts in engineering applications, assisting with wastewater treatment, resource recovery, and bioenergy production. However, this framing entails a risk: engineers frequently treat microbial systems as controllable machines, expecting predictable results from defined inputs. Real microbial communities resist linear control. They change, adapt, and can sabotage the intended process by selecting faster but less efficient routes. This makes long-term administration a matter of directing community structure rather than imposing rigid conditions (Cerruti et al., 2021; Sun et al., 2025). Another important consideration is context dependency. A microbial consortia that thrives in the lab may fail in the environment due to temperature changes, nutritional constraints, or competition from native species. Restoration and bioengineering must consequently adopt an ecological engineering perspective, building with microbes rather than simply exploiting them. Effective techniques integrate metagenomic understanding with environmental feedback loops, allowing populations to self-organize toward stability rather than being driven to an artificial equilibrium (Rebello et al., 2021; Sharma, 2022).

Finally, microbial communities are important because they demonstrate the idea that function develops from interaction. A single strain cannot replicate the durability, flexibility, or biochemical diversity of a well-structured community. Treating microorganisms as discrete tools ignores the underlying truth: ecosystems are cooperative metabolic systems, and restoration or engineering can only succeed if those cooperative networks are understood and harnessed, rather than oversimplified (Shade et al., 2012; Bengtsson-Palme, 2020). Culture-based microbiology taught us the fundamentals of microbial physiology, metabolism, and taxonomy. Without it, we would not understand the fundamental growth requirements or functional roles of even common microorganisms. However, there is a limitation: culture methods only reveal a subset of microbial diversity, and they are often biased toward fast-growing, easy-to-manage organisms. The premise that what we can culture accurately represents the genuine microbial world is wrong. Most environmental microorganisms resist normal cultivation; thus, our initial assessment was skewed (Lagier et al., 2015; Wu et al., 2020).

By directly sequencing DNA from ambient samples, metagenome-driven techniques broke down that barrier. This enabled us to identify organisms that cannot be grown, discover new metabolic pathways, and map community relationships. However, this approach has several limitations. Sequencing depth, assembly problems, and reliance on reference databases can all skew ecological interpretations. Calling metagenomics "complete" is a little arrogant (Pettersen et al., 2022; Abdul Aziz & Masmoudi, 2025).Another issue to consider is that culture and metagenomics are not mutually exclusive. New technologies, like high-throughput culturing driven by genetic predictions, are narrowing the gap. This questions the basic binary of old vs new.

9.2 METAGENOMICS IN BIOREMEDIATION

9.2.1 Concept of Bioremediation

Bioremediation is the use of living organisms, primarily bacteria, that degrade, detoxify, or convert environmental contaminants into less dangerous forms. This can include chemicals found in soil, water, or air (Singh et al., 2016).Bioremediation processes are typically categorized as *in situ* or *ex situ*. *In situ* bioremediation occurs directly at the source of contamination, when treatments such as nutrient solution injection are utilized to encourage indigenous microbial communities, such as those found in groundwater. While this strategy eliminates the need to transport contaminated material, it has drawbacks, such as limited control over environmental conditions and uneven degradation rates (Singh et al., 2016). *Ex situ* bioremediation, on the other hand, entails removing contaminated material and treating it elsewhere, such as digging up soil for microbial composting. Although this technology allows for more controlled circumstances, it is frequently costly, logistically complicated, and may fail to imitate natural microbial interactions, thereby reducing overall effectiveness (Bala et al., 2022; Kuppan et al., 2024). Bioremediation occurs via different approaches, each dealing with pollutants in a distinct way.

Bioaugmentation involves the introduction of certain microorganisms into contaminated environments to expedite breakdown processes. The premise is that these bacteria will flourish and metabolize contaminants effectively; however, their survival frequently hinges on local conditions, competition with indigenous microbes, and the availability of nutrients (Muter, 2023).

Biostimulation aims to augment the activity of native microbial communities by the provision of nutrients, oxygen, or other conducive conditions. This approach is effective but assumes that native bacteria have the necessary metabolic pathways to degrade the pollutants, which is not assured (Tribedi et al., 2018; Li et al., 2025). Phytoremediation employs plants and their symbiotic bacteria to absorb, accumulate, or convert contaminants in soil or water. This approach seems refined and sustainable; yet, it may be sluggish and significantly reliant on plant species, pollutant type, and environmental circumstances (Sharma et al., 2023). These mechanisms reveal that bioremediation is seldom a straightforward solution; its efficacy necessitates a meticulous alignment of the biological agent, environmental context, and contaminant attributes, and it frequently benefits from the integration of many strategies. All three processes are depicted in Figure 9.1.

The fundamental basis of efficient bioremediation lies in microbial diversity and metabolic plasticity. A diversified microbial population enhances the likelihood that certain species can withstand and decompose a broad spectrum of contaminants, including those that are chemically intricate or resistant (Kanwar et al., 2024; Singh & Chand Kumawat, 2025). Metabolic flexibility enables microorganisms to alternate between several energy sources or degradation pathways based on environmental variables, hence enhancing the resilience of bioremediation to variations in pH, temperature, or nutrient supply (Qattan, 2025). Diversity alone is insufficient; a proper combination of species with complementary metabolic capabilities is essential. Exaggerating the influence of a solitary "superbug" is a frequent error, as actual ecosystems are predominantly influenced by relationships, competition, and environmental stressors, rather than the conditions observed in laboratory settings (Cunningham et al., 2020; Koch et al., 2021). The greater the diversity of the microbial toolbox, the more resistant and adaptable the bioremediation process can become; however, leveraging this diversity necessitates a thorough comprehension of ecological dynamics rather than the indiscriminate addition of microbes.

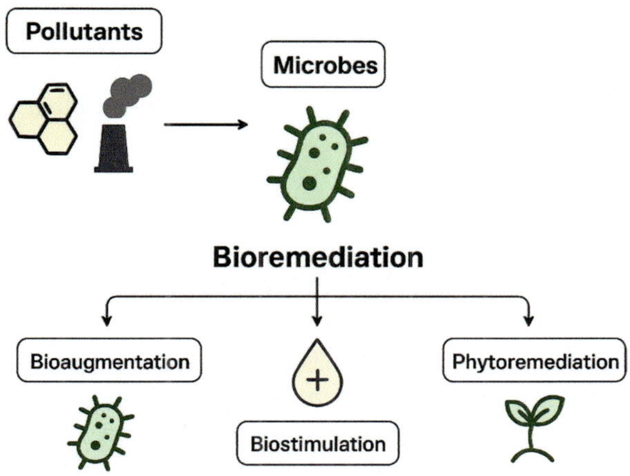

FIGURE 9.1 Integrated Bioremediation Technologies for Pollutant Degradation and Environmental Restoration.

9.2.2 Pollutant Degradation Pathways

9.2.2.1 Hydrocarbon and Oil Spill Degradation

The breakdown of hydrocarbons and oil spills is predominantly dependent on bacteria capable of metabolizing intricate combinations of alkanes, aromatics, and polycyclic molecules. Natural attenuation can be effective; nevertheless, the degradation process is sometimes sluggish due to the resistance of heavier hydrocarbons to microbial decomposition (Xu et al., 2018). Biostimulation or bioaugmentation can expedite the process; nevertheless, success is contingent upon oxygen availability, nutritional equilibrium, and the indigenous microbial community. There exists a potential that partial degradation generates intermediate molecules that are more hazardous than the original hydrocarbons, a fact frequently neglected (Atlas & Hazen, 2011).

9.2.2.2 Heavy Metal Detoxification

Heavy metal detoxification, encompassing arsenic, mercury, cadmium, and lead, functions distinctly, as metals are indestructible; microorganisms convert them into less bioavailable or less poisonous forms. For example, certain bacteria methylate mercury or convert hexavalent chromium to trivalent chromium (Monachese et al., 2012; Naziębło & Dobrzyński, 2025). The issue lies in the reversibility of these conversions and the potential for environmental changes to remobilize metals; thus, bioremediation focuses more on containment and stabilization rather than complete elimination. Assuming that microbial detoxification is irreversible is an oversimplification that may yield adverse consequences.

9.2.2.3 Pesticide and Herbicide Biodegradation

The biodegradation of pesticides and herbicides relies on bacteria with enzymes that can decompose organophosphates, carbamates, or chlorinated chemicals. The issue is that numerous synthetic pesticides were engineered to withstand natural breakdown, necessitating adaptation or augmentation of microbial communities (Vermelho et al., 2024). Moreover, breakdown pathways may produce secondary metabolites that maintain toxicity or bioaccumulate; thus, monitoring the end products is essential rather than presuming the pollutant merely vanishes.

9.2.2.4 Plastic and Microplastic Degradation

The degradation of plastic and microplastics is becoming a significant area of emphasis, especially with polyethylene, PET, and polystyrene. Some bacteria and fungi can excrete enzymes such as PETases or laccases to degrade polymer chains; nevertheless, the rates are exceedingly slow in normal environments (Dhali et al., 2024). Environmental factors such as ultraviolet exposure, temperature, and microbial competition significantly influence outcomes. Additionally, microplastics provide a complex substrate containing additives that may impede microbial activity, necessitating tailored solutions or synergistic microbial consortia for breakdown (Sutkar et al., 2023). All major pathways are discussed in Table 9.1.

9.2.3 Metagenomic Insights

9.2.3.1 Identification of Key Taxa and Functional Genes Involved in Pollutant Degradation

Identifying key taxa and functional genes implicated in pollutant degradation necessitates more than just DNA sequencing and database alignment. The procedure commences with careful sampling across environmental gradients, including variations in depth, redox potential, or pollutant concentration, succeeded by the extraction of DNA and RNA. Amplicon sequencing identifies the current bacteria, whereas shotgun metagenomics elucidates the extensive gene pool and possible metabolic functions (Chettri et al., 2024). Metatranscriptomics and metaproteomics are employed to correlate these genes with functional activity by identifying the pathways activated in response

TABLE 9.1
Major Microbial Pathways Involved in Pollutant Degradation across Different Contaminant Types

Pollutant Type	Primary Mechanisms	Key Microorganisms	Main Enzymes	Degradation Products	Environmental Conditions	Key Citations/References
Hydrocarbon and oil spills	Aerobic biodegradation, oxidation, mineralization	*Acinetobactersp., Pseudomonassp., Sphingomonassp., Gordonia, Mycobacterium*	Alkane hydroxylases, cytochrome P450, oxygenases, dehydrogenases	CO_2, H_2O, biomass, fatty acids	Aerobic conditions, optimal temperature 15–45°C, neutral pH	Das and Chandran (2011); Kostka et al. (2011); Ron and Rosenberg (2014)
Heavy metal detoxification	Biosorption, bioaccumulation, enzymatic detoxification, chelation	*Pseudomonassp., Bacillus sp., microalgae, fungi*	Metallothioneins, glutathione peroxidase, superoxide dismutase	Metal-protein complexes, precipitates, sequestered metals	Variable pH, presence of chelating agents, reducing conditions	Balali-Mood et al. (2021); Ayilara and Babalola (2023); Patil et al. (2024)
Pesticide and herbicide biodegradation	Hydrolysis, oxidation, conjugation, ring cleavage	*Sphingomonas sp., Novosphingobium sp., Bacillus sp., Pseudomonas sp.*	Hydrolases, oxidases, dehalogenases, cytochrome P450	Metabolites, amino acids, CO_2, H_2O, inorganic salts	Aerobic/anaerobic conditions, neutral pH, adequate moisture	Guerrero Ramírez et al. (2023); Wang et al. (2024); Shah et al. (2024)
Plastic and microplastic degradation	Enzymatic hydrolysis, oxidation, biofilm formation	*Aspergillus sp., Penicillium sp., Fusarium sp., Pseudomonas sp.*	PETases, cutinases, lipases, esterases, oxidases	Oligomers, monomers, CO_2, H_2O, organic acids	Elevated temperature (30–70°C), specific pH ranges, UV exposure	Cai et al. (2023); Zaini et al. (2024); Li and Zhang (2025)

to pollutant exposure. Stable isotope probing identifies active degraders by tracking labeled substrates, such as 13C-tagged contaminants, into microbial DNA or proteins, whereas metagenome-assembled genomes associate individual genes with their microbial hosts (Chandran et al., 2020). It is essential to recognize that the existence of a gene does not necessarily indicate its function; several studies have suggested microbial involvement by conflating DNA detection with degrading activity. Relative abundance measures can be misleading; a gene that seems seldom may nonetheless catalyze essential processes if expressed at high levels (Lee & Kalia, 2020). Furthermore, horizontal gene transfer typically obscures taxon-function boundaries, as degradative genes commonly exist on plasmids that transmit between unrelated hosts (Wang et al., 2014). To draw strong results, researchers must use absolute quantification by qPCR, multi-omics data, and microscale environmental context, preferably over time rather than relying on isolated snapshots (Aminov, 2011).

9.2.3.2 Functional Metagenomics for Enzyme Discovery

Functional metagenomics provides an effective means to identify novel enzymes, including oxygenases, hydrolases, and reductases, that decompose contaminants. Two primary strategies exist. The initial method is sequence-driven mining, wherein researchers employ bioinformatic techniques to forecast enzyme families and active sites, emphasizing genes produced in response to pollution stress. The second method is function-based screening, which entails cloning ambient DNA into expression hosts and assessing real activity on substrates. Each possesses drawbacks. Sequence-based approaches may inaccurately classify enzymes that deviate from established patterns, whereas function-based screens frequently overlook enzymes that do not express adequately in model organisms such as *E. coli* (Ngara & Zhang, 2018). Oxygenases specifically necessitate particular cofactors and companion proteins that several hosts lack, whereas reductases may rely on atypical electron donors. An intelligent strategy integrates both methodologies by utilizing metagenomic and transcriptome data to identify candidates, then expressing them in various systems such as *Pseudomonas*, *Rhodococcus*, or yeast (Nagayama et al., 2015). Hits from these screens must undergo biochemical validation by verifying substrate range, cofactor dependency, kinetics, and product synthesis under realistic conditions (Sharma et al., 2018; Fenner et al., 2021). Activity-based protein profiling can directly capture native enzyme activity from environmental materials, offering functional evidence. The overarching lesson is to reject the allure of novelty assertions predicated merely on sequence similarity, as discovery is only significant when the enzyme functions effectively in relevant contexts (Ninck et al., 2024).

9.2.3.3 Case Studies in Soil, Groundwater, and Marine Ecosystems

Case studies in soil, groundwater, and marine systems demonstrate how environmental context alters microbial functions in bioremediation. Soil degradation frequently entails bacteria such as *Pseudomonas* and *Rhodococcus*, which possess ring-hydroxylating dioxygenases, and fungi that generate laccases and peroxidases. Soil variability, microscale redox zones, unequal pollutant distribution, and sorption to organic matter result in degradation occurring in spatially localized hotspots. Metal and nutrient constraints also influence microbial activity, necessitating detailed geochemical mapping (Liu et al., 2025). Groundwater systems impose many constraints, including low biomass, restricted electron donors, and sluggish flow, which hinder anaerobic degraders such as *Dehalococcoides*, whose reductive dehalogenases eliminate chlorine atoms from solvents like TCE and PCE. Nonetheless, incomplete reactions may produce hazardous intermediates such as vinyl chloride if electron donors are inadequately controlled (van der Zaan et al., 2010; Dutta et al., 2022). In marine environments, the breakdown of hydrocarbons following oil spills is mostly facilitated by taxa such as *Alcanivorax* and *Cycloclasticus*, which possess alkane monooxygenases and aromatic ring-cleaving enzymes (Naether et al., 2013). Conversely, plastics breakdown occurs at a significantly slower rate; even the promising enzyme PETase, identified in terrestrial bacteria, exhibits diminished efficacy in saline, nutrient-deficient aquatic environments. Environmental pressures, including salinity, nutrient limitation, and dilution, determine the dominant taxa and their response rates (Kim et al., 2020). Controlled mesocosm investigations and metagenomic analyses of particle-attached vs free-living communities demonstrate that the majority of degradation occurs inside confined biofilms rather than in the bulk water.

9.3 METAGENOMICS IN BIOENGINEERING

9.3.1 Microbial Engineering and Synthetic Biology

Utilizing metagenomic data for the development of engineered microbial strains has emerged as a fundamental aspect of contemporary bioremediation and biocatalytic advancements. The core concept is to convert environmental genetic potential into functional, designed systems that execute specific tasks. Effectively utilizing metagenomic knowledge necessitates transcending the rudimentary concept that a sequence equates to a solution. Environmental metagenomes disclose a considerable diversity of genes, although only a subset is functionally expressed under particular conditions. Consequently, gene selection should depend not solely on sequence similarity but also on contextual information, including gene clusters, promoter structures, cofactor dependencies, and regulatory networks obtained from metatranscriptomic or metaproteomic studies (Zhang et al., 2021; Sinha et al., 2025). The design of engineered strains entails the reconstruction of whole functional modules, incorporating accessory proteins—such as electron transport components, chaperones, and membrane transporters—that maintain route integrity. A prevalent error in synthetic design is the presumption that genes derived from metagenomes may be directly produced in model organisms such as *Escherichia coli* without modification. Codon bias, folding assistance, and cofactor compatibility frequently constrain effective expression (Bhatwa et al., 2021). To tackle this issue, various expression systems, cell-free synthesis, and adaptive laboratory evolution can be utilized to evaluate gene performance and stability. Moreover, the ethical and ecological aspects of utilizing modified strains, including the risk of horizontal gene transfer or inadvertent ecological impact, must be integrated into design considerations from the beginning.

Pathway reconstruction and metabolic modeling offer a systems-level framework for predicting and optimizing the behavior of modified microbial pathways. Reconstruction commences with the identification of enzymes that may catalyze each reaction step, the validation of reaction directionality, and the integration of cofactor dependencies. Genome-scale metabolic models and flux balance analysis are prevalent methodologies employed to assess metabolic fluxes and pinpoint potential bottlenecks or hazardous intermediates (Qiu et al., 2023). Nevertheless, these models rely on simplifying assumptions, such as steady-state circumstances and fixed cellular objectives, which seldom reflect the dynamic and adaptive characteristics of microbial systems. Gap-filling algorithms, although beneficial for completeness, may generate implausible reactions, and an overdependence on model predictions might result in unwarranted confidence. Consequently, computer modeling needs to serve as a generator of hypotheses rather than a conclusive prediction. Iterative design–build–test–learn cycles that incorporate experimental data, isotope labeling investigations, and metabolomics yield a more robust empirical foundation. Special emphasis must be placed on cofactor equilibrium, thermodynamic viability, enzyme kinetics, and proteome allocation expenses, as these factors substantially affect pathway efficacy (Xeon, 2024). In specific instances, partitioning pathways among other strains or utilizing modular designs might mitigate metabolic burden and improve overall system resilience.

Enzyme mining for industrial biocatalysis exemplifies a significant use of metagenomics and synthetic biology, granting access to innovative catalytic functionalities capable of transforming industrial and environmental processes. Conventional sequence-based mining depends on homology searches and conserved motifs to forecast enzyme families, whereas functional metagenomics directly examines environmental DNA for quantifiable activity on specific substrates. Both strategies possess constraints. Sequence-based methodologies may neglect distantly related or novel enzymes, whereas function-based techniques rely on effective expression in heterologous hosts, which frequently fails for complicated or multicomponent enzymes. A synergistic strategy is thus more efficacious, amalgamating *in silico* predictions with empirical screening (Ngara & Zhang, 2018). Upon identifying potential enzymes, their industrial applicability must be assessed through biochemical characterization, emphasizing criteria such as substrate range, catalytic turnover, stability, solvent tolerance, and byproduct generation. Many natural enzymes necessitate additional optimization to satisfy industrial standards, achievable via directed evolution, structure-guided mutagenesis, or machine learning-assisted design. The utilization of high-throughput screening platforms, such as microfluidic devices and cell-free assays, expedites the discovery and optimization process (Sharma et al., 2024). Moreover, techniques such as enzyme immobilization, cofactor recycling, and incorporation into synthetic consortia can enhance process efficiency and scalability.

A critical viewpoint is vital when implementing microbial engineering and synthetic biology to address real-world issues. Not all functions identified in metagenomes are suitable for transfer into a single designed strain. Synthetic consortia, modular route distribution, and cell-free catalytic systems provide safer and more controllable options in certain instances. Furthermore, whereas synthetic biology offers the potential for precise design, the inherent unpredictability of biological systems requires ongoing validation, feedback, and modification. Through the integration of metagenomic insights, meticulous modeling, and iterative experimentation, microbial engineering can transition from potential to performance, thereby connecting environmental discovery with manufactured functionality (Grandel et al., 2021).

9.3.2 Industrial Applications

9.3.2.1 Production of Biofuels

Metagenomics has become a central tool for industrial biotechnology, providing direct access to the genetic potential of microbial communities without the need for culturing. One of the most prominent applications is in the production of biofuels, including ethanol, butanol, methane, and hydrogen. Environmental samples, such as anaerobic digesters, hot springs, and termite guts, harbor microbial enzymes capable of efficiently degrading lignocellulosic biomass and converting it into fermentable sugars or energy-rich gases. Metagenomic analysis allows the identification of robust cellulases, xylanases, hydrogenases, and dehydrogenases, which can then be incorporated into engineered strains or consortia to improve yield and process efficiency (Sartaj et al., 2022).

9.3.2.2 Production of Bioplastics and Biodegradable Polymers

Another major application is the development of bioplastics and biodegradable polymers. Microbes naturally produce polyhydroxyalkanoates (PHAs) and related polymers as storage compounds, and metagenomic mining of soil, compost, and marine environments has revealed novel PHA synthases, depolymerases, and accessory enzymes with enhanced activity and substrate versatility. These enzymes can be harnessed to create microbial platforms for sustainable polymer production, offering renewable alternatives to petroleum-based plastics while allowing control over polymer properties such as melting temperature, elasticity, and degradation rate (Paloyan et al., 2025).

9.3.2.3 Secondary Metabolites: Antibiotics, Pigments, and Novel Biomolecules

Metagenomics is crucial for the identification of secondary metabolites, such as antibiotics, pigments, and other bioactive chemicals. A significant number of these chemicals are encoded by biosynthetic gene clusters that are either silent or inadequately represented in cultivated organisms. Through the sequencing of ambient DNA and the reconstruction of these clusters, researchers can discern novel enzymes and routes for antibiotic synthesis, pigment creation, and specialized compounds (Lema et al., 2025). Synthetic biology techniques provide heterologous expression, route optimization, and modification of metabolites for commercial applications, hence enabling the synthesis of chemicals that are challenging or unattainable via conventional fermentation or chemical synthesis (Kotopka et al., 2018).

Moreover, metagenomic insights facilitate the development of microbial consortia for intricate industrial processes, such as multistep degradation of waste streams or co-fermentation of varied feedstocks. By recognizing complementary metabolic abilities among several species, it is feasible to construct communities that execute synergistic functions more effectively than individual strains. This concept facilitates applications from wastewater treatment to biorefinery operations and enhances circular economy initiatives (Tao et al., 2022).

9.3.3 Host–Microbe Systems

9.3.3.1 Engineering Microbial Consortia for Synergistic Processes

Engineering microbial consortia for synergistic activities signifies a transition from optimizing single strains to designing at the community level. Natural ecosystems illustrate that collaboration, specialization, and metabolic interdependence frequently surpass solitary monocultures. By emulating these dynamics, designed consortia can accomplish intricate transformations such as pollutant degradation, waste valorization, or multistep biosynthesis that would normally exceed the capabilities of a single organism (Duncker et al., 2021). Metagenomic discoveries inform the selection of complementing species by elucidating natural relationships, metabolite exchange pathways, and intercellular communication networks. The difficulty resides in converting ecological theory into reliable engineering outcomes. Consortia are intrinsically dynamic systems, responsive to nutritional variations, spatial arrangement, and competitive interactions. Synthetic biology instruments, including synthetic promoters, quorum-sensing circuits, and modular plasmid systems, facilitate the stabilization of population ratios and the coordination of gene expression among species (McCarty & Ledesma-Amaro, 2019). Nonetheless, an unresolved tension persists between control and adaptability: enforcing stringent regulations may diminish resilience, whereas allowing autonomy poses a risk of community disintegration. The most effective designs incorporate computational modeling, microfluidic coculture systems, and adaptive evolution experiments to optimize collaboration under variable settings.

9.3.3.2 CRISPR-Cas Systems Informed by Metagenomics

Metagenomics-informed CRISPR–Cas systems have greatly enhanced the accuracy and adaptability of genome engineering. Metagenomic data sets have revealed a broad array of CRISPR–Cas variations beyond the traditional Cas9 and Cas12 systems, encompassing small enzymes, multisubunit complexes, and RNA-guided nucleases with innovative recognition methods. These findings enable researchers to customize editing techniques for particular applications, including base editing, transcriptional regulation, and protection against mobile genetic elements (Azeez et al., 2024). Metagenomics not only uncovers novel Cas proteins but also provides contextual information regarding their natural regulatory systems, spacers, and anti-CRISPR interactions. This ecological comprehension facilitates the logical adaptation of CRISPR systems to novel hosts and settings. However, enthusiasm for CRISPR innovation must be tempered with realities. Not every newly found Cas variation is biochemically manageable or suitable for conventional laboratory expression platforms. Functional validation, structural characterization, and optimization of PAM compatibility are essential procedures before deployment (Hu et al., 2022). Moreover, ethical considerations with genome editing, particularly gene drive applications and environmental release, must be approached with transparency and prudence.

9.3.3.3 Application of Microbial Chassis Organisms

The utilization of microbial chassis organisms, including *Escherichia coli*, *Pseudomonas*, and *Bacillus*, underpins the majority of synthetic biology initiatives. These species are preferred due to their well-defined genetics, accessibility of molecular techniques, and vigorous growth under diverse environments (Calero & Nikel, 2019). *Escherichia coli* is the predominant host for heterologous expression owing to its rapid proliferation and comprehensive toolset; yet, it frequently has difficulties expressing intricate pathways necessitating membrane-associated proteins, cofactors, or secretion systems. *Pseudomonas* exhibits extensive metabolic diversity and solvent tolerance, rendering it suitable for the degradation of hydrophobic substrates and the production of secondary metabolites. *Bacillus* species serve as superior platforms for protein secretion, especially in the synthesis of industrial enzymes. The strategic selection of a chassis must evaluate growth kinetics, genetic accessibility, metabolic compatibility, and biosafety. Recently, nontraditional chassis, including *Corynebacterium*, *Streptomyces*, and extremophiles, have garnered interest for specific purposes (Qattan, 2025). Notwithstanding these advancements, the presumption that a single chassis can fulfill all functions is fundamentally erroneous. Metagenomic data promote a distributed design philosophy wherein various chassis fulfill specific functions within a modular or consortium-based framework.

9.4 TOOLS AND APPROACHES

9.4.1 Shotgun Metagenomics and Targeted Approaches (16S/ITS)

Understanding complex microbial systems and their biotechnological potential depends on the careful selection and integration of analytical tools. One foundational choice lies between shotgun metagenomics and targeted approaches such as 16S rRNA or ITS sequencing. Targeted methods offer cost-effective community profiling, revealing taxonomic composition and relative abundance patterns across samples. They are particularly useful for monitoring shifts in microbial diversity under environmental or experimental stresses (Mani, 2023; Kumar, Gangani, et al., 2024). However, they fall short when the goal is to uncover functional potential, since marker genes provide limited insight into metabolic pathways or adaptive traits. Shotgun metagenomics, by contrast, sequences all genomic material in a sample, enabling the reconstruction of entire genomes, metabolic pathways, and gene clusters. The trade-off lies in cost, computational complexity, and the challenge of assembling fragmented sequences from mixed communities. An in-depth analysis identifies these approaches as complementary: targeted sequencing provides ecological resolution, while shotgun metagenomics delivers functional depth (Sharpton, 2014; Kumar, Shukla, et al., 2024). The key is not choosing one over the other but designing workflows that integrate both, balancing resolution, resource constraints, and research objectives.

9.4.2 Functional Metagenomics Libraries for Novel Gene Discovery

Functional metagenomics libraries constitute a robust experimental approach for identifying novel genes and enzymes independent of prior sequence annotations. This method involves cloning environmental DNA into appropriate hosts, typically *E. coli* or *Bacillus*, and screening for phenotypic characteristics such as pollutant breakdown, antibiotic synthesis, or substrate utilization. This approach connects sequence data with biochemical activity, uncovering functional advances that sequence homology may miss. Numerous advancements in enzyme discovery, including novel oxygenases, hydrolases, and reductases, have arisen from the functional screening of metagenomic libraries sourced from harsh or contaminated environments (Murray et al., 2023; Rich et al., 2023). This approach has intrinsic obstacles: expression biases in heterologous hosts, inadequate library coverage, and constrained screening throughput may hinder discovery. Integrating functional metagenomics with cell-free expression methods, microfluidic droplet screening, or machine learning-assisted candidate selection can alleviate these challenges, improving hit rates and expanding enzyme diversity (Conchouso et al., 2021; Robinson et al., 2021).

9.4.3 Bioinformatics Pipelines for Pathway Reconstruction

Bioinformatics pipelines for pathway reconstruction constitute the foundation of contemporary metagenomic analysis. Instruments such as MEGAHIT, MetaBAT, Prokka, and KEGG Mapper enable researchers to construct contigs, bin genomes, annotate genes, and integrate them into metabolic frameworks (Pita-Galeana et al., 2025). Pathway reconstruction connects genetic potential to ecological function, demonstrating how microbial communities collaborate to cycle resources, destroy contaminants, or synthesize beneficial substances. Nevertheless, reconstruction is far from simple. The precision of annotation is significantly influenced by reference databases, which are sometimes insufficient and skewed toward cultivated species (Chorlton, 2024). Furthermore, inferring pathway activity solely from DNA may lead to overinterpretation, as the mere presence of genes does not ensure their expression or catalytic significance. Integrating pathway reconstruction with expression-based data, such as metatranscriptomics or metaproteomics, yields a more accurate representation of active metabolism under particular conditions (Pita-Galeana et al., 2025).

9.4.4 Integration with Multi-Omics (Metat-Ranscriptomics, Meta-Proteomics, Metabolomics)

Integration with multi-omics approaches such as meta-transcriptomics, meta-proteomics, and metabolomics provides a comprehensive understanding of microbial systems at several biological levels. While metagenomics shows potential, meta-transcriptomics discloses active gene expression, meta-proteomics identifies functional proteins, and metabolomics records biochemical outputs (Aguiar-Pulido et al., 2016; Arıkan & Muth, 2023). Together, these layers allow for dynamic tracking of community responses to environmental changes, nutritional alterations, and engineering interventions. To avoid misleading connections, multi-omics data must be carefully normalized, calibrated across platforms, and supported by robust statistical frameworks. The promise of multi-omics resides not just in data collection but also in developing cohesive narratives that connect genetics, phenotype, and environment (Zheng et al., 2024).

9.4.5 Systems Biology and Metabolic Network Modeling

Systems biology and metabolic network modeling provide a conceptual framework for combining these discoveries to create predictive designs. By combining experimental data and computational models, researchers may simulate flow distributions, identify bottlenecks, and forecast how perturbations affect overall system behavior (Srinak et al., 2025). Constraint-based modeling, flux balance analysis, and genome-scale metabolic reconstructions enable the rational engineering of microbial metabolism. However, models are only as reliable as their assumptions. Simplified stoichiometries, missing regulatory layers, and context-specific interactions can all reduce predictive power. Iterative validation is crucial, as it refines models and closes the loop between hypothesis and observation (Heinken et al., 2021).

9.5 CASE STUDIES

9.5.1 Metagenomic Discovery of Enzymes for Oil Spill Cleanup

Metagenomic methods have revolutionized environmental biotechnology by revealing new enzymes, pathways, and adaptive traits within complex ecosystems. An exemplary case is the metagenomic identification of enzymes that facilitate oil spill remediation (Liu et al., 2024). Conventional bioremediation strategies have typically depended on culturable hydrocarbon-degrading bacteria, including *Pseudomonas* and *Alcanivorax*; however, these organisms constitute merely a small portion of the microbial diversity present in contaminated environments (Tedesco et al., 2024). Researchers have identified previously unknown alkane hydroxylases, monooxygenases, and dioxygenases through shotgun metagenomics, which are capable of degrading a wide range of hydrocarbons under various conditions, including high salinity and low temperature. These findings have broadened the enzymatic resources available for petroleum degradation and demonstrated synergistic interactions within microbial consortia, wherein one organism initiates partial oxidation while others finalize mineralization (Zhang & Zhang, 2022). Translating these findings into large-scale cleanup operations presents significant challenges. Enzymes that are optimized for natural environments may exhibit reduced stability in engineered systems, and ecological interactions observed *in situ* are challenging to replicate *ex situ* (Ndochinwa et al., 2024). This highlights the significance of integrating metagenomic discovery with synthetic biology methods for enzyme stabilization and the design of controlled communities.

9.5.2 Plastic-Degrading Enzymes from Marine Microbiomes

Marine microbiomes containing plastic-degrading enzymes represent another major discovery. Global plastic pollution favors bacteria that use synthetic polymers as carbon sources. Metagenomic investigations of ocean gyres, coastal sediments, and deep-sea settings have identified PETases, MHETases, and cutinases that depolymerize PET and other plastics (Dhali et al., 2024; Méndez-Zamora et al., 2024). In particular, these enzymes are structurally adapted to cold or saline conditions, making them suited for environmental applications. Enthusiasm, however, must remain realistic. These enzymes function slowly in nature; hence, plastic polymers seldom mineralize completely. To improve catalytic efficiency and substrate selectivity, directed evolution, rational mutagenesis, and hybrid enzyme design are being investigated. This case study demonstrates that natural adaptation is only a starting point and that effective remediation requires integrated enzymatic and mechanical techniques within circular economy frameworks.

9.5.3 Heavy-Metal Resistant Genes from Mine Tailings

Heavy-metal-resistant genes identified in mine tailings exemplify a significant application of metagenomics in elucidating environmental resilience. Mining sites frequently contain microbial communities that can endure high levels of metal toxicity, utilizing mechanisms including efflux pumps, metal-binding proteins, and enzymatic reduction pathways. Metagenomic analyses of sites contaminated with arsenic, mercury, cadmium, and lead have identified a wide range of genes linked to detoxification and sequestration processes. Arsenate reductases (*arsC*), mercuric reductases (*merA*), and metallothionein-like proteins have been identified in previously uncultured, novel taxa. This research enhances our comprehension of microbial adaptation and indicates possible biotechnological applications, including biofilters, biosensors, and engineered strains for heavy-metal remediation. The assumption that these mechanisms can be readily integrated into industrial systems is frequently overly optimistic. Metal resistance is often contingent upon contextual factors, including cofactors, redox conditions, and community-level buffering effects. Effective biotechnological translation necessitates engineering strategies informed by ecosystem dynamics rather than mere gene transfer.

9.5.4 Metagenome-Guided Biofuel Production Pathways

Metagenome-guided biofuel production pathways demonstrate the potential of utilizing microbial diversity for renewable energy generation. Samples from anaerobic digesters, termite guts, and hot springs have yielded novel enzymes for lignocellulose degradation, syngas fermentation, and methanogenesis. Metagenomic screening has identified thermostable cellulases and xylanases that decompose agricultural residues into fermentable sugars with greater efficiency than current commercial enzymes (Love, 2022). The identification of hydrogenases and dehydrogenases from extremophiles has contributed to the design of microbial systems for biohydrogen and bioethanol production. Biofuel synthesis continues to be limited by metabolic bottlenecks, cofactor imbalances, and substrate inhibition. Metagenomic analysis offers alternative pathways, including mixed-acid fermentation and syntrophic interactions, which allocate metabolic responsibilities among community members (Santos Correa et al., 2023). The essential inquiry is not the existence of new enzymes but their potential integration into scalable, cost-effective systems capable of competing with petrochemical processes.

9.6 CHALLENGES AND LIMITATIONS

9.6.1 Complexity of Environmental Samples (Heterogeneity, Low Biomass)

Metagenomics and microbial engineering have many limitations despite their potential. Environmental sample complexity represents a major challenge. Nutrient distribution, pH, temperature, and microbial composition vary greatly in soil, groundwater, and marine sediments. Such heterogeneity complicates sampling and often underrepresents rare but ecologically important taxa. Deep subterranean aquifers and oligotrophic oceans yield low amounts of DNA, reducing sequencing efficiency and increasing contamination risk (Suttner et al., 2020).

9.6.2 Limitations in DNA Extraction and Sequencing Depth

Limitations in DNA extraction and sequencing depth are also critical. Standard extraction procedures may favor readily lysed cells over microorganisms with robust cell walls or extensive extracellular matrices. Obtaining broad functional insight from shotgun metagenomic sequencing requires sufficient depth to capture low-abundance genes and reconstruct near-complete genomes. In many complex communities, even deep sequencing fails to resolve genomes adequately, resulting in fragmented data that hinder pathway reconstruction and functional inference (Sharpton, 2014).

9.6.3 Annotation Gaps and Uncharacterized Genes ("Microbial Dark Matter")

Another limitation involves annotation gaps and "microbial dark matter," referring to uncharacterized genes. Functional prediction remains uncertain for many genes derived from environmental samples that lack close homologs in reference databases. Annotations sometimes rely on distant homology or inferred motifs, which can misrepresent enzyme activity, substrate selectivity, or regulatory context. This knowledge gap reduces the reliability of downstream applications, such as enzyme discovery and metabolic engineering (Varadi et al., 2025).

9.6.4 Scalability and Regulatory Hurdles in Applying Engineered Microbes

Scalability and regulatory challenges hinder the application of engineered microbes. Competition, predation, and environmental stresses often diminish the performance of laboratory-optimized strains outside controlled conditions. Large-scale bioprocess adoption requires technical optimization, biosafety compliance, environmental risk assessment, and social acceptance. Engineering microorganisms for field deployment may face stringent containment, licensing, and monitoring requirements that hinder or prevent implementation despite promising laboratory results (Singh, 2025).

9.7 FUTURE PERSPECTIVES

9.7.1 AI/ML-Driven Predictions of Microbial Functions and Interactions

The future of metagenomics and microbial engineering will be shaped by the integration of advanced computational tools, precision biology, and ethical governance. Artificial intelligence and machine learning offer substantial potential for predicting microbial functions, interactions, and emergent community behaviors. By analyzing large-scale omics datasets, AI models can detect patterns that may escape human observation, including co-occurrence networks, functional redundancies, and novel enzyme candidates (Pandey et al., 2019; Trivedi et al., 2025). These predictions can guide targeted experimental validation, thereby accelerating the discovery of effective biocatalysts and informing rational community design. Nevertheless, caution is warranted. Machine-learning outputs are contingent upon the quality and completeness of input data, and models may exhibit overfitting or misinterpret correlation as causation. Continuous experimental feedback remains essential to ensure alignment between computational predictions and biological reality (Akintola, 2024).

9.7.2 Genome-Resolved Metagenomics for Precision Bioremediation

Genome-resolved metagenomics provides a targeted approach to bioremediation. Reconstructing near-complete genomes from environmental samples allows researchers to associate specific metabolic pathways with individual taxa, thereby identifying organisms actively involved in pollutant degradation. This resolution supports the design of customized interventions, including stimulation of indigenous populations, engineering of complementary strains, or assembly of synthetic consortia with defined functional roles (Kayani et al., 2021; Chettri et al., 2024). Translating genomic insights into field-scale impact remains challenging. Environmental heterogeneity, variable physicochemical conditions, and microbial competition may diminish predicted outcomes, underscoring the need for ongoing field validation and adaptive management strategies (Kuppan et al., 2024).

9.7.3 Synthetic Microbiomes for Targeted Pollutant Degradation

Synthetic microbiomes represent an emerging research frontier. The deliberate assembly of multiple microbial species with complementary functions enables the targeting of complex pollutants or the execution of multi-step transformations unattainable by individual organisms. Metagenomic and multi-omics data guide species selection, assess metabolic compatibility, and inform the design of intercellular signaling networks (Contreras-Salgado et al., 2024). Key challenges include maintaining community stability, preventing dominance by rapidly proliferating species, and ensuring resilience under environmental stress. Modular design principles, quorum-sensing circuits, and feedback-regulated gene-expression systems are being employed to address these challenges; however, successful implementation requires careful monitoring and control (Jaiswal & Shukla, 2020).

9.7.4 Engineering Novel Bio-Based Materials and Sustainable Bioprocesses

The development of innovative bio-based materials and sustainable bioprocesses will be driven by continued advances in environmental gene discovery. Metagenome-guided enzyme mining identifies catalysts for polymer synthesis, degradation, and modification, enabling the production of bioplastics, biofuels, and specialty chemicals from renewable feedstocks (Xing et al., 2012; Rahmati et al., 2024). Integrating metabolic engineering with systems biology supports the optimization of yield, efficiency, and scalability. Achieving industrial feasibility requires technical innovation alongside economic and environmental assessments to ensure sustainable implementation (Weiland et al., 2024).

9.7.5 Policy and Ethics in Deploying Metagenome-Guided Bioengineering

Policy and ethical considerations will increasingly influence the deployment of metagenome-guided bioengineering. Engineering microbes and synthetic consortia for environmental or industrial use raises concerns related to biosafety, ecological impact, intellectual property, and public perception (Saukshmya & Chugh, 2010; Hoffmann et al., 2023). Establishing clear regulatory frameworks, conducting transparent risk assessments, and engaging stakeholders are essential for balancing innovation with responsibility. Addressing these concerns proactively during early research and development stages will shape the safe and widespread adoption of metagenomic technologies (Wheeler, 2025).

9.8 CONCLUSION

Metagenomics has emerged as a transformative tool for understanding and harnessing microbial diversity in environmental and industrial contexts. By providing access to the complete genetic repertoire of complex microbial communities, it enables the identification of key taxa, functional genes, and metabolic pathways governing pollution degradation, biofuel production, and the synthesis of valuable biomolecules. This knowledge underpins microbial engineering, synthetic biology, and the rational design of microbial consortia, facilitating targeted interventions that were previously unattainable.

The range of potential applications is extensive. Metagenome-guided discovery and engineering can enhance environmental sustainability through precision bioremediation, plastic degradation, and heavy-metal detoxification. In industrial biotechnology, these approaches advance the development of sustainable biofuels, bioplastics, and novel bio-based materials, offering alternatives to conventional chemical processes and fossil-derived resources. By leveraging environmental microbial intelligence, metagenomics opens new avenues for efficient and sustainable bioprocesses aligned with global ecological and economic goals.

Realizing this potential requires integrative, multidisciplinary research. Genomics, bioinformatics, systems biology, metabolic engineering, and synthetic ecology must be integrated with ethical, regulatory, and economic considerations. Translational efforts must prioritize not only laboratory success but also scalability, stability, and ecological compatibility in real-world settings. Ultimately, metagenomics provides both a conceptual framework and a practical toolkit, with the central challenge being the translation of discovery into effective, responsible, and sustainable solutions that bridge scientific insight and tangible impact.

ACKNOWLEDGMENT

I acknowledge the use of ChatGPT [https://chatgpt.com/] in assisting with the review of my writing during the final stage of chapter preparation and figure development.

REFERENCES

Abdul Aziz M, Masmoudi K (2025) Molecular breakthroughs in modern plant breeding techniques. *Horticultural Plant Journal* 11:15–41. https://doi.org/10.1016/j.hpj.2024.01.004

Aguiar-Pulido V, Huang W, Suarez-Ulloa V, et al. (2016) Metagenomics, metatranscriptomics, and metabolomics approaches for microbiome analysis. *Evol Bioinform Online* 12:5–16. https://doi.org/10.4137/EBO.S36436

Akintola AA (2024) AI-driven monitoring systems for bioremediation: Real-time data analysis and predictive modelling. *World J Adv Res Rev* 24:788–803. https://doi.org/10.30574/wjarr.2024.24.1.3099

Aminov RI (2011) Horizontal gene exchange in environmental microbiota. *Front Microbiol* 2:158. https://doi.org/10.3389/fmicb.2011.00158

Arıkan M, Muth T (2023) Integrated multi-omics analyses of microbial communities: A review of the current state and future directions. *Mol Omics* 19:607–623. https://doi.org/10.1039/D3MO00089C

Atlas RM, Hazen TC (2011) Oil biodegradation and bioremediation: A tale of the two worst spills in U.S. history. *Environ Sci Technol* 45:6709–6715. https://doi.org/10.1021/es2013227

Ayilara MS, Babalola OO (2023) Bioremediation of environmental wastes: The role of microorganisms. *Front Agron* 5:1183691. https://doi.org/10.3389/fagro.2023.1183691

Azeez SS, Hamad RS, Hamad BK, et al. (2024) Advances in CRISPR-Cas technology and its applications: Revolutionising precision medicine. *Front Genome Ed* 6:1509924. https://doi.org/10.3389/fgeed.2024.1509924

Bala S, Garg D, Thirumalesh BV, et al. (2022) Recent strategies for bioremediation of emerging pollutants: A review for a green and sustainable environment. *Toxics* 10:484. https://doi.org/10.3390/toxics10080484

Balali-Mood M, Naseri K, Tahergorabi Z, et al. (2021) Toxic mechanisms of five heavy metals: Mercury, lead, chromium, cadmium, and arsenic. *Front Pharmacol* 12:643972. https://doi.org/10.3389/fphar.2021.643972

Bengtsson-Palme J (2020) Microbial model communities: To understand complexity, harness the power of simplicity. *Comput Struct Biotechnol J* 18:3987–4001. https://doi.org/10.1016/j.csbj.2020.11.043

Bhatwa A, Wang W, Hassan YI, et al. (2021) Challenges associated with the formation of recombinant protein inclusion bodies in *Escherichia coli* and strategies to address them for industrial applications. *Front Bioeng Biotechnol* 9:630551. https://doi.org/10.3389/fbioe.2021.630551

Cai Z, Li M, Zhu Z, et al. (2023) Biological degradation of plastics and microplastics: A recent perspective on associated mechanisms and influencing factors. *Microorganisms* 11:1661. https://doi.org/10.3390/microorganisms11071661

Calero P, Nikel PI (2019) Chasing bacterial chassis for metabolic engineering: A perspective review from classical to non-traditional microorganisms. *Microb Biotechnol* 12:98–124. https://doi.org/10.1111/1751-7915.13292

Cerruti M, Guo B, Delatolla R, et al. (2021) Plant-wide systems microbiology for the wastewater industry. *Environ Sci: Water Res Technol* 7:1687–1706. https://doi.org/10.1039/D1EW00231G

Chandran H, Meena M, Sharma K (2020) Microbial biodiversity and bioremediation assessment through Omics approaches. *Front Environ Chem* 1:570326. https://doi.org/10.3389/fenvc.2020.570326

Chettri D, Verma AK, Chirania M, Verma AK (2024) Metagenomic approaches in bioremediation of environmental pollutants. *Environmental Pollution* 363:125297. https://doi.org/10.1016/j.envpol.2024.125297

Chorlton SD (2024) Ten common issues with reference sequence databases and how to mitigate them. *Front Bioinform* 4:1278228. https://doi.org/10.3389/fbinf.2024.1278228

Conchouso D, Al-Ma'abadi A, Behzad H, et al. (2021) Integration of Droplet microfluidic tools for single-cell functional metagenomics: An engineering head start. *Genomics, Proteomics & Bioinformatics* 19:504–518. https://doi.org/10.1016/j.gpb.2021.03.010

Contreras-Salgado EA, Sánchez-Morán AG, Rodríguez-Preciado SY, et al. (2024) Multifaceted applications of synthetic microbial communities: Advances in biomedicine, bioremediation, and industry. *Microbiology Research* 15:1709–1727. https://doi.org/10.3390/microbiolres15030113

Cunningham CJ, Kuyukina MS, Ivshina IB, et al. (2020) Potential risks of antibiotic resistant bacteria and genes in bioremediation of petroleum hydrocarbon contaminated soils. *Environ Sci: Processes Impacts* 22:1110–1124. https://doi.org/10.1039/C9EM00606K

Das N, Chandran P (2011) Microbial degradation of petroleum hydrocarbon contaminants: An overview. *Biotechnol Res Int* 2011:941810. https://doi.org/10.4061/2011/941810

Dhali SL, Parida D, Kumar B, Bala K (2024) Recent trends in microbial and enzymatic plastic degradation: A solution for plastic pollution predicaments. *Biotechnol Sustain Mater* 1:11. https://doi.org/10.1186/s44316-024-00011-0

Duncker KE, Holmes ZA, You L (2021) Engineered microbial consortia: Strategies and applications. *Microb Cell Fact* 20:211. https://doi.org/10.1186/s12934-021-01699-9

Dutta N, Usman M, Ashraf MA, et al. (2022) A critical review of recent advances in the bio-remediation of chlorinated substances by microbial dechlorinators. *Chemical Engineering Journal Advances* 12:100359. https://doi.org/10.1016/j.ceja.2022.100359

Fenner K, Elsner M, Lueders T, et al. (2021) Methodological advances to study contaminant biotransformation: New prospects for understanding and reducing environmental persistence? *ACS EST Water* 1:1541–1554. https://doi.org/10.1021/acsestwater.1c00025

Grandel NE, Reyes Gamas K, Bennett MR (2021) Control of synthetic microbial consortia in time, space, and composition. *Trends in Microbiology* 29:1095–1105. https://doi.org/10.1016/j.tim.2021.04.001

Guerrero Ramírez JR, Ibarra Muñoz LA, Balagurusamy N, et al. (2023) Microbiology and biochemistry of pesticides biodegradation. *Int J Mol Sci* 24:15969. https://doi.org/10.3390/ijms242115969

Heinken A, Basile A, Thiele I (2021) Advances in constraint-based modelling of microbial communities. *Current Opinion in Systems Biology* 27:100346. https://doi.org/10.1016/j.coisb.2021.05.007

Hoffmann SA, Diggans J, Densmore D, et al. (2023) Safety by design: Biosafety and biosecurity in the age of synthetic genomics. *iScience* 26:106165. https://doi.org/10.1016/j.isci.2023.106165

Hu Y, Chen Y, Xu J, et al. (2022) Metagenomic discovery of novel CRISPR-Cas13 systems. *Cell Discov* 8:107. https://doi.org/10.1038/s41421-022-00464-5

Jaiswal S, Shukla P (2020) Alternative strategies for microbial remediation of pollutants via synthetic biology. *Front Microbiol* 11:808. https://doi.org/10.3389/fmicb.2020.00808

James AS, Adil NA, Goltz D, et al. (2024). Abnormalities in gut virome signatures linked with cognitive impairment in older adults. *Gut Microbes* 16(1):2431648. https://doi.org/10.1080/19490976.2024.2431648

Kanwar P, Yadav N, Srivastava S (2024) Microbial biodiversity and bioremediation: A systematic, biological and metabolic engineering tool. In: Gupta J, Verma A (eds) *Microbiology–2.0 Update for a Sustainable Future*. Springer Nature, Singapore, pp. 77–93.

Kayani MUR, Huang W, Feng R, Chen L (2021) Genome-resolved metagenomics using environmental and clinical samples. *Brief Bioinform* 22:bbab030. https://doi.org/10.1093/bib/bbab030

Kim JW, Park S-B, Tran Q-G, et al. (2020) Functional expression of polyethylene terephthalate-degrading enzyme (PETase) in green microalgae. *Microb Cell Fact* 19:97. https://doi.org/10.1186/s12934-020-01355-8

Koch N, Islam NF, Sonowal S, et al. (2021) Environmental antibiotics and resistance genes as emerging contaminants: Methods of detection and bioremediation. *Current Research in Microbial Sciences* 2:100027. https://doi.org/10.1016/j.crmicr.2021.100027

Kostka JE, Prakash O, Overholt WA, et al. (2011) Hydrocarbon-degrading bacteria and the bacterial community response in gulf of Mexico beach sands impacted by the deepwater horizon oil spill. *Appl Environ Microbiol* 77:7962–7974. https://doi.org/10.1128/AEM.05402-11

Kotopka BJ, Li Y, Smolke CD (2018) Synthetic biology strategies toward heterologous phytochemical production. *Nat Prod Rep* 35:902–920. https://doi.org/10.1039/C8NP00028J

Kumar V, Gangani S, Shukla R, Prajapati SK, Shekhar H, Shukla V (2024) Next-generation sequencing and genomic data analysis. In: Chaudhary A, Sethi SK, Verma A (eds) *Unraveling New Frontiers and Advances in Bioinformatics*. Springer Nature, pp. 49–76. https://doi.org/10.1007/978-981-97-7123-3_3

Kumar V, Sharma JK, Kumar S, et al. (2025) Metagenomic and proteomic approach for bioremediation of environmental pollutants. *In: Development in Waste Water Treatment Research and Processes*. Elsevier, pp. 635–647.

Kumar V, Shukla R, Gangani S, Shekhar H, Shukla V (2024) Metagenomics and microbiome analysis. In: Chaudhary A, Sethi SK, Verma A (eds) *Unraveling New Frontiers and Advances in Bioinformatics*. Springer Nature, Singapore, pp. 77–92. https://doi.org/10.1007/978-981-97-7123-3_4

Kuppan N, Padman M, Mahadeva M, et al. (2024) A comprehensive review of sustainable bioremediation techniques: Eco friendly solutions for waste and pollution management. *Waste Management Bulletin* 2:154–171. https://doi.org/10.1016/j.wmb.2024.07.005

Lagier J-C, Hugon P, Khelaifia S, et al. (2015) The rebirth of culture in microbiology through the example of culturomics to study human gut microbiota. *Clin Microbiol Rev* 28:237–264. https://doi.org/10.1128/CMR.00014-14

Lee J-K, Kalia VC (2020) Mapping microbial capacities for bioremediation: Genes to genomics. *Indian J Microbiol* 60:45–53. https://doi.org/10.1007/s12088-019-00842-w

Lema NK, Gemeda MT, Woldesemayat AA (2023) Recent advances in metagenomic approaches, applications, and challenges. *Curr Microbiol* 80:347. https://doi.org/10.1007/s00284-023-03451-5

Lema NK, Gemeda MT, Woldesemayat AA (2025) Secondary metabolite biosynthetic gene clusters and microbial diversity as predicted by shotgun metagenomic sequencing approach in hospitals and pharmaceutical industry wastes. *Sci Rep* 15:25500. https://doi.org/10.1038/s41598-025-03467-w

Li S, Zhang W (2025) Computational identification of plastic-degrading enzymes in ocean microbiomes. *Sci Rep* 15:15332. https://doi.org/10.1038/s41598-025-99275-3

Li Y, Yang J, Song Y, Wei M (2025) Progress in biostimulation-based remediation of TPH-contaminated soils: A comprehensive review. *PeerJ* 13:e19991. https://doi.org/10.7717/peerj.19991

Liu P, Wen S, Zhu S, et al. (2025) Microbial degradation of soil organic pollutants: Mechanisms, challenges, and advances in forest ecosystem management. *Processes* 13:916. https://doi.org/10.3390/pr13030916

Liu Q, He W, Zhang W, et al. (2024) Metagenomic analysis reveals the microbial response to petroleum contamination in oilfield soils. *Science of the Total Environment* 912:168972. https://doi.org/10.1016/j.scitotenv.2023.168972

Love J (2022) Microbial pathways for advanced biofuel production. *Biochem Soc Trans* 50:987–1001. https://doi.org/10.1042/BST20210764

Madigan KE, Maffei VJ, Shukla R, Kumar V, Yadav H, Weinberg RB (2025) Fecal microbiome richness and evenness are reduced and the relative abundance of firmicutes species is increased in patients with intestinal methanogen overgrowth. *Gastroenterology* 169(Supplement_1):S–1263. https://doi.org/10.1016/S0016-5085(25)03787-4

Mani I (2023) Cutting-edge tools to assess microbial diversity and their function in land remediation. In: Pandey VC (ed) *Bio-Inspired Land Remediation*. Springer International Publishing, Cham, pp. 115–132.

McCarty NS, Ledesma-Amaro R (2019) Synthetic biology tools to engineer microbial communities for biotechnology. *Trends in Biotechnology* 37:181–197. https://doi.org/10.1016/j.tibtech.2018.11.002

Méndez-Zamora A, Diaz-Vidal T, Robles-Machuca M, et al. (2024) Unveiling potential PET degrading eukaryotes through in silico bioprospecting of PETases. *Science of The Total Environment* 948:174978. https://doi.org/10.1016/j.scitotenv.2024.174978

Mishra S, Buddendorff L, Hoover J, Shukla R, Kumar V, Jain S, Yadav H (2024) Microbiome role in cognitive impairment: MiaGB Cohort insights. *Innovation in Aging* 8(Supplement_1):811. https://doi.org/10.1093/geroni/igae098.2629

Monachese M, Burton JP, Reid G (2012) Bioremediation and tolerance of humans to heavy metals through microbial processes: A potential role for probiotics? *Appl Environ Microbiol* 78:6397–6404. https://doi.org/10.1128/AEM.01665-12

Murray AK, Zhang L, Snape J, Gaze WH (2023) Functional metagenomic libraries generated from anthropogenically impacted environments reveal importance of metabolic genes in biocide and antibiotic resistance. *Curr Res Microb Sci* 4:100184. https://doi.org/10.1016/j.crmicr.2023.100184

Muter O (2023) Current trends in bioaugmentation tools for bioremediation: A critical review of advances and knowledge gaps. *Microorganisms* 11:710. https://doi.org/10.3390/microorganisms11030710

Naether DJ, Slawtschew S, Stasik S, et al. (2013) Adaptation of the hydrocarbonoclastic bacterium Alcanivorax borkumensis SK2 to alkanes and toxic organic compounds: A physiological and transcriptomic approach. *Appl Environ Microbiol* 79:4282–4293. https://doi.org/10.1128/AEM.00694-13

Nagayama H, Sugawara T, Endo R, et al. (2015) Isolation of oxygenase genes for indigo-forming activity from an artificially polluted soil metagenome by functional screening using Pseudomonas putida strains as hosts. *Appl Microbiol Biotechnol* 99:4453–4470. https://doi.org/10.1007/s00253-014-6322-2

Nair SU, Lin Y, Jain S, Chaudhari DS, Shukla R, Kumar V, Mankowski RT, Meng H, Buford TW, Kuchel GA, Yadav H, Consortium M (2024). Unique gut microbiome signatures associated with physical and cognitive function. *Alzheimer's & Dementia* 20(S2):e092044. https://doi.org/10.1002/alz.092044

Nam NN, Do HDK, Loan Trinh KT, Lee NY (2023) Metagenomics: An effective approach for exploring microbial diversity and functions. *Foods* 12:2140. https://doi.org/10.3390/foods12112140

Naziębło A, Dobrzyński J (2025) Biotransformation of As, Cr, Hg, and Mn by Pseudomonadota: Chances and risks. *Biodegradation* 36:60. https://doi.org/10.1007/s10532-025-10157-x

Ndochinwa OG, Wang Q-Y, Amadi OC, et al. (2024) Current status and emerging frontiers in enzyme engineering: An industrial perspective. *Heliyon* 10:e32673. https://doi.org/10.1016/j.heliyon.2024.e32673

Ngara TR, Zhang H (2018) Recent advances in function-based metagenomic screening. *Genomics Proteomics Bioinformatics* 16:405–415. https://doi.org/10.1016/j.gpb.2018.01.002

Ninck S, Klaus T, Kochetkova TV, et al. (2024) Environmental activity-based protein profiling for function-driven enzyme discovery from natural communities. *Environmental Microbiome* 19:36. https://doi.org/10.1186/s40793-024-00577-2

Paloyan A, Tadevosyan M, Ghevondyan D, et al. (2025) Biodegradation of polyhydroxyalkanoates: Current state and future prospects. *Front Microbiol* 16:1542468. https://doi.org/10.3389/fmicb.2025.1542468

Pandey A, Tripathi PH, Tripathi AH, et al. (2019) Omics technology to study bioremediation and respective enzymes. *In: Smart Bioremediation Technologies*. Elsevier, pp. 23–43.

Patil A, Chakraborty S, Yadav Y, et al. (2024) Bioremediation strategies and mechanisms of bacteria for resistance against heavy metals: A review. *Bioremediation Journal* 1–33. https://doi.org/10.1080/10889868.2024.2375204

Pettersen VK, Antunes LCM, Dufour A, Arrieta M-C (2022) Inferring early-life host and microbiome functions by mass spectrometry-based metaproteomics and metabolomics. *Computational and Structural Biotechnology Journal* 20:274–286. https://doi.org/10.1016/j.csbj.2021.12.012

Pita-Galeana MA, Ruhle M, López-Vázquez L, et al. (2025) Computational metagenomics: State of the art. *Int J Mol Sci* 26:9206. https://doi.org/10.3390/ijms26189206

Prajapati SK, Kumar V, Shukla R, Szekeres C, Ma Y, Jain S, Yadav H (2024). Microbiome-immune interplay in aging brains: A pilot investigation from the MiaGB Cohort. *Alzheimer's & Dementia* 20(S2):e090734. https://doi.org/10.1002/alz.090734

Qattan SYA (2025) Harnessing bacterial consortia for effective bioremediation: Targeted removal of heavy metals, hydrocarbons, and persistent pollutants. *Environ Sci Eur* 37:85. https://doi.org/10.1186/s12302-025-01103-y

Qiu S, Yang A, Zeng H (2023) Flux balance analysis-based metabolic modeling of microbial secondary metabolism: Current status and outlook. *PLoS Comput Biol* 19:e1011391. https://doi.org/10.1371/journal.pcbi.1011391

Rahmati F, Sethi D, Shu W, et al. (2024) Advances in microbial exoenzymes bioengineering for improvement of bioplastics degradation. *Chemosphere* 355:141749. https://doi.org/10.1016/j.chemosphere.2024.141749

Rebello S, Nathan VK, Sindhu R, et al. (2021) Bioengineered microbes for soil health restoration: Present status and future. *Bioengineered* 12:12839–12853. https://doi.org/10.1080/21655979.2021.2004645

Rich MH, Sharrock AV, Mulligan TS, et al. (2023) A metagenomic library cloning strategy that promotes high-level expression of captured genes to enable efficient functional screening. *Cell Chemical Biology* 30:1680–1691.e6. https://doi.org/10.1016/j.chembiol.2023.10.001

Robinson SL, Piel J, Sunagawa S (2021) A roadmap for metagenomic enzyme discovery. *Nat Prod Rep* 38:1994–2023. https://doi.org/10.1039/d1np00006c

Ron EZ, Rosenberg E (2014) Enhanced bioremediation of oil spills in the sea. *Curr Opin Biotechnol* 27:191–194. https://doi.org/10.1016/j.copbio.2014.02.004

Santos Correa S, Schultz J, Lauersen KJ, Soares Rosado A (2023) Natural carbon fixation and advances in synthetic engineering for redesigning and creating new fixation pathways. *J Adv Res* 47:75–92. https://doi.org/10.1016/j.jare.2022.07.011

Sarfraz H, Czerniecki BJ, Stankiewicz K, Jaglal MV, Jain S, Yadav H, Liu M, Shukla R, Kumar V, Alhomsi M, Hathout M, Tuz Zahra F, Darwin A, Joshi U, Sen A, Muzaffar J, Bari S (2024) Evaluating gut microbiome as a biomarker of pathological complete response in patients with early locally advanced triple negative breast cancer (LA TNBC): A pilot study. *Journal of Clinical Oncology* 42(Supplement_16):593–593. https://doi.org/10.1200/JCO.2024.42.16_suppl.593

Sartaj K, Patel A, Matsakas L, Prasad R (2022) Unravelling metagenomics approach for microbial biofuel production. *Genes (Basel)* 13:1942. https://doi.org/10.3390/genes13111942

Saukshmya T, Chugh A (2010) Commercializing synthetic biology: Socio-ethical concerns and challenges under intellectual property regime. *J Commer Biotechnol* 16:135–158. https://doi.org/10.1057/jcb.2009.28

Shade A, Peter H, Allison SD, et al. (2012) Fundamentals of microbial community resistance and resilience. *Front Microbiol* 3:417. https://doi.org/10.3389/fmicb.2012.00417

Shah BA, Malhotra H, Papade SE, et al. (2024) Microbial degradation of contaminants of emerging concern: Metabolic, genetic and omics insights for enhanced bioremediation. *Front Bioeng Biotechnol* 12:1470522. https://doi.org/10.3389/fbioe.2024.1470522

Sharma B, Dangi AK, Shukla P (2018) Contemporary enzyme based technologies for bioremediation: A review. *Journal of Environmental Management* 210:10–22. https://doi.org/10.1016/j.jenvman.2017.12.075

Sharma JK, Kumar N, Singh NP, Santal AR (2023) Phytoremediation technologies and their mechanism for removal of heavy metal from contaminated soil: An approach for a sustainable environment. *Front Plant Sci* 14:1076876. https://doi.org/10.3389/fpls.2023.1076876

Sharma N, Ahlawat YK, Stalin N, et al. (2024) Microbial enzymes in industrial biotechnology: Sources, production, and significant applications of lipases. *J Ind Microbiol Biotechnol* 52:kuaf010. https://doi.org/10.1093/jimb/kuaf010

Sharma P (2022) Role and significance of biofilm-forming microbes in phytoremediation–a review. *Environmental Technology & Innovation* 25:102182. https://doi.org/10.1016/j.eti.2021.102182

Sharpton TJ (2014) An introduction to the analysis of shotgun metagenomic data. *Front Plant Sci* 5:209. https://doi.org/10.3389/fpls.2014.00209

Singh A, Chand Kumawat K (2025) Unraveling the potential of microbial diversity in pesticide remediation: An eco-friendly approach for environmental sustainability. *Journal of Agriculture and Food Research* 21:101832. https://doi.org/10.1016/j.jafr.2025.101832

Singh NP, Sharma JK, Santal AR (2016) Biotechnological approaches to remediate soil and water using plant–microbe interactions. In: Ansari AA, Gill SS, Gill R, et al. (eds) *Phytoremediation*. Springer International Publishing, Cham, pp. 131–152.

Singh S (2025) The role of environmental microbiota in bioremediation: Harnessing bacteria for pollutant degradation. *Clin Med Helt Res J* 5:1228–1240. https://doi.org/10.18535/cmhrj.v5i02.464

Sinha L, Sharma JK, Lath A, et al. (2025) Novel tools in phytoremediation technology. *In: Green Biotechnology for Herbal and Medicinal Plant Metabolites*, 1st edn. CRC Press, Boca Raton, pp. 163–180.

Srinak N, Krüger F, Kaleta C, Taubenheim J (2025) Metabolic modeling of host-microbe interactions. *Computational and Structural Biotechnology Journal* 27:4304–4319. https://doi.org/10.1016/j.csbj.2025.10.006

Sun Y, Li D, Li B, Zhao K (2025) Editorial: The role of microbial communities in environmental engineering systems. *Front Microbiol* 16:1644856. https://doi.org/10.3389/fmicb.2025.1644856

Sutkar PR, Gadewar RD, Dhulap VP (2023) Recent trends in degradation of microplastics in the environment: A state-of-the-art review. *Journal of Hazardous Materials Advances* 11:100343. https://doi.org/10.1016/j.hazadv.2023.100343

Suttner B, Johnston ER, Orellana LH, et al. (2020) Metagenomics as a public health risk assessment tool in a study of natural creek sediments influenced by agricultural and livestock runoff: Potential and limitations. *Appl Environ Microbiol* 86:e02525-19. https://doi.org/10.1128/AEM.02525-19

Tao J, Chen Q, Chen S, et al. (2022) Metagenomic insight into the microbial degradation of organic compounds in fermented plant leaves. *Environmental Research* 214:113902. https://doi.org/10.1016/j.envres.2022.113902

Tedesco P, Balzano S, Coppola D, et al. (2024) Bioremediation for the recovery of oil polluted marine environment, opportunities and challenges approaching the blue growth. *Marine Pollution Bulletin* 200:116157. https://doi.org/10.1016/j.marpolbul.2024.116157

Tribedi P, Goswami M, Chakraborty P, et al. (2018) Bioaugmentation and biostimulation: A potential strategy for environmental remediation. *JMEN* 6:. https://doi.org/10.15406/jmen.2018.06.00219

Trivedi P, Kumar A, Gupta N, Patel CN (2025) Artificial intelligence and machine learning in microbial degradation of pollutants and toxins. In: Kaur S, Dhiman S, Tripathi M (eds) *Microbial Metabolomics*. Springer Nature, Singapore, pp. 377–400.

van der Zaan B, Hannes F, Hoekstra N, et al. (2010) Correlation of Dehalococcoides 16S rRNA and chloroethene-reductive dehalogenase genes with geochemical conditions in chloroethene-contaminated groundwater. *Appl Environ Microbiol* 76:843–850. https://doi.org/10.1128/AEM.01482-09

Varadi M, Tsenkov M, Velankar S (2025) Challenges in bridging the gap between protein structure prediction and functional interpretation. *Proteins* 93:400–410. https://doi.org/10.1002/prot.26614

Vermelho AB, Moreira JV, Akamine IT, et al. (2024) Agricultural pest management: The role of microorganisms in biopesticides and soil bioremediation. *Plants* 13:2762. https://doi.org/10.3390/plants13192762

Wang H, Wang Q, Lv M, et al. (2024) Study on the degradation and metabolic mechanism of four quinolone antibiotics by mixed strains. *Front Environ Chem* 5:1326206. https://doi.org/10.3389/fenvc.2024.1326206

Wang Y, Jiang Q, Zhou C, et al. (2014) *In-situ* remediation of contaminated farmland by horizontal transfer of degradative plasmids among rhizosphere bacteria. *Soil Use and Management* 30:303–309. https://doi.org/10.1111/sum.12105

Weiland F, Kohlstedt M, Wittmann C (2024) Biobased de novo synthesis, upcycling, and recycling—the heartbeat toward a green and sustainable polyethylene terephthalate industry. *Current Opinion in Biotechnology* 86:103079. https://doi.org/10.1016/j.copbio.2024.103079

Wheeler NE (2025) Responsible AI in biotechnology: Balancing discovery, innovation and biosecurity risks. *Front Bioeng Biotechnol* 13:1537471. https://doi.org/10.3389/fbioe.2025.1537471

Wu X, Spencer S, Gushgari-Doyle S, et al. (2020) Culturing of "unculturable" subsurface microbes: Natural organic carbon source fuels the growth of diverse and distinct bacteria from groundwater. *Front Microbiol* 11:610001. https://doi.org/10.3389/fmicb.2020.610001

Xeon G (2024) Computational approaches in bioremediation leveraging bioinformatics for the biodegradation of environmental pollutants. *Journal of Bioremediation & Biodegradation* 15:1–3. https://doi.org/10.4172/2155-6199.1000654

Xing M-N, Zhang X-Z, Huang H (2012) Application of metagenomic techniques in mining enzymes from microbial communities for biofuel synthesis. *Biotechnology Advances* 30:920–929. https://doi.org/10.1016/j.biotechadv.2012.01.021

Xu X, Liu W, Tian S, et al. (2018) Petroleum hydrocarbon-degrading bacteria for the remediation of oil pollution under aerobic conditions: A perspective analysis. *Front Microbiol* 9:2885. https://doi.org/10.3389/fmicb.2018.02885

Zaini N, Kasmuri N, Mojiri A, et al. (2024) Plastic pollution and degradation pathways: A review on the treatment technologies. *Heliyon* 10:e28849. https://doi.org/10.1016/j.heliyon.2024.e28849

Zhang L, Chen F, Zeng Z, et al. (2021) Advances in metagenomics and its application in environmental microorganisms. *Front Microbiol* 12:766364. https://doi.org/10.3389/fmicb.2021.766364

Zhang T, Zhang H (2022) Microbial consortia are needed to degrade soil pollutants. *Microorganisms* 10:261. https://doi.org/10.3390/microorganisms10020261

Zheng Y, Liu Y, Yang J, et al. (2024) Multi-omics data integration using ratio-based quantitative profiling with quartet reference materials. *Nat Biotechnol* 42:1133–1149. https://doi.org/10.1038/s41587-023-01934-1

10 Metagenomics and Antimicrobial Resistance (AMR)

Ankita Hulekal[1#], K. Aadira[1#], and Prashant Singh[2]*
[1]JSS Academy of Higher Education and Research, Mysuru, Karnataka, India
[2]Indian Institute of Technology Hyderabad, Sangareddy, Telangana, India
[#]Equal Contribution
[*]Corresponding Author

10.1 INTRODUCTION

Alexander Fleming discovered penicillin in 1928, and it was often described as a medical milestone, with good reason (Ventola, 2015). By early 1940s, it had progressed into clinical use, and overnight, infections that were once frequently fatal became treatable (Salam et al., 2023; Ventola, 2015). Pneumonia, septicaemia, and even syphilis could be managed with a single drug. For many patients, this was nothing short of miraculous[1] (Ventola, 2015). Yet this breakthrough did not go acknowledged for long. By early 1940s, doctors were already reporting strains of Staphylococcus aureus that could produce penicillinase, an enzyme which destroyed the drug before it had any effect (Pitiot et al., 2025; Kumar et al., 2023; Yadav & Kapley, 2021). In hospital wards specifically, penicillin often lost its power almost as quickly as it had arrived. What had begun as a medical revolution was already shifting into something far less certain, a constant race between new antibiotics and the bacteria learning how to resist them (Pitiot et al., 2025; Ventola, 2015; Yadav & Kapley, 2021).

Methicillin was introduced in 1959 as a targeted approach to penicillin-resistant infections, and for a brief period it seemed to be highly effective (Ventola, 2015). This approach did not last long. In 1960, Professor Patricia Jevons reported the first case of methicillin-resistant Staphylococcus aureus (MRSA), a finding that altered how resistance was understood by both clinicians and scientists (Pitiot et al., 2025; Ventola, 2015). This issue grew more rapidly over the following decades. During the 1970s, resistance genes spread across bacteria, and by the late 1980s vancomycin-resistant Enterococcus (VRE) had appeared as a new concern (Salam et al., 2023; Ventola, 2015). By the 1990s, multidrug-resistant (MDR) pathogens were rising more frequently, including strains displaying resistance to carbapenems[5] (Woolhouse & Ward, 2013). Another turning point in 2008 was led with the discovery of the NDM-1 gene, which gave bacteria the capacity to withstand nearly all available beta-lactam antibiotics and spread quickly on a global scale (Salam et al., 2023). What started as a local issue in the middle of the 20th century has, over several decades, evolved to become one of the health emergencies of our time (Salam et al., 2023; Ventola, 2015; Woolhouse & Ward, 2013).

Antibiotic resistance in bacteria is achieved through different means. Some of them secrete enzymes like b-lactamases which neutralize the drug (Anjum et al., 2017; Galhano et al., 2021). In some situations, one mutation of a target enzyme, for example, DNA gyrase, is also sufficient to

DOI: 10.1201/9781003670155-10

prevent the effect of the drug (Galhano et al., 2021; Salam et al., 2023). Efflux pumps also aid bacteria to expel the drugs out of the cell (Galhano et al., 2021; Mendes et al., 2023). The fast diffusion of resistance characteristics is also a significant issue. Most of the genes are housed on portable components like plasmids or transposons and it can be transferred between one bacterium and another (Woolhouse & Ward, 2013). Because of this, resistance is not confined to a single strain but is very easily transmitted among groups of people, thus the use of the term superbugs was coined (Salam et al., 2023). The misuse of antibiotics in hospitals and clinical practices is a part and parcel issue, yet the farm has played a significant role too (Das et al., 2024; Salam et al., 2023). The amount of drugs administered to livestock is large and accumulates the sources of resistance in animal microbes (Das et al., 2024; Munk et al., 2017). The worry is that these genes fail to remain in animal species, they are passed to humans in food chains, directly through handling, or even by waste released into water and soil (Hendriksen et al., 2019; Munk et al., 2017; Williams et al., 2023). Calculating the interrelationship of all these paths is a pain in itself, as they are overlapped with the resistance between human beings themselves (Woolhouse & Ward, 2013). In order to calculate it, researchers are turning to such modern methods as whole-genome sequencing (Crofts et al., 2017; Oniciuc et al., 2018; Sherry et al., 2025). As an example, the opportunity to compare Salmonella Typhimurium DT104 of animal and human outbreaks has allowed establishing common patterns in various environments (Munk et al., 2017).

A vast number of microorganisms that exist in nature cannot be brought into culture with standard methods. This long-standing problem is often termed as the "great plate count anomaly," and it has made it difficult to understand the true extent of microbial diversity (Hugenholtz & Tyson, 2008; Sleator et al., 2008; Kumar, Shukla, et al., 2024). Even organisms that are common in their environments often refuse to grow in the laboratory. This disadvantage delayed any potential advancement by decades (Hugenholtz & Tyson, 2008; Thomas et al., 2012). Metagenomics swooped in and transformed the entire picture. Scientists can now sequence DNA directly from soil, water, or clinical samples (Quince et al., 2017; Thomas et al., 2012). This advancement made it possible to assemble near complete genomes, even whole ones, without having to culture the cells. These assemblies contribute new information to that of reference strains and give a solid basis to comparative and functional studies as well as ecological models (Garza & Dutilh, 2015; Quince et al., 2017). The technique extends into a concept known as microbial dark matter. It consists of new lines of descent, novel metabolic forces and new systems of regulation. These findings are not mere facts but make a better model of community structure to clarify host-microbe interactions and attract focus to new reservoirs of essential genes, including those related to antimicrobial resistance (Garza & Dutilh, 2015; Nam et al., 2023). Over time, this work has shifted microbiology out of a very limited perspective of laboratory cultures. It demonstrates that microbial life is much larger and diverse than the past methods had enabled us to observe (Garza & Dutilh, 2015; Hugenholtz & Tyson, 2008).

Metagenomics shifted the antimicrobial resistance research by offering a comprehensive procedure that studies the resistome. Instead of focusing on known genome resistance determinants, it captures known as well as unknown elements (Crofts et al., 2017; De, 2019). Unlike gene-targeted methods, shotgun metagenomics provide a different perspective by identifying the presence of resistance genes in a community and displaying their organization in respect to its neighbors and mobile genetic platforms (Ko et al., 2022; Lanza et al., 2018; Kumar, Gangani, et al., 2024). This type of analysis is more than that of single-gene detection. It provides an image of microbial communities on a systems level and visualizes the impact of human actions and environmental stress on the abundance, maintenance, and spread of resistance (Hendriksen et al., 2019; Vallejo-Espin et al., 2025). In this regard, metagenomics explains not only the existence of the genes themselves but the spread and adaptation to the contexts (Ko et al., 2022). The other advantage of this approach is its encouragement of reference independent discovery relying on assembly-based approaches particularly in combination with a more recent development in the field of machine learning (Anahtar et al., 2021; Roy et al., 2024). The ability to identify much more divergent or hidden resistance genes, which would remain hidden with database-dependent tools, is now possible to researchers (Alcock et al., 2022; Crofts et al., 2017). With a lowering cost of sequencing and an increasing computational pipeline, metagenomics is starting to be used in a real-time surveillance capacity (Ko et al., 2022; Oniciuc et al., 2018). Historically it has been used in a vast number of fields that involve resistance monitoring in hospitals and clinical samples, gauging dissemination by wastewater and the effects of its impacts in agricultural settings (Hendriksen et al., 2019; Munk et al., 2017; Williams et al., 2023). All this research yield data that can potentially inform the public about general health policies, enhanced interventions, and create awareness into the evolution of AMR within ecosystems (Djordjevic et al., 2023; Vallejo-Espin et al., 2025).

This specific chapter investigates the application of metagenomics in the study of antimicrobial resistance (AMR) corresponding to its problem in terms of its methodological framework and its practical aspects. Metagenomics offer a channel to culture-independent analysis of microbial communities thoroughly (Quince et al., 2017; Thomas et al., 2012; Kumar et al., 2025). Direct sequencing of clinical, agricultural or environmental samples of DNA allows the detection of the presence of resistance genes without the biases that comes with cultivation (De, 2019; Ko et al., 2022). The chapter contains the typical workflow of a metagenomic study that typically starts with the collection and preparation of samples, followed by generating data on sequencing. This data undergoes bioinformatic pipelines to identify and quantify AMR genes. At every stage, various layers of information are piled on, and altogether, they give us a holistic view of the resistome that is better than what traditional techniques deliver (Hendriksen et al., 2019; Ko et al., 2022). The chapter ends by highlighting new sequencing platforms, evolved curated databases of genes, and inclusion of metagenomics into global systems of surveillance and public health. These factors illustrate not only the growing capabilities of studying resistance but also the transformed methods with which resistance is contained and monitored.

10.2 METAGENOMICS: CONCEPTS AND METHODOLOGIES

10.2.1 Metagenomics: Definition and History

One of the biggest problems of microbiology is that the prokaryotic life on Earth, estimated between 4 and 6 x 1030 cells, cannot be cultivated in a laboratory (Hugenholtz & Tyson, 2008; Sleator et al., 2008). This enormous silent majority is an enormous genetic repository. This is a window to this hidden world made available by metagenomics. As opposed to studying the separate species, it studies the entire genetic makeup of entire microbial communities in their natural habitats (Quince et al., 2017; Thomas et al., 2012). This culture-free approach provides us with the information that would otherwise be unreachable, allows not only to discover new genes during sequencing, but also new metabolic functions (Garza & Dutilh, 2015). Metagenomic surveys of environments as diverse as ocean waters and insect guts have altered our concept of the microbial ecology, symbiosis, and even the biochemical potential of the planet over the past two decades (Ko et al., 2022; Quince et al., 2017).

Classical genomics had followed a very different direction. It was intended to map the entire genome of one organism, such as the Human Genome Project (Slatko et al., 2018). By contrast, in metagenomics, all the DNA in a community is analyzed in the entirety recovered on a single sample (Thomas et al., 2012). Generally speaking, it is sequence-based as well as functional analysis of the collective microbiome (Hugenholtz & Tyson, 2008; Quince et al., 2017). This change provides us with a comparatively objective picture of the community structure in terms of richness, distribution, and diversity and an understanding of the metabolic potential that is coded in the gene pool (Quince et al., 2017). Its versatility is evident in applications ranging from small, extreme environments such as acid mine biofilms to global initiatives like the Ocean Sampling Expedition and the Human Microbiome Project (Ko et al., 2022; Quince et al., 2017). Although we think of antibiotics as inventions of modern medicine, they are in fact ancient molecules. For billions of years, microbes in soil and other habitats have produced them as chemical weapons to outperform rivals (Yadav & Kapley, 2021).

At the same time, they have served as communication signals, influencing microbial behavior such as biofilm formation (Mendes et al., 2023). Resistance follows where antibiotics exist. The ability to withstand these compounds is as old as the molecules themselves (De, 2019). Bacteria achieve resistance in two main ways. The first is through spontaneous mutation, for instance, altering the binding site of a drug. The second pathway involves horizontal gene transfer (HGT)—essentially, bacteria sharing genetic material with their neighbors like trading cards. When this happens, organisms can pick up resistance genes that help them survive antibiotic attacks (Galhano et al., 2021; Woolhouse & Ward, 2013). These borrowed genes might produce enzymes capable of breaking down the drug, altering the drug's intended target, or even pumping the antibiotic right back out of the cell. Through HGT, an organism can acquire ready-made resistance genes that provide defenses such as enzymes to break down the antibiotic, proteins to modify the drug's target, or efflux pumps to eject the compound from the cell (Anjum et al., 2017; Galhano et al., 2021; Mendes et al., 2023).

What is particularly fascinating is that many of these resistance genes weren't originally designed for fighting antibiotics at all. They started out as ordinary housekeeping genes helping with routine cell wall maintenance but just happened to have structures similar enough to antibiotic targets that they could interfere (De, 2019). Once these genes found their way onto mobile genetic elements like plasmids or transposons, they became shareable currency in the bacterial world, easily passed to dangerous pathogens where they fuel clinical resistance (Woolhouse & Ward, 2013). The human gut provides a perfect example of how this all plays out. Our intestinal tract, especially the colon, hosts one of the most crowded microbial neighborhoods on Earth, which is over a trillion bacterial cells per milliliter (Nam et al., 2023; van Schaik, 2015; Gupta et al., 2025). Most of these residents are helpful anaerobes from groups like Bacteroidetes and Firmicutes. They're the good bacteria, breaking down nutrients and churning out vitamins we need. But living alongside these beneficial bacteria are smaller populations from the Enterobacteriaceae and Enterococcaceae families (van Schaik, 2015). This is where things get concerning. These groups harbor harmful microbes like E. coli, Klebsiella pneumoniae, and various Enterococcus species, the very bacteria that have become infamous for causing hard-to-treat, multidrug-resistant infections in hospitals over the past few decades (Crofts et al., 2017; van Schaik, 2015).

In essence, our gut serves as a breeding ground for potentially dangerous pathogens just waiting for their chance. This makes the gut a key reservoir of opportunistic pathogens (van Schaik, 2015). Usually, beneficial microbes keep pathogenic groups in check. Antibiotic treatment, especially in hospitalized patients can disrupt this balance (De, 2019). Commensal populations are suppressed, giving resistant opportunists a chance to expand. Once established, they might escape into the bloodstream or colonize medical devices, causing severe infections (Liu & Ma, 2024). The density and diversity of the gut microbiota also create a perfect environment for horizontal gene transfer to occur. With trillions of cells in close proximity, genes can move readily between commensals and pathogens (Woolhouse & Ward, 2013). This has made characterization of the "gut resistome" which is the totality of resistance genes present in the gut microbiota, a central research priority (Crofts et al., 2017; De, 2019; van Schaik, 2015). Understanding its composition and tracking how genes move within this ecosystem is critical for addressing AMR (Djordjevic et al., 2023). Here, metagenomics has become indispensable as sequencing entire microbial communities directly from samples, researchers can now capture not only the well-known genes but also the uncertain or novel ones (Ko et al., 2022). This generates a comprehensive picture of resistance potential in the community. Given these realities, metagenomics has emerged as an invaluable approach. It helps us spot new threats before they explode, diagnose tricky infections, and develop smart, evidence-based strategies for both treating individual patients and protecting public health at large (Djordjevic et al., 2023; Olsen & Riber, 2025; Oniciuc et al., 2018).

10.2.2 High-Throughput Sequencing Technologies

Advancements in DNA sequencing technologies have repeatedly transformed genomic research capabilities. The first generation included Sanger and Maxam-Gilbert methods (Pareek et al., 2011; Slatko et al., 2018). These techniques enabled the earliest genomic breakthroughs, but limitations were severe. Electrophoresis-based single-fragment sequencing made the process extremely slow. It was expensive too. Whole-genome projects were nearly impossible for most labs. Next-generation sequencing (NGS) arrived in the mid-2000s and changed everything (Levy & Boone, 2019; Slatko et al., 2018). The key innovation was parallelization—processing millions of DNA fragments simultaneously rather than one at a time. Multiple platforms emerged during this period. Roche/454 used pyrosequencing. Illumina developed sequencing-by-synthesis. SOLiD and Ion Torrent explored different chemistries, including semiconductor-based detection methods. The impact across biology was immediate. Population genetics studies that once took years could be completed in weeks. Transcriptome profiling became standard practice. Even personalized medicine started becoming feasible (Levy & Boone, 2019; Kumar et al., 2019; Slatko et al., 2018).

Research that had been limited to major consortiums was suddenly possible in regular university labs. The democratization of genomics had begun. Yet NGS had its own problems. Read lengths were extremely short, typically under 300 bases. This created major challenges. De novo assembly was particularly difficult. Repetitive regions couldn't be resolved properly. Structural variants were often missed. Complex genomic regions remained problematic to sequence accuracy (Levy & Boone, 2019; Slatko et al., 2018). Third-generation platforms were developed specifically to address these short-read limitations. Two main technologies emerged. Pacific Biosciences created SMRT sequencing which uses zero-mode waveguides to watch DNA synthesis in real-time (Kim et al., 2024). Oxford Nanopore took a completely different approach. Their system passes DNA through protein pores and measures electrical current changes. The results from both platforms are remarkable. Read lengths now extend tens of thousands of bases. Some reads even exceed 100,000 bases. This changes everything for genome assembly. Previously impossible regions become readable. Structural variants are easily detected. Haplotype phasing works properly.

Even highly repetitive sequences can be resolved. Plus, real-time sequencing is now possible. The MinION device from Oxford Nanopore is pocket-sized, enabling sequencing anywhere from hospitals to field sites (Kim et al., 2024; Willbanks et al., 2022). For metagenomics, these advances have been transformative. High-throughput sequencing made large-scale microbiome studies practical (Ko et al., 2022; Quince et al., 2017). But it wasn't just about raw sequencing power. Many incremental improvements mattered too. Better chemistries reduced error rates. Improved basecalling algorithms increased accuracy. Library preparation became more efficient (Maghini et al., 2020; Purushothaman et al., 2024). One crucial development was hosting DNA depletion. Clinical samples like blood contain mostly human DNA, sometimes over 99%. Early metagenomic studies struggled with this contamination. New protocols can now effectively remove host DNA. This allows detection of low-abundance pathogens (Purushothaman et al., 2024; Zheng et al., 2021). Even challenging samples like respiratory fluids yield useful data. The computational side has evolved in parallel. Better assembly programs handle metagenomic data more effectively. Binning algorithms can now separate closely related species (Navgire et al., 2022; Quince et al., 2017).

Resistance gene databases have grown larger and more accurate. Long-read data has been particularly helpful here. It's much easier to determine which organism carries which resistance gene when you have 20,000-base reads instead of 200-base ones (Kim et al., 2024; Olsen & Riber, 2025). Clinical applications have matured significantly. We now have mNGS (metagenomic Next-Generation Sequencing) as a diagnostic tool. Complete workflows exist from sample to result. Collection protocols are standardized. DNA extraction is optimized for different sample types. Library preparation follows validated procedures. Sequencing parameters are well-established. Bioinformatic pipelines can process data automatically. The clinical benefits are substantial. Difficult infections can be diagnosed without culture. Multiple pathogens in the same sample are detected simultaneously. Even novel, unexpected pathogens can be identified. Turn-around times keep improving. What once took days or weeks can sometimes be done in hours (Willbanks et al., 2022). Looking at the broader picture, HTS has fundamentally altered how we study microbial communities. Complete genome recovery from complex samples is now routine (Garza & Dutilh, 2015; Quince et al., 2017). We can maintain ecological context while examining individual organisms. Real-time analysis enables rapid public health responses. Both basic research and applied work benefit. Understanding of the microbiome continues to deepen. Meanwhile, practical tools for combating antimicrobial resistance keep improving. The technology serves multiple purposes effectively (Djordjevic et al., 2023; Olsen & Riber, 2025).

10.2.3 Metagenomics Approaches

The culture-independent approach of metagenomics has fundamentally altered how antimicrobial resistance (AMR) is studied. It goes beyond the tiny fraction of microbes that can be grown, which has been estimated as less than 1% and furthermore accesses the entire genetic pool of microbial ecologies (Hugenholtz & Tyson, 2008; Nam et al., 2023). There are two strategies that prevail in this field of research. Them being sequence-based metagenomics, which maps genetic content, and functional metagenomics which directly experimentally measures the functioning of genes. They are complementary in nature and not substitutes but rather have their strengths in the bigger battle against AMR (Duarte et al., 2021). The core of today's AMR research is shotgun sequencing. This method involves analyzing total DNA from a microbial community without cultivation, producing an extensive snapshot of the genes carried by the organisms within (Ko et al., 2022; Quince et al., 2017). The principle is simple but technologically significant where DNA is randomly fragmented, sequenced, and compiled into a dataset of short reads that together represent the collective metagenome. With this data, several fundamental questions can be addressed.

i. *Community composition:* Taxonomic marker genes or reference mapping provide insight into the different microbial groups present followed by their relative proportions (Quince et al., 2017).
ii. *Functional potential:* Genes encoding certain functions such as metabolic functions, virulence factors and resistance mechanisms can be detected by analysis (Crofts et al., 2017; Hendriksen et al., 2019).
iii. *Profiling of the resistome:* comparing reads against curated AMR databases provide a detailed map of resistance mechanisms, such as β-lactamases, target-modifying enzymes, or efflux systems along with their abundance across the community (Alcock et al., 2022).

The interpretation of this data depends on two main strategies: the gene-centric stratergy which maps reads directly to resistance gene databases that provides a rapid overview of content and the second one being a genome-centric strategy which assembles reads into longer fragments or near-complete genomes, giving crucial context (Navgire et al., 2022). The second approach shows which organisms accommodate which resistance genes, and whether those genes sit on mobile genetic elements. This is a key factor in judging their transmission capabilities (Olsen & Riber, 2025). Collectively, shotgun metagenomics produce quantitative, high-throughput data on resistomes in humans, animals, and the environment. It has become the foundation of AMR surveillance and ecological awareness (Ko et al., 2022; Vallejo-Espín et al., 2025).

Shotgun methods characterize the genetic potential, but to a great extent they depend on sequence similarity and prediction (Crofts et al., 2017). They can suggest that a gene looks like a resistance gene but cannot prove that it acts like one. To resolve this limitation, researchers use functional metagenomics, which follows a different logic that starts with phenotype, then moves to sequence. Here, DNA from environmental samples is cloned into a laboratory host, commonly E. coli (De, 2019; Mullany, 2014). This produces large libraries in which each clone carries a fragment of foreign DNA. When the library is exposed to an antibiotic, only those clones that received a functional AMR gene will survive (Mullany, 2014). These surviving inserts can then be sequenced to reveal their identity. The great strength of this approach is its independence from prior knowledge because the selection is based on function rather than sequence comparison. It can also reveal genes that share little to no similarity with those already catalogued (Duarte et al., 2021).

This allows truly novel resistance determinants to be discovered. In other words, it uncovers resistome diversity that would otherwise remain hidden, while also warning us of the mechanisms that might later emerge in pathogens (Olsen & Riber, 2025). It is important to stress that functional screening complements shotgun sequencing rather than replacing it. Sequence-based methods provide the ecological overview. Function-based screens validate biological activity and bring light to unexpected resistance genes. Once identified, these genes are incorporated into reference databases, where they strengthen future sequence-based surveillances. This cycle of discovery and integration enhances both approaches (Alcock et al., 2022; Crofts et al., 2017). The most powerful insights come from combining both techniques. Shotgun sequencing delivers breadth, capturing the overall genetic repertoire of communities, while functional analysis supplies depth by testing which genes actually confer resistance (Crofts et al., 2017; Mullany, 2014). The improvement in long-read sequencing and its integration with other omics data make it feasible to link the resistance gene with its hosts and to their broader biochemical networks (Kim et al., 2024; Olsen & Riber, 2025). It made a foundation for tackling the AMR crisis by providing a holistic view of the resistome and bringing the field closer (Djordjevic et al., 2023; Ko et al., 2022).

10.3 APPLICATIONS OF METAGENOMICS IN AMR

10.3.1 Detection of Antibiotic Resistance Genes (ARGs)

The development of the antimicrobial resistance genes (ARGs) detection techniques has suffered significant evolution. Microbiologists over the decades resorted to the use of phenotypic tests that were essentially asking the question of whether or not an organism could grow when subjected to an antibiotic (Anjum et al., 2017; Galhano et al., 2021). Although these assays were useful, they had apparent limitations. They were long, in most cases taking days, and limited to a small number of microorganisms that could be grown in the laboratory (Anjum et al., 2017). As a result, some organisms and their resistance determinants could not be seen. With the advent of molecular techniques, a whole new dimension was now created. Instead of waiting until growth could be seen, scientists started to study genetic material directly. This direct inspection of DNA or RNA made the study of resistant genes that provide survival an ease of achievement, which is a revolutionary change (Anjum et al., 2017; Banerjee & Patel, 2023). Quicker responses enabled clinicians to commence specific treatments much earlier on and, in some cases, save lives of patients with severe infections. Of equal importance is the scope of the information that such methods can give.

Only the growing organisms that grow well in vitro can be evaluated using the culture-based techniques (Galhano et al., 2021). Conversely, molecular tests have been shown to be able to detect ARGs even in cells that have been killed or in microorganisms that cannot be grown at all (Anjum et al., 2017). Multiplex PCR is one such format. It uses panels of primers to target a specific set of pathogens and their resistance determinants, which provide results in the case of hours (Banerjee & Patel, 2023). The tests offer extremely actionable information to the clinicians, but the trade-off was reduced to a small scope. They will only sense what they are designed to sense. New resistance genes, or surprise variants, will always escape these specific assays (Anjum et al., 2017). This is where metagenomics has been a necessity. Unlike PCR panels, metagenomic sequencing is not restricted by the search space of predefined targets. Rather, it sequences the whole set of DNA in a sample, creating a picture of the entire resistome (Crofts et al., 2017).

Shotgun metagenomics can sequence the entire set of genetic fragments, providing community composition and resistance profiles in a single set. It is especially useful in the identification of resistance in uncultivable organisms, the characterization of new allelic variants, or the determination of resistance determinants in complex system settings, such as soil, wastewater, or the human gut (Liu & Ma, 2024; Munk et al., 2017). Functional metagenomics brings in additional strength. Instead of using similarity in sequence it filters DNA fragments based on their resistance conferring ability when introduced into a laboratory host. In doing so, it can discover any resistance genes that are not like the previously known ones, showing entirely new mechanisms (Crofts et al., 2017; Mullany, 2014). Outcomes of these experiments enlarge reference databases on which sequence-based approaches are based, forming a cycle of continuous enhancement (Alcock et al., 2022). Surveillance feeds discovery, discovery identifies new targets of surveillance.

Direct genetic detection has numerous benefits compared to more conventional culture-based techniques (Galhano et al., 2021). The first is the time issue through the reduction of the diagnostic window, which might be days, to hours, which is a basic clinical achievement. Second is the extent of discovery through investigation of the whole resistome. These techniques broaden the perspective to incorporate genes that are contained in organisms that would be difficult to examine otherwise (Hendriksen et al., 2019). Finally, there are surveillance applications. Direct sampling of hospitals, farms, wastewater systems, and even natural ecosystems provide real-time genetic snapshots of resistance on a population level (Djordjevic et al., 2023; Hendriksen et al., 2019; Munk et al., 2017; Olsen & Riber, 2025).

Such data currently informs policy on antimicrobial stewardship, and this information is used to drive interventions within the area of public health (Salam et al., 2023; Tang et al., 2023). Specifically, molecular and metagenomic methodologies allow identifying resistance that has not been identified before. These technologies discover the antibiotic resistance genes in unculturable species, inform about emerging genetic threats, and enable real-time tracking of the spread of resistance among human, animal, and environmental reservoirs, thus bypassing the cultivation bottleneck (Djordjevic et al., 2023; Olsen & Riber, 2025). This results in the researchers and clinicians being spared of the technicality that defined the past. The shift to fast, culture-free genomic profiling, compared to traditional slow phenotypic methods, is not only an incremental change but also a paradigm shift in the nature of the method of antimicrobial resistance in scientific research and clinical practice (Ko et al., 2022).

10.3.2 Profiling Microbial Communities

The two linked perspective of metagenomics foundation reveals the both, the compositional and functional capacity of the microbial community and its genes, respectively (Hugenholtz & Tyson, 2008; Quince et al., 2017). This dual view is particularly significant for AMR research. The presence of a resistance gene is inseparable from the ecological and taxonomic context in which it resides (Ko et al., 2022).

10.3.2.1 Diversity and Community Structure

The first stage of most metagenomic studies is a taxonomic survey. Sequencing all DNA from a sample makes it possible to identify major groups of bacteria, archaea, viruses, fungi, and also measure their relative abundance (Quince et al., 2017; Sleator et al., 2008). There are several databases available for the resolving this problem as shown in Table 10.1. Understanding this baseline structure is essential for AMR studies for several reasons.

- *Reservoirs of pathogens:* Metagenomics capture the full range of organisms present, including rare members usually overlooked by culturing methods (De, 2019; Ko et al., 2022). In the human gut, this involves opportunistic pathogens such as Escherichia coli, Klebsiella pneumoniae, and Enterococcus faecium, which may persist at low levels among dominant commensals. Documenting their existence helps establish a baseline measure of risk (van Schaik, 2015).

TABLE 10.1
Most Commonly Used AMR Databases

Database	Key features	Data Types Provided	Original Publication
CARD (Comprehensive Antibiotic Resistance Database)	Curated reference database with ontology-based classifications of resistance mechanisms and AMR gene variants. Widely used in clinical and research pipelines	DNA/Protein sequences	Alcock et al. (2022)
ResFinder	Detects acquired AMR genes via sequence alignment. Widely used for bacterial isolate analysis	DNA sequences	Florensa et al. (2022)
MEGARes	Hierarchically structured resistance gene database optimized for metagenomic read classification	DNA sequences	Lakin et al. (2017)
ARG-ANNOT	Manually curated database focused on acquired resistance genes from pathogens	DNA sequences	Gupta et al. (2014)
BacMet	Manually curated entries of experimentally validated genes and a large set of predicted resistance genes	DNA/Protein sequences	Pal et al. (2014)
ARDB (Antibiotic Resistance Genes Database)	Earliest AMR gene repositories that is not updated regularly but historically important	DNA/Protein sequences	Liu & Pop (2009)
NDARO (National Database of Antibiotic Resistant Organisms)	Integrated AMR reference database within NCBI with links to pathogen genomes and metadata	DNA/Protein sequences	National Institutes of Health (2022)
SARG (Structured ARG Database)	Combines CARD, ARDB and literature-based curation for environmental resistome analysis	DNA sequences	Yang et al. (2016)
PATRIC/BV-BRC (PAThosystems Integration Resource Centre/ Bacterial and Viral Bioinformatics Resource Centre)	Includes resistance gene annotation integrated with pathogen genome browser and ML prediction tools	Genomic and phenotypic data	Olson et al. (2023)

- *Indicators of dysbiosis:* In ecological theory, greater diversity is directly proportional to stability. Within the gut, diverse microbial groups restrict pathogen overgrowth through constant competition for resources and space (van Schaik, 2015). Antibiotics disrupt this balance as they reduce diversity and eliminate sensitive commensals. The resulting dysbiosis creates an empty niche that resistant organisms can occupy. In practice, metagenomics allows these shifts to be observed and quantified directly (Liu & Ma, 2024).
- *Comparisons across contexts:* A community profile also enables comparisons between environments and health states. Hospital surfaces, community dwellings, agricultural soils, and forest ecosystems often show distinct microbial indicators (Djordjevic et al., 2023; Vallejo-Espín et al., 2025). Similar comparisons between healthy individuals and patients suffering from recurrent infections can reveal which conditions are associated with higher resistance burdens (Zheng et al., 2021).

10.3.2.2 Functional Potential

The structure of the community is not enough. The shared genes are also cataloged by metagenomic sequencing, and the functional repertoire coded by the community (Quince et al., 2017). The latter aspect of analysis is the central issue in AMR studies.

- *Resistome characterization:* The dataset of all antibiotic resistance genes (ARGs) in a sample is referred to as its resistome. Using the sequences as a point of comparison with specific databases, such as CARD or ResFinder, researchers can track down resistance genes, starting with 2-lactamases (degrading 2-lactam antibiotics), and ending with the efflux pump and antibiotic target modifying enzymes (Alcock et al., 2022; Crofts et al., 2017). This method also aids in the discovery of the existence of resistance genes as well as the broad spectrum of mechanisms that they employ.
- *Estimating resistance load:* Resistome analysis is not only descriptive but also quantitative. The sequencing data can be used to estimate the abundance of various ARGs in comparison to the full metagenome. A big and complicated resistome may imply a great deal of antibiotic consumption or exposure in that setting. Hospital input into wastewater will generally contain a dense resistome (Hendriksen et al., 2019; Munk et al., 2017), whereas in remote soils or mountain streams, there are likely to be much fewer resistance markers.
- *Relations to other traits:* Resistance genes are frequently present with other clinically relevant phenotypes. The virulence factors, plasmids, and transposons may be listed concurrently, and the researchers are able to assess the risk of mobility, pathogenic potential, and resistance (Alcock et al., 2022; Crofts et al., 2017). Of special concern is the finding of ARGs on mobile genetic elements, which is an indication of an apparent ability to undergo horizontal transfer.

It is important to stress that sequence-based analyses describe potential rather than activity. A gene detected in metagenomic data may not be expressed under natural conditions. The transcriptome level research is required to verify its functionality, including meta-transcriptomics (Liu & Ma, 2024; Zheng et al., 2021). Therefore, metagenomics provides the sketch of the available genetic characteristics, and integrated methods of omics strategies narrow this image into a comprehensive perspective of resistance ecology.

10.3.3 Tracking Resistance across the One Health Spectrum

Metagenomics does give us the opportunity to see beyond individual AMR incidences and have a feel of the bigger interconnected system. Regarding AMR, it is not only a health concern but a complicated problem that connects humans, animals, and the environment in terms of the One Health lens (Djordjevic et al., 2023; Olsen & Riber, 2025). The standard one-point techniques simply fail miserably in the sheer magnitude of this international issue. Rather, metagenomics is a paradigm-shifting technology that allows us to simultaneously investigate the entire resistome of various ecosystems (Elton et al., 2025; Vallejo-Espin et al., 2025). Another highly documented and direct route of ARG transmission is the human-animal interface. Extensive application of antibiotics in livestock, usually to stimulate growth or to prevent diseases, is expected to generate a strong selection pressure that accelerates the growth of ARGs in the gut microbiomes of animals (Munk et al., 2017; Williams et al., 2023). They can be directly passed to humans by close contact with animals, for example, between farmers and their livestock or veterinarians and their patients.

Its transfer is more often indirectly by the food chain, when ARGs in the contaminated meat or produce may be ingested by human beings. Such genes can subsequently be transferred into the human gut microbiome, which might be transferred to pathogenic bacteria (van Schaik, 2015). Human population is one of the major contributors of ARGs introduced into an environment. One of the most prominent avenues of this is via wastewater which can carry a wide variety of ARGs due to human waste, hospitals, and pharmaceutical production (Hendriksen et al., 2019). Although the wastewater treatment plant (WWTPs) is created to eliminate the contamination, it is not specifically geared at eradicating ARGs. Rather, they become hotspots where all kinds of bacteria of diverse origins are combined and share genetic material and are, in essence, a breeding ground of resistance. Wastewater containing ARGs and the bacteria that carry them is often released into rivers, oceans, and soil, spreading clinically important resistance genes into the wider environment (Hendriksen et al., 2019; Munk et al., 2017). In agriculture, the use of animal manure as fertilizer provides another major route for ARGs to enter soil and groundwater. These genes can be picked up by environmental bacteria or even contaminate crops directly. Once in the environment, they create a large reservoir of resistance that can be passed to wild animals, birds, and insects, helping ARGs spread even further across the globe (Vallejo-Espín et al., 2025). Such a complex network of distribution underlines the need to adopt a comprehensive, metagenomics-centered surveillance strategy to monitor ARGs in all three areas as well as formulate combined ways of containing them (Djordjevic et al., 2023; Olsen & Riber, 2025).

One of the studies conducted at a veterinary center in the UK used metagenomics to study the resistome of animal patients and the hospital environment. The researchers wiped down communal locations in the hospital and sampled animal patients. The metagenomic revealed high abundance and diversity of ARGs like genes of tetracyclines and β-lactamases resistance, in places that are regularly visited by animals, for example, dog wards and surgical preparation rooms. This study gave first hand proof that the animal communal areas serve as reservoirs of ARGs (Elton et al., 2025). It has also reported how resistance could be present in the environment of a veterinary facility and possibly be passed between sick animals, and to those who work in these facilities, highlighting the essential connection between the health of animals and that of humans (Djordjevic et al., 2023; Elton et al., 2025).

Another interesting study was centered on a city setting in Bangladesh, a highly populated area with a high number of humans and animals and a poor sanitation system. The samples of the human population, poultry and common environment (i.e., wastewater) were collected and have been analyzed using metagenomic techniques. The findings indicated that there was great diversity in ARGs across the three areas. The wastewater treatment plants has been identified as a major hotspot for antibiotic resistance genes (ARGs) and a key pathway for their spread (Hendriksen et al., 2019; Munk et al., 2017; Williams et al., 2023). Research shows that while direct clonal transmission of bacteria between humans and animals may be limited, the sharing of ARGs is common. Wastewater acts as a shared environment—or "super-highway"—where these genetic elements of resistance are exchanged. This case study highlights how metagenomics can uncover the complex movement of resistance in crowded urban areas, showing how shared environments tightly connect human and animal health (Djordjevic et al., 2023; Vallejo-Espín et al., 2025).

10.4 METAGENOMICS STUDY PIPELINE

10.4.1 Sample Collection and Preparation

A successful metagenomics study unfolds in several phases, each building on the previous one (see Figure 10.1). The quality and reliability of the final bioinformatic results depend entirely on the care and precision applied in the early stages. Put simply, "garbage in, garbage out"—poor sample collection or preparation will undermine the entire pipeline. This is why designing a solid workflow starts with getting the basics, like sampling and preparation, exactly right (Quince et al., 2017; Navgire et al., 2022). The first step is, perhaps, the most important step of the whole metagenomic workflow. It aims at capturing a sample that is a true reflection of the microbial community in vivo and to maintain the integrity of its nucleic acids until the time it can be extracted in the laboratory. Any bias and contamination that is introduced at this point is almost intractable later in the analysis and will be transduced throughout the analysis with consequences of possible inaccurate conclusions (Thomas et al., 2012).

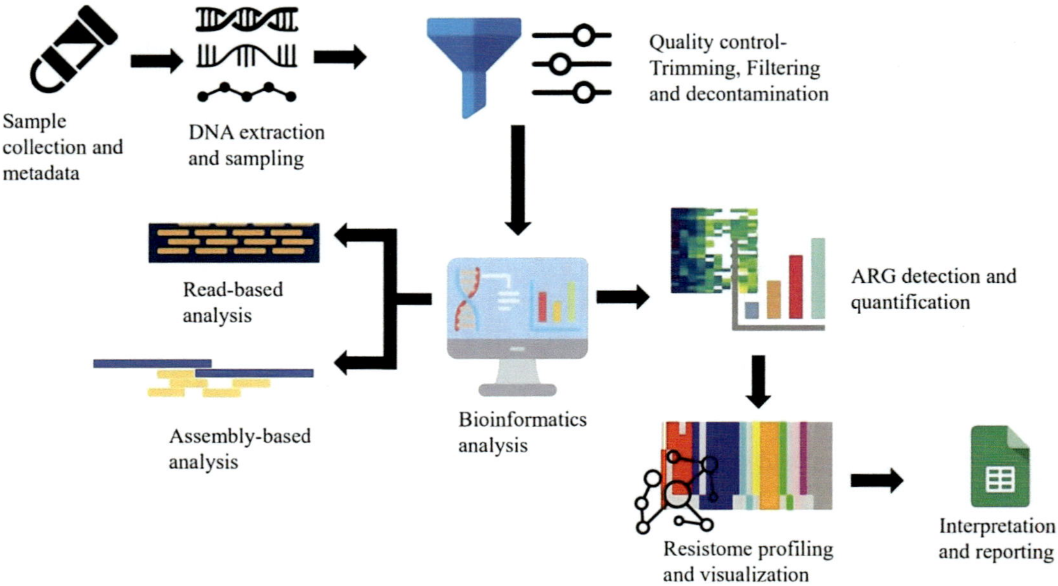

FIGURE 10.1 Metagenomics Analysis Pipeline.

10.4.1.1 Types of Samples

The research question determines the selection of the sample. In the AMR research, important reservoirs and transmission environments are usually sampled (Djordjevic et al., 2023). Common sample types include:

- *Fecal/Stool Samples:* This is the sample type that is most frequently used to study the gut microbiome. It is non-invasive and a great proxy of the composition and resistome of the distal gut, one of the primary reservoirs of ARGs and opportunistic pathogens. Fecal samples are the best in monitoring AMR in farm environments as evidenced in the case of swine herds (Munk et al., 2017).
- *Oral Samples:* Oral samples like Saliva, Swab are used to investigate the microbiome of the mouth and its contribution to reservoir existence.
- Skin Swabs: These are essential in the research on the skin microbiota and its contribution to hospital-acquired infections (Djordjevic et al., 2023; Nam et al., 2023).
- *Environmental Samples:* These are important in the monitoring of AMR spreading outside a host. River, lake and ocean samples are used to monitor the background levels of AMR. Sewage and wastewater are unique sources and a great use for surveillance of health to the population because it gives a community-wide view of the resistome (Hendriksen et al., 2019; Munk et al., 2017).
- *Soil and Sediment:* These living environments are regarded as great reservoirs of new AMR genes.
- *Low-Biomass Samples:* These pose special challenges because of low content of microbial DNA as compared to possible contaminants.
- *Air Samples:* Air samples are used in the investigation of transmission of microbes in the air.
- *Clinical Fluids:* Samples such as blood or cerebrospinal fluid must have a very sensitive and sterile method of sample collection to prevent subduing the real signal by the contaminants (Liu & Ma, 2024).

10.4.1.2 Best Practices and Considerations

A good sampling strategy would be characterized by thorough preparation and keen execution. These are some of the considerations that need to be made to produce high-quality metagenomic information.

- *Sampling must be preceded by study design:* It is the study design that is the most important factor. The sampling strategy is determined by the research question. A cross-sectional study, giving a picture needs to sample a large number of individuals or areas at a time. A longitudinal study, which is meant to follow changes (e.g., before, during and after the antibiotic treatment) needs repeated sampling using the same sources (Quince et al., 2017).
- *Replication and Representativeness:* Adequate biological replicates are non-negotiable to use to make statistically valid conclusions. It is not sufficient to sample just one animal or a liter of water in order to describe a whole farm or river. The plan should be used to acquire the natural fluctuation of the system. With large numbers, such as a swine herd, a typical approach includes pooling of samples, in which several separate samples of a compartment are pooled together to create one composite sample. This gives an economical and representative mean of the group which minimizes variation between animals (Munk et al., 2017).
- *Aseptic Technique and Contamination Control:* Contamination is one of the major threats to data integrity. All samples should be collected using sterile equipment such as gloves, tubes, and swabs. This is especially crucial for samples with low biomass, where even small amounts of contamination from reagents, lab surfaces, or the collector's skin can easily overpower the true microbial signal. Hence, it is a significant best practice to incorporate negative controls. Including a negative control—such as running a blank sample of sterile water through the entire extraction and sequencing process—helps identify and measure background contamination. This contamination, often called the "kitome," comes from the DNA extraction kit itself (Trigodet et al., 2022).
- *Immediate Sample Preservation:* After the sample has been collected, immediate stoppage of biological activity of the sample is necessary to avoid alteration of microbial community in the sample and deterioration of the DNA by inactivating enzymes. Preservation using liquid nitrogen (-196 degree Celsius) freezing followed by storage at -80 degree Celsius is the gold standard of preservation. Where this is not practical, a -20 degree Celsius freezer or dry ice can be used as an alternate for immediate storage. To do remote fieldwork, the chemical preservation buffers (e.g., RNAlater, ethanol) can stabilize the nucleic acids at ambient temperature so that they can be safely stored at a later date (Maghini et al., 2020).
- *Extensive Metadata Collection:* Raw sequence data has no meaning. An elaborate metadata should be carefully documented on each sample. It contains the date, time, and exact place of collection and other pertinent environmental parameters (e.g., temperature, pH, chemical concentrations) or host data (e.g., age, diet, recent antibiotic use). It is this metadata that can be used by researchers in the future to relate changes in the metagenome to certain external factors (Thomas et al., 2012).

10.4.2 DNA Extraction and Sequencing

The study of the collective genomes of whole microbial communities in their natural conditions has transformed our knowledge of the microbial life. In contrast to the conventional methods founded on a culture-based understanding of the biodiversity, which is usually restricted to the minority of cultured microorganisms, metagenomics provides a holistic perspective of the biosphere in terms of both biodiversity and functionality. Nonetheless, it will depend on the data sampling quality, and the

DNA extraction and sequencing, the most crucial operations in the workflow. According to recent studies, methodology during this stage is a critical decision because it can induce serious biases and artifacts, which can invalidate the latter bioinformatic analyses (Navgire et al., 2022; Rastogi & Sani, 2011). This section will serve as a handy guide to the core principles of the protocols and quality checks needed to generate high-quality data for metagenomic research.

10.4.2.1 DNA Extraction: From Sample to DNA

The main problem with metagenomic DNA extraction is, in essence, to efficiently lyse any type of cell gram-positive and gram-negative bacteria, fungi, archaea, and so on; at the same time, eliminate any contaminants of the sample matrix. This is to collect that required amount of DNA purity and quantity and leave the molecules long, which reduces bias in its turn.

- *The Challenge of Metagenomic DNA*: Each sample type—whether soil, stool, or water—presents its own unique challenges. For example, both wet and dry materials may contain large amounts of humic acids and other substances that can inhibit enzymes and disrupt later processing steps. Additionally, Gram-positive bacteria have particularly tough cell walls, which makes them resistant to mild lysis methods (Elton et al., 2025; Vallejo-Espín et al., 2025).
- *Mechanical and Chemical Lysis Methods:* A combination of mechanical and chemical lysis is a common and effective technology. Physically perturbing a sample with glass or ceramic beads is known as bead-beating and forms the foundation of numerous commercial kits and in-house protocols, in particular with environmental samples. As an example, the comparison of methods of DNA extraction revealed that bead-beating systems, including the KingFisher system, can offer a high-throughput, high-read answer to environmental samples of varied types without taxonomic or abundance details being compromised. Mechanical lysis is followed by the use of chemical degrading factors such as SDS and enzymes such as proteinase K, to further disperse cell membranes and proteins to guarantee a more complete and impartial yield of DNA (Crofts et al., 2017; Nam et al., 2023).
- *Commercial Kits* versus *In-house Protocols:* In-house protocols mean that the protocol can be tailored but commercial kits have a standardized and often more reproducible answer. FastDNA Spin Kit of Soil is a popular kit that has shown to work across most environmental matrices. It should be highlighted though, that the kit efficacy is sometimes affected by the type of sample and the desired sequencing platform. As an example, certain bead-beating systems can shear DNA so those are not more appropriate with long-read sequencing systems that need high-molecular-weight fragments like DNA (Crofts et al., 2017; Nam et al., 2023).
- Extracted DNA Quality Control: The process of the extraction of the DNA would not be effective without quality control. The DNA extracted has to be evaluated in terms of quantity, purity and integrity.
  - *Quantity:* Typical DNA quantification methods are spectrophotometry (e.g., Nanodrop) and fluorometry (e.g., Qubit). To give a rapid estimate, a Nanodrop can be used, however, a Qubit fluorometer is preferable as it is specific to double-stranded DNA, which is vital in preparing a library.
  - *Purity:* Evaluation of purity is done based on the A260/A280 and A260/A230 ratios on a spectrophotometer. A A260/A280 ratio of ~1.8–2.0 shows a clean sample with no formation of serious protein contamination, and an A260/A230 ratio of ~2.0–2.2 indicates that the humic acids or other organic substances were not significantly contaminated (Elton et al., 2025; Vallejo-Espín et al., 2025).
  - *Integrity:* The integrity of the DNA, or degree of fragmentation, is important, particularly on long-read sequencing. It is possible to visualize the integrity of DNA with

gel electrophoresis or more sophisticated equipment, such as a Fragment Analyzer. The goal is to obtain high-molecular-weight DNA, as fragmented DNA can negatively affect the accuracy and success of later genome assembly steps (Liu & Ma, 2024).

10.4.2.2 Sequencing

Sequencing use to transforms the extracted DNA into raw digital data. Choosing the right sequencing platform depends on practical factors like the research goals, budget, and the type and quality of data you want to generate. It is an important factor to consider include the length of the reads, which influences how well genomes can be assembled, as well as the sequencing throughput and cost per base, which determine how large and affordable the study can be. There are point listed below which mention feature of know sequencing platforms:

- *Short-Read Sequencing (e.g., Illumina):* Illumina is the dominant platform for metagenomics due to its high throughput, low cost, and low error rate. Its "Sequencing by Synthesis" principle generates billions of short reads (up to ~300 bp), which are ideal for abundance profiling and gene-centric analyses (Duarte et al., 2021; Galhano et al., 2021). However, the short reads can make it challenging to assemble contiguous genomes from a complex mixture of species, especially when dealing with repetitive regions.
- *Long-Read Sequencing (e.g., PacBio, Oxford Nanopore):* In effort to address the reducibility of short reads, long-read sequencing technologies are turning more popular. Such platforms, including Oxford Nanopore, are capable of generating reads that are tens of thousands base pairs long, and are able to span repetitive genomic components to give a better picture of genomes in a community. Though their error rates were more in the past, this is steadily getting better. New research has shown the possibility of using long-read sequencing to detect taxonomic and antimicrobial resistance in complex samples in a relatively short period of time (Liu & Ma, 2024; Willbanks et al., 2022).
- *Quality Check on Sequencing Data:* The raw data from a sequencing run cannot be analyzed at all. It has to be implemented through a rather strict quality control process in order to guarantee reliability.
  - *Pre-processing:* It starts with sharing tools such as FastQC to produce a report on the raw reads indicating potential problems such as adapter contamination, poorly sequenced ends of reads, and non-matching GC content (Duarte et al., 2021).
  - *Filtering and Trimming:* Filtering and trimming of reads based on the FastQC report is performed by software like Trimmomatic which is used to remove adapters and poor-quality reads. The quality of downstream analyses is largely dependent on how the strand is sequenced and often bases of low quality can generate issues during assemblies and fall into wrong taxonomies.
  - *Quality Metrics:* The quality of every base call is determined using key metrics like Phred Quality Score (Q-score). The probability of missing a base call is equal to 1 in 1000 base calls and Q-score should be used to estimate this and one should aim at getting high Q-scores (Galhano et al., 2021).

10.4.3 Bioinformatics Analyses

The sequencing process has no biological knowledge output unless it is subsequently processed with bioinformatics tools (or mechanisms that transform raw data). The algorithmic process of converting this raw data into controlled meaning revealing the microbial community and resistome is known as the bioinformatics pipeline and it is complicated and computationally costly (Khan et al., 1 C.E.; Roy et al., 2024). There are some available tools for the ease of the analysis work as shown in Table 10.2. This pipeline can be generally broken down into three main phases, namely, initial data

TABLE 10.2
Bioinformatics Tools and Pipelines Used for Metagenomic AMR Detection

Tool/Pipeline	Primary Function	Method Type	Input Type	Features	Tool/Pipeline Citation
MEGAHIT/ metaSPAdes	Metagenome assembly	Assembly based	Short reads	Fast de novo assembly for large metagenomes	Li et al. (2016)
Kraken2/Bracken	Taxonomic classification	K-mer based	Reads	High speed composition profiling	Lu et al. (2017)
QIIME/DADA2	Microbiome data analysis	Quality control	Raw data files	Denoising and filtering	Estaki et al. (2020)
AMRFinderPlus	ARG annotation	Alignment based	Reads/ contigs	NCBI-supported tool with phenotype integration	Feldgarden et al. (2022)
DeepARG	AMR prediction using deep learning	Machine learning	Reads/ORFs	Identifies diverse or cryptic AMR genes	Arango-Argoty et al. (2018)
Abricate	Batch screening of assemblies	Alignment based	Assembled contigs	Scan multiple databases simultaneously	Seemann T, Abricate, GitHub
MetaPhlAn3	Species level microbial profiling	Marker gene based	Reads	Substantial for human microbiome studies	Francesco Beghini et al. (2021)
Resistance Gene Identifier	Detects AMR genes using CARD	Alignment and ontology	Contigs/ proteins	Industry standard for resistome analysis	Kwatra (2021)
ShortBRED	Marker based AMR quantification	Profile based	Reads	High specificity for functional markers	Kaminski et al. (2015)
SURPI/IDseq	Diagnostic metagenomics	Cloud/hybrid	Reads	Hospital and outbreak surveillance	Naccache et al. (2014); Kalantar et al. (2020)

processing, identification and quantification of AMR genes, and the need of overcoming the inevitable obstacles of the methodology. The complete metagenomics pipeline is depicted in Figure 10.1.

10.4.3.1 Data Processing

It all depends on the quality of the input data to generate a good quality of analysis. Raw sequencing reads are not perfect; they have technical artifacts of the sequencing process, and usually contain a significant fraction of non-microbial DNA. Thus the initial step is intensive data processing and quality control (QC) workflow (Angers-Loustau et al., 2018; Navgire et al., 2022).

- *Quality Filtering and Trimming:* The first step is to check the quality of the raw sequencing reads. Bases with low confidence scores from the sequencer are trimmed from the ends, and any reads that remain too short or have poor overall quality are removed. This process is often carried out using tools such as Trimmomatic or Fastp (Navgire et al., 2022). During library preparation, small adapter sequences are attached to DNA fragments to help with sequencing. Since these adapters are artificial and not part of the original DNA, they need to be removed before further analysis.
- *Host DNA Decontamination:* This step is especially important when working with samples from host organisms, such as the human gut, swine feces, or clinical specimens. In such cases, the host's DNA can make up more than 90% of the total sequencing data, overwhelming the microbial signal and making analysis more difficult. To overcome this, reads are aligned to the reference genome of the host (e.g., human, pig). Any reads that coincide

are deemed as contaminants of the host and excluded out of the dataset. Specialized equipment such as KneadData can be fitted to carry this quality control and decontamination pipeline out effectively (Hendriksen et al., 2019; Munk et al., 2017).

It is after such cleaning procedures that the final product of high-quality, non-host, microbial reads are prepared to be analyzed in the downstream.

10.4.3.2 Molecular Detection and Quantification of AMR Genes

Having a clean dataset, the most important objective is to find and count AMR genes in it. This is usually done by mapping the metagenomic sequences to existing curated collections of known AMR genes (which include, but are not limited to, the Comprehensive Antibiotic Resistance Database (CARD), ResFinder or MEGARes) (Alcock et al., 2022; Galhano et al., 2021; Mullany, 2014). There are two primary analytical strategies to accomplish this:

- *Read-Based Analysis (Mapping Approach):*
  It is an immediate and quick way. These high-quality short reads are aligned or mapped to an AMR gene sequences reference database. Aligning short reads to the database is performed using tools such as BWA or Bowtie2 (Navgire et al., 2022; Roy et al., 2024). A read is deemed as a hit in case of alignment with an adequate similarity (e.g., >90% sequence identity). To compare resistomes of two or more samples, raw hit counts are to be normalized. It is necessary to consider discrepancies in sequencing depth (number of reads in each sample), and even length of resistance genes themselves. One popular measure of abundance is the RPKM (Reads Per Kilobase of gene length per Million sample reads) or other measures such as TPM (Transcripts Per Million) which give a proportion of the abundance of a gene. Specialized versions of this method have been developed such as ShortBRED, which employs unique protein markers to measure gene families to minimize false positives (Lanza et al., 2018).
- *Assembly-Based:*
  This procedure is more computationally costly, and yet offers priceless genomic context. They are initially assembled into longer more contiguous sequences (contigs) with assemblers such as MEGAHIT or SPAdes (Navgire et al., 2022). Through the process, this is that of solving a giant jigsaw puzzle. As contigs have been constructed, the Prodigal gene prediction algorithm is employed to recognize possible protein-coding genes in those lengthier sequences (Navgire et al., 2022). The predicted genes are subsequently compared with the AMR databases. This method is typically also more precise, since full length gene prediction accuracy is less obscure than that of short reads. Most importantly, this maintains the genomic context, enabling scientists to be aware of other genes that are in the vicinity of an ARG, and most importantly, is able to identify whether the ARG is a chromosome or a mobile genetic element, such as a plasmid, which would be of significant importance to understand its risk of mobilization (Ko et al., 2022; Olsen & Riber, 2025).

10.4.3.3 Challenges in Bioinformatic Analysis

Metagenomic AMR analysis is a powerful tool, but it comes with several challenges that researchers need to carefully address. Some of them are pointed bellow:

- *Computational Requirements:* The sheer volume of data generated by modern sequencers is a major hurdle. A terabytes of data is generated for a single metagenomics study. A high-performance computing (HPC) clusters, large amounts of RAM, and massive storage capacity are need to processing, assembling, and analyzing these kind of datasets (Khan et al., 1 C.E.). The assembly step is often the most demanding bottleneck.

- *Sensitivity:* This refers to the ability to detect rare ARGs from the datasets. The abundance of genes in a community can span many orders of magnitude. If an emerging or low-abundance resistance gene is present, it may not be detected unless the sequencing is sufficiently "deep" (i.e., a very high number of total reads are generated). This trade-off between sequencing depth and cost is a constant consideration (Hendriksen et al., 2019).
- *Specificity:* This is the ability of ARG-identification and false positives. Secondary reads are not always good. The read of a conserved portion of a housekeeping gene may be incorrectly identified as an ARG in the event that the conserved portion is shared between the genes. Methods and tools that exploit assembly-based techniques that rely on strict, full length alignment of genes contribute towards creating greater specificity and minimizing these errors (Olsen & Riber, 2025).
- *Database Dependency:* All sequence-based methods are fundamentally limited by the knowledge contained within the reference databases. They can only identify what is already known. This means that truly novel or highly divergent resistance genes will be missed, highlighting why functional metagenomics remains a crucial, complementary approach for AMR discovery (Ko et al., 2022; Mullany, 2014).

10.5 CHALLENGES AND LIMITATIONS

Metagenomic sequencing has emerged as a key method to analyze the overall repertoire of antimicrobial resistance (AMR) genes, called the resistome. It offers novel methods of studying the microbial communities, yet this method is limited. Technical, analytical and interpretive issues have to be carefully addressed in order to produce data that are reliable and meaningful (Ko et al., 2022; Quince et al., 2017). Without addressing them, the estimates of AMR prevalence and its effect on the health of the population are misleading. The journey towards resistome profiling is initiated by DNA extraction and sequencing, and issues may arise during the first step. Acquiring a sample that will effectively represent the whole community is one of the greatest challenges (Rastogi & Sani, 2011; Trigodet et al., 2022). Cell envelopes of microorganisms are very diverse. Thick peptidoglycan layers of gram-positive bacteria, such as those of gram-positive bacteria, resist lysis, whereas the cell walls of many Gram-negative bacteria are relatively weak and readily ruptured. They do not respond to just one extraction method. This leads to the under-representation of some organisms in the final dataset (Purushothaman et al., 2024; Trigodet et al., 2022). This gives an inaccurate image of the resistome. A further complication in host-associated samples is the presence of the host DNA dominating the extract (Kim et al., 2024; Munk et al., 2017).

This background can be expensive to eliminate but is an essential step when detection of low-abundance resistance genes is required. Selection of sequencing platform is an added problem. Illumina and other short-read systems are still affordable, however, they cannot readily reassemble complete genomes of complex mixtures of species (Duarte et al., 2021; Galhano et al., 2021). The piecemeal information can hardly be used to reproduce the genetic background of a gene that can either be embedded in a chromosome or a mobile plasmid. Conversely, long-read sequencing shows read lengths long enough to cover an entire gene or even large plasmids, for example, the Oxford Nanopore (Liu & Ma, 2024; Willbanks et al., 2022). This enables us to position the resistance genes in a better position which is one of the crucial steps in getting to know their mobility. There is a trade-off, though, namely that long-read data are commonly more costly, and may contain higher error rates, which becomes a challenge when analyzing such data (Liu & Ma, 2024). After getting the reads there are more problems of interpretation. A resistance gene detected in a dataset does not indicate that the gene is expressed, or even functional. It can be either silent or truncated, or it can be conveyed under specific circumstances. In the absence of evidence in the form of phenotypic techniques, there is a genuine danger of overestimating the level of resistance (Anjum et al., 2017; Mullany, 2014).

This is a big weakness of sequence-based studies because it fills this gap between what is in the DNA and what is actually expressed in living cells as resistance. Another problem yet to be solved is the association of genes with their hosts. It is important to know whether a specific gene is of a harmless soil bacterium or of a clinically significant pathogen to determine its risk. This connection is hardly ever confidently made using a short-read data (Galhano et al., 2021; Willbanks et al., 2022). Long-read techniques, and the creation of metagenome-assembled genomes, are just starting to bridge this gap, but in complex communities the issue is still difficult (Willbanks et al., 2022). Another challenge is to resolve allelic variants. Minor alterations in a resistance gene could affect its functionality and the amount of resistance it generates. These variants are often lumped together into short reads resulting in a loss of resolution and an incomplete picture of functional diversity (Mullany, 2014). Another issue is sensitivity. In rare cases, the resistance genes at low frequencies can avoid detection unless the depth of sequencing is extremely high, increasing the cost (Hendriksen et al., 2019). It is important to meticulously optimize the sequencing depth to both the type of sample, and the study question. Lastly, any analysis relies on comparison to reference databases. Blind spots have their place here, as well. Development of resistance occurs quickly and novel variants emerge at a higher rate than the databases can be revised. Popular tools like CARD or ResFinder are priceless, yet they are not complete (Alcock et al., 2022; Galhano et al., 2021). Unsubmitted genes will just be overlooked. There are numerous entries of bare sequence data only. However, context plays a role in understanding that presence or absence of a gene on a plasmid or transposon is essential in determining its transferability (Ko et al., 2022; Olsen & Riber, 2025). Long-read sequencing is well adapted to this kind of genomic detail and can be used to fill gaps that are not yet present in databases (Willbanks et al., 2022).

10.6 FUTURE PERSPECTIVES

Metagenomics is a field that is fast developing, it is no longer a descriptive science but a predictive science that can be directly applied in clinical and community health practice. As the sequencing platforms evolve and new analytical tools are created, the convenience of metagenomics will be more defined by its speed, reliability, and building larger surveillance networks (Djordjevic et al., 2023; Olsen & Riber, 2025; Rastogi & Sani, 2011; Trigodet et al., 2022; Vallejo-Espin et al., 2025). The sequencing could only be done in well-endowed centers that are equipped with large and expensive machines throughout the 20th century. This has begun to shift with the advent of portable and long-read sequencing technology, the most notable one being the MinION by Oxford Nanopore (Elton et al., 2025; Mullany, 2014). These handheld devices are not only sequencing outside the centralized laboratories, but also into new areas like the clinic, the bedside and even the farm field.

In effect, diagnostics can now be done at the point of need rather than wait to receive a result in remote laboratories. Its implications on antimicrobial resistance (AMR) are massive. In life threatening infections such as sepsis or meningitis, traditional culture may take two to three days to realize the pathogen and its resistance profile, which can be fatal. Portable metagenomic sequencing of blood or other clinical materials can be brought down to hours (Crofts et al., 2017; Mullany, 2014). This allows the clinicians to get timely reports that contain the causative agent and its resistance genes that lead to early and specific therapy. It is a more accurate treatment that is advantageous to both the patient and the decrease in the waste of broad-spectrum antibiotics, which is one of the pillars of stewardship (Duarte et al., 2021). Sequencing in the future will be portable, real time and the ability to provide genomic context in a manner never observed previously. Democratizing genomics is by point-of-need sequencing, which is opening genomics to the outside of the core facilities to hospitals, farms and the environment.

Such a shift in AMR research means that now sample-based diagnostics (such as urine or blood) can be conducted on-site, and the complete picture of the resistome can be obtained in hours rather than days, which can potentially transform the approach to acute infection treatment. Another development of significance is long-read sequencing technologies. Among the most widespread weaknesses of the short-read shotgun metagenomics is that it is more likely to produce fragmented data that obscures the relationship between resistance genes, their hosts and mobile genetic elements. Newer long-read methods can resolve this issue, both of which are Oxford Nanopore and PacBio (Galhano et al., 2021; van Schaik, 2015). A single read may be either a whole plasmid or chromosomal fragment and thus, it can be beyond doubt that a resistance gene exists on the genome of a specific pathogen, and it is situated on a mobile element. This increased resolution is critical to the interpretation of the risks of gene mobilization and the larger mechanics of transmission.

Even with these developments there are still problems. Even the sequence-based methods remain to be largely reliant on homology since they are only capable of identifying genes that are similar to those already present in the database, and in many cases, novel forms of resistance are not even in sight. Machine learning (ML) and other methods of artificial intelligence (AI) are now starting to overcome this limitation not by traditional alignment alone, but by deep learning, which can be trained to learn biochemical signatures or structural motifs (Banerjee & Patel, 2023). As an example, the catalytic site of b-lactamase. By doing so, even low-sequence similarity of potential resistance genes to known ARGs can be detected. This change in the mode of pattern matching to predictive modelling should enhance the process of finding cryptic genes in environmental reservoirs and enhance the reliability of resistome profiling. The problem of the quality of annotation is also crucial. The phrase "garbage in, garbage out" is true to bioinformatics.

The quality of a metagenomic profile is limited by the quality of the underlying database, although most AMR databases are either inconsistent, out-of-date, or not functionally validated (Willbanks et al., 2022; Zheng et al., 2021). Issues such as errors in gene entries and absence of common rules in naming and classification are problems. To overcome these problems, international efforts should be organized to deal with expert curation, functional validation, and standard ontology (Kim et al., 2024). This would enhance the quality of data, the interoperability of tools and enable the comparison of results of various studies with greater reliability. Altogether, these developments may bring genomic discoveries to actual benefits to society. Rapid metagenomic diagnostics provide an avenue in clinical settings that avoids empirical broad-spectrum prescribing and instead provides precision therapy. This shift would enhance the time to recovery, decrease the length of stay, and minimize the evolutionary pressure which drives resistance (Levy & Boone, 2019).

On the scale of the public health policy, metagenomic surveillance can shift the practice towards responding to outbreaks instead of anticipating risks. Unstable environments can be identified sooner and actions to address the problem can be initiated before the situation becomes widespread (Pareek et al., 2011; Slatko et al., 2018). Finally, metagenomic data can be truly valued only when it is put into consonance with the action of the public health. Future directions would be to incorporate metagenomic surveillance as a direct part of international systems like the Global Antimicrobial Resistance and Use Surveillance System (GLASS) of the World Health Organization. Currently, GLASS predominantly uses phenotypic resistance data that is retrospective in nature which is based on cultured isolates. This would be supplemented with wastewater based and environmental metagenomics to provide a visionary aspect (Kumar et al., 2023). Genomic data would be provided by countries in conjunction with clinical reports and this would make it feasible to provide early warnings on emerging genes, track the flow of resistance between human, animal and environmental sources and also assess the effects of new policies such as restrictions on farm antibiotics (Quince et al., 2017; Zhao et al., 2024). This would bring about a paradigm shift. It would also substitute a reactive approach to resistance with a proactive strategy, which would incorporate metagenomics into the heart of global AMR policy (Hugenholtz & Tyson, 2008).

10.7 CONCLUSION

Metagenomics is a potential research paradigm and it is among the most important tools to examine antimicrobial resistance. It provides strengths to access resistance genes by the use of new sequencing technologies which retain a decent accuracy level that was never possible with the help of previous culture technologies. It does not only as a tool to detect the existence of resistance genes, but also shows the entrenchment of resistance genes in microbial communities which would not be available to targeted assays. It not only changed the technical but also brought a conceptual alteration to examine microbial communities and surveillance by the researcher and clinicians, respectively. Metagenomics has not just limited implications of clinical research as far as its public health is concerned. It has been made potential to develop resistance turns in which various sources are involved like hospitals, farm-sites, sewage treatment plants and natural ecosystems. Integrating such profiles in risk-assessment models, the placement of resistance pressures and the circulation of the same can be predicted easier. This foresight would recommend more precise interventions in the area of sanitation or the prescription practices. By this way the data generated by metagenomics are directly applicable to the stewardship programs, which support precise use of drugs and reduce selective pressure. The speedy discovery of resistant strains can be the difference between success and failure of an outbreak and its control. With further enhancement in sequencing technology and computational analysis becoming more and more common and metagenomics will probably be a part of global AMR strategy which will fill the gap between the technical development, clinical care and the health of the population policy in general.

ABBREVIATIONS

AMR	Anti Microbial Resistance
ARGs	Antimicrobial Resistant Genes
HGT	Horizontal Gene Transfer
HTS	High-Throughput Screening
NGS	Next Generation Sequencing
VRE	Vancomycin-Resistant Enterococci
MDR	Multi Drug Resistant
MRSA	Methicillin-resistant Staphylococcus aureus
SMRT	Strengthening Antimicrobial Resistance Surveillance
NMD-1	New Delhi metallo-beta-lactamase-1 enzyme
PCR	Polymerase Chain Reaction
SDS	Sodium Dodecyl Sulphate
HTC	High-throughput Conjugation
WWTPs	Wastewater Treatment Plants
RPKM	Reads Per Kilobase Million
TPM	Transcripts Per Million
HPC	High-Performance Computing
BWA	Burrows-Wheeler Aligner
WHO GLASS	World Health Organization Global Antimicrobial Resistance and Use Surveillance System

REFERENCES

Alcock, B. P., Huynh, W., Chalil, R., Smith, K. W., Raphenya, A. R., Wlodarski, M. A., Edalatmand, A., Petkau, A., Syed, S. A., Tsang, K. K., Baker, S. J. C., Dave, M., Mccarthy, M. C., Mukiri, K. M., Nasir, J. A., Golbon, B., Imtiaz, H., Jiang, X., Kaur, K., . . . Mcarthur, A. G. (2022). CARD 2023: Expanded curation, support for machine learning, and resistome prediction at the comprehensive antibiotic resistance database. *Nucleic Acids Research*, *51*(D1), D690. https://doi.org/10.1093/NAR/GKAC920

Anahtar, M. N., Yang, J. H., & Kanjilal, S. (2021). Applications of machine learning to the problem of antimicrobial resistance: An emerging model for translational research. *Journal of Clinical Microbiology*, *59*(7). https://doi.org/10.1128/JCM.01260-20/ASSET/D6BD3115-0DFD-4335-BC91-2E4D9AC64511/ASSETS/IMAGES/LARGE/JCM.01260-20-F0002.JPG

Angers-Loustau, A., Petrillo, M., Bengtsson-Palme, J., Berendonk, T., Blais, B., Chan, K. G., Coque, T. M., Hammer, P., Heß, S., Kagkli, D. M., Krumbiegel, C., Lanza, V. F., Madec, J. Y., Naas, T., O'Grady, J., Paracchini, V., Rossen, J. W. A., Ruppé, E., Vamathevan, J., . . . Van den Eede, G. (2018). The challenges of designing a benchmark strategy for bioinformatics pipelines in the identification of antimicrobial resistance determinants using next generation sequencing technologies. *F1000Research*, *7*, ISCB Comm J-459. https://doi.org/10.12688/F1000RESEARCH.14509.2

Anjum, M. F., Zankari, E., & Hasman, H. (2017). Molecular methods for detection of antimicrobial resistance. *Microbiology Spectrum*, *5*(6). https://doi.org/10.1128/MICROBIOLSPEC.ARBA-0011-2017

Banerjee, R., & Patel, R. (2023). Molecular diagnostics for genotypic detection of antibiotic resistance: Current landscape and future directions. *JAC-Antimicrobial Resistance*, *5*(1), dlad018. https://doi.org/10.1093/JACAMR/DLAD018

Crofts, T. S., Gasparrini, A. J., & Dantas, G. (2017). Next-generation approaches to understand and combat the antibiotic resistome. *Nature Reviews Microbiology*, *15*(7), 422–434. https://doi.org/10.1038/nrmicro.2017.28

Das, R., Burnase, N., Suryawanshi, N., Barapatre, A., & Kumar, B. (2024). Infliction of antibiotic resistance in the environment: Challenges and outlook. *Antimicrobials in Environment*, 76–105. https://doi.org/10.1201/9781003268925-7/INFLICTION-ANTIBIOTIC-RESISTANCE-ENVIRONMENT-REENA-DAS-NISHANT-BURNASE-NISHA-SURYAWANSHI-ANAND-BARAPATRE-BIKASH-KUMAR

De, R. (2019). Metagenomics: Aid to combat antimicrobial resistance in diarrhea. In *Gut Pathogens* (Vol. 11). BioMed Central Ltd. https://doi.org/10.1186/s13099-019-0331-8

Djordjevic, S. P., Jarocki, V. M., Seemann, T., Cummins, M. L., Watt, A. E., Drigo, B., Wyrsch, E. R., Reid, C. J., Donner, E., & Howden, B. P. (2023). Genomic surveillance for antimicrobial resistance—a one health perspective. *Nature Reviews Genetics*, *25*(2), 142–157. https://doi.org/10.1038/s41576-023-00649-y

Duarte, A. S. R., Röder, T., Van Gompel, L., Petersen, T. N., Hansen, R. B., Hansen, I. M., Bossers, A., Aarestrup, F. M., Wagenaar, J. A., & Hald, T. (2021). Metagenomics-based approach to source-attribution of antimicrobial resistance determinants—identification of reservoir resistome signatures. *Frontiers in Microbiology*, *11*, 601407. https://doi.org/10.3389/FMICB.2020.601407/BIBTEX

Elton, L., Mateos, A. D., Frosini, S. M., Jepson, R., Rofael, S., McHugh, T. D., & Wey, E. Q. (2025). A metagenomic approach to one health surveillance of antimicrobial resistance in a UK veterinary centre. *BioRxiv*. https://doi.org/10.1101/2025.02.12.636430

Florensa, A. F., Kaas, R. S., Clausen, P. T. L. C., Aytan-Aktug, D., & Aarestrup, F. M. (2022). ResFinder—an open online resource for identification of antimicrobial resistance genes in next-generation sequencing data and prediction of phenotypes from genotypes. *Microbial Genomics*, *8*(1). https://doi.org/10.1099/MGEN.0.000748

Galhano, B. S. P., Ferrari, R. G., Panzenhagen, P., de Jesus, A. C. S., & Conte-Junior, C. A. (2021). Antimicrobial resistance gene detection methods for bacteria in animal-based foods: A brief review of highlights and advantages. *Microorganisms*, *9*(5), 923. https://doi.org/10.3390/MICROORGANISMS9050923

Garza, D. R., & Dutilh, B. E. (2015). From cultured to uncultured genome sequences: Metagenomics and modeling microbial ecosystems. *Cellular and Molecular Life Sciences: CMLS*, *72*(22), 4287. https://doi.org/10.1007/S00018-015-2004-1

Gupta, A., Pahal, S., Kumar, V., & Chaudhary, A. (2025). Role of monoclonal antibodies in alzheimer's disease: Successes, failures, and future directions. *Neurological Sciences*. https://doi.org/10.1007/s10072-025-08456-5

Gupta, S. K., Padmanabhan, B. R., Diene, S. M., Lopez-Rojas, R., Kempf, M., Landraud, L., & Rolain, J. M. (2014). ARG-ANNOT, a new bioinformatic tool to discover antibiotic resistance genes in bacterial genomes. *Antimicrobial Agents and Chemotherapy*, *58*(1), 212–220. https://doi.org/10.1128/AAC.01310-13

Hendriksen, R. S., Munk, P., Njage, P., van Bunnik, B., McNally, L., Lukjancenko, O., Röder, T., Nieuwenhuijse, D., Pedersen, S. K., Kjeldgaard, J., Kaas, R. S., Clausen, P. T. L. C., Vogt, J. K., Leekitcharoenphon, P., van de Schans, M. G. M., Zuidema, T., de Roda Husman, A. M., Rasmussen, S., Petersen, B., . . . Aarestrup, F. M. (2019). Global monitoring of antimicrobial resistance based on metagenomics analyses of urban sewage. *Nature Communications*, *10*. https://doi.org/10.1038/s41467-019-08853-3

Hugenholtz, P., & Tyson, G. W. (2008). Microbiology: Metagenomics. *Nature*, *455*(7212), 481–483. https://doi.org/10.1038/455481A;KWRD=SCIENCE

The Human Microbiota and Chronic Disease: Dysbiosis as a Cause of Human . . .—Google Books. (n.d.). Retrieved August 21, 2025, from https://books.google.co.in/books?hl=en&lr=&id=Ik3ODAAAQBAJ&

oi=fnd&pg=PA3&dq=+historical+perspectives+of+metagenomics&ots=Uf4xZKXdC0&sig=y0DfHgiLBNgYLHoukXRcVsL9Nfg&redir_esc=y#v=onepage&q&f=false

Khan, N. J., Kami Reza, K., Adil, A., & Jeelani, A. (1 C.E.). Big data in bioinformatics: Exploring antimicrobial resistance (AMR). In *AI-driven breakthroughs in antimicrobial resistance* (pp. 115–130). IGI Global Scientific Publishing. https://Services.Igi-Global.Com/Resolvedoi/Resolve.Aspx?Doi=10.4018/979-8-3693-7550-1.Ch005; https://doi.org/10.4018/979-8-3693-7550-1.CH005

Kim, C., Pongpanich, M., & Porntaveetus, T. (2024). Unraveling metagenomics through long-read sequencing: A comprehensive review. *Journal of Translational Medicine*, *22*(1), 111. https://doi.org/10.1186/S12967-024-04917-1

Ko, K. K. K., Chng, K. R., & Nagarajan, N. (2022). Metagenomics-enabled microbial surveillance. *Nature Microbiology*, *7*, 486–496. https://doi.org/10.1038/s41564-022-01089-w

Kumar, R., Yadav, G., Kuddus, M., Ashraf, G. M., & Singh, R. (2023). Unlocking the microbial studies through computational approaches: How far have we reached? *Environmental Science and Pollution Research International*, *30*(17), 48929. https://doi.org/10.1007/S11356-023-26220-0

Kumar, V., Gangani, S., Shukla, R., Prajapati, S. K., Shekhar, H., & Shukla, V. (2024). Next-generation sequencing and genomic data analysis. In A. Chaudhary, S. K. Sethi, & A. Verma (Eds.), *Unraveling new frontiers and advances in bioinformatics* (pp. 49–76). Springer Nature. https://doi.org/10.1007/978-981-97-7123-3_3

Kumar, V., Pandey, A., Arora, A., Gautam, P., Bisht, D., Gupta, S., Chaurasia, A., & Sachan, M. (2023). Diagnostics and therapeutic potential of miR-205 and miR-34a in ovarian cancer management: A miRNA-target-based analysis. *DNA and Cell Biology*, *42*(3), 151–162. https://doi.org/10.1089/dna.2022.0487

Kumar, V., Sharma, J. K., Kumar, S., Gulia, J., Singh, N. P., & Santal, A. R. (2025). Metagenomic and proteomic approach for bioremediation of environmental pollutants. In M. P. Shah (Eds.), *Development in waste water treatment research and processes*. Elsevier. https://doi.org/10.1016/B978-0-443-13615-3.00020-0

Kumar, V., Shukla, R., Gangani, S., Shekhar, H., & Shukla, V. (2024). Metagenomics and microbiome analysis. In A. Chaudhary, S. K. Sethi, & A. Verma (Eds.), *Unraveling new frontiers and advances in bioinformatics* (pp. 77–92). Springer Nature Singapore. https://doi.org/10.1007/978-981-97-7123-3_4

Kumar, V., Singh, A., Gautam, P., & Sachan, M. (2019). Techniques/tools to study epigenetic biomarkers in human cancer detection. In S. Paul (Ed.), *Biomedical engineering and its applications in healthcare*. Springer Singapore. https://doi.org/10.1007/978-981-13-3705-5_14

Lakin, S. M., Dean, C., Noyes, N. R., Dettenwanger, A., Ross, A. S., Doster, E., Rovira, P., Abdo, Z., Jones, K. L., Ruiz, J., Belk, K. E., Morley, P. S., & Boucher, C. (2017). MEGARes: An antimicrobial resistance database for high throughput sequencing. *Nucleic Acids Research*, *45*(D1), D574–D580. https://doi.org/10.1093/NAR/GKW1009

Lanza, V. F., Baquero, F., Martínez, J. L., Ramos-Ruíz, R., González-Zorn, B., Andremont, A., Sánchez-Valenzuela, A., Ehrlich, S. D., Kennedy, S., Ruppé, E., van Schaik, W., Willems, R. J., de la Cruz, F., & Coque, T. M. (2018). In-depth resistome analysis by targeted metagenomics. *Microbiome*, *6*(1). https://doi.org/10.1186/S40168-017-0387-Y

Levy, S. E., & Boone, B. E. (2019). Next-generation sequencing strategies. *Cold Spring Harbor Perspectives in Medicine*, *9*(7), a025791. https://doi.org/10.1101/CSHPERSPECT.A025791

Li, D., Luo, R., Liu, C. M., Leung, C. M., Ting, H. F., Sadakane, K., Yamashita, H., & Lam, T. W. (2016). MEGAHIT v1.0: A fast and scalable metagenome assembler driven by advanced methodologies and community practices. *Methods*, *102*, 3–11. https://doi.org/10.1016/j.ymeth.2016.02.020

Liu, B., & Pop, M. (2009). ARDB—antibiotic resistance genes database. *Nucleic Acids Research*, *37*(Database issue). https://doi.org/10.1093/NAR/GKN656

Liu, Y., & Ma, Y. (2024). Clinical applications of metagenomics next-generation sequencing in infectious diseases. *Journal of Zhejiang University. Science. B*, *25*(6), 471. https://doi.org/10.1631/JZUS.B2300029

Lu, J., Breitwieser, F. P., Thielen, P., & Salzberg, S. L. (2017). Bracken: Estimating species abundance in metagenomics data. *PeerJ Computer Science*, *3*, e104. https://doi.org/10.7717/peerj-cs.104

Maghini, D. G., Moss, E. L., Vance, S. E., & Bhatt, A. S. (2020). Improved high-molecular-weight DNA extraction, nanopore sequencing and metagenomic assembly from the human gut microbiome. *Nature Protocols*, *16*(1), 458. https://doi.org/10.1038/S41596-020-00424-X

Mendes, S. G., Combo, S. I., Allain, T., Domingues, S., Buret, A. G., & Da Silva, G. J. (2023). Co-regulation of biofilm formation and antimicrobial resistance in acinetobacter baumannii: From mechanisms to therapeutic strategies. *European Journal of Clinical Microbiology and Infectious Diseases*, *42*(12), 1405–1423. https://doi.org/10.1007/S10096-023-04677-8

Mullany, P. (2014). Functional metagenomics for the investigation of antibiotic resistance. *Virulence*, *5*(3), 443–447. https://doi.org/10.4161/VIRU.28196

Munk, P., Andersen, V. D., de Knegt, L., Jensen, M. S., Knudsen, B. E., Lukjancenko, O., Mordhorst, H., Clasen, J., Agersø, Y., Folkesson, A., Pamp, S. J., Vigre, H., & Aarestrup, F. M. (2017). A sampling and metagenomic sequencing-based methodology for monitoring antimicrobial resistance in swine herds. *Journal of Antimicrobial Chemotherapy*, *72*(2), 385–392. https://doi.org/10.1093/JAC/DKW415

Nam, N. N., Do, H. D. K., Loan Trinh, K. T., & Lee, N. Y. (2023). Metagenomics: An effective approach for exploring microbial diversity and functions. *Foods*, *12*(11), 2140. https://doi.org/10.3390/FOODS12112140

Navgire, G. S., Goel, N., Sawhney, G., Sharma, M., Kaushik, P., Mohanta, Y. K., Mohanta, T. K., & Al-Harrasi, A. (2022). Analysis and interpretation of metagenomics data: An approach. *Biological Procedures Online*, *24*(1), 1–22. https://doi.org/10.1186/S12575-022-00179-7

Olsen, N. S., & Riber, L. (2025). Metagenomics as a transformative tool for antibiotic resistance surveillance: Highlighting the impact of mobile genetic elements with a focus on the complex role of phages. *Antibiotics*, *14*(3), 296. https://doi.org/10.3390/ANTIBIOTICS14030296

Olson, R. D., Assaf, R., Brettin, T., Conrad, N., Cucinell, C., Davis, J. J., Dempsey, D. M., Dickerman, A., Dietrich, E. M., Kenyon, R. W., Kuscuoglu, M., Lefkowitz, E. J., Lu, J., Machi, D., Macken, C., Mao, C., Niewiadomska, A., Nguyen, M., Olsen, G. J., . . . Stevens, R. L. (2023). Introducing the bacterial and viral bioinformatics resource center (BV-BRC): A resource combining PATRIC, IRD and ViPR. *Nucleic Acids Research*, *51*(D1), D678–D689. https://doi.org/10.1093/NAR/GKAC1003

Oniciuc, E. A., Likotrafiti, E., Alvarez-Molina, A., Prieto, M., Santos, J. A., & Alvarez-Ordóñez, A. (2018). The present and future of whole genome sequencing (WGS) and whole metagenome sequencing (WMS) for surveillance of antimicrobial resistant microorganisms and antimicrobial resistance genes across the food chain. *Genes*, *9*(5), 268. https://doi.org/10.3390/GENES9050268

Pal, C., Bengtsson-Palme, J., Rensing, C., Kristiansson, E., & Larsson, D. G. J. (2014). BacMet: Antibacterial biocide and metal resistance genes database. *Nucleic Acids Research*, *42*(Database issue). https://doi.org/10.1093/NAR/GKT1252

Pareek, C. S., Smoczynski, R., & Tretyn, A. (2011). Sequencing technologies and genome sequencing. *Journal of Applied Genetics*, *52*(4), 413. https://doi.org/10.1007/S13353-011-0057-X

Pitiot, A., Rolin, C., Seguin-Devaux, C., & Zimmer, J. (2025). Fighting antibiotic resistance: Insights into human barriers and new opportunities: Antibiotic resistance constantly rises with the development of human activities. We discuss barriers and opportunities to get it under control. *BioEssays*, *47*(6), e70001. https://doi.org/10.1002/BIES.70001;JOURNAL:JOURNAL:15211878;WGROUP:STRING:PUBLICATION

Purushothaman, S., Meola, M., Roloff, T., Rooney, A. M., & Egli, A. (2024). Evaluation of DNA extraction kits for long-read shotgun metagenomics using Oxford Nanopore sequencing for rapid taxonomic and antimicrobial resistance detection. *Scientific Reports*, *14*(1), 29531. https://doi.org/10.1038/S41598-024-80660-3

Quince, C., Walker, A. W., Simpson, J. T., Loman, N. J., & Segata, N. (2017). Shotgun metagenomics, from sampling to analysis. *Nature Biotechnology*, *35*(9), 833–844. https://doi.org/10.1038/nbt.3935

Rastogi, G., & Sani, R. K. (2011). Molecular techniques to assess microbial community structure, function, and dynamics in the environment. *Microbes and Microbial Technology: Agricultural and Environmental Applications*, 29–57. https://doi.org/10.1007/978-1-4419-7931-5_2

Roy, G., Prifti, E., Belda, E., & Zucker, J. D. (2024). Deep learning methods in metagenomics: A review. *Microbial Genomics*, *10*(4), 001231. https://doi.org/10.1099/MGEN.0.001231

Salam, M. A., Al-Amin, M. Y., Salam, M. T., Pawar, J. S., Akhter, N., Rabaan, A. A., & Alqumber, M. A. A. (2023). Antimicrobial resistance: A growing serious threat for global public health. *Healthcare*, *11*(13), 1946. https://doi.org/10.3390/HEALTHCARE11131946

Sherry, N. L., Lee, J. Y. H., Giulieri, S. G., Connor, C. H., Horan, K., Lacey, J. A., Lane, C. R., Carter, G. P., Seemann, T., Egli, A., Stinear, T. P., & Howden, B. P. (2025). Genomics for antimicrobial resistance—progress and future directions. *Journals.Asm.Org*, *69*(5). https://doi.org/10.1128/AAC.01082-24

Slatko, B. E., Gardner, A. F., & Ausubel, F. M. (2018). Overview of next generation sequencing technologies. *Current Protocols in Molecular Biology*, *122*(1), e59. https://doi.org/10.1002/CPMB.59

Sleator, R. D., Shortall, C., & Hill, C. (2008). Metagenomics. *Letters in Applied Microbiology*, *47*(5), 361–366. https://doi.org/10.1111/J.1472-765X.2008.02444.X

Tang, K. W. K., Millar, B. C., & Moore, J. E. (2023). Antimicrobial resistance (AMR). *British Journal of Biomedical Science*, *80*, 11387. https://doi.org/10.3389/BJBS.2023.11387/TEXT

Thomas, T., Gilbert, J., & Meyer, F. (2012). Metagenomics—a guide from sampling to data analysis. *Microbial Informatics and Experimentation*, *2*. https://doi.org/10.1186/2042-5783-2-3

Trigodet, F., Lolans, K., Fogarty, E., Shaiber, A., Morrison, H. G., Barreiro, L., Jabri, B., & Eren, A. M. (2022). High molecular weight DNA extraction strategies for long-read sequencing of complex metagenomes. *Molecular Ecology Resources*, *22*(5), 1786. https://doi.org/10.1111/1755-0998.13588

Vallejo-Espín, D., Galarza-Mayorga, J., Lalaleo, L., & Calero-Cáceres, W. (2025). Beyond clinical genomics: Addressing critical gaps in one health AMR surveillance. *Frontiers in Microbiology*, *16*, 1596720. https://doi.org/10.3389/FMICB.2025.1596720

van Schaik, W. (2015). The human gut resistome. *Philosophical Transactions of the Royal Society B: Biological Sciences*, *370*(1670), 20140087. https://doi.org/10.1098/RSTB.2014.0087

Ventola, C. L. (2015). The antibiotic resistance crisis: Part 1: Causes and threats. *Pharmacy and Therapeutics*, *40*(4), 277. https://doi.org/Article

Willbanks, A., Leary, M., Greenshields, M., Tyminski, C., Heerboth, S., Lapinska, K., Haskins, K., & Sarkar, S. (2022). Metagenomic sequencing for microbial DNA in human samples: Emerging technological advances. *International Journal of Molecular Sciences*, *23*(4), 2181. https://doi.org/10.3390/IJMS23042181

Williams, A. D., Rousham, E., Neal, A. L., Amin, M. B., Hobman, J. L., Stekel, D., & Islam, M. A. (2023). Impact of contrasting poultry exposures on human, poultry, and wastewater antibiotic resistomes in Bangladesh. *Microbiology Spectrum*, *11*(6), e01763–23. https://doi.org/10.1128/SPECTRUM.01763-23

Woolhouse, M. E. J., & Ward, M. J. (2013). Sources of antimicrobial resistance. *Science*, *341*(6153), 1460–1461. https://doi.org/10.1126/SCIENCE.1243444/ASSET/A2F3C26A-C55E-4D7F-B671-4A3AC136673E/ASSETS/GRAPHIC/341_1460_F1.JPEG

Yadav, S., & Kapley, A. (2021). Antibiotic resistance: Global health crisis and metagenomics. *Biotechnology Reports*, *29*, e00604. https://doi.org/10.1016/J.BTRE.2021.E00604

Yang, Y., Jiang, X., Chai, B., Ma, L., Li, B., Zhang, A., Cole, J. R., Tiedje, J. M., & Zhang, T. (2015). ARGs-OAP: Online analysis pipeline for antibiotic resistance genes detection from metagenomic data using an integrated structured ARG-database. *Bioinformatics*, *32*(15), 2346–2351. https://doi.org/10.1093/BIOINFORMATICS/BTW136

Zhao, Y., Zhang, W., & Zhang, X. (2024). Application of metagenomic next-generation sequencing in the diagnosis of infectious diseases. *Frontiers in Cellular and Infection Microbiology*, *14*, 1458316. https://doi.org/10.3389/FCIMB.2024.1458316

Zheng, Y., Qiu, X., Wang, T., & Zhang, J. (2021). The diagnostic value of metagenomic next–generation sequencing in lower respiratory tract infection. *Frontiers in Cellular and Infection Microbiology*, *11*, 694756. https://doi.org/10.3389/FCIMB.2021.694756

11 Computational Challenges and Solutions in Metagenomics

Vivek Kumar[1], *Shaile Thakur*[2], *Mansi Gautam*[3], *and Rohit Shukla*[1]*
[1]USF Center for Microbiome Research, Microbiomes Institute, University of South Florida, Tampa, FL, USA.
[2]Department of Forensic Sciences, PIAS, Parul University, Vadodara, Gujrat, India
[3]Department of Biomedical Sciences, Galgotia University, Greater Noida, Uttar Pradesh, India
*Corresponding author

11.1 INTRODUCTION

Recent metagenomics advancements have transformed the environmental and clinical microbiology fields. These advancements have now made possible the direct assessment of environmental and host-associated samples without culturing. Metagenomics has made it possible to reduce the cultural biases that arise from traditional methods. Pooled genetic material can now be sequenced through metagenomics and analyzed for microbial diversity, taxonomic organization, and functional potential (Handelsman, 2005). Recent research has established the breadth of the human microbiome, dealing with uncharacterized microbial consortia that influence nutrient cycles, ecosystems, and human health (Wooley et al., 2010; Kumar, Shukla, et al., 2024). Recent research has demonstrated how gut microbiota affect immunity, metabolism, and neurocognitive function. Marine microorganisms control biogeochemical fluxes, and soil bacteria drive carbon and nitrogen turnover.

A major factor in this change has been the rapid advancement of sequencing technologies. Technological advancements such as long-read systems (PacBio, Oxford Nanopore) allow for the reassembly of whole genomes and mobile genetic elements, which improves the resolution of complex microbial communities, while short-read Illumina sequencing provides excellent depth and accuracy (Quince et al., 2017; Kumar, Gangani, et al., 2024). However, each method presents unique computation difficulties; for example, short reads frequently result in fragmented assemblies, whereas long reads demand substantial error correction and significant computational power. Recently, the scale of metagenomic research has exploded with data. Initiatives such as the Earth Microbiome Project and the Human Microbiome Project have produced information on a petabyte scale, thus highlighting the necessity of integrated techniques, standardized workflows, and scalable algorithms for understanding datasets. In addition to basic documentation, functional inference (such as gene expression, pathway activity, and metabolite synthesis) and multi-omics integration with meta-transcriptomics, meta-proteomics, and metabolomics are becoming increasingly important.

This chapter provides an accessible summary of the main computational issues in metagenomics and the new approaches being developed to address them. Additionally, we highlight case studies in translational, clinical, and environmental microbiome research. Moreover, the chapter concludes with future directions that help us understand the field's transition to predictive, systems-level

 DOI: 10.1201/9781003670155-11

microbiome modeling by emphasizing expansion, accuracy, simplified access, and integrative multi-omics.

11.2 EXPERIMENTAL BIASES AND THEIR COMPUTATIONAL IMPLICATIONS

Even though metagenomics is primarily a computational field, the laboratory environment is significantly influenced by upstream experimental decisions. Several variations are introduced during sample collection, DNA extraction, library preparation, and sequencing, and these variations must be minimized by computational techniques. To prevent misunderstandings about microbial diversity and function, it is crucial to acknowledge these factors. Some of the challenges are given below.

- *Sampling and Extraction Biases.* The organisms found in various collection settings (soil, marine, gut, and sediments) range from spore-forming archaea to Gram-positive bacteria and differ widely in their cell wall characteristics. The composition of ecosystems is frequently enriched or depleted at the species and taxa levels by DNA extraction techniques that depend on chemical lysis, enzymatic digestion, or mechanical disruption (such as bead beating) (Handelsman, 2005). This increases the possibility of persistent underrepresentation of certain groups, which leads to complications in downstream computational analysis.
- *Sequencing Technology Biases.* Additionally, computational complications and costs are introduced by sequencing platform selection. Despite producing high-accuracy reads, short-read technologies (such as Illumina) fragment complex genomes, making resource-intensive assembly methods more necessary. Although long-read systems (PacBio, Oxford Nanopore) enhance genome reconstruction and contiguity, error correction, and computational requirements increases the cost (Quince et al., 2017; Kumar et al., 2023). Despite the several strength of hybrid sequencing, it drastically increases the amount and complexity of data analysis (Nicholls et al., 2019)
- *Batch and Contamination Effects.* Cross-contamination, reagent-derived DNA, and technical batch effects are also major sources of false signals. The application of computational denoising techniques, such as statistical correction (e.g., "decontam") and prevalence-based filtering, is growing to detect and reduce these contaminants. Overlooking these variables may cause errors in downstream network analysis, functional annotation, and taxonomic classification.
- *Implications for Computational Analysis.* All computational stages are affected by these upstream biases. Downstream analyses such as functional annotation pipelines need to account for potentially contaminated contigs, taxonomic classifiers need to manage data with unequal abundance distributions, and assembly algorithms need to handle highly fragmented or error-prone reads. To provide accuracy, bioinformatic procedures now often incorporate statistical correction for batch effects, quality-control checkpoints, and benchmarking against mock communities (Sczyrba et al., 2017).

11.3 COMPUTATIONAL CHALLENGES IN TAXONOMIC CLASSIFICATION

The foundation of metagenomic research is based on taxonomic categorization, which connects raw sequencing data to ecological and clinical interpretation. Function prediction, diversity measurements, and downstream modeling are highly dependent upon precise classification. However, achieving higher precision is not simple. There are several ambiguities, such as database limitations, strain-level ambiguities, and methodological needs, that must be resolved by researchers using computational methods (Quince et al., 2017)

11.3.1 Database Bias and Gaps

The accuracy of reference-based classification tools is dependent upon databases. There are millions of microbial sequences stored in databases such as NCBI RefSeq, SILVA, and GTDB that overwhelmingly reflect pathogens, model organisms, and human-associated microbes. These databases have several shortcomings; for example, some whole clades from soil-, host-associated, and marine environments are either missing or underrepresented. This results in the association of new taxa with nearest neighbors, which undervalues diversity and increases the apparent abundance of well-studied species (Pasolli et al., 2019). However, metagenome-assembled genome (MAG) coverage has been improved by adding thousands of uncultivated lineages. Yet, variation in MAG quality is substantial, and their use in downstream assays might spread contaminants and incomplete assemblies (Kumar et al., 2019; Parks et al., 2020). Even with the growth of databases in recent years, problems such as bias and unequal representation remain.

11.3.2 Resolution Constraints

Even after recent growth in comprehensive catalogs, differentiation between closely related strains remains challenging. Short-read sequencing is highly accurate, but its limited read length undermines strain resolution. It has been observed that strains differ by only a few SNPs or mobile elements, and the generated reads are insufficient to resolve them. A recent study showed that *Escherichia coli* and *Shigella* species commonly group together, making discrimination clinically irrelevant (Parks et al., 2018). In contrast, long-read technologies such as PacBio HiFi and Oxford Nanopore offer improved consistency in strain separation. However, these techniques increase error rates, require more computational power, and raise the cost of analysis (Wick et al., 2023). To overcome these issues, hybrid sequencing (short and long reads) is being used more frequently; however, data volume and analysis complexity remain major concerns (Nicholls et al., 2019).

11.3.3 Misclassifications: False Positives and Negatives

Another major issue is misclassification during analysis. Current analyses show many false-positive and false-negative classifications. False-positive assignments arise from incorrect classification of conserved regions such as ribosomal proteins or plasmid elements. Similarly, false-negative assignments occur when no close reference genome exists, resulting in unclassified organisms. Both errors lead to altered downstream relative abundance estimates and false interpretation of function (Sczyrba et al., 2017). These changes are particularly significant in clinical settings, where misclassification can lead to incorrect associations or dissociations with disease. Recent benchmarking assessments revealed that the sensitivity and specificity of commonly used methods vary significantly based on strain overlap and dataset complexity (Meyer et al., 2022).

11.3.4 Emerging Computational Solutions

Recent extensive studies have revealed the importance of taxonomic classification, which is considered the most error-prone step in metagenomics analysis. Therefore, several novel techniques have been developed to overcome these limitations and enhance accuracy and robustness. These methods are summarized below.

- *Marker gene profiling:* MetaPhlAn3 focuses on clade-specific marker genes, improving precision in complex communities (Beghini et al., 2021).
- *K-mer classifiers:* Kraken2 and Centrifuge use exact k-mer matches for ultrafast classification but depend heavily on database completeness (Wood et al., 2019)

- *Reference-free methods:* mOTUs2 detects species directly from universal single-copy marker genes, enabling robust profiling even in the absence of close references (Milanese et al., 2019).
- *Ensemble approaches*: Combining marker-, k-mer-, and phylogeny-based strategies reduces tool-specific biases and improves robustness (Sczyrba et al., 2017)

11.3.5 Benchmarking and Standards

With the increasing development of computational tools, the process of benchmarking has emerged as a key component in metagenomic analysis. Recently, the Critical Assessment of Metagenome Interpretation (CAMI) project was launched to provide realistic reference datasets that help in the evaluation of classifiers, assemblers, and binning approaches under uniform conditions. The CAMI study reported significant variations in recall and precision when datasets contain closely related microbial strains. (Sczyrba et al., 2017). In addition, the CAMI II study was expanded by including long-read sequencing and more intricate community structures, but major obstacles persisted (Meyer et al., 2022). However, in the current scenario, benchmarking is essential for evaluating performance, promoting database transparency, and strengthening reproducibility. More importantly, these efforts provide developers and end users with productive feedback loops that accelerate advancements in taxonomic categorization and associated analyses.

11.4 COMPUTATIONAL CHALLENGES IN FUNCTIONAL ANNOTATION

Whereas taxonomic profiling focuses on identifying "who is present," functional annotation asks, "what can they do?" This step is central to microbiome ecology, biotechnology, and medical applications, but remains one of the most error-prone stages in metagenomics. There are several obstacles; among them, the main obstacles are the frequency of uncharacterized genes, the spread of incorrect annotations, and the discrepancy between actual activity and genetic potential. Additionally, there are other challenges for functional annotation, which include the integration of heterogeneous omics datasets, database growth, and algorithmic limitations.

11.4.1 The "Dark Matter" of Uncharacterized Genes

The biggest challenge in metagenomics analysis is the excessive number of genes with unknown roles that hinder functional annotation. Around 30–60% of predicted coding sequences coming from metagenomic datasets are not linked to characterized proteins and are annotated as "hypothetical proteins" (Steinegger et al., 2017). These uncharacterized proteins are termed the functional dark matter of genomes. This issue is more pronounced when samples from new environments are sequenced. For instance, sequenced genes from deep-sea sediments, desert soils, or harsh settings such as hydrothermal vents are frequently shown to have nearly half of their sequences without any functional classification. Recent studies have reported that even highly characterized human gut microbiomes contain a high fraction of unknown proteins. This finding implies that the diversity of microorganisms is growing far more quickly than our capacity to verify gene functions through experimentation. The lack of functional annotation creates lacunae, which include incomplete metabolic modeling and fragmented ecological networks. As large-scale studies such as the Earth Microbiome Project and Tara Oceans continue to generate vast amounts of data, the "dark fraction" continues to expand. This trend highlights the urgent need for new computational and experimental approaches that can link sequences to functions.

11.4.2 Propagation of Annotation Errors

Homology-based transfer is the most often used method for the assignment of function. This approach characterizes bacteria based on sequence similarity indexes. This method is efficient but not error-free. Sometimes, reference sequences are incorrectly annotated, and these errors are further propagated across databases. Thus, functional misclassification of homolog systems leads to bias in functional annotation databases, which complicates downstream analysis (Schnoes et al., 2009). A recent study assessed this problem and reported misannotation rates of up to 63% in enzyme superfamilies. These mistakes are particularly prevalent in protein families with substantial sequence homology or multifunctional domains, leading to inaccurate functional assignment.

The computational implications are severe. Misannotations distort comparative analyses, yield false positives in pathway enrichment, and may falsely suggest metabolic capabilities. For instance, biosynthetic gene cluster predictions may appear in communities where those metabolites are not produced. Recent computational solutions include confidence scoring, probabilistic annotation frameworks, and stricter curation of reference sequences, but large public repositories still contain legacy errors that continue to propagate.

11.4.3 Gene Presence versus Functional Activity

Metagenomics offers an understanding of gene potential; however, this does not mean that these genes are functionally active. For instance, Zhu et al. (2015) showed that several antibiotic-resistance genes are present in the soil microbiome; however, they are active only under selective environmental pressure. This issue is particularly important for ecological and clinical research. Metagenomes of species can indicate the existence of resistance genes, metabolic pathways, or virulence factors; however, phenotypic validation is required to confirm actual characteristics. This gap is addressed by meta-transcriptomics, meta-proteomics, and metabolomics, which reveal gene expression, metabolic activity, and protein translation. Nevertheless, these data types present unique computational difficulties; for instance, proteomics suffers from inadequate coverage, metabolomics from unclear molecule identification, and transcriptomic datasets from dynamic range and normalization problems (Cheng et al., 2024). Thus, DNA-based annotation is crucial but provides only a partial picture of microbial activity. More accurate functional interpretation now depends on integrative multi-omics techniques.

11.4.4 Emerging Computational Solutions

The growth and volume of metagenomics data are expanding rapidly; therefore, traditional annotation techniques need to be advanced for higher sensitivity, scalability, and comprehension. To overcome these obstacles, several recent research efforts have been conducted to create novel computational strategies. These strategies are employed to make use of integrative multi-omics frameworks, sophisticated machine learning, and deeper biological context. Several key directions are outlined below:

- *Expanded Functional Databases*
  Databases such as eggNOG, KEGG Orthology, and Pfam are continuously updated with curated orthologous groups, domain architectures, and pathway information. The latest eggNOG release includes over 4.4 million orthologous groups, providing far broader functional coverage than earlier versions (Huerta-Cepas et al., 2019).
- *Machine Learning Approaches*
  Deep learning and graph-based models are increasingly used to predict function beyond simple sequence homology. For instance, Huerta-Cepas et al. (2019) demonstrated that graph neural networks can assign Gene Ontology terms to proteins with no known homologs, significantly improving recall for novel sequences. These methods can mine structural and contextual features that traditional alignment-based approaches overlook.

- *Context-Aware Annotations*
  Emerging pipelines use gene neighborhood analysis, synteny, and co-occurrence networks to predict functional roles. For example, genes located within biosynthetic gene clusters or operons can be functionally inferred even without direct homology. Such approaches improve annotation for pathways involving secondary metabolites, antibiotic resistance, and mobile genetic elements.
- ***Multi-omics Integration***
  By combining DNA, RNA, protein, and metabolite data, researchers can directly validate whether predicted genes are expressed and functional. This integrative approach has revealed, for example, that butyrate-production pathways predicted in gut microbiomes are not only present but also actively expressed, with measurable effects on host physiology (Cheng et al., 2024).
- *Benchmarking Frameworks*
  Similar to taxonomic profiling, functional annotation pipelines are beginning to undergo systematic benchmarking. Salerno et al. (2023) highlighted the importance of robust annotation and reproducibility, calling for community standards and shared evaluation datasets. Although no equivalent to CAMI exists yet for function, efforts are underway to provide gold-standard datasets.

11.4.5 Implications for Metagenomic Studies

The functional annotation of genes is crucial for the interpretation of microbial communities. Inadequate annotation of new enzymes, natural products, or metabolic pathways results in incorrect functional assignment. Misannotations may imply incorrect ecological interactions or divert clinical diagnoses. The overinterpretation of gene presence without circumstantial evidence is very harmful. Incorrect inferences concerning the role of microbes in illness, the environment, or biotechnology lead to misleading conclusions. Going forward, progress will depend on three parallel efforts: (1) expanding and curating databases to minimize the dark fraction and reduce error propagation, (2) integrating AI-driven function prediction for proteins with no homologs, and (3) embracing multi-omics validation to ground predictions in biological activity. Together, these strategies will help close the gap between metagenomic sequence and microbial function.

11.5 CHALLENGES IN ASSEMBLY AND BINNING

Assembly and binning are critical steps in metagenomic workflows. Assembly transforms short- or long-sequencing reads into longer contiguous sequences (contigs), while binning organizes these contigs into MAGs. These processes enable genome-resolved insights into microbial communities, facilitating the discovery of novel taxa, strain-level variation, and metabolic potential. Despite major advances, assembly and binning remain computationally challenging due to the scale of datasets, the complexity of microbial communities, and technical limitations of sequencing platforms.

11.5.1 Computational Intensity of De Novo Assembly

Metagenome *de novo* assembly contains billions of reads with distorted abundance and is computationally very intensive. To detect and fix sequencing problems such as repeat removal and to create de Bruijn or overlap graphs, assemblers require high memory and CPU time. Memory use can reach hundreds of terabytes for high-diversity settings such as soil or ocean microbiomes, making them inaccessible to entities without cloud or HPC resources (Li et al., 2015). Furthermore, community structure adds additional complexity. Organisms with higher abundance show higher coverage and produce dense coverage, dominating graph paths, while taxa with lower abundance produce sparse and fragmented contigs. This unevenness leads to challenges for assemblers in balancing sensitivity and specificity. The computational cost is further increased by hybrid assemblies that blend Illumina short reads and long reads because assemblers must balance widely disparate error models and read lengths.

11.5.2 Fragmentation from Short Reads

The most widely used method is still short-read sequencing (100–250 bp) because of its excellent accuracy and low cost. However, short reads result in fragmented assemblies and incomplete MAGs due to their inability to navigate conserved genomic regions, mobile elements, or long repeats. This fragmentation leads to the loss of gene order, regulatory regions, and structural variations that are essential for comprehending microbial adaptability. For instance, results from short-read assemblies often fail to resolve plasmids, phages, and genomic islands that are crucial for horizontal gene transfer and antibiotic resistance. The accuracy of downstream annotations decreases when long contigs are absent because they are unable to properly reconstruct pathway completeness or synteny. Recent research by Quince et al. highlighted that short-read metagenomics frequently underestimates microbial genome sizes and misses ecologically significant features (Quince et al., 2017).

11.5.3 Chimeric Bins and Contamination

After contigs are assembled, binning techniques use coverage profiles across samples, compositional characteristics (such as GC content and tetranucleotide frequencies), and, occasionally, similarity to reference genomes to categorize them into draft genomes. The binning process is valuable, but it also comes with a significant error rate. Chimeric bins are created by the inadvertent combination of contigs from multiple species into a single MAG. This usually occurs when species covary in abundance across samples or have homologous sequence compositions. In addition, contamination can skew gene content estimations, increase MAG size, and confuse phylogenetic placement. The CAMI benchmarking project showed that the presence of closely related strains causes a dramatic decline in binning accuracy and that there is frequently significant disagreement among binning methods (Sczyrba et al., 2017). Mis-binned genomes have cascading effects: they alter estimates of microbial diversity, bias functional annotation, and confound evolutionary inferences.

11.5.4 Emerging Computational Solutions

The process of rebuilding genomes from complicated metagenomic combinations presents enormous computational challenges, as highlighted by advances in sequencing and assembly. New algorithms and hybrid approaches that improve the accuracy, repeatability, and contiguity of MAGs are being explored to address these issues. Key emerging approaches include:

- *Hybrid Assemblies:* Combining Illumina short reads with PacBio or Oxford Nanopore long reads yields more contiguous assemblies. Long reads span repeats, while short reads provide base accuracy. Nicholls et al. (2019) demonstrated that hybrid strategies can recover near-complete genomes from mock communities.
- *Graph-Based Assemblers:* New tools employ repeat graphs and strain-aware algorithms to disentangle closely related organisms and improve contiguity. Kolmogorov et al. (2019) introduced metaFlye, a long-read assembler that resolves strain variation and produces higher-quality assemblies compared to short-read methods.
- *Ensemble Binning Approaches:* Instead of relying on a single algorithm, ensemble methods combine outputs from multiple binners (MetaBAT, MaxBin, and CONCOCT). DAS Tool integrates these results to improve both precision and recall, producing higher-quality MAGs (Kolmogorov et al., 2019).

- *Quality Standards:* To harmonize reporting, the Minimum Information about MIMAG standard provides guidelines for MAG quality, recommending thresholds for completeness, contamination, and strain heterogeneity. These standards promote reproducibility and comparability across studies (Bowers et al., 2017)

11.5.5 Implications for Genome-Resolved Metagenomics

The success of genome-resolved metagenomics depends heavily on the quality of assembly and binning. When assemblies are fragmented, metabolic pathways may appear incomplete, while chimeric bins can give a misleading impression of microbial diversity or wrongly assign critical functions. Such errors inevitably carry through to downstream analyses, shaping ecological interpretations, functional predictions, and even potential clinical or biotechnological applications. To address these issues, the field is increasingly adopting methods that balance accuracy with practical usability. Hybrid sequencing approaches help to reduce fragmentation, graph-based algorithms are better at untangling strain-level complexity, and ensemble binning tools are improving confidence in genome recovery. Alongside these methodological advances, quality standards such as MIMAG now provide clear guidelines for evaluating and reporting metagenome-assembled genomes. Collectively, these developments are shifting assembly and binning from persistent obstacles into reliable strategies that can illuminate the vast microbial diversity still hidden in complex communities.

11.6 CHALLENGES IN DATA INTEGRATION AND REPRODUCIBILITY

Contemporary metagenomic studies demand far more than simply assembling and annotating sequences. Researchers are now tasked with integrating information across multiple samples, diverse omics layers, and a wide array of computational pipelines. This integrative perspective promises a more complete understanding of microbial ecology and its biomedical implications, but it also brings significant technical and methodological hurdles. In parallel, reproducibility has become a pressing concern in bioinformatics. Differences in software tools, parameter choices, and database versions can yield divergent results, raising questions about consistency, transparency, and the reliability of published findings.

11.6.1 High Dimensionality and Multi-Sample Integration

Metagenomic datasets are extremely high dimensional in nature. Even a single cohort can encompass thousands of microbial taxa, tens of thousands of genes, and hundreds of individual samples. The addition of other omics layers, such as meta-transcriptomics, metabolomics, meta-proteomics, or host genomics, increases the complexity. These "wide but shallow" datasets present significant statistical challenges. The chances of overfitting and incorrect findings are always associated with having higher feature counts than sample numbers. To address these challenges, the mixOmics package has integrated frameworks such as sparse partial least squares discriminant analysis (sPLS-DA) and multi-block methods. These integrations enable the identification of more informative features by reducing dimensionality (Rohart et al., 2017). Similarly, another approach involves network-based strategies that link correlations and co-occurrences among different data types. However, the applicability of these methods requires caution, validation, and parameter optimization.

11.6.2 Tool and Version Dependence

A major issue in metagenomics data reproducibility is the selection of computational tools and database versions. On the same dataset, different read trimmers, assemblers, taxonomic classifiers, and functional annotators can yield divergent results. Moreover, even upgrades of the same pipeline can alter outcomes because of algorithmic refinements or changes in reference databases. For example, recent comparative studies have shown that different alignment tools, such as Kraken2, MetaPhlAn, or Centrifuge, have significant differences in taxonomic (Wood et al., 2019) The CAMI benchmarking initiative has repeatedly highlighted that pipelines performing well with mock datasets can show different accuracy on complex samples (Meyer et al., 2022). Such variability complicates longitudinal studies and meta-analyses, where consistent comparisons are essential.

11.6.3 Interpretability of Statistical Outputs

Even when reproducibility is achieved, interpreting complex, multilayered results is nontrivial. Statistical models such as random forests, neural networks, or network inference frameworks can achieve high predictive accuracy but often function as "black boxes." This lack of interpretability reduces biological insight and limits translational applications in medicine. Recent work has emphasized the importance of interpretability in machine learning for microbiome science. Machine learning approaches have employed network visualization, SHAP values, and feature importance measures to elucidate biological meaning (Topçuoğlu et al., 2020). Nevertheless, interpretability remains challenging for datasets with strong confounding factors or batch effects.

11.6.4 Solutions for Reproducible and FAIR Workflows

Reproducibility is one of the most pressing challenges in metagenomic research. Several factors affect the reproducibility of metagenomics results. These factors include the complexity of datasets, diversity in computational tools, and variability across computing infrastructures. To address these issues, the community has increasingly embraced approaches that emphasize reproducibility, portability, and adherence to shared standards. These solutions not only minimize technical discrepancies but also foster transparency, comparability, and long-term usability of data and workflows. Key strategies include:

- *Workflow Managers:* Pipelines orchestrated by workflow managers such as Snakemake or Nextflow facilitate reproducible execution across datasets and platforms (Köster & Rahmann, 2012).
- *Containerization:* Environments encapsulated with Docker or Singularity ensure that dependencies, software versions, and libraries are preserved, thus minimizing discrepancies across computing infrastructures (Kurtzer et al., 2017)
- *FAIR Principles:* The adoption of FAIR (Findable, Accessible, Interoperable, Reusable) data standards promotes transparency and interoperability in microbiome research (Wilkinson et al., 2016).
- *Visualization Tools:* Platforms such as Krona and Anvi'o facilitate reproducibility by offering interactive and transparent visualizations of taxonomic and functional data, thus allowing researchers to explore rather than obscure variability (Ondov et al., 2011; Eren et al., 2015).

11.6.5 Implications for Metagenomics

Challenges in data integration and reproducibility are not minor details; instead, they directly influence how reliable and meaningful downstream analyses will be. If left unaddressed, inconsistencies across software tools, parameter settings, or database versions can compromise comparative studies, lead to misleading functional interpretations, and ultimately weaken confidence in microbiome research and its applications. On the other hand, workflows that embrace FAIR principles, containerization, and systematic benchmarking offer clear benefits: they enhance reproducibility, promote transparency, and make advanced bioinformatics more accessible to smaller research groups. The future of metagenomics is moving toward increasingly integrative studies, spanning multiple omics layers and large, diverse cohorts. Ensuring reproducibility in this landscape will require not only robust community standards and ongoing benchmarking efforts but also sustained investment in user-friendly, scalable computational infrastructure. With this support in place, the field will be better positioned to unlock the full potential of genome-resolved, multi-omic metagenomics.

11.7 ADVANCED COMPUTATIONAL FRONTIERS

Although metagenomic analysis has advanced considerably, the field is now pushing into new and more ambitious computational directions. Emerging priorities include the integration of multiple omics layers, the application of machine learning and artificial intelligence (AI), and the deployment of scalable platforms such as cloud-based resources and high-performance computing (HPC) clusters. Each of these frontiers holds the potential to deliver transformative insights into microbial communities, yet they also bring their own technical and methodological challenges that must be addressed for the field to move forward.

11.7.1 Multi-Omics Integration

Analysis of a single omics layer only offers a limited view of microbial communities. In metagenomics, data analysis can only reveal the genetic potential of a community. However, the addition of complementary layers such as meta-transcriptomics, meta-proteomics, or metabolomics will provide more insight into biological processes. The real challenge occurs when these diverse data types are combined. Each of these has its own distribution patterns, sparsity, and noise characteristics. A recently developed mixOmics suite has integrated a computational framework known as DIABLO. This integration helps with supervised integration of multiple omics blocks and can predict features that jointly discriminate between phenotypes (F et al., 2017). Likewise, another network-based approach was developed that allows researchers to connect heterogeneous datasets. This connection is based on correlation or causal graphs that highlight the interactions spanning across molecular layers (Bodein et al., 2022). Despite these advances, applying multi-omics integration at the population-scale remains a formidable task. The main hurdles in integrating cross-omics data are batch effects, the selection of features within high-dimensional data, and the development of standardized formats. Without rigorous preprocessing and validation, there is a high risk that spurious associations may overwhelm genuine biological signals.

11.7.2 Machine Learning and Artificial Intelligence

AI and machine learning (ML) are increasingly shaping the way microbiome research is conducted. These methods have been applied to everything from taxonomic classification and functional annotation to disease risk prediction, uncovering patterns in high-dimensional data that traditional approaches often miss. Despite their promise, ML models face two recurring challenges: they can be difficult to interpret, and they are prone to overfitting when applied to small or noisy datasets. Recent advances in deep learning have shown how powerful these tools can be. AlphaFold2, for example, has revolutionized protein structure prediction and opened new possibilities for inferring function within the "dark proteome" (Jumper et al., 2021). Beyond protein structures, graph neural networks and embedding-based models are now being used to predict microbial interactions and enzyme activities, even in cases where sequence homology offers little guidance (Gligorijević et al., 2021). Still, the full integration of AI into microbiome science is limited by a lack of large, well-annotated datasets and by the demand for outputs that can be trusted in clinical and translational contexts. This has motivated the development of explainable AI (XAI) tools, which aim to balance predictive accuracy with interpretability. Such tools help ensure that computational predictions remain biologically meaningful and experimentally testable.

11.7.3 Cloud and High-Performance Computing

The scale of metagenomic data continues to grow exponentially, with projects like the Earth Microbiome Project generating petabytes of data. Traditional desktop computing is insufficient for such workloads, making scalable infrastructure a necessity. HPC clusters and cloud computing platforms (e.g., AWS, Google Cloud, and Microsoft Azure) offer elastic resources for assembly, binning, and multi-omics integration. Cloud-native platforms such as MG-RAST (Wilke et al., 2016; Afgan et al., 2018) have democratized access to computational pipelines, enabling smaller labs to analyze terabyte-scale datasets without dedicated infrastructure. Nonetheless, cost, data transfer bottlenecks, and the need for user training remain barriers to adoption. Moreover, ensuring reproducibility across heterogeneous computing environments requires strict adherence to containerization and workflow standards.

11.7.4 Implications for Future Research

The convergence of multi-omics integration, artificial intelligence, and scalable computing is pushing microbiome research toward a new era—one where studies move beyond descriptive catalogs and into predictive, mechanistic modeling. Achieving this vision will depend on overcoming key limitations: the need for interpretable AI, standardized and reproducible pipelines for multi-omics integration, and equitable access to advanced computing resources across the globe. These computational frontiers are as challenging as they are promising. If the field can successfully navigate them, the outcome will be transformative: real-time, clinically actionable microbiome analytics, innovative biotechnological applications, and an unprecedented ability to unravel the complexity of microbial ecosystems at scale.

11.8 CASE STUDIES

The practical impact of computational challenges in metagenomics becomes clearest when examined through real-world examples. Whether in clinical diagnostics, environmental monitoring, or public health surveillance, case studies reveal how the accuracy or inaccuracy of analytical approaches can profoundly influence biological interpretations and translational applications.

11.8.1 Human Microbiome Surveillance and Antimicrobial Resistance

Antimicrobial resistance (AMR) monitoring has been revolutionized by metagenomic sequencing, which makes it possible to identify resistance genes in intricate microbial communities without culturing them. Yet, accurate detection is confounded by several problems, including reference bias, homologous genes, and fragmented assemblies. A recent groundbreaking global analysis studied national patterns of antibiotic consumption, and their findings suggest that the composition of the human gut resistome mirrors antibiotic consumption (Forslund et al., 2013). Curated repositories such as the Comprehensive Antibiotic-Resistance Database (CARD) now underpin AMR analytics by providing standardized gene models and ontologies for robust surveillance pipelines (Alcock et al., 2020).

In parallel, wastewater metagenomics has emerged as a scalable proxy for tracking AMR at the community level. A multi-country survey revealed globally structured resistomes and highlighted how wastewater sequencing can serve as a sentinel system for both human and environmental health (Hendriksen et al., 2019). More recently, harmonized workflows applied to hundreds of cities demonstrated reproducibility and predictive power, reinforcing wastewater monitoring as a cost-effective public health strategy (Munk et al., 2022).

Key computational takeaways: Accurate AMR surveillance requires version-pinned curated databases, hybrid read/assembly-based pipelines to minimize false positives, and benchmarking across multiple sample types before scaling.

11.8.2 Soil Metagenomics and Novel Biosynthetic Gene Clusters

Soil is an unmatched source of microbial secondary metabolites; however, the immense diversity of microbial populations complicates assembly and functional inference. Several biosynthetic gene clusters (BGCs) remain undiscovered due to fragmented assemblies and chimeric bins that obscure their genomic context. A study by Donia et al. demonstrated that methodical computational mining can aid in the identification of BGCs and uncovered thiopeptide antibiotic clusters (Donia et al., 2014; Kumar et al., 2025). Recent automated platforms such as antiSMASH can provide annotation and comparative frameworks for BGC prediction; however, this platform remains dependent on high-quality assemblies (Donia et al., 2014). Linking predicted BGCs to chemical outputs remains a bottleneck.

Key computational takeaways: Robust BGC discovery relies on hybrid assemblies, stringent binning and quality control, and integration with orthogonal data layers.

11.8.3 Clinical Applications: Microbiome-Based Cancer Diagnostics

Metagenomics is now utilized in clinical settings and shows significant promise for cancer diagnostics. Recent studies showed clinical translatability in uncovering microbial signatures that are consistently linked to tumors. However, the potential use of metagenomics in clinical setups faces several computational hurdles, including rigorous contamination control, effective removal of host DNA, and reliable feature selection. A recent independent meta-analysis of fecal metagenomes in colorectal cancer showed reproducible microbial signatures across diverse populations. These studies reveal that generalization of pipelines to different cohorts is possible and depends upon standardization of pipelines that carefully manage contamination (Cuskin et al., 2015; Wirbel et al., 2019). Similarly, another study identified dysregulated microbiome abundance in breast cancer progression (Sarfraz et al., 2024).

Another study highlights the predictive accuracy of colorectal cancer using machine learning approaches. In addition, techniques such as SHAP (SHapley Additive exPlanations) are frequently incorporated into microbiome-based classifiers. These methods highlight prediction models based on microbial taxa or pathways, further enhancing transparency and clinical decision-making. For clinical applications, computational requirements focus on stringent decontamination protocols, interpretable machine learning frameworks, and validation across independent cohorts.

11.8.4 Metagenomics and the Microbiome in Alzheimer's Disease

Alzheimer's disease (AD) is shaped by genetic and neurodegenerative factors. However, a recent study revealed that AD progression is also influenced by interactions between the host and gut microbiome, referred to as the "gut–brain axis." Recent advancements in metagenomics have allowed researchers to map microbial, viral, and immune signatures linked to cognitive decline. Computational challenges are evident in these studies. A recent work from the MiaGB consortium and collaborators provided insights into how the gut virome differs in cognitively impaired individuals. Findings suggest that cognitive impairment is associated with altered viral community structure (James et al., 2024). Beyond virome alterations, integrative studies suggest that the diet–microbiome–inflammation axis can impact cognitive impairment (Arikawa et al., 2025). Similarly, another study from the same cohort showed that immune–microbiome interactions in aging brains distinguish cognitively impaired individuals. The integration of such heterogeneous datasets requires advanced frameworks, such as multi-block modeling and network inference, to avoid spurious associations (Prajapati et al., 2024).

Other metagenomics studies reveal the potential use of microbial abundance as a biomarker. For example, a recent study in intestinal disorders showed that reduced richness of taxa, alongside shifts in Firmicutes, overlap with cognitive symptoms and can be used for predictive modeling (Madigan et al., 2025; Gupta et al., 2025). In addition, recent work reveals linkage of gut microbiome signatures with physical and cognitive function, strengthening the case for microbial biomarkers in AD diagnosis (Nair et al., 2024). Despite several obstacles, the convergence of microbiome, virome, and immune profiling through NGS and metagenomics is opening new avenues for biomarker discovery and mechanistic insights into Alzheimer's disease. By grounding these approaches in reproducible computational frameworks, researchers are beginning to translate microbial signatures into meaningful clinical applications.

11.9 FUTURE PERSPECTIVES

Recent advancements have moved metagenomics from cataloging microbial diversity to an integrated discipline that supports medicine, biotechnology, and environmental research. However, persistent computational obstacles remain, ranging from upstream biases to reproducibility and interpretability. Future advances in the field will be determined by the intersection of NGS, AI, and multi-omics integration.

- *Real-Time and Portable Metagenomics*

  The rapid real-time pathogen identification enabled by portable nanopore sequencing now allows rapid pathogen detection and outbreak surveillance. For example, the recent Ebola outbreak was monitored using a handheld sequencer with cloud analytics (Quick et al., 2016). However, the applicability of such approaches requires robust base calling, error correction, and standardized workflows to minimize false positives in noisy field data.
- *Artificial Intelligence and Interpretable Models*

  Recent advancements in AI are increasingly used for applications ranging from taxonomic profiling to protein function prediction. AI techniques, such as graph neural networks and deep learning, now extend beyond sequence similarity. However, interpretability remains a critical component. Recently, a machine learning-based framework was developed for microbiome classification that balances accuracy and transparency (Topçuoğlu et al., 2020). Similarly, deep learning has been advanced using graph convolutional networks for protein structure-based function prediction (Gligorijević et al., 2021). Embedding such interpretable approaches into containerized workflows will enhance reproducibility and translational reliability.
- *Multi-Omics and Systems Integration*

  Recent advancements in multi-omics integration have revealed its usefulness in analyzing complex, integrated datasets. Multi-omics integration of metagenomics, transcriptomics, proteomics, metabolomics, and host genomics now provides more robust results. However, this shift requires high-end computational frameworks capable of managing high dimensionality and reducing spurious associations. A recently developed platform, the MixOmics platform, has pioneered multi-block modeling strategies for such integration, providing reproducible pipelines for feature selection (Rohart et al., 2017; https://doi.org/10.1371/journal.pcbi.1005752). Applications to Alzheimer's disease and cancer microbiome studies already illustrate how cross-omic integration can reveal mechanistic insights unavailable to single-omic analyses.
- *Toward Global Standards and Equity*

  Another key future priority for metagenomics analysis is global standardization. Factors such as sequencing depth, pipeline choice, and database selection are major hurdles for the reproducibility. FAIR principles provide a roadmap for harmonized data practices (Wilkinson et al., 2016). At the same time, open-access platforms such as MG-RAST (Rigden et al., 2016) and Galaxy (Afgan et al., 2018) are democratizing access to large-scale metagenomic analyses, lowering barriers for smaller labs and under-resourced regions.

11.10 CONCLUSION

Metagenomics now allows us to move far beyond simple species lists. We can ask what microbes are doing, how they interact, and why that matters for health and ecosystems. However, turning raw reads into trustworthy. Sampling and extraction choices bias what we see; reference databases remain incomplete; closely related strains blur together; assemblies fragment; and a large share of genes still have no known function. As projects scale to many cohorts and multiple omics layers, the biggest challenges increasingly involve integration, reproducibility, and making complex models understandable. The good news is that practical solutions are already helping. Hybrid short- and long-read strategies, graph-based assemblers, and ensemble binning produce cleaner MAGs, while standards like MIMAG improve clarity in quality assessment and reporting. On the functional side, expanded databases, context-aware methods, and machine learning and AI models validated against transcriptomic, proteomic, and metabolomic evidence are steadily reducing the "unknowns" in gene function. Simultaneously, workflow managers, containerized environments, and FAIR principles are making analytical pipelines more portable, transparent, and easier to share.

Case studies in antimicrobial resistance, cancer, and Alzheimer's disease demonstrate that when these practices are applied, results replicate across cohorts and begin to translate into meaningful insights for both research and clinical decision-making. Where does this leave us? The field is heading toward integrative, population-scale studies and models that predict behavior, not just describe it. Achieving this vision requires three things: (1) interpretable AI that generates testable, biologically sensible hypotheses; (2) standardized, benchmarked pipelines with versioned databases so comparisons remain fair over time; and (3) equitable, scalable computing so rigor is not a luxury. If progress continues along these lines, metagenomics will evolve from simply cataloging microbes to uncovering mechanisms that guide clinical care, accelerate biotechnological discovery, and reveal how microbial worlds shape human health and ecosystems.

REFERENCES

Afgan, E., Baker, D., Batut, B., Van Den Beek, M., Bouvier, D., Ech, M., Chilton, J., Clements, D., Coraor, N., Grüning, B. A., Guerler, A., Hillman-Jackson, J., Hiltemann, S., Jalili, V., Rasche, H., Soranzo, N., Goecks, J., Taylor, J., Nekrutenko, A., & Blankenberg, D. (2018). The Galaxy platform for accessible, reproducible and collaborative biomedical analyses: 2018 update. *Nucleic Acids Research*, *46*(W1), W537–W544. https://doi.org/10.1093/nar/gky379

Alcock, B. P., Raphenya, A. R., Lau, T. T. Y., Tsang, K. K., Bouchard, M., Edalatmand, A., Huynh, W., Nguyen, A. L. V., Cheng, A. A., Liu, S., Min, S. Y., Miroshnichenko, A., Tran, H. K., Werfalli, R. E., Nasir, J. A., Oloni, M., Speicher, D. J., Florescu, A., Singh, B., . . . McArthur, A. G. (2020). CARD 2020: Antibiotic resistome surveillance with the comprehensive antibiotic resistance database. *Nucleic Acids Research*, *48*(D1), D517–D525. https://doi.org/10.1093/nar/gkz935

Arikawa, A. Y., Waterman, A., Mishra, S. P., Labyak, C., Williams, C., Chaudhari, D. S., Shukla, R., Kumar, V., Masternak, M., Holland, P., Golden, A., Dangiolo, M., Kociolek, J., Fraser, A., Agronin, M., Aymat, M., Yadav, H., & Jain, S. (2025). Cognitive impairment is associated with alterations in diet quality and inflammatory biomarkers in older adults: A cross-sectional analysis of data collected from the Microbiome in Aging Gut and Brain (MiaGB) consortium cohort. *The Journal of Nutrition*. https://doi.org/10.1016/j.tjnut.2025.06.026

Beghini, F., McIver, L. J., Blanco-Míguez, A., Dubois, L., Asnicar, F., Maharjan, S., Mailyan, A., Manghi, P., Scholz, M., Thomas, A. M., Valles-Colomer, M., Weingart, G., Zhang, Y., Zolfo, M., Huttenhower, C., Franzosa, E. A., & Segata, N. (2021). Integrating taxonomic, functional, and strain-level profiling of diverse microbial communities with biobakery 3. *ELife*, *10*, 1–42. https://doi.org/10.7554/eLife.65088

Bodein, A., Scott-Boyer, M. P., Perin, O., Lê Cao, K. A., & Droit, A. (2022). Interpretation of network-based integration from multi-omics longitudinal data. *Nucleic Acids Research*, *50*(5), E27. https://doi.org/10.1093/nar/gkab1200

Bowers, R. M., Kyrpides, N. C., Stepanauskas, R., Harmon-Smith, M., Doud, D., Reddy, T. B. K., Schulz, F., Jarett, J., Rivers, A. R., Eloe-Fadrosh, E. A., Tringe, S. G., Ivanova, N. N., Copeland, A., Clum, A., Becraft, E. D., Malmstrom, R. R., Birren, B., Podar, M., Bork, P., . . . Woyke, T. (2017). Minimum information about a single amplified genome (MISAG) and a metagenome-assembled genome (MIMAG) of bacteria and archaea. *Nature Biotechnology*, *35*(8), 725–731. https://doi.org/10.1038/nbt.3893

Cheng, M., Xu, B., Sun, Y., Wang, J., Lu, Y., Shi, C., Pan, T., & Zhao, W. (2024). By targeting TRAF6 protein stability and affecting the. *15*(9), 1–23.

Cuskin, F., Baslé, A., Ladevèze, S., Day, A. M., Gilbert, H. J., Davies, G. J., Potocki-Véronèse, G., & Lowe, E. C. (2015). The GH130 family of mannoside phosphorylases contains glycoside hydrolases that target β-1,2-mannosidic linkages in Candida mannan. *Journal of Biological Chemistry*, *290*(41), 25023–25033. https://doi.org/10.1074/jbc.M115.681460

Donia, M. S., Cimermancic, P., Schulze, C. J., Wieland Brown, L. C., Martin, J., Mitreva, M., Clardy, J., Linington, R. G., & Fischbach, M. A. (2014). A systematic analysis of biosynthetic gene clusters in the human microbiome reveals a common family of antibiotics. *Cell*, *158*(6), 1402–1414. https://doi.org/10.1016/j.cell.2014.08.032

Eren, A. M., Esen, O. C., Quince, C., Vineis, J. H., Morrison, H. G., Sogin, M. L., & Delmont, T. O. (2015). Anvi'o: An advanced analysis and visualization platformfor Omics data. *PeerJ*, *2015*(10), 1–29. https://doi.org/10.7717/peerj.1319

Forslund, K., Sunagawa, S., Kultima, J. R., Mende, D. R., Arumugam, M., Typas, A., & Bork, P. (2013). Country-specific antibiotic use practices impact the human gut resistome. *Genome Research*, *23*(7), 1163–1169. https://doi.org/10.1101/gr.155465.113

Gligorijević, V., Renfrew, P. D., Kosciolek, T., Leman, J. K., Berenberg, D., Vatanen, T., Chandler, C., Taylor, B. C., Fisk, I. M., Vlamakis, H., Xavier, R. J., Knight, R., Cho, K., & Bonneau, R. (2021). Structure-based protein function prediction using graph convolutional networks. *Nature Communications*, *12*(1). https://doi.org/10.1038/s41467-021-23303-9

Gupta, A., Pahal, S., Kumar, V., & Chaudhary, A. (2025). Role of monoclonal antibodies in alzheimer's disease: Successes, failures, and future directions. *Neurological Sciences*. https://doi.org/10.1007/s10072-025-08456-5

Handelsman, J. (2005). Metagenomics: Application of genomics to uncultured microorganisms. *Microbiology and Molecular Biology Reviews*, *69*(1), 195–195. https://doi.org/10.1128/mmbr.69.1.195.2005

Hendriksen, R. S., Munk, P., Njage, P., van Bunnik, B., McNally, L., Lukjancenko, O., Röder, T., Nieuwenhuijse, D., Pedersen, S. K., Kjeldgaard, J., Kaas, R. S., Clausen, P. T. L. C., Vogt, J. K., Leekitcharoenphon, P., van de Schans, M. G. M., Zuidema, T., de Roda Husman, A. M., Rasmussen, S., Petersen, B., . . . Aarestrup, F. M. (2019). Global monitoring of antimicrobial resistance based on metagenomics analyses of urban sewage. *Nature Communications*, *10*(1). https://doi.org/10.1038/s41467-019-08853-3

Huerta-Cepas, J., Szklarczyk, D., Heller, D., Hernández-Plaza, A., Forslund, S. K., Cook, H., Mende, D. R., Letunic, I., Rattei, T., Jensen, L. J., Von Mering, C., & Bork, P. (2019). EggNOG 5.0: A hierarchical, functionally and phylogenetically annotated orthology resource based on 5090 organisms and 2502 viruses. *Nucleic Acids Research*, *47*(D1), D309–D314. https://doi.org/10.1093/nar/gky1085

James, A. S., Adil, N. A., Goltz, D., Tangudu, D., Chaudhari, D. S., Shukla, R., Kumar, V., . . . Yadav, H. (2024). Abnormalities in gut virome signatures linked with cognitive impairment in older adults. *Gut Microbes*, *16*(1). https://doi.org/10.1080/19490976.2024.2431648

Jumper, J., Evans, R., Pritzel, A., Green, T., Figurnov, M., Ronneberger, O., Tunyasuvunakool, K., Bates, R., Žídek, A., Potapenko, A., Bridgland, A., Meyer, C., Kohl, S. A. A., Ballard, A. J., Cowie, A., Romera-Paredes, B., Nikolov, S., Jain, R., Adler, J., . . . Hassabis, D. (2021). Highly accurate protein structure prediction with AlphaFold. *Nature*, *596*(7873), 583–589. https://doi.org/10.1038/s41586-021-03819-2

Kolmogorov, M., Yuan, J., Lin, Y., & Pevzner, P. A. (2019). Assembly of long, error-prone reads using repeat graphs. *Nature Biotechnology*, *37*(5), 540–546. https://doi.org/10.1038/s41587-019-0072-8

Köster, J., & Rahmann, S. (2012). Snakemake-a scalable bioinformatics workflow engine. *Bioinformatics*, *28*(19), 2520–2522. https://doi.org/10.1093/bioinformatics/bts480

Kumar, V., Gangani, S., Shukla, R., Prajapati, S. K., Shekhar, H., & Shukla, V. (2024). Next-generation sequencing and genomic data analysis. In A. Chaudhary, S. K. Sethi, & A. Verma (Eds.), *Unraveling new frontiers and advances in bioinformatics* (pp. 49–76). Springer Nature. https://doi.org/10.1007/978-981-97-7123-3_3

Kumar, V., Pandey, A., Arora, A., Gautam, P., Bisht, D., Gupta, S., Chaurasia, A., & Sachan, M. (2023). Diagnostics and therapeutic potential of miR-205 and miR-34a in ovarian cancer management: A miRNA-target-based analysis. *DNA and Cell Biology*, *42*(3), 151–162. https://doi.org/10.1089/dna.2022.0487

Kumar, V., Sharma, J. K., Kumar, S., Gulia, J., Singh, N. P., & Santal, A. R. (2025). Metagenomic and proteomic approach for bioremediation of environmental pollutants. In M. P. Shah (Ed.), *Development in waste water treatment research and processes*. Elsevier. https://doi.org/10.1016/B978-0-443-13615-3.00020-0

Kumar, V., Shukla, R., Gangani, S., Shekhar, H., & Shukla, V. (2024). Metagenomics and microbiome analysis. In A. Chaudhary, S. K. Sethi, & A. Verma (Eds.), *Unraveling new frontiers and advances in bioinformatics* (pp. 77–92). Springer Nature. https://doi.org/10.1007/978-981-97-7123-3_4

Kumar, V., Singh, A., Gautam, P., & Sachan, M. (2019). Techniques/tools to study epigenetic biomarkers in human cancer detection. In S. Paul (Ed.), *Biomedical engineering and its applications in healthcare* (pp. 327–351). Springer, Singapore. https://doi.org/10.1007/978-981-13-3705-5_14

Kurtzer, G. M., Sochat, V., & Bauer, M. W. (2017). Singularity: Scientific containers for mobility of compute. *PLoS ONE, 12*(5), 1–20. https://doi.org/10.1371/journal.pone.0177459

Li, D., Liu, C. M., Luo, R., Sadakane, K., & Lam, T. W. (2015). MEGAHIT: An ultra-fast single-node solution for large and complex metagenomics assembly via succinct de Bruijn graph. *Bioinformatics*, *31*(10), 1674–1676. https://doi.org/10.1093/bioinformatics/btv033

Madigan, S., Thiemann, R., Deneault, A.-A., Fearon, R. M. P., Racine, N., Park, J., Lunney, C. A., Dimitropoulos, G., Jenkins, S., Williamson, T., & Neville, R. D. (2025). Prevalence of adverse childhood experiences in child population samples: A systematic review and meta-analysis. JAMA Pediatrics, 179(1), 19–33. https://doi.org/10.1001/jamapediatrics.2024.4385

Meyer, F., Fritz, A., Deng, Z. L., Koslicki, D., Lesker, T. R., Gurevich, A., Robertson, G., Alser, M., Antipov, D., Beghini, F., Bertrand, D., Brito, J. J., Brown, C. T., Buchmann, J., Buluç, A., Chen, B., Chikhi, R., Clausen, P. T. L. C., Cristian, A., . . . McHardy, A. C. (2022). *Nature Methods*, *19*(4), 429–440. https://doi.org/10.1038/s41592-022-01431-4

Milanese, A., Mende, D. R., Paoli, L., Salazar, G., Ruscheweyh, H. J., Cuenca, M., Hingamp, P., Alves, R., Costea, P. I., Coelho, L. P., Schmidt, T. S. B., Almeida, A., Mitchell, A. L., Finn, R. D., Huerta-Cepas, J., Bork, P., Zeller, G., & Sunagawa, S. (2019). Microbial abundance, activity and population genomic profiling with mOTUs2. *Nature Communications*, *10*(1). https://doi.org/10.1038/s41467-019-08844-4

Munk, P., Brinch, C., Møller, F. D., Petersen, T. N., Hendriksen, R. S., Seyfarth, A. M., Kjeldgaard, J. S., Svendsen, C. A., van Bunnik, B., Berglund, F., Bego, A., Power, P., Rees, C., Lambrinidis, D., Neilson, E. H. J., Gibb, K., Coventry, K., Collignon, P., Cassar, S., . . . Aarestrup, F. M. (2022). Genomic analysis of sewage from 101 countries reveals global landscape of antimicrobial resistance. *Nature Communications*, *13*(1), 1–16. https://doi.org/10.1038/s41467-022-34312-7

Nair, S. U., Lin, Y., Jain, S., Chaudhari, D. S., Shukla, R., Kumar, V., Mankowski, R. T., Meng, H., Buford, T. W., Kuchel, G. A., Yadav, H., & Consortium, M. (2024), Unique gut microbiome signatures associated with physical and cognitive function. *Alzheimer's Dementia*, *20*, e092044. https://doi.org/10.1002/alz.092044

Nicholls, S. M., Quick, J. C., Tang, S., & Loman, N. J. (2019). Ultra-deep, long-read nanopore sequencing of mock microbial community standards. *GigaScience*, *8*(5), 1–9. https://doi.org/10.1093/gigascience/giz043

Ondov, B. D., Bergman, N. H., & Phillippy, A. M. (2011). Interactive metagenomic visualization in a Web browser. *BMC Bioinformatics*, *12*(1), 385. https://doi.org/10.1186/1471-2105-12-385

Parks, D. H., Chuvochina, M., Chaumeil, P. A., Rinke, C., Mussig, A. J., & Hugenholtz, P. (2020). A complete domain-to-species taxonomy for bacteria and Archaea. *Nature Biotechnology*, *38*(9), 1079–1086. https://doi.org/10.1038/s41587-020-0501-8

Parks, D. H., Chuvochina, M., Waite, D. W., Rinke, C., Skarshewski, A., Chaumeil, P. A., & Hugenholtz, P. (2018). A proposal for a standardized bacterial taxonomy based on genome phylogeny. *Nature Biotechnology*, *36*, 996–1004.

Pasolli, E., Asnicar, F., Manara, S., Zolfo, M., Karcher, N., Armanini, F., Beghini, F., Manghi, P., Tett, A., Ghensi, P., Collado, M. C., Rice, B. L., DuLong, C., Morgan, X. C., Golden, C. D., Quince, C., Huttenhower, C., & Segata, N. (2019). Extensive unexplored human microbiome diversity revealed by over 150,000 genomes from metagenomes spanning age, geography, and lifestyle. *Cell*, *176*(3), 649–662.e20. https://doi.org/10.1016/j.cell.2019.01.001

Prajapati, S. K., Kumar, V., Shukla, R., Szekeres, C., Ma, Y., Jain, S., & Yadav, H. (2024). Microbiome-immune interplay in aging brains: A pilot investigation from the MiaGB Cohort. *Alzheimer's Dementia*, *20,* e090734. https://doi.org/10.1002/alz.090734

Quick, J., Loman, N. J., Duraffour, S., Simpson, J. T., Severi, E., Cowley, L., Bore, J. A., Koundouno, R., Dudas, G., Mikhail, A., Ouédraogo, N., Afrough, B., Bah, A., Baum, J. H. J., Becker-Ziaja, B., Boettcher, J. P., Cabeza-Cabrerizo, M., Camino-Sánchez, Á., Carter, L. L., . . . Carroll, M. W. (2016). Real-time, portable genome sequencing for Ebola surveillance. *Nature*, *530*(7589), 228–232. https://doi.org/10.1038/nature16996

Quince, C., Walker, A. W., Simpson, J. T., Loman, N. J., & Segata, N. (2017). Shotgun metagenomics, from sampling to analysis. *Nature Biotechnology*, *35*(9), 833–844. https://doi.org/10.1038/nbt.3935

Rigden, D. J., Fernández-Suárez, X. M., & Galperin, M. Y. (2016). The 2016 database issue of nucleic acids research and an updated molecular biology database collection. *Nucleic Acids Research*, *44*(D1), D1–D6. https://doi.org/10.1093/nar/gkv1356

Rohart, F., Gautier, B., Singh, A., & Lê Cao, K. A. (2017). mixOmics: An R package for Omics feature selection and multiple data integration. *PLoS Computational Biology*, *13*(11), 1–14. http://www.embase.com/

search/results?subaction=viewrecord&from=export&id=L619520782%0Ahttp://dx.doi.org/10.1371/journal.pcbi.1005752

Salerno, F., Howden, A. J. M., Matheson, L. S., Gizlenci, Ö., Screen, M., Lingel, H., Brunner-Weinzierl, M. C., & Turner, M. (2023). An integrated proteome and transcriptome of B cell maturation defines poised activation states of transitional and mature B cells. *Nature Communications*, *14*(1), 1–18. https://doi.org/10.1038/s41467-023-40621-2

Sarfraz, H., Czerniecki, J. B., Stankiewicz, K., Michael Jaglal, V., Jain, S., Yadav, H., Liu, M., Shukla, R., Kumar, V., Alhomsi, M., Hathout, M., Zahra, T. F., Darwin, A., Joshi, U., Sen, A., Muzaffar, J., & Bari, S. (2024) Evaluating gut microbiome as a biomarker of pathological complete response in patients with early locally advanced triple negative breast cancer (LA TNBC): A pilot study. *JCO*, *42*(2024), 593–593. http://https://doi.org/10.1200/JCO.2024.42.16_suppl.593

Schnoes, A. M., Brown, S. D., Dodevski, I., & Babbitt, P. C. (2009). Annotation error in public databases: Misannotation of molecular function in enzyme superfamilies. *PLoS Computational Biology*, *5*(12). https://doi.org/10.1371/journal.pcbi.1000605

Sczyrba, A., Hofmann, P., Belmann, P., Koslicki, D., Janssen, S., Dröge, J., Gregor, I., Majda, S., Fiedler, J., Dahms, E., Bremges, A., Fritz, A., Garrido-Oter, R., Jørgensen, T. S., Shapiro, N., Blood, P. D., Gurevich, A., Bai, Y., Turaev, D., . . . McHardy, A. C. (2017). Critical assessment of metagenome interpretation—a benchmark of metagenomics software. *Nature Methods*, *14*(11), 1063–1071. https://doi.org/10.1038/nmeth.4458

Steinegger, M., & Söding, J. (2017). MMseqs2 enables sensitive protein sequence searching for the analysis of massive data sets. *Nat Biotechnol*, *35*, 1026–1028. https://doi.org/10.1038/nbt.3988

Topçuoğlu, B. D., Lesniak, N. A., Ruffin, M. T., Wiens, J., & Schloss, P. D. (2020). A framework for effective application of machine learning to microbiome-based classification problems. *MBio*, *11*(3). https://doi.org/10.1128/mBio.00434-20

Wick, R. R., Judd, L. M., & Holt, K. E. (2023, March 2). Assembling the perfect bacterial genome using Oxford Nanopore and illumina sequencing. *PLoS Comput Biol*, *19*(3), e1010905. https://doi.org/10.1371/journal.pcbi.1010905. PMID: 36862631; PMCID: PMC9980784

Wilke, A., Bischof, J., Gerlach, W., Glass, E., Harrison, T., Keegan, K. P., Paczian, T., Trimble, W. L., Bagchi, S., Grama, A., Chaterji, S., & Meyer, F. (2016). The MG-RAST metagenomics database and portal in 2015. *Nucleic Acids Research*, *44*(D1), D590–D594. https://doi.org/10.1093/nar/gkv1322

Wilkinson, M. D., Dumontier, M., Aalbersberg, Ij. J., Appleton, G., Axton, M., Baak, A., Blomberg, N., Boiten, J. W., da Silva Santos, L. B., Bourne, P. E., Bouwman, J., Brookes, A. J., Clark, T., Crosas, M., Dillo, I., Dumon, O., Edmunds, S., Evelo, C. T., Finkers, R., . . . Mons, B. (2016). Comment: The FAIR guiding principles for scientific data management and stewardship. *Scientific Data*, *3*, 1–9. https://doi.org/10.1038/sdata.2016.18

Wirbel, J., Pyl, P. T., Kartal, E., Zych, K., Kashani, A., Milanese, A., Fleck, J. S., Voigt, A. Y., Palleja, A., Ponnudurai, R., Sunagawa, S., Coelho, L. P., Schrotz-King, P., Vogtmann, E., Habermann, N., Niméus, E., Thomas, A. M., Manghi, P., Gandini, S., . . . Zeller, G. (2019). Meta-analysis of fecal metagenomes reveals global microbial signatures that are specific for colorectal cancer. *Nature Medicine*, *25*(4), 679–689. https://doi.org/10.1038/s41591-019-0406-6

Wood, D. E., Lu, J., & Langmead, B. (2019). Improved metagenomic analysis with Kraken 2. *Genome Biology*, *20*(1), 1–13. https://doi.org/10.1186/s13059-019-1891-0

Wooley, J. C., Godzik, A., & Friedberg, I. (2010). A primer on metagenomics. *PLoS Computational Biology*, *6*(2). https://doi.org/10.1371/journal.pcbi.1000667

Zhu, Y.-G. G., Johnson, T. a, Su, J.-Q. Q., Qiao, M., Guo, G.-X., Stedtfeld, R. D., Hashsham, S. a, Tiedje, J. M., Landers, T. F., Cohen, B., Wittum, T. E., Larson, E. L., Pal, C., Bengtsson-Palme, J., Kristiansson, E., Larsson, D. G. J. G. J. J., Salyers, A. A., Amábile-Cuevas, C. F., Bengtsson-Palme, J., . . . Fitzgerald, G. F. (2015). Glazer, Evan InterMath[14]—professional and cognitive development through problem solving with technology. *Nature Reviews Microbiology*, *13*(1), 1224–1232. https://doi.org/10.1126/science.1220761

12 Single-Cell and Long-Read Metagenomics

Ashish Kumar[1], *Shruti Shukla*[1], *and Ashutosh Mani*[1*]
[1]Department of Biotechnology, Motilal Nehru National Institute of Technology, Prayagraj,
*Corresponding author

12.1 INTRODUCTION

Microbes are ubiquitous, inhabiting the guts of humans and other organisms, and play vital roles in health, environmental balance, and agricultural productivity. Studying these microorganisms allows us to understand their functions and interactions. Until about two decades ago, genome sequencing was primarily restricted to microbes that could be cultured in the lab. However, with the development of next-generation sequencing technologies, metagenomics, and single-cell genomics, researchers can now characterize microbial communities more comprehensively, capturing the full diversity of the global microbiome without cultivation bias (Koonin et al., 2021). It has been estimated that more than 90–99% of microbes are nonculturable using available techniques. The introduction of metagenomics has revolutionized our comprehension of microbial communities by facilitating the examination of microorganisms without the necessity of culturing (Kumar, Shukla, et al., 2024). Traditional short-read sequencing and long-read sequencing platforms have offered valuable insights into microbial diversity, taxonomic categorization, functional annotations, and the formation of complete microbial genomes. However, these platforms face critical limitations in resolving genomes, repetitive elements, and rare microbial populations. This challenge becomes more pronounced because of genome variability, high mutation rates, and differing levels of microbial complexity. Hence, new cultivation-independent methods for understanding the function and diversity of microorganisms have been much-needed technologies.

12.2 SINGLE-CELL METAGENOMICS

Single-cell metagenomics represents a cutting-edge approach for the genomic characterization of individual microbial cells. Its principle lies in directly analyzing the genome of a single cell isolated from a complex microbial community without the need for cultivation. The methodology overcomes one of the significant limitations of traditional microbiology, in which more than 90–99% of microbes remain unculturable under laboratory conditions. It uncovers the hidden diversity within complex microbial ecosystems. The workflow involves key steps such as cell isolation, genome amplification, sequencing, and downstream analysis. Beginning with environmental sample collection and extending through sequencing and downstream analysis, this approach allows researchers to capture the genomic blueprint of individual uncultured microbes precisely.

DOI: 10.1201/9781003670155-12

12.3 SINGLE-CELL METAGENOMICS WORKFLOW

Single-cell metagenomics exploration is a very tough process; hence, sample collection and sequence analysis involving different steps are illustrated in Figure 12.1.

i. Cell isolation techniques
ii. Library preparation for single-cell sequencing
iii. Quality control
iv. Genome amplification strategies
iv. Downstream analysis

12.3.1 Single-Cell Isolation

In single-cell metagenomics, samples are obtained from various sources, including environmental samples, human- or animal-associated microbiomes, industrial and applied systems, clinical and pathogen surveillance, and plant-associated microbiomes. To analyze genomic content, individual microorganism cell isolation is the initial step, laying the foundation for identifying and characterizing microbes in these complex communities. Almost 90–99% of microbial species remain uncharacterized due to their unculturable properties in the laboratory. Micromanipulation, laser-assisted cell isolation, fluorescence *in situ* hybridization (FISH), microarrays, flow cytometry, microfluidic systems, and fluorescence-activated cell sorting (FACS) are techniques used for isolating single microbes from samples. Notably, flow cytometry and FACS stand out for their adaptability, as they are suitable for analyzing both culturable and nonculturable microbial populations (Guo et al., 2017). Fluorescence-activated cell sorting (FACS) takes advantage of fluorescent signals emitted by fluorescently labeled cells. The mechanism involves laser excitation of the fluorescently labeled cells; the resulting fluorescence is detected and measured by a detector. Although this method is highly efficient and rapid, the need for fluorescent labeling can sometimes introduce stress or damage to the

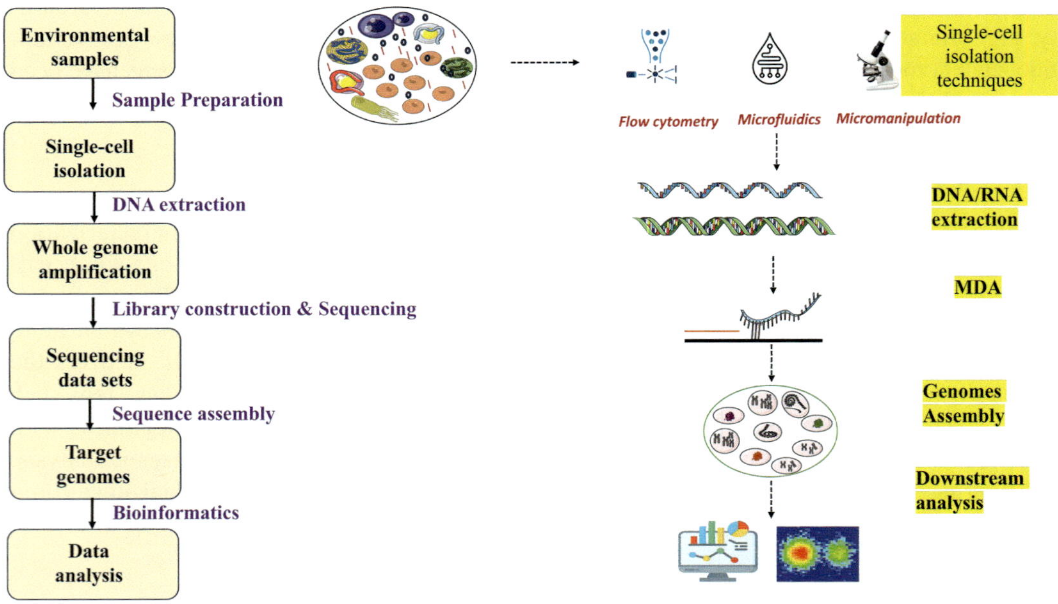

FIGURE 12.1 General overview of the scRNA-seq metagenomics process. It begins with sample collection and cell isolation, followed by DNA extraction and long-read sequencing (LRS), and concludes with downstream analysis.

cells. It is also noted that fluorescent dyes have potential toxicity to cells, nonspecific labeling, and disruption of normal cellular function (Jensen, 2012). To overcome these challenges, microfluidic-based methods for single-cell isolation are used. This technology has significantly enhanced single microbial cell isolation with high precision, efficiency, and minimal risk of contamination. It has multiple types based on procedures, such as droplet-based systems, microtraps, dielectrophoresis, and fully integrated on-chip workflows. It uses electric or magnetic forces to isolate cells from mixtures based on cell morphology and composition. This makes it possible to capture single microbial cells in a highly controlled way. Microfluidic technology helps reduce experimental variability and improve reproducibility, offering a more reliable way to explore microbial diversity at the single-cell level. Importantly, microfluidics has become central to single-cell metagenomics, allowing the recovery and genomic decoding of rare, unculturable, and functionally significant microbes previously inaccessible through conventional methods (Khan & Lin, 2019).

12.3.1.1 Micromanipulation

In micromanipulation, individual microbial cells are isolated by physical methods using high-powered microscopes and micropipettes or microneedles, or optical tweezers. Micromanipulation provides visual control, allowing researchers to pick single cells of interest based on morphology, fluorescence, or labeling.

12.3.1.2 Raman Spectroscopy

Raman spectroscopy is a widely accepted technique that has the advantage of preserving the fundamental nature of individual cells (Yuan et al., 2018; Lee et al., 2019). as it allows a label-free and nondestructive method that elucidates the inherent biochemical properties of cells and facilitates phenotypic analysis through inelastic light scattering (Butler et al., 2016; Suhito et al., 2018). Raman spectroscopy has the advantage of preserving the fundamental nature of individual cells (Yuan et al., 2018; Lee et al., 2019).

12.3.1.3 epicPCR

epicPCR (emulsion, paired isolation, and concatenation PCR), first introduced by Spencer and colleagues, is a breakthrough technique designed to link functional genes with phylogenetic markers in uncultured single microbial cells. The method works by compartmentalizing single cells within water-in-oil emulsions, where each droplet acts as an independent microreactor. As a result, it cannot reliably detect novel or uncharacterized genes, limiting its potential to uncover previously unknown microbial functions. Despite these challenges, epicPCR remains a valuable tool for linking microbial identity with function at the single-cell level, especially in studies of uncultured microbes.

12.3.2 Library Preparation for Single-Cell Sequencing

Library preparation for sequencing is the most important step in single-cell metagenomics, as further processes rely on this step. Since the starting material for a single cell comes from a single microbial genome, the DNA quantity is minimal, often only in the range of femtograms to picograms. To make this material suitable for sequencing, well-stabilized workflows are used to carefully capture, preserve, and amplify the genomic information while minimizing bias and contamination. The prepared library consists of larger DNA fragments equipped with sequencing adapters and barcodes. These fragments are then loaded onto sequencing platforms, generating reads that can later be assembled into single-cell genomes for detailed downstream analysis.

12.3.3 Long Read Sequencing Platform

Oxford Nanopore Technologies (ONT) and Pacific Biosciences (PacBio) are major long-read sequencing platforms, each offering unique technological innovations and advantages, which are discussed below.

12.3.3.1 PacBio

PacBio uses single-molecule real-time (SMRT) sequencing, which monitors DNA polymerase activity within zero-mode waveguides (ZMWs). DNA fragments are circularized with hairpin adapters, allowing polymerase to sequence the same templates multiple times. This process generates multiple subreads from single DNA fragments. As these fragments are sequenced, they are combined into circular consensus sequences (CCS), or HiFi reads. The PacBio can produce sequences with read lengths of 10–25 kb, which can exceed 100 kb with up to 99.99% accuracy. We can detect and find the de novo genome assembly, full-length transcript analysis, variant calling, and metagenomics with these long-read sequencing properties. The older versions have moderate throughput compared to short reads, but the newly introduced Revio system has significantly improved its throughput and lowered the cost of sequencing.

12.3.3.2 Oxford Nanopore Sequencing (ONT)

ONT relies on the nanopore sequencing principle, in which single DNA or RNA fragments pass through biological nanopores present in a membrane. These nanopores are made of protein or synthetic material. A steady electric current is supplied to the nanopore membrane as the tagged adapter sample strand travels through the nanopore until the sequencing adapter reaches the top. Once the adapter is engaged, it acts like a helicase, unwinding DNA and allowing single nucleotides to pass through the nanopore. An electrical signal fluctuates, resulting in changes to the ion current. This denotes each nucleotide type capable of impeding ion movement into the pore for distinct time intervals. The sequence of the utilized strand can be determined by measuring shifts in ion current generated by these impediments. ONT produces the longest sequencing reads, ranging from a few kilobases to over 100 kb, with a maximum read length of 2.273 megabases (Mb) (Amarasinghe et al., 2020). This sequencing capability makes it especially useful for resolving complex genomic regions, structural variants, and repetitive elements. ONT's most recent tool, PromethION, released in 2018, incorporates 48 flow cells, with a maximum run time of 68 hours, a maximum productivity of 8.6 to 15 Tb, and up to 144,000 possible channels (Hall et al., 2020)

12.3.4 Quality Control

Quality control in single-cell metagenomics requires multilayered filtering and contamination management, including the removal of low-quality reads and adapters to identify chimeras, trimming of adapters and overlapping reads, standardizing coverage, and evaluating assemblies. LongQC and NanoPack are best suited to present visual output, with optional filtering and trimming, as adapters are sometimes required in downstream processing because of the intrinsic variations in long-read sequencing by the two foremost sequencers, PacBio and ONT. A single quality control approach may not adequately cover every sequenced data point. For example, PacBio's long-read data from the Sequel sequencing platform may lack meaningful Phred scores, whereas ONT's sequenced data show meaningful Phred scores. As a result, platform-specific quality control tools are required to enable accurate assessment and processing of the relevant long-read datasets. Filtlong, LongQC, Porechop, and NanoPack are standard tools used to check the quality of long-read sequencing reads.

LongQC was designed to check the quality of long-read sequences generated by PacBio and ONT. Quality checks by LongQC provide read length distribution, read quality score distribution, GC content profiles, coverage with a reference genome, k-mer-based error checks, and detection of adapter sequences. (Fukasawa et al., 2020). LongQC is an easy-to-use interface and includes a variety of modules and algorithms for extensive analysis. It works by statistical sampling of reads and k-mer-based analysis for data quality. Quality control checks for reference-free datasets evaluate read length, error signature, and base composition. For reference-based quality control, it maps reads to the reference genome to measure coverage distribution. It uses the FASTA file format for its assessment. It generates interactive HTML and statistical plots summarizing read length histograms, percent GC content, quality score distribution, and sequence read coverage heatmaps (only for reference-based quality checks).

NanoPack is another reliable tool for long-read sequencing (LRS) quality control. It comprises a variety of subtools, including Cramino, NanoPlot, NanoStat, NanoComp, NanoFilt, and NanoLyse, each specialized for a separate purpose, such as NanoPlot, which is used to generate read length distribution, quality scores, cumulative yields, and other purposes. (Feng et al., 2022). NanoStat provides total read length, N50, mean and median quality, and the longest read. NanoComp is used to compare multiple sequencing runs and barcodes. NanoFilt filters and trims sequences and filters reads based on quality, length, and GC content. In addition, NanoFilt supports streaming mode and real-time sequencing processing. NanoLyse removes contaminant DNA sequences by applying lightweight alignment. NanoPack facilitates statistical summaries and visualization of quality checks across all these subtools. All these steps make it a complete suite for quality evaluation and downstream data analysis (De Coster et al., 2018). NanoPack2 is the sequel to NanoPack, with several advancements to its modules, such as increased plot production options, code optimization, and more dynamic HTML charts than NanoPlot and NanoComp. Along with these advancements, NanoPack2 integrates NanoLyse and NanoFilt functions into the Chopper tool, allowing complete filtering based on read score, read duration, contamination level, and other criteria. In addition, various minor improvements and adjustments have been made throughout the system (De Coster & Rademakers, 2023). It is a complete application created exclusively for working with PacBio data, making it one of the best options for PacBio sequencing data management.

12.3.5 Genome Amplification Strategies

One of the major technical barriers in single-cell metagenomics is the minute quantity of genomic DNA obtained from a single microbial cell, usually in the femtogram (~1–10 fg) range. Low-abundance bacteria are difficult to detect and are frequently overlooked using current cultivation and gene retrieval techniques.

12.3.5.1 Multiple Displacement Amplification (MDA)

Multiple Displacement Amplification (MDA) generates many DNA fragments using random hexamer primers and ϕ29 DNA polymerase in an isothermal environment (~30 °C). MDA provides an accurate, uniform representation across the genome. DNA polymerase ϕ29 has strong strand displacement capacity and high fidelity, which enable it to generate long DNA fragments. MDA is now widely used in single-cell whole-genome amplification. MDA involves cell lysis and genome extraction, primer binding, isothermal amplification, and amplified DNA recovery, typically yielding micrograms of DNA. Once the genomic contents are extracted, short random hexamer nucleotides are added to each part of the genome to ensure amplification starts at many positions covering the whole genome. After primer integration with sample sequences, the next step is polymerization by DNA polymerase ϕ29, which has exceptional displacement capacity and synthesizes long strands of DNA without dissociating from the template DNA. The ϕ29 polymerase has high processivity and high fidelity, which result in fewer error rates. During new strand synthesis, dNTPs (deoxynucleotide triphosphates) act as building blocks for new DNA synthesis.

To enhance MDA reaction specificity, the reaction volume (Hutchison et al., 2005) can be reduced, which decreases contaminating DNA and the strand synthesis of primer dimers (Marcy et al., 2007). Chimaeras can be resolved during genome assembly by achieving sufficient sequencing depth; however, removing them before sequencing may simplify the process. Chimeras in samples that have completed MDA processes can be treated with S1 nuclease to eliminate single-stranded DNA forms, resulting in an 80% reduction in chimeras (Zhang et al., 2006). To evaluate biases in MDA, a few mL aliquots from a series of DNA dilutions extracted from 12CH_3OH-exposed seawater were used as templates for amplification with the GenomiPhi V2 DNA Amplification Kit (GE Healthcare). Amplification reactions were conducted in 20 μL volumes using template inputs with different concentrations and incubated for several hours at 30 °C. Additional tested templates included control Lambda DNA supplied with the kit, heavy DNA from the 13CH_3OH-exposed seawater SIP gradient, and the corresponding light DNA from the same SIP gradient. Following amplification, aliquots of each product were analyzed on an agarose gel. DNA concentrations were further quantified using a NanoDrop ND-1000 spectrophotometer. Amplified DNA samples were diluted tenfold, and aliquots of the original DNA samples were subsequently used as templates for PCR and DGGE. In addition, aliquots of amplified DNA were subjected to pulsed-field gel electrophoresis (PFGE) using a fraction of low-melt agarose in TBE buffer on a CHEF Mapper system. The run was performed under default Auto Algorithm parameters, configured to achieve separation of fragments in the range of 10–100 kb. As controls, 300 ng of Lambda DNA and 300 ng of a heavy (ETBR) DNA marker were included in neighboring lanes. The gels were subsequently stained with ETBR and visualized using a suitable scanner (Neufeld et al., 2008).

12.3.5.2 Multiple Annealing and Looping-Based Amplification Cycles (MALBAC)

MALBAC is an advanced whole-genome amplification (WGA) method developed to improve the widely used multiple displacement amplification (MDA) method. Although MDA has been the standard for single-cell genome amplification, it often introduces strong amplification bias, resulting in uneven genome coverage. MALBAC addresses this issue through a quasi-linear amplification strategy designed to minimize bias and generate a more uniform genome representation. MALBAC amplifies the genome through its distinctive primer design and amplification strategy. Special primers have a shared sequence tag and random nucleotides to initiate DNA synthesis. Initially, the primer anneals at random sites across the genome, synthesizing new DNA strands. After generating complementary strands, their ends fold into looped structures, effectively blocking them from acting as templates in later cycles. This mechanism prevents the repeated amplification of the same types of DNA fragments. Following the initial linear amplification phase, the process shifts to exponential amplification, in which the shared primer sequences serve as binding sites for PCR. This two-step approach, combining linear preamplification with controlled exponential amplification, results in more uniform genome coverage than traditional MDA methods. By minimizing overamplification of certain regions, MALBAC effectively detects genetic variations, including single-nucleotide variants (SNVs), copy number, and structural rearrangements. Nonetheless, its combination of linear and exponential phases represents a significant advancement in genome amplification, making MALBAC one of the reliable methods available for single-cell genomics.

12.3.5.3 Microfluidics-Based MDA

Using the microfluidics method for MDA, single cells are encapsulated, and the MDA reaction is carried out in each microdroplet. This approach reduces contamination and cross-sample bias and provides higher throughput with high reproducibility. This technique is preferably used for marine microbiome studies. MDA has to be performed with altered tagged primers, which are sometimes used instead of random hexamers to reduce bias. This method significantly improves amplification of GC-rich or low-complexity regions, which conventional MDA cannot achieve. Another method for MDA involves the use of commercial kits, such as the REPLI-g Single Cell Kit (Qiagen) and GenomiPhi V2/V3, which are widely used kits standardizing MDA reactions for reproducibility and optimized for microbial single-cell genomics and are commonly used for high-fidelity DNA amplification. These kits incorporate proprietary lysis and amplification buffers to maximize yield and fidelity.

12.3.5.4 Challenges in Amplification

Amplification bias refers to the preferential amplification of specific genomic regions over others during genomic amplification. As whole-genome amplification (WGA) of microorganisms starts with tiny amounts of DNA, sometimes as little as a single microbial genome, even subtle biases in primer annealing or enzyme activity can have significant effects. One of the major challenges linked to amplification bias in whole-genome amplification (WGA) is uneven genome coverage. Instead of providing a balanced representation of the entire genome, WGA often produces data in which some regions are sequenced at extremely high depth while others receive little or no coverage. This imbalance creates gaps in genome assemblies and complicates the reconstruction of complete sequences. It also reduces the reliability of downstream analyses, including the detection of single-nucleotide variations (SNVs) and the identification of copy-number variations (CNVs). In single-cell studies of *Prochlorococcus*, a dominant marine microorganism, uneven coverage frequently leads to only partial genome reconstructions. As a result, important functional pathways may remain incomplete, and strain-specific adaptations are often overlooked. This limitation highlights the need for improved amplification strategies to deliver uniform genome representation in single-cell genomics. As WGA methods amplify DNA from extremely small quantities, they are highly susceptible to contamination during sample handling and from reagents, or from laboratory environmental microbes that may be amplified along with the target DNA. This challenge is particularly severe in environmental metagenomics, as distinguishing genuine microbial sequences from contaminants is already difficult due to the complexity of microbial communities.

12.3.6 Downstream Analysis

Downstream analysis refers to cleaning and evaluating a single-cell genome assembly. It begins with sequencing the long reads. The downstream processing starts with assembling the sequenced reads with the available reference database. The assembly quality is then checked. This is followed by binning similar sequences and then taxonomic and functional annotation of known and unknown sequences.

12.3.6.1 Genome Assembly

Genomic assembly assembles fragmented sequencing reads to reassemble the original sequence. Genomic assembly in metagenomics is complex due to the large number of microbial species in the sample. The metagenomic collection includes various species with different abundance levels, resulting in uneven read coverage throughout the genomes. Metagenomic samples often suffer from limited representation of particular species compared to cultured samples, leading to fragmented or incorrect assemblies (Lapidus & Korobeynikov, 2021). There are two main approaches for assembling long-read sequencing (LRS) data: HiFi assembly, explicitly designed for LRS, and hybrid assembly, which combines HiFi sequencing with short-read sequencing (SRS) methods. These approaches typically rely on three key assembly algorithms: the greedy algorithm, overlap layout consensus (OLC), and De Bruijn graph methods (Lapidus & Korobeynikov, 2021). The greedy algorithm assembles contigs by connecting reads based on overlaps, stopping once no further connections can be made. In contrast, OLC calculates pairwise overlaps between reads, forming an overlap graph where nodes represent individual reads and edges represent the overlaps. The consensus step then determines the most probable sequence by aligning reads along the various paths in the graph (Ayling et al., 2020) Meanwhile, the De Bruijn graph method uses the k-mer algorithm, which breaks down sequences into small overlapping segments (k-mers) and builds a graph based on their co-occurrence across reads (Ghurye et al., 2016). Metagenomic assemblers then use these algorithms to generate contigs, construct assembly graphs, and identify paths that help link the sequence data. The metagenomic assembly method can also result in a metagenome-assembled genome (MAG). MAGs are especially useful when there are not enough high-quality reference genomes (Hugerth et al., 2015) Table 12.1 summarizes the genome assembly algorithms, tools, and assembly strategies.

12.3.6.2 Assembly Quality Assessment

Metagenomes are complex due to the presence of unidentified microbes in samples. For de novo metagenomics, assembly quality can be highly compromised due to the presence of inversions, relocations, and interspecies translocations. MetaQUAST (Mikheenko et al., 2016) can perform reference-based and de novo assembly assessments. Another tool based on deep learning approaches,

TABLE 12.1
Genome Assembly Algorithms, Tools, and Strategies Used in Metagenomic Sequencing

Genome assembly methodology	Approach/ Tool	Key Features	Limitations	References
Genome assembly algorithms	Greedy algorithm	Sequentially connects overlapping reads until no more connections are possible	Produces fragmented assemblies, less accurate for complex genomes	Ayling et al. (2020)
	Overlap layout consensus (OLC)	Builds an overlap graph of reads, finds the best-supported path, and then constructs the consensus	Computationally intensive for large datasets	Ayling et al. (2020)
	De Bruijn graph	Breaks read into k-mers, builds a graph of overlaps, and reconstructs the genome from paths	Struggles with uneven coverage and repeats	Ghurye et al. (2016)
Genome assembly tools	MetaFlye	Modified Flye assembler; global and local k-mer analysis; repeat detection with bubbles/superbubbles	Needs high read depth for rare species	Kolmogorov et al. (2020); Cuscó et al. (2021)
	Canu	Uses tf-idf weighted MinHash and MinHash Alignment Process (MAP); sparse overlap graph	Computationally heavy for very large metagenomes	Koren et al. (2017)
	Hifiasm-meta	Designed for PacBio HiFi reads; overlap graph and haplotype-aware assembly	Newer tool; less tested on highly diverse communities	Lapidus & Korobeynikov (2021)
Assembly strategies	HiFi-only	Uses highly accurate PacBio HiFi reads	May miss low-abundance organisms	Lapidus & Korobeynikov (2021)
	Hybrid assembly	Combines long reads (LRS) with short reads (SRS)	Requires both data types	Lapidus & Korobeynikov (2021)

Deep Metagenome Assembly Error Detection (DeepMAsED), is used to assess quality without needing a reference genome. DeepMAsED provides an *insilico* pipeline for generating a large, realistic metagenome assembly for comprehensive model training and testing. In addition, DeepMAsED is flexible and can be applied to a wide, diversified metagenome assembly (Mineeva et al., 2020).

12.3.6.3 Binning

Binning refers to grouping assembled sequenced data into clusters, known as bins. Ideally, each bin represents a single genome or a Metagenome assembled genome (MAG). Binning tools are often based on short DNA sequences found within the contigs to identify similarity that suggests the contigs or DNA sequence came from which organism. (Setubal, 2021). Based on the availability of reference genomes, binning is divided into two types: taxonomic-dependent binning and taxonomic-independent binning.

12.3.6.3.1 Taxonomic Dependent Binning

Taxonomy-dependent binning works by comparing unknown microbial DNA sequences against reference databases. Using alignment tools, it assigns contigs to known microbial taxa based on sequence similarity (Yue et al., 2020). This method is effective when close reference genomes are available, but its accuracy may drop when dealing with novel or poorly characterized species.

12.3.6.3.2 Taxonomic Independent Binning

This method does not rely on external references. Instead, it groups bins based on patterns of GC content, tetranucleotide repetitions, and the abundance of sequences across different samples. This approach is beneficial for analyzing complex microbial communities with an undefined number of microbes. These methods are further divided into composition-based binning and abundance-based binning. In composition-based binning, microbial genomes from the same organism have unique signature sequences that share similar features, such as GC content, k-mer distribution, and the presence of conserved single-copy genes. Longer bins produce more reliable patterns, making it easier to assign them more accurately (Sedlar et al., 2017). Meanwhile, in abundance-based binning, the same genome appears at similar coverage levels. For multiple samples, this method looks for consistent coverage profiles across them. As complexity increases, abundance-based methods provide more reliable binning results (Yue et al., 2020). A hybrid method can be used for binning to obtain high accuracy and high-quality metagenome-assembled genomes (MAGs). Table 12.2 presents binning tools applied in metagenomics.

TABLE 12.2
Overview of Binning Tools Used in Metagenomics

Genome Binner	Parameters	Model	References
BASALT	Post binning refinement and Deredundancy for higher-quality, non-redundant MAGs	Strong bin cleanup and consolidation	Qiu et al. (2024)
SMeta	Integrates single-cell sequences with metagenomic data to guide contig binning	Explicitly leverages SAG reads to improve MAG recovery	Zhang et al. (2024)
SemiBin2	Self-supervised constructive learning for binning; suitable for short/Long/hybrid	Uses must link and cannot link to constraints for learning	Pan et al. (2023)
Autometa	sequence homology, single-copy genes, 5-mer frequency, coverage, single-copy genes	Lowest standard ancestor analysis, DBSCAN algorithm, supervised decision tree classifier, and recruit unclustered contigs	Miller et al. (2019)
SolidBin-naive	single-copy mark genes, tetranucleotide frequencies, coverage, pairwise constraints	Semi-supervised spectral Normalized cut	Wang et al. (2019)
BMC3C	frequency variation of oligonucleotides, coverage, and codon usage	Ensemble k-means, construct a weigh graph and partition it by Normalized cuts	Yu et al. (2018)
Vamb	tetranucleotide frequencies, coverage	Variational autoencoders, iterative medoid clustering algorithm	Nissen et al. (2021)
DAS Tool	original binner output bin sets	Refine bins according to shared contigs between the two original binner results	Sieber et al. (2018)
MetaWrap	original binner output bin sets	Separating every pair of contigs into different bins, selecting the best bin sets according to completion and contamination	Uritskiy et al. (2018)
Binsanity	coverage, tetranucleotide frequency, percent GC content	Affinity propagation	Graham et al. (2017)
COCACOLA	k-mer frequency, coverage, co-alignment, paired-end read linkage	K-means based on L1 distance, non-negative matrix factorization with sparse regularization, and hierarchical clustering	Lu et al. (2017)
Binning refiner	original binner output bin sets, single-copy genes	Scoring bins based on single-copy genes and picking up high-score bins iteratively	Song & Thomas (2017)

TABLE 12.3
Taxonomic Annotation and Software

Name	Algorithm	Reference Database	Developed Year
MEGAN-LR	Translation+LCA algorithm	NCBI nt	2018
	Nucleotide+LCA algorithm	NCBI nt	
MMseq2	Translation+LCA algorithm	NCBI nt	2017
BugSeq-v2	Nucleotide+LCA algorithm	NCBI nt	2021
CDKAM	Approximate matching+kmer	NCBI nt	2019
MetaMaps	Approximate mapping+EM	Miniseq+H	2016

12.3.6.4 Genome Annotation

Genome annotation is a computational process of identifying and functionally characterizing genes and other functional genomic elements in complex microbes. Further, microbes can be classified based on taxonomy and functional genome annotation.

12.3.6.4.1 Taxonomic Annotation

The main objective is determining which organisms the sequences belong to and at what taxonomic level. It refers to categorizing microbes based on their short-read sequenced or assembled sample genomes. For an overview of all microorganisms and their relative abundances in a sample, a comprehensive approach is to obtain taxonomic annotations for all possible sequencing reads. The outcome profiles suggest the number of genomes related to taxonomic categories (Hauptfeld et al., 2024).

Two principal methods are used for taxonomic annotation: marker gene-based methods and whole sequence-based methods. The first, a marker gene-based approach, relies on conserved genes in bacteria and the internal transcribed spacer (ITS) for fungi. These markers are matched with reference databases to find similarity with already known taxa. Whole sequence-based methods involve screening sequenced reads against large genomic databases. The software used in taxonomic annotation is listed in Table 12.3.

12.3.6.4.2 Functional Annotation

In metagenomics, functional annotation helps us go beyond simply knowing who is present in a microbial community to understanding what they can do. By studying the genes, researchers can predict which metabolic processes and biological pathways the microbes may be involved in. This reveals their potential activities and shows the important roles they might play in maintaining or influencing their environment. Table 12.4 presents the functional annotation tools and pipelines.

12.4 APPLICATIONS OF SINGLE-CELL METAGENOMICS

Since advanced technologies have emerged, single-cell metagenomics involves numerous applications.

12.4.1 Identification of Rare and Uncultivable Microbes

Researchers have applied metagenomic strategies to detect unknown and uncultivable microbes in marine environments. In one such approach, surface-water samples were incubated with near *in situ* concentrations of methanol and carefully labeled. Only picogram amounts of labeled DNA were recovered from these samples using cesium chloride gradient centrifugation. The DNA was then amplified through MDA, producing microgram quantities of high-molecular-weight DNA fragments

TABLE 12.4
Functional Annotation Tools and Pipelines

Name of Tool and Pipelines	Feature	Developed Year	Database	References
Mettannotator	A scalable Nextflow pipeline for annotating prokaryotic assemblies (isolates & MAGs). Predicts coding & non-coding genes, antimicrobial resistance, biosynthetic gene clusters, etc.	2024	Uses multiple reference databases	Gurbich et al. (2025)
MetaTP	For meta transcriptomics: automated workflows for functional annotation, expression quantification, and enrichment. Includes co-expression network analysis	2024	Uses eggnog, KEGG, GO	He et al. (2025)
FGBERT	A gene-language model pre-trained on metagenomic data. Helps capture contextual and structural relationships among genes, improving functional predictions	2024		Duan et al. (2024)
PhaGO	Specialized for bacteriophages. Leverages genomic context + protein foundation models to annotate phage proteins with rare or divergent functions	2024	Phage genome & protein databases	Guan et al. (2024)
EggNOG-mapper v2	Rapid functional annotation, orthology inference, domain prediction, metabolic pathway reconstruction	2021	eggnog, KEGG, FOAM	Cantalapiedra et al. (2021)
MEGAN LR	Long-read binning and annotation; taxonomic and functional analysis; interactive visualization	2021	SwissProt, SILVA SSU	Huson et al (2018)
Nanopore (pipeline)	Functional annotation pipeline for long reads; CRISPR-Cas detection, enzyme classification, metabolic profiling	2021	RefSeq, EC	Wick et al. (2019)
MetaERG	Profile HMMs and blast-based searches to detect similarity	2019	Eggnog, InterPro	Dong & Strous (2019)

(<40 kb), which were cloned to generate a fosmid library of over 10,000 clones. Denaturing Gradient Gel Electrophoresis (DGGE) confirmed minimal amplification bias during the MDA step and linked *Methylophaga*-like phylotypes to methanol metabolism in marine systems.

Further screening of 1,500 clones by PCR identified one carrying a methanol dehydrogenase (MDH)-encoding insert. Shotgun sequencing of this insert yielded a 9-kb DNA segment containing a cluster of genes involved in MDH production, regulation, and assembly. This integrated approach demonstrates how combining stable isotope probing, multiple MDA, and metagenomic cloning allows researchers to link specific functions with the microbes actively living in their natural environments, a connection that has long been sought after in microbial ecology. (Neufeld et al., 2008). MDA plays a crucial role by uniformly amplifying genomes and enabling single-cell genomics (SCG), which provides reference genomes for microorganisms that have never been cultured. These reference genomes are essential for making sense of community-wide omics datasets (Woyke et al., 2009). For instance, SCG has been used to explore the worldwide presence of uncultured marine *Flavobacterium* strains (Mason et al., 2012).

12.4.2 Metabolic Heterogeneity Within Communities

Understanding the wide range of metabolic capabilities within microbial communities is essential for linking different organisms to their roles in the environment. Single-cell metagenomics is a powerful tool in this effort. For example, study on *Methylophaga*-like bacteria, known for their role in methanol metabolism, used MDA along with PCR screening to identify around 1,500 clones that contained the methanol dehydrogenase (MDH) gene. By performing shotgun sequencing on these clones, scientists identified a 9-kilobase DNA fragment encoding a group of enzymes responsible for MDH biosynthesis, regulation, and assembly. This targeted metagenomic approach highlights how studying microbes in conditions close to their natural environment can effectively link specific microbial populations to their metabolic functions (Neufeld et al., 2008).

12.4.3 Strain-Level Resolution of Gene Functions and Interactions

Single-cell metagenomics adds to this by allowing researchers to recover whole genomes from nonculturable microorganisms. This capability makes it possible to reconstruct complete metabolic pathways, examine gene functions, and even support efforts to cultivate unculturable microbes. Early applications include the recovery of polyketide biosynthesis pathways from sponge symbionts (Siegl & Hentschel, 2010; Siegl et al., 2011) and marine cyanobacteria, as well as genomes of uncultured microbes degrading macromolecules (Hess et al., 2011) or fixing CO_2 via chemoautotrophy (Swan et al., 2011). Looking ahead, integrating single-cell metagenomics-based pathway discovery with synthetic biology promises innovative solutions in clean energy, bioremediation, and the production of nutritional and medicinal natural products (Yadav et al., 2019).

12.5 LONG READ SEQUENCING (LRS)

Long-read sequencing (LRS) represents a form of third-generation sequencing technology, with significant advancements pioneered by companies such as Pacific Biosciences (PacBio) and Oxford Nanopore Technologies (ONT) (Bleidorn, 2016). The sequence length of LRS varies depending on the sequencing platform and the technologies used. The PacBio Sequel II system generally produces reads with an average length of 10–20 kilobases (kb), with some extending beyond 1100 kb. In contrast, the Oxford Nanopore platform, such as MinION and GridION, can generate reads ranging from 10 kb to 100 kb, with maximum read lengths reported up to 2.27 megabases (Mb) (Amarasinghe et al., 2020). Longer sequencing reads produce fewer but longer fragments, which makes genome assembly less complicated and reduces the chances of errors. In contrast, short reads often struggle to assemble genomes accurately, especially in highly repetitive regions (Tørresen et al., 2019).

This extended sequence read length improves the ability to detect diverse genomic variants, such as large insertions and deletions, complex rearrangements, structural variations, repetitive elements, regions with high GC content, and epigenetic information, including modifications such as 5-methylcytosine (5mC). Despite these advantages, LRS has traditionally been associated with lower accuracy compared to short-read sequencing (SRS). In ONT, many errors are due to difficulties in regulating the speed at which DNA molecules move through the nanopore. At the same time, PacBio systems are more prone to random error (Delahaye & Nicolas, 2021). To overcome this issue, PacBio added circular consensus sequencing techniques, which increase read accuracy to about

99.8%. Similarly, Oxford Nanopore improved its performance by adding new flow cell technology, raising its accuracy to 99.5% (MacKenzie & Argyropoulos, 2023). Additionally, the incorporation of phi29 DNA polymerase in ONT has reduced DNA movement through the nanopore, thereby contributing to a reduction in error rates (Wang et al., 2021).

12.5.1 Short Read Sequencing (SRS)

In the early development of sequencing technologies, researchers often faced a trade-off between read length and accuracy. Short reads, usually up to 300 bases with around 99% accuracy, were widely adopted because of their reliability. These highly precise reads became the foundation for applications such as germline and somatic variant detection (Krøigård et al., 2016), exome sequencing, liquid biopsy analysis (Plagnol et al., 2018), noninvasive prenatal testing (Chitty et al., 2015), as well as transcriptome profiling (Cloonan et al., 2008) and single-cell studies (Bentley et al., 2008). On the other hand, long-read platforms could produce sequences spanning tens to hundreds of kilobases but with lower accuracy, typically in the range of 75–90% (Eid et al., 2009). Short-read sequencing (SRS), which commonly generates 600–700 base-pair reads, is limited, particularly when applied to complex metagenomic samples. Rodrigue and colleagues introduced a novel experimental and computational pipeline to overcome these challenges that assembles longer, high-quality composite reads from overlapping mate-pair data. This advancement expanded the potential of low-cost, high-throughput sequencing technologies, making them more suitable for large-scale metagenomic investigations. Table 12.5 presents a comparison between SRS and LRS.

TABLE 12.5
Comparison Between SRS and LRS

Feature	Short-Read Sequencing (SRS)	Long-Read Sequencing (LRS)
Sequencing platforms	Illumina (MiSeq, HiSeq, NovaSeq)	PacBio HiFi, Oxford Nanopore Technologies (ONT)
Read length	50–300 bp (very short)	PacBio HiFi: 10–25 kb ONT: up to >2 Mb (Megabase)
Accuracy	Very high raw accuracy (≥99.9%)	PacBio HiFi: very high (~99.9%) ONT: (95–99%)
Cost efficiency	Low cost per base; suitable for large-scale studies	Higher per-base cost
Assembly quality	Struggles with repetitive or complex regions; often produces fragmented assemblies	Produces contiguous assemblies; resolves repeats, structural variants, plasmids, and phage elements
Genome resolution	Difficult to resolve at the strain level; often collapses similar genomes	High-resolution, strain-level discrimination and accurate MAG (Metagenome-Assembled Genome) reconstruction
Throughput	Extremely high (billions of reads per run)	Moderate throughput; fewer reads but much longer
Epigenetic detection	Not possible (requires separate assays like bisulfite sequencing)	Direct detection of DNA modifications (e.g., 5mC, 6mA) during sequencing
Error profile	Minimal errors (mostly substitution errors, easily corrected)	ONT: insertion/deletion (indel) errors; PacBio HiFi: very low error rates
Applications in metagenomics	Taxonomic profiling, gene abundance estimation, and comparative metagenomics	High-quality genome assembly, plasmid/phage detection, structural variant discovery, epigenetic studies
Computational requirements	Lower; well-established pipelines (e.g., MetaPhlAn, HUMAnN)	Higher; requires error correction, hybrid assembly, and advanced tools (e.g., Canu, Flye, MetaWRAP)
Best use cases	Large-scale surveys of microbial diversity, abundance profiling, and cost-sensitive studies	High-resolution genome reconstruction, strain-level variation, mobile genetic element detection, epigenomics

12.6 APPLICATIONS OF LONG-READ SEQUENCING

LRS has the ability to make considerable advances and provide valuable insights in areas where short-read sequencing and 16S rRNA fail. Its capacity to generate longer and more contiguous reads enables improved characterization of complicated genomic areas, resolving repetitive elements, and detecting novel sequences.

12.6.1 Reconstruction of Metagenome-Assembled Genomes (MAGs)

MAGs are regarded as a valuable technique for studying all microbes. MAGs are important in identifying new microbes and strains from complicated microbial populations and providing insights into how they interact with others in their habitats. Importantly, each MAG often represents a unique microbial species, allowing for high-resolution characterization of microbial communities.

12.6.2 Identification of Structural Variants, Plasmids

To find large-scale variations, assembly-driven techniques combine reads into larger sequences (contigs) and then align them to another contig or a reference. MetaSVs, an assembly-driven technique, was intended for detecting types of structural variants and utilizes the benefits of long-read technology (Li et al., 2023). Long-read sequencing (LRS) has opened new possibilities for epigenetic studies by enabling the detection of DNA methylation while simultaneously capturing DNA sequences that span several kilobases. This capacity has allowed researchers to explore chromatin features, including DNA methylation, histone modifications, chromatin accessibility, and transcription factor binding. Both major long-read platforms, Oxford Nanopore Technologies (ONT) and Pacific Biosciences (PacBio), now extend their detection capabilities beyond the common 5-methylcytosine (5mC) to include other modifications such as N6-methyladenine (6mA) and 4-methylcytosine (4mC). PacBio identifies these modifications in microbial genomes using its SMRT Link microbial genome analysis suite. At the same time, ONT applies its Dorado software, which can detect multiple modification types and even combinations (e.g., 5mC with 5-hydroxymethylcytosine [5hmC] or 4mC with 5mC). Integrating deep learning-based methylation callers has further enhanced PacBio's accuracy, achieving strong concordance with traditional bisulfite sequencing, the gold standard for methylation detection. (Tse et al., 2021). Comparative studies indicate that both ONT and PacBio platforms provide reliable and high-resolution detection of CpG methylation. Typically, chromatin accessibility analysis with LRS begins by identifying methylation sites at the single-molecule level. This requires a base-calling algorithm tailored to the methyltransferase used in the experiment, targeting either 6mA or 5mC.

12.7 APPLIED CASE STUDIES OF INTEGRATION OF SINGLE-CELL METAGENOMICS AND LRS

12.7.1 Considerations in Sampling the Human Microbiome

The human body has an uncountable number of microorganisms that together encode ten times as many genes as the human genome (Zeng et al., 2016). Through this extensive genetic diversity, gut microbes play critical roles in our well-being by extracting energy from dietary nutrients (Schwalm & Groisman, 2017), synthesizing vitamins (Goodman et al., 2009), offering protection against pathogens, and boosting immunological homeostasis (Peterson et al., 2015). Changes in these microbial populations have been connected with various illnesses, including inflammatory bowel disease (Vich Vila et al., 2018), metabolic disorder, and cancer (Kostic et al., 2013). The existence of microbial taxa in every area of the human body has been the primary focus of these connections. Beyond the level of individual taxa, however, most microbially mediated human diseases necessitate a far more profound understanding of the molecular and functional characteristics of the gut microbial community. Because microbial populations differ greatly depending on anatomical location, collection period, and handling techniques, an accurate sample is crucial for human microbiome research. Fecal samples, for instance, are commonly employed to depict the gut microbiome, although they cannot accurately depict the microbial populations that adhere to the intestinal mucosa. Likewise, site-specific collection methods are required for oral, cutaneous, and vaginal microbiomes in order to avoid contamination. Factors such as sample storage, methods of DNA extraction, and sequencing depth can greatly influence metagenomic results. To gain reproducibility and comparable outcomes across studies, sampling procedures need to be standardized (Zeng et al., 2016; James et al., 2024; Nair et al., 2024), Moreover, the gut microbiome stimulates the immune system (Peterson et al., 2015). Gut bacteria are essential to human health. Inflammatory bowel disease, metabolic disorders like type 2 diabetes, and cancer (Kostic et al., 2013; Sarfraz et al., 2024). Moreover, the gut microbiome stimulates the immune system (Peterson et al., 2015). These gut bacteria are responsible for many disorders linked to changes in these microbial ecosystems. The existence or lack of microbial taxa in the human gut has been the primary focus of these connections (Yue et al., 2020).

12.7.2 Matching Phylogeny and Function of the Uncultured Microbial Community

SCG analysis of microbes provides information about the connection between the cell's metabolic markers and phylogenetic information with the help of sequencing in the cell embryo. Conversely, the dark ocean's carbon budget provides a convincing example of chemolithoautotrophy pathways in uncultured proteobacteria responsible for biomass production in the dark ocean (Swan et al., 2011; Garcia et al., 2013). The 100-year-old argument concerning the textbook's phylogenetic positioning was resolved when it was found that members of the poorly understood phylum Verrucomicrobia are most likely important contributors by breaking polysaccharides and iron by iron oxidizer *Ochracea leptothrix* (Blainey et al., 2011). It provided more evidence of the function of archaea in the nitrogen cycle under abnormal conditions. Numerous freshwater (Garcia et al., 2013) and marine taxonomic groups (Not et al., 2007) have confirmed the presence of rhodopsin and bacteriochlorophyll genes in aquatic bacteria, demonstrating that photoheterotrophy is common. Additionally, Picobiliphytes, a novel phylum of marine protists, are heterotrophic rather than phototrophic. These instances demonstrate how SCG may successfully address the long-standing problem of determining the metabolic capability of the uncultured microbial communities that control a variety of habitats and biochemical cycles.

12.7.3 Human Microbiome from Correlations to Mechanistic Understanding

Because of the human microbiome's complexity and variety, most connections in microbiome investigations are correlative rather than causative. Differences in microbial composition are frequently associated with disorders such as obesity, diabetes, and inflammatory bowel disease, but these findings do not always imply that microorganisms are to blame. Additional specific assays are necessary to determine causation and underlying processes. Standard procedures include transferring human microbiome samples to gnotobiotic animals, particularly germ-free mice. This allows researchers to determine whether a human microbial population can cause specific phenotypes in a controlled host. Another option is to conduct perturbation experiments, such as providing antibiotics, probiotics, or nutritional manipulations, to purposely modify the microbiome and examine the resulting impact on host physiology.

12.7.4 Understanding the Human Microbiome by Using Model Organisms

Model organisms provide an environment for studying how hosts interact with their causative organisms. For example, gnotobiotic mice are the gold standard. They enable scientists to explore cause-and-effect relationships by introducing specific microbial communities or even the entire human microbiota in a controlled environment. In addition to mice, other models like zebrafish, fruit flies (*Drosophila melanogaster*), and worms (*Caenorhabditis elegans*) offer unique advantages due to their ease of handling, short life cycles, and the availability of genetic tools. These systems allow researchers to examine how microbes influence host immunity, metabolism, and behavior, questions that are often difficult or impossible to study directly in humans.

12.7.5 Culture Systems for Characterizing the Human Microbiome

Despite the dominance of sequencing-based methods, culture-based approaches remain invaluable for human microbiome research. Traditional culture methods have been limited because many microbes are fastidious or unculturable under standard laboratory conditions. However, advances in high-throughput culturing (culturomics) and specialized growth media, anaerobic chambers, and microfluidic droplet-based isolation have dramatically expanded the range of cultivable human-associated microbes. Culturing isolates enables functional testing and metabolite profiling and allows for the construction of defined microbial consortia, which are useful in both mechanistic studies and therapeutic development.

12.7.6 Assaying the Microbiome in an Organism

Researchers employ a wide range of molecular and observational assays to characterize the microbiome. The most common is nucleotide sequencing, which includes DNA-based approaches (such as 16S rRNA gene sequencing for taxonomic profiling and shotgun metagenomics for functional gene content) and RNA-based approaches (meta-transcriptomics, which reveals active microbial functions). Beyond sequencing, other molecular profiling techniques such as metabolomics (identifying small molecules produced by microbes), proteomics (profiling microbial proteins), and lipidomics provide complementary insights into microbial activity. Direct observation methods such as fluorescence *in situ* hybridization (FISH), confocal microscopy, and flow cytometry enable spatial and quantitative analysis of microbial populations within host tissues. Integrating these multi-omics and imaging approaches is key to moving from descriptive correlations toward a mechanistic understanding of microbiome function.

12.7.7 Monitoring Bioreactor Function, Composition, and Stability

Bioreactors frequently experience changes in temperature, pH, or the amount of substrate added, which can disrupt the microbial communities' balance. Metagenomics helps identify key species and the backup roles certain microbes play, providing valuable insights into how these communities stay resilient. For instance, monitoring nitrifying and denitrifying microbes in wastewater treatment plants allows operators to maintain steady nitrogen removal even when conditions fluctuate. Beyond taxonomic identification, metagenomic approaches reveal the functional gene repertoire of microbial communities. This allows operators to assess the metabolic potential of a bioreactor, such as genes associated with lignocellulose degradation, nitrogen cycling, or stress tolerance. For instance, detecting antibiotic resistance genes or toxin-producing pathways in industrial fermentation systems provides early warning signals for contamination or functional shifts.

12.7.8 Monitoring of Environmental Conditions

Environmental monitoring offers important information about the baseline levels of antibiotic resistance in human and wildlife populations and in pristine and human-impacted settings. Identifying high-risk locations for establishing, selecting, and disseminating antibiotic-resistant bacteria is another benefit of this surveillance (Berendonk et al., 2015; Niemi & Rosenberry, 1976). This information is especially crucial for directing the creation of focused policies and mitigation plans. (Aarestrup & Woolhouse, 2020). Recent developments in wastewater-based surveillance have made it possible to identify collective antibiotic resistance genes in human populations, primarily by monitoring influent sewage to wastewater treatment facilities (Prieto Riquelme et al., 2022; Hendriksen et al., 2019), often mirroring local clinical resistance patterns (Pärnänen et al., 2019). Evaluations of ARG and ARB removal efficiency are crucial since the WWTP is a hotspot for resistance dynamics and a barrier to additional diffusion (Majeed et al., 2021). Additionally, contaminated surface waters and wastewater reuse serve as possible pathways for transmission between human populations and environmental reservoirs (Garner et al., 2018) In this context, new technologies like long-read sequencing improve the capacity to reconstruct plasmids and mobile genetic elements that promote horizontal gene transfer. In contrast, single-cell sequencing offers the resolution to connect resistance determinants to their microbial hosts directly. Combined, these methods can improve risk assessment and surveillance systems by expanding our mechanistic understanding of how ARB and ARGs move across clinical and environmental interfaces.

12.8 SYNERGY BETWEEN METAGENOMICS OF SINGLE-CELL AND LONG-READ SEQUENCING

The combination of both single-cell and long-read sequencing has immense applications, as single-cell technology enables the isolation of genomic information from rare and nonculturable microbes. Moreover, LRS resolves structural variation and repetitive regions and improves genome contiguity. Combining both technologies improves the recovery of high-quality MAGs, minimizes assembly breaks, and strengthens the identification of plasmids, mobile genetic elements, and epigenetic biomarkers.

12.8.1 Integration of Single-Cell and Long-Read Sequencing

Integrating single-cell approaches with long-read sequencing has reshaped strategies in metagenomic research. Single-cell techniques, which include microfluidics, micromanipulation, and FACS, allow researchers to separate individual microbial cells from samples of complex microbial communities. These isolation techniques help overcome challenges due to diversity in microbial communities and uneven species representation. This approach, together with long-read sequencing technologies, enables the recovery of near-complete genomes from single cells. In addition, the combined strategy simplifies genome assembly compared to short-read sequencing data, and it provides access to rare or uncultivable microorganisms that often remain undetected in bulk metagenomic sequencing data.

12.8.2 Enhanced Resolution of Microbial Communities

Long-read sequencing has dramatically advanced our ability to study complex microbial communities, providing a level of resolution that short-read platforms cannot match. Short-read sequence analysis methods often merge closely related strains, resulting in mixed or incomplete assemblies. Long-read sequencing technologies preserve genomic continuity and allow researchers to clearly distinguish between strains and even sub-strains within the same community. When paired with single-cell genomics, the power of long-read sequencing becomes more sophisticated. This combination makes it possible to identify fine-scale variations, including strain-level diversity, the presence and spread of mobile genetic elements, and evolutionary changes within microbial groups. Such detailed information is invaluable for uncovering ecological interactions, competition, and cooperative relationships across various environments, from the complex ecosystem of the human gut to extreme habitats like hydrothermal vents and polar regions.

12.8.3 Linking Mobile Genetic Elements to Host Genomes: Toward Multi-Omics Integration

The future of metagenomics lies in its integration with other omics disciplines, aiming to build an accurate systems-level view of microbial communities. Single-cell genomics and LRS provide the structural and functional frameworks of microbial genomes, meta-transcriptomics captures real-time gene expression patterns, metabolomics maps the chemical products of microbial activity, and proteomics identifies the enzymes and molecular interactions that drive these processes. Combining this information provides a holistic perspective linking genetic potential with biological functions and ecological roles. For example, a bioreactor system can be examined simultaneously at genomic, transcriptional, and metabolic levels, yielding more profound insights into how microbial communities maintain stability and productivity under different operating conditions.

12.8.4 Implications for Precision Microbiome Engineering

Metagenomics, along with other "omics" approaches, aims to gain a complete systems-level understanding of microbial communities. Single-cell genomics and long-read sequencing offer detailed insights into the structure and function of microbial genomes. At the same time, meta-transcriptomics reveals which genes are actively expressed, metabolomics tracks the chemical products generated by microbial activity, and proteomics uncovers the enzymes and molecular interactions driving these processes. By combining all these approaches, scientists can form a comprehensive picture that connects genetic potential to fundamental biological functions and ecological roles. For example, in a bioreactor, researchers can study the genomic, gene expression, and metabolic activity of the microbial community simultaneously. This integrated approach helps explain how microbial populations remain stable and productive even as environmental or operational conditions change.

12.9 CONCLUSION AND FUTURE PERSPECTIVES

The combination of single-cell technologies with advances in sequencing is driving significant progress toward higher portability, accuracy, and comprehensive analysis in microbiome research. Single-cell metagenomics and long-read sequencing significantly change traditional methods of exploring microbial communities by breaking through long-standing barriers related to cultivation, genome assembly, and taxonomic and functional interpretation. This integration allows researchers to examine rare microbes individually, which is impossible without single-cell metagenomics and LRS. Looking forward, blending these methods with multi-omics data, AI-powered analysis from single-cell isolation to downstream processing and synthetic biology holds great promise for breakthroughs in medicine, agriculture, and environmental sustainability. These cutting-edge tools are reshaping the field of metagenomics and paving the way for a new era of microbiome science where precision meets innovation.

REFERENCES

Aarestrup, F. M., & Woolhouse, M. E. J. (2020). Using sewage for surveillance of antimicrobial resistance. *Science*, 367(6478), 630–632. https://doi.org/10.1126/science.aba3432

Adan, A., Alizada, G., Kiraz, Y., Baran, Y., & Nalbant, A. (2017). Flow cytometry: Basic principles and applications. *Critical Reviews in Biotechnology*, 37(2), 163–176. https://doi.org/10.3109/07388551.2015.1128876

Amarasinghe, S. L., Su, S., Dong, X., Zappia, L., Ritchie, M. E., & Gouil, Q. (2020). Opportunities and challenges in long-read sequencing data analysis. *Genome Biology,* 21(1), 30. https://doi.org/10.1186/s13059-020-1935-5

Ayling, M., Clark, M. D., & Leggett, R. M. (2020). New approaches for metagenome assembly with short reads. *Briefings in Bioinformatics*, 21(2), 584–594. https://doi.org/10.1093/bib/bbz020

Bentley, D. R., Balasubramanian, S., Swerdlow, H. P., Smith, G. P., Milton, J., Brown, C. G., Hall, K. P., Evers, D. J., Barnes, C. L., Bignell, H. R., Boutell, J. M., Bryant, J., Carter, R. J., Keira Cheetham, R., Cox, A. J., Ellis, D. J., Flatbush, M. R., Gormley, N. A., Humphray, S. J., . . . Smith, A. J. (2008). Accurate whole human genome sequencing using reversible terminator chemistry. *Nature*, 456(7218), 53–59. https://doi.org/10.1038/nature07517

Berendonk, T. U., Manaia, C. M., Merlin, C., Fatta-Kassinos, D., Cytryn, E., Walsh, F., Bürgmann, H., Sørum, H., Norström, M., Pons, M.-N., Kreuzinger, N., Huovinen, P., Stefani, S., Schwartz, T., Kisand, V., Baquero, F., & Martinez, J. L. (2015). Tackling antibiotic resistance: The environmental framework. *Nature Reviews Microbiology*, 13(5), 310–317. https://doi.org/10.1038/nrmicro3439

Blainey, P. C., Mosier, A. C., Potanina, A., Francis, C. A., & Quake, S. R. (2011). Genome of a low-salinity Ammonia-oxidizing Archaeon determined by single-cell and metagenomic analysis. *PLoS ONE,* 6(2), e16626. https://doi.org/10.1371/journal.pone.0016626

Bleidorn, C. (2016). Third generation sequencing: Technology and its potential impact on evolutionary biodiversity research. *Systematics and Biodiversity*, 14(1), 1–8. https://doi.org/10.1080/14772000.2015.1099575

Butler, H. J., Ashton, L., Bird, B., Cinque, G., Curtis, K., Dorney, J., Esmonde-White, K., Fullwood, N. J., Gardner, B., Martin-Hirsch, P. L., Walsh, M. J., McAinsh, M. R., Stone, N., & Martin, F. L. (2016). Using Raman spectroscopy to characterize biological materials. *Nature Protocols*, 11(4), 664–687. https://doi.org/10.1038/nprot.2016.036

Cantalapiedra, C. P., Hernández-Plaza, A., Letunic, I., Bork, P., & Huerta-Cepas, J. (2021). eggNOG-mapper v2: Functional annotation, orthology assignments, and domain prediction at the metagenomic scale. *Molecular Biology and Evolution*, 38(12), 5825–5829. https://doi.org/10.1093/molbev/msab293

Chitty, L. S., Mason, S., Barrett, A. N., McKay, F., Lench, N., Daley, R., & Jenkins, L. A. (2015). Non-invasive prenatal diagnosis of achondroplasia and thanatophoric dysplasia: Next-generation sequencing allows for a safer, more accurate, and comprehensive approach. *Prenatal Diagnosis*, 35(7), 656–662. https://doi.org/10.1002/pd.4583

Cloonan, N., Forrest, A. R. R., Kolle, G., Gardiner, B. B. A., Faulkner, G. J., Brown, M. K., Taylor, D. F., Steptoe, A. L., Wani, S., Bethel, G., Robertson, A. J., Perkins, A. C., Bruce, S. J., Lee, C. C., Ranade, S. S., Peckham, H. E., Manning, J. M., McKernan, K. J., & Grimmond, S. M. (2008). Stem cell transcriptome profiling via massive-scale mRNA sequencing. *Nature Methods*, 5(7), 613–619. https://doi.org/10.1038/nmeth.1223

Cuscó, A., Pérez, D., Viñes, J., Fàbregas, N., & Francino, O. (2021). Long-read metagenomics retrieves complete single-contig bacterial genomes from canine feces. *BMC Genomics,* 22(1), 330. https://doi.org/10.1186/s12864-021-07607-0

De Coster, W., D'Hert, S., Schultz, D. T., Cruts, M., & Van Broeckhoven, C. (2018). NanoPack: Visualizing and processing long-read sequencing data. *Bioinformatics*, 34(15), 2666–2669. https://doi.org/10.1093/bioinformatics/bty149

De Coster, W., & Rademakers, R. (2023). NanoPack2: Population-scale evaluation of long-read sequencing data. *Bioinformatics,* 39(5), btad311. https://doi.org/10.1093/bioinformatics/btad311

Delahaye, C., & Nicolas, J. (2021). Sequencing DNA with nanopores: Troubles and biases. *PLOS ONE,* 16(10), e0257521. https://doi.org/10.1371/journal.pone.0257521

Dong, X., & Strous, M. (2019). An integrated pipeline for annotation and visualization of metagenomic contigs. *Frontiers in Genetics*, 10, 999. https://doi.org/10.3389/fgene.2019.00999

Duan, C., Zang, Z., Xu, Y., He, H., Liu, Z., Li, S., Song, Z., Zheng, J.-S., & Li, S. Z. (2024). FGBERT: Function-driven pre-trained gene language model for metagenomics (version 2). *arXiv*. https://doi.org/10.48550/ARXIV.2402.16901

Eid, J., Fehr, A., Gray, J., Luong, K., Lyle, J., Otto, G., Peluso, P., Rank, D., Baybayan, P., Bettman, B., Bibillo, A., Bjornson, K., Chaudhuri, B., Christians, F., Cicero, R., Clark, S., Dalal, R., deWinter, A., Dixon, J., . . . Turner, S. (2009). Real-time DNA sequencing from single polymerase molecules. *Science*, 323(5910), 133–138. https://doi.org/10.1126/science.1162986

Feng, X., Cheng, H., Portik, D., & Li, H. (2022). Metagenome assembly of high-fidelity long reads with hifiasm-meta. *Nature Methods*, 19(6), 671–674. https://doi.org/10.1038/s41592-022-01478-3

Fukasawa, Y., Ermini, L., Wang, H., Carty, K., & Cheung, M.-S. (2020). LongQC: A quality control tool for third generation sequencing long read data. *G3 Genes|Genomes|Genetics*, 10(4), 1193–1196. https://doi.org/10.1534/g3.119.400864

Garcia, S. L., McMahon, K. D., Martinez-Garcia, M., Srivastava, A., Sczyrba, A., Stepanauskas, R., Grossart, H.-P., Woyke, T., & Warnecke, F. (2013). Metabolic potential of a single cell belonging to one of the most abundant lineages in freshwater bacterioplankton. *The ISME Journal*, 7(1), 137–147. https://doi.org/10.1038/ismej.2012.86

Garner, A. J., Weiss, J. L., Parris, A., Kopp, R. E., Horton, R. M., Overpeck, J. T., & Horton, B. P. (2018). Evolution of 21st century sea level rise projections. *Earth's Future*, 6(11), 1603–1615. https://doi.org/10.1029/2018EF000991

Ghurye, J. S., Cepeda-Espinoza, V., & Pop, M. (2016). Metagenomic assembly: Overview, challenges and applications. *The Yale Journal of Biology and Medicine*, 89(3), 353–362.

Goodman, A. L., McNulty, N. P., Zhao, Y., Leip, D., Mitra, R. D., Lozupone, C. A., Knight, R., & Gordon, J. I. (2009). Identifying genetic determinants needed to establish a human Gut Symbiont in its Habitat. *Cell Host & Microbe*, 6(3), 279–289. https://doi.org/10.1016/j.chom.2009.08.003

Graham, E. D., Heidelberg, J. F., & Tully, B. J. (2017). BinSanity: Unsupervised clustering of environmental microbial assemblies using coverage and affinity propagation. *PeerJ*, 5, e3035. https://doi.org/10.7717/peerj.3035

Guan, J., Ji, Y., Peng, C., Zou, W., Tang, X., Shang, J., & Sun, Y. (2024). PhaGO: Protein function annotation for bacteriophages by integrating the genomic context (version 2). *arXiv*. https://doi.org/10.48550/ARXIV.2408.06402

Guo, F., Li, L., Li, J., Wu, X., Hu, B., Zhu, P., Wen, L., & Tang, F. (2017). Single-cell multi-omics sequencing of mouse early embryos and embryonic stem cells. *Cell Research*, 27(8), 967–988. https://doi.org/10.1038/cr.2017.82

Gurbich, T. A., Beracochea, M., De Silva, N. H., & Finn, R. D. (2025). Mettannotator: A comprehensive and scalable Nextflow annotation pipeline for prokaryotic assemblies. *Bioinformatics,* 41(2), btaf037. https://doi.org/10.1093/bioinformatics/btaf037

Hall, C. L., Zascavage, R. R., Sedlazeck, F. J., & Planz, J. V. (2020). Potential applications of nanopore sequencing for forensic analysis. *Forensic Science Review*, 32(1), 23–54.

Hauptfeld, E., Pappas, N., Van Iwaarden, S., Snoek, B. L., Aldas-Vargas, A., Dutilh, B. E., & Von Meijenfeldt, F. A. B. (2024). Integrating taxonomic signals from MAGs and contigs improves read annotation and taxonomic profiling of metagenomes. *Nature Communications,* 15(1), 3373. https://doi.org/10.1038/s41467-024-47155-1

He, L., Zou, Q., & Wang, Y. (2025). metaTP: A meta-transcriptome data analysis pipeline with integrated automated workflows. *BMC Bioinformatics,* 26(1), 111. https://doi.org/10.1186/s12859-025-06137-w

Hendriksen, R. S., Munk, P., Njage, P., Van Bunnik, B., McNally, L., Lukjancenko, O., Röder, T., Nieuwenhuijse, D., Pedersen, S. K., Kjeldgaard, J., Kaas, R. S., Clausen, P. T. L. C., Vogt, J. K., Leekitcharoenphon, P., Van De Schans, M. G. M., Zuidema, T., De Roda Husman, A. M., Rasmussen, S., Petersen, B., . . . Aarestrup, F. M. (2019). Global monitoring of antimicrobial resistance based on metagenomics analyses of urban sewage. *Nature Communications,* 10(1), 1124. https://doi.org/10.1038/s41467-019-08853-3

Hess, M., Sczyrba, A., Egan, R., Kim, T.-W., Chokhawala, H., Schroth, G., Luo, S., Clark, D. S., Chen, F., Zhang, T., Mackie, R. I., Pennacchio, L. A., Tringe, S. G., Visel, A., Woyke, T., Wang, Z., & Rubin, E. M. (2011). Metagenomic discovery of biomass-degrading genes and genomes from Cow Rumen. *Science*, 331(6016), 463–467. https://doi.org/10.1126/science.1200387

Hugerth, L. W., Larsson, J., Alneberg, J., Lindh, M. V., Legrand, C., Pinhassi, J., & Andersson, A. F. (2015). Metagenome-assembled genomes uncover a global brackish microbiome. *Genome Biology,* 16(1), 279. https://doi.org/10.1186/s13059-015-0834-7

Huson, D. H., Albrecht, B., Bağcı, C., Bessarab, I., Górska, A., Jolic, D., & Williams, R. B. H. (2018). MEGAN-LR: New algorithms allow accurate binning and easy interactive exploration of metagenomic long reads and contigs. *Biology Direct,* 13(1), 6. https://doi.org/10.1186/s13062-018-0208-7

Hutchison, C. A., Smith, H. O., Pfannkoch, C., & Venter, J. C. (2005). Cell-free cloning using φ29 DNA polymerase. *Proceedings of the National Academy of Sciences*, 102(48), 17332–17336. https://doi.org/10.1073/pnas.0508809102

James, A. S., Adil, N. A., Goltz, D., Tangudu, D., Chaudhari, D. S., Shukla, R., Kumar, V., Kumar, A., Masternak, M. M., Holland, P., Labyak, C., Golden, A., Dangiolo, M., Arikawa, A. Y., Kociolek, J., Fraser, A., Williams, C., Agronin, M., Aymat, M., Jain, S., & Yadav, H. (2024). Abnormalities in gut virome signatures linked with cognitive impairment in older adults. *Gut Microbes*, *16*(1). https://doi.org/10.1080/19490976.2024.2431648

Jensen, E. C. (2012). Use of fluorescent probes: Their effect on cell biology and limitations. *The Anatomical Record*, 295(12), 2031–2036. https://doi.org/10.1002/ar.22602

Khan, M., & Lin, J.-M. (2019). Microfluidics for single-cell genomics. In J.-M. Lin (Ed.), *Microfluidics for single-cell analysis* (pp. 143–161). Springer, Singapore. https://doi.org/10.1007/978-981-32-9729-6_6

Kolmogorov, M., Bickhart, D. M., Behsaz, B., Gurevich, A., Rayko, M., Shin, S. B., Kuhn, K., Yuan, J., Polevikov, E., Smith, T. P. L., & Pevzner, P. A. (2020). MetaFlye: Scalable long-read metagenome assembly using repeat graphs. *Nature Methods*, 17(11), 1103–1110. https://doi.org/10.1038/s41592-020-00971-x

Koonin, E. V., Makarova, K. S., & Wolf, Y. I. (2021). Evolution of microbial genomics: Conceptual shifts over a quarter century. *Trends in Microbiology*, 29(7), 582–592. https://doi.org/10.1016/j.tim.2021.01.005

Koren, S., Walenz, B. P., Berlin, K., Miller, J. R., Bergman, N. H., & Phillippy, A. M. (2017). Canu: Scalable and accurate long-read assembly via adaptive k-mer weighting and repeat separation. *Genome Research*, 27(5), 722–736. https://doi.org/10.1101/gr.215087.116

Kostic, A. D., Chun, E., Robertson, L., Glickman, J. N., Gallini, C. A., Michaud, M., Clancy, T. E., Chung, D. C., Lochhead, P., Hold, G. L., El-Omar, E. M., Brenner, D., Fuchs, C. S., Meyerson, M., & Garrett, W. S. (2013). Fusobacterium nucleatum potentiates intestinal tumorigenesis and modulates the tumor-immune microenvironment. *Cell Host & Microbe*, 14(2), 207–215. https://doi.org/10.1016/j.chom.2013.07.007

Krøigård, A. B., Thomassen, M., Lænkholm, A.-V., Kruse, T. A., & Larsen, M. J. (2016). Evaluation of nine somatic variant callers for detection of somatic mutations in exome and targeted deep sequencing data. *PLOS ONE,* 11(3), e0151664. https://doi.org/10.1371/journal.pone.0151664

Kumar, V., Gangani, S., Shukla, R., Prajapati, S. K., Shekhar, H., & Shukla, V. (2024). Next-generation sequencing and genomic data analysis. In A. Chaudhary, S. K. Sethi, & A. Verma (Eds.), *Unraveling new frontiers and advances in bioinformatics* (pp. 49–76). Springer Nature. https://doi.org/10.1007/978-981-97-7123-3_3

Kumar, V., Shukla, R., Gangani, S., Shekhar, H., & Shukla, V. (2024). Metagenomics and microbiome analysis. In A. Chaudhary, S. K. Sethi, & A. Verma (Eds.), *Unraveling new frontiers and advances in bioinformatics* (pp. 77–92). Springer Nature. https://doi.org/10.1007/978-981-97-7123-3_4

Lapidus, A. L., & Korobeynikov, A. I. (2021). Metagenomic data assembly—the way of decoding unknown microorganisms. *Frontiers in Microbiology*, 12, 613791. https://doi.org/10.3389/fmicb.2021.613791

Lee, K. S., Palatinszky, M., Pereira, F. C., Nguyen, J., Fernandez, V. I., Mueller, A. J., Menolascina, F., Daims, H., Berry, D., Wagner, M., & Stocker, R. (2019). An automated Raman-based platform for the sorting of live cells by functional properties. *Nature Microbiology*, 4(6), 1035–1048. https://doi.org/10.1038/s41564-019-0394-9

Li, Y., Cao, J., & Wang, J. (2023). MetaSVs: A pipeline combining long and short reads for analysis and visualization of structural variants in metagenomes. *iMeta,* 2(4), e139. https://doi.org/10.1002/imt2.139

Lu, Y. Y., Chen, T., Fuhrman, J. A., & Sun, F. (2017). COCACOLA: Binning metagenomic contigs using sequence COmposition, read CoverAge, CO-alignment and paired-end read LinkAge. *Bioinformatics*, 33(6), 791–798. https://doi.org/10.1093/bioinformatics/btw290

MacKenzie, M., & Argyropoulos, C. (2023). An introduction to nanopore sequencing: Past, present, and future considerations. *Micromachines,* 14(2), 459. https://doi.org/10.3390/mi14020459

Majeed, A., Rehman, M., Hussain, I., Imran, I., Saleem, M. U., Saeed, H., Hashmi, F. K., Akbar, M., Abrar, M. A., Ramzan, B., Chaudhry, M. O., Islam, M., Nisar, N., & Rasool, M. F. (2021). The impact of treatment adherence on quality of life among type 2 diabetes mellitus patients—findings from a cross-sectional study. *Patient Preference and Adherence*, 15, 475–481. https://doi.org/10.2147/PPA.S295012

Marcy, Y., Ishoey, T., Lasken, R. S., Stockwell, T. B., Walenz, B. P., Halpern, A. L., Beeson, K. Y., Goldberg, S. M. D., & Quake, S. R. (2007). Nanoliter reactors improve multiple displacement amplification of genomes from single cells. *PLoS Genetics,* 3(9), e155. https://doi.org/10.1371/journal.pgen.0030155

Martinez-Garcia, M., Brazel, D. M., Swan, B. K., Arnosti, C., Chain, P. S. G., Reitenga, K. G., Xie, G., Poulton, N. J., Gomez, M. L., Masland, D. E. D., Thompson, B., Bellows, W. K., Ziervogel, K., Lo, C.-C., Ahmed, S., Gleasner, C. D., Detter, C. J., & Stepanauskas, R. (2012). Capturing single cell genomes of active polysaccharide degraders: An unexpected contribution of verrucomicrobia. *PLoS ONE,* 7(4), e35314. https://doi.org/10.1371/journal.pone.0035314

Martinez-Garcia, M., Swan, B. K., Poulton, N. J., Gomez, M. L., Masland, D., Sieracki, M. E., & Stepanauskas, R. (2012). High-throughput single-cell sequencing identifies photoheterotrophs and chemoautotrophs in freshwater bacterioplankton. *The ISME Journal*, 6(1), 113–123. https://doi.org/10.1038/ismej.2011.84

Mason, O. U., Hazen, T. C., Borglin, S., Chain, P. S. G., Dubinsky, E. A., Fortney, J. L., Han, J., Holman, H.-Y. N., Hultman, J., Lamendella, R., Mackelprang, R., Malfatti, S., Tom, L. M., Tringe, S. G., Woyke, T., Zhou, J., Rubin, E. M., & Jansson, J. K. (2012). Metagenome, metatranscriptome and single-cell sequencing reveal microbial response to Deepwater Horizon oil spill. *The ISME Journal*, 6(9), 1715–1727. https://doi.org/10.1038/ismej.2012.59

Mikheenko, A., Saveliev, V., & Gurevich, A. (2016). MetaQUAST: Evaluation of metagenome assemblies. *Bioinformatics*, 32(7), 1088–1090. https://doi.org/10.1093/bioinformatics/btv697

Miller, I. J., Rees, E. R., Ross, J., Miller, I., Baxa, J., Lopera, J., Kerby, R. L., Rey, F. E., & Kwan, J. C. (2019). Autometa: Automated extraction of microbial genomes from individual shotgun metagenomes. *Nucleic Acids Research,* 47(10), e57–e57. https://doi.org/10.1093/nar/gkz148

Mineeva, O., Rojas-Carulla, M., Ley, R. E., Schölkopf, B., & Youngblut, N. D. (2020). DeepMAsED: Evaluating the quality of metagenomic assemblies. *Bioinformatics*, 36(10), 3011–3017. https://doi.org/10.1093/bioinformatics/btaa124

Nair, S. U., Lin, Y., Jain, S., Chaudhari, D. S., Shukla, R., Kumar, V., Mankowski, R. T., Meng, H., Buford, T. W., Kuchel, G. A., Yadav, H., & Consortium, M. (2024). Unique gut microbiome signatures associated with physical and cognitive function. *Alzheimer's Dementia*, 20, e092044. https://doi.org/10.1002/alz.092044

Neufeld, J. D., Chen, Y., Dumont, M. G., & Murrell, J. C. (2008). Marine methylotrophs revealed by stable-isotope probing, multiple displacement amplification and metagenomics. *Environmental Microbiology*, 10(6), 1526–1535. https://doi.org/10.1111/j.1462-2920.2008.01568.x

Niemi, W. D., & Rosenberry, T. L. (1976). Veratridine enhancement of agonist-induced synaptic membrane depolarization in electroplax. *Life Sciences*, 19(5), 707–718. https://doi.org/10.1016/0024-3205(76)90168–5

Nissen, J. N., Johansen, J., Allesøe, R. L., Sønderby, C. K., Armenteros, J. J. A., Grønbech, C. H., Jensen, L. J., Nielsen, H. B., Petersen, T. N., Winther, O., & Rasmussen, S. (2021). Improved metagenome binning and assembly using deep variational autoencoders. *Nature Biotechnology*, 39(5), 555–560. https://doi.org/10.1038/s41587-020-00777-4

Not, F., Valentin, K., Romari, K., Lovejoy, C., Massana, R., Töbe, K., Vaulot, D., & Medlin, L. K. (2007). Picobiliphytes: A marine picoplanktonic algal group with unknown affinities to other eukaryotes. *Science*, 315(5809), 253–255. https://doi.org/10.1126/science.1136264

Pan, S., Zhao, X.-M., & Coelho, L. P. (2023). SemiBin2: Self-supervised contrastive learning leads to better MAGs for short- and long-read sequencing. *Bioinformatics,* 39(Supplement_1), i21–i29. https://doi.org/10.1093/bioinformatics/btad209

Pärnänen, K. M. M., Narciso-da-Rocha, C., Kneis, D., Berendonk, T. U., Cacace, D., Do, T. T., Elpers, C., Fatta-Kassinos, D., Henriques, I., Jaeger, T., Karkman, A., Martinez, J. L., Michael, S. G., Michael-Kordatou, I., O'Sullivan, K., Rodriguez-Mozaz, S., Schwartz, T., Sheng, H., Sørum, H., . . . Manaia, C. M. (2019). Antibiotic resistance in European wastewater treatment plants mirrors the pattern of clinical antibiotic resistance prevalence. *Science Advances,* 5(3), eaau9124. https://doi.org/10.1126/sciadv.aau9124

Peterson, C. T., Sharma, V., Elmén, L., & Peterson, S. N. (2015). Immune homeostasis, dysbiosis and therapeutic modulation of the gut microbiota. *Clinical and Experimental Immunology*, 179(3), 363–377. https://doi.org/10.1111/cei.12474

Plagnol, V., Woodhouse, S., Howarth, K., Lensing, S., Smith, M., Epstein, M., Madi, M., Smalley, S., Leroy, C., Hinton, J., De Kievit, F., Musgrave-Brown, E., Herd, C., Baker-Neblett, K., Brennan, W., Dimitrov, P., Campbell, N., Morris, C., Rosenfeld, N., . . . Forshew, T. (2018). Analytical validation of a next generation sequencing liquid biopsy assay for high sensitivity broad molecular profiling. *PLOS ONE,* 13(3), e0193802. https://doi.org/10.1371/journal.pone.0193802

Prieto Riquelme, M. V., Garner, E., Gupta, S., Metch, J., Zhu, N., Blair, M. F., Arango-Argoty, G., Maile-Moskowitz, A., Li, A., Flach, C.-F., Aga, D. S., Nambi, I. M., Larsson, D. G. J., Bürgmann, H., Zhang, T., Pruden, A., & Vikesland, P. J. (2022). Demonstrating a comprehensive wastewater-based surveillance approach that differentiates globally sourced resistomes. *Environmental Science & Technology*, 56(21), 14982–14993. https://doi.org/10.1021/acs.est.1c08673

Qiu, Z., Yuan, L., Lian, C.-A., Lin, B., Chen, J., Mu, R., Qiao, X., Zhang, L., Xu, Z., Fan, L., Zhang, Y., Wang, S., Li, J., Cao, H., Li, B., Chen, B., Song, C., Liu, Y., Shi, L., . . . Yu, K. (2024). BASALT refines binning from metagenomic data and increases resolution of genome-resolved metagenomic analysis. *Nature Communications,* 15(1), 2179. https://doi.org/10.1038/s41467-024-46539-7

Sarfraz, H., Czerniecki, J. B., Stankiewicz, K., Michael Jaglal, V., Jain, S., Yadav, H., Liu, M., Shukla, R., Kumar, V., Alhomsi, M., Hathout, M., Zahra, T. F., Darwin, A., Joshi, U., Sen, A., Muzaffar, J., & Bari, S. (2024). Evaluating gut microbiome as a biomarker of pathological complete response in patients with early locally advanced triple negative breast cancer (LA TNBC): A pilot study. *JCO*, 42(2024), 593–593. http://https://doi.org/10.1200/JCO.2024.42.16_suppl.593

Schwalm, N. D., & Groisman, E. A. (2017). Navigating the Gut Buffet: Control of polysaccharide utilization in bacteroides spp. *Trends in Microbiology*, 25(12), 1005–1015. https://doi.org/10.1016/j.tim.2017.06.009

Sedlar, K., Kupkova, K., & Provaznik, I. (2017). Bioinformatics strategies for taxonomy independent binning and visualization of sequences in shotgun metagenomics. *Computational and Structural Biotechnology Journal*, 15, 48–55. https://doi.org/10.1016/j.csbj.2016.11.005

Setubal, J. C. (2021). Metagenome-assembled genomes: Concepts, analogies, and challenges. *Biophysical Reviews*, 13(6), 905–909. https://doi.org/10.1007/s12551-021-00865-y

Sieber, C. M. K., Probst, A. J., Sharrar, A., Thomas, B. C., Hess, M., Tringe, S. G., & Banfield, J. F. (2018). Recovery of genomes from metagenomes via a dereplication, aggregation and scoring strategy. *Nature Microbiology*, 3(7), 836–843. https://doi.org/10.1038/s41564-018-0171-1

Siegl, A., & Hentschel, U. (2010). PKS and NRPS gene clusters from microbial symbiont cells of marine sponges by whole genome amplification. *Environmental Microbiology Reports*, 2(4), 507–513. https://doi.org/10.1111/j.1758-2229.2009.00057.x

Siegl, A., Kamke, J., Hochmuth, T., Piel, J., Richter, M., Liang, C., Dandekar, T., & Hentschel, U. (2011). Single-cell genomics reveals the lifestyle of Poribacteria, a candidate phylum symbiotically associated with marine sponges. *The ISME Journal*, 5(1), 61–70. https://doi.org/10.1038/ismej.2010.95

Song, W.-Z., & Thomas, T. (2017). Binning_refiner: Improving genome bins through the combination of different binning programs. *Bioinformatics*, 33(12), 1873–1875. https://doi.org/10.1093/bioinformatics/btx086

Suhito, I. R., Han, Y., Min, J., Son, H., & Kim, T.-H. (2018). In situ label-free monitoring of human adipose-derived mesenchymal stem cell differentiation into multiple lineages. *Biomaterials*, 154, 223–233. https://doi.org/10.1016/j.biomaterials.2017.11.005

Swan, B. K., Martinez-Garcia, M., Preston, C. M., Sczyrba, A., Woyke, T., Lamy, D., Reinthaler, T., Poulton, N. J., Masland, E. D. P., Gomez, M. L., Sieracki, M. E., DeLong, E. F., Herndl, G. J., & Stepanauskas, R. (2011). Potential for chemolithoautotrophy among ubiquitous bacteria lineages in the dark ocean. *Science*, 333(6047), 1296–1300. https://doi.org/10.1126/science.1203690

Tørresen, O. K., Star, B., Mier, P., Andrade-Navarro, M. A., Bateman, A., Jarnot, P., Gruca, A., Grynberg, M., Kajava, A. V., Promponas, V. J., Anisimova, M., Jakobsen, K. S., & Linke, D. (2019). Tandem repeats lead to sequence assembly errors and impose multi-level challenges for genome and protein databases. *Nucleic Acids Research*, 47(21), 10994–11006. https://doi.org/10.1093/nar/gkz841

Tse, O. Y. O., Jiang, P., Cheng, S. H., Peng, W., Shang, H., Wong, J., Chan, S. L., Poon, L. C. Y., Leung, T. Y., Chan, K. C. A., Chiu, R. W. K., & Lo, Y. M. D. (2021). Genome-wide detection of cytosine methylation by single molecule real-time sequencing. *Proceedings of the National Academy of Sciences*, 118(5), e2019768118. https://doi.org/10.1073/pnas.2019768118

Uritskiy, G. V., DiRuggiero, J., & Taylor, J. (2018). MetaWRAP—a flexible pipeline for genome-resolved metagenomic data analysis. *Microbiome,* 6(1), 158. https://doi.org/10.1186/s40168-018-0541-1

Vich Vila, A., Imhann, F., Collij, V., Jankipersadsing, S. A., Gurry, T., Mujagic, Z., Kurilshikov, A., Bonder, M. J., Jiang, X., Tigchelaar, E. F., Dekens, J., Peters, V., Voskuil, M. D., Visschedijk, M. C., Van Dullemen, H. M., Keszthelyi, D., Swertz, M. A., Franke, L., Alberts, R., . . . Weersma, R. K. (2018). Gut microbiota composition and functional changes in inflammatory bowel disease and irritable bowel syndrome. *Science Translational Medicine,* 10(472), eaap8914. https://doi.org/10.1126/scitranslmed.aap8914

Wang, Y., Zhao, Y., Bollas, A., Wang, Y., & Au, K. F. (2021). Nanopore sequencing technology, bioinformatics and applications. *Nature Biotechnology*, 39(11), 1348–1365. https://doi.org/10.1038/s41587-021-01108-x

Wang, Z., Wang, Z., Lu, Y. Y., Sun, F., & Zhu, S. (2019). SolidBin: Improving metagenome binning with semi-supervised normalized cut. *Bioinformatics*, 35(21), 4229–4238. https://doi.org/10.1093/bioinformatics/btz253

Wick, R. R., Judd, L. M., & Holt, K. E. (2019). Performance of neural network basecalling tools for Oxford Nanopore sequencing. *Genome Biology,* 20(1), 129. https://doi.org/10.1186/s13059-019-1727-y

Woyke, T., Xie, G., Copeland, A., González, J. M., Han, C., Kiss, H., Saw, J. H., Senin, P., Yang, C., Chatterji, S., Cheng, J.-F., Eisen, J. A., Sieracki, M. E., & Stepanauskas, R. (2009). Assembling the marine metagenome, one cell at a time. *PLoS ONE,* 4(4), e5299. https://doi.org/10.1371/journal.pone.0005299

Yadav, B. S., Yadav, A. K., Singh, S., Singh, N. K., & Mani, A. (2019). Methods in metagenomics and environmental biotechnology. In K. M. Gothandam, S. Ranjan, N. Dasgupta, & E. Lichtfouse (Eds.), *Nanoscience and biotechnology for environmental applications* (Vol. 22, pp. 85–113). Springer International Publishing. https://doi.org/10.1007/978-3-319-97922-9_4

Yu, G., Jiang, Y., Wang, J., Zhang, H., & Luo, H. (2018). BMC3C: Binning metagenomic contigs using codon usage, sequence composition and read coverage. *Bioinformatics*, 34(24), 4172–4179. https://doi.org/10.1093/bioinformatics/bty519

Yuan, X., Song, Y., Song, Y., Xu, J., Wu, Y., Glidle, A., Cusack, M., Ijaz, U. Z., Cooper, J. M., Huang, W. E., & Yin, H. (2018). Effect of laser irradiation on cell function and its implications in Raman spectroscopy. *Applied and Environmental Microbiology*, 84(8), e02508–17. https://doi.org/10.1128/AEM.02508-17

Yue, Y., Huang, H., Qi, Z., Dou, H.-M., Liu, X.-Y., Han, T.-F., Chen, Y., Song, X.-J., Zhang, Y.-H., & Tu, J. (2020). Evaluating metagenomics tools for genome binning with real metagenomic datasets and CAMI datasets. *BMC Bioinformatics,* 21(1), 334. https://doi.org/10.1186/s12859-020-03667-3

Zeng, M. Y., Cisalpino, D., Varadarajan, S., Hellman, J., Warren, H. S., Cascalho, M., Inohara, N., & Núñez, G. (2016). Gut Microbiota-induced immunoglobulin G controls systemic infection by symbiotic bacteria and pathogens. *Immunity*, 44(3), 647–658. https://doi.org/10.1016/j.immuni.2016.02.006

Zhang, K., Martiny, A. C., Reppas, N. B., Barry, K. W., Malek, J., Chisholm, S. W., & Church, G. M. (2006). Sequencing genomes from single cells by polymerase cloning. *Nature Biotechnology*, 24(6), 680–686. https://doi.org/10.1038/nbt1214

Zhang, Y., Cheng, M., & Ning, K. (2024). SMeta, a binning tool using single-cell sequences to aid in reconstructing species from metagenome accurately. *Bioinformatics*. https://doi.org/10.1101/2024.08.25.609542

13 Multi-Omics Integration

Shruti Shukla[1], *Ashish Kumar*[1], *Ashutosh Mani*[1]*, *and Ravindra Pratap Singh*[2]

[1]Department of Biotechnology, Motilal Nehru National Institute of Technology, Prayagraj, Uttar Pradesh, India.

[2]Department of Biochemistry, Uttar Pradesh Rajarshi Tandon Open University, Prayagraj, Uttar Pradesh, India.

*Corresponding author

13.1 INTRODUCTION

Multi-omics is an integrative approach in systems biology that combines different types of biological data. like genes, RNA, Proteins, and metabolites to study living systems in a complete way. It integrates multiple "omics"—genome, transcriptome, proteome, metabolome, epigenome, and microbiome, all measured from the same group of samples. The central dogma layers (genomics, transcriptomics, proteomics) help us to understand DNA variation, RNA expression, and protein levels or state, while other fields like lipidomics and glycomics add more detail by capturing important biochemical information, as shown in Figure 13.1. Compared with single-omics, multi-omics or vertical integration combines signals from multiple omics layers, improving power for discovery and interpretation (e.g., disease genes/loci, regulatory programs). However, the data are large, heterogeneous, and often platform-specific, with added complexity from clinical covariates and site effects, plus concerns around standards, sharing, and reproducibility.

Biology functions across multiple interconnected layers. DNA variants can influence chromatin structure, which in turn regulates gene expression; transcripts are then processed and translated into proteins that undergo post-translational modifications; metabolism reflects both the activity of these proteins and the surrounding environment; and the microbiome further shapes many of these processes. Due to these complex interactions, using single-omics data alone is not enough; it cannot find causal connections across layers, detect ways for them to work together to make things better, or accurately group together complicated and different patient groups. To solve these limitations, multi-omics integration combines different types of biological data, such as DNA methylation and gene expression, or microbial metabolites and host pathways. This helps capture interactions between them, improves prediction accuracy, and provides a better understanding of how biological processes work together (Shukla et al., 2024; Kumar et al., 2024, 2025).

13.2 OMICS DATA TYPES AND REPOSITORIES

Multi-omics data include information about genes, proteins, RNA, metabolites, and other molecules like lipids and modified proteins like phosphorylated proteins and sugar-linked proteins. When these data are collected from the same samples, they help researchers to understand how biological processes work at different levels and understand the causes of diseases. Some public databases provide such patient-based multi-omics datasets for research.

DOI: 10.1201/9781003670155-13

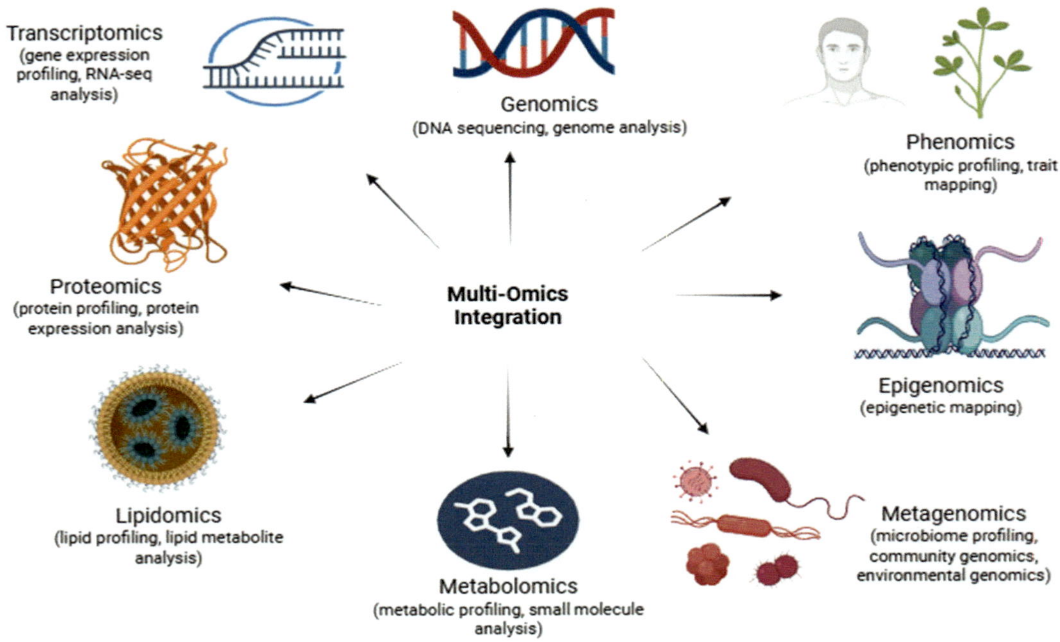

FIGURE 13.1 Integration of diverse omics layers in systems biology.

13.2.1 NCBI Gene Expression Omnibus (GEO)

The NCBI GEO is a public database developed by the National Centre for Biotechnology Information to store and share high-throughput gene expression and functional genomics data. It includes data from microarray, RNA-sequencing and other sequencing technologies covering a wide range of species, tissues and experimental conditions. GEO data can be explored through tools such as GEO DataSets, GEO Profiles, and GEO2R, and can be integrated with other resources like ArrayExpress and ENSEMBL for multi-omics studies (Clough & Barrett, 2016).

13.2.2 Quartet Data Portal

The Quartet Data Portal provides high-quality reference materials and multi-omics datasets designed for benchmarking and improving data accuracy across laboratories. It includes genomics, transcriptomics, proteomics, metabolomics, epigenomics and metagenomics data from B-lymphoblastoid cell lines (LCLs) made from blood samples of four family members: a father (F7), a mother (M8), and identical twin daughters (D5 and D6). Samples of DNA, RNA, proteins, and metabolites from the same batch were sent to different labs. Most labs analyzed three replicates of each sample in the same batch, except for long-read sequencing, which used one replicate per sample (Zheng et al., 2024). To ensure fair testing, labs received blinded samples so they wouldn't know which donor they came from. In total, data from six types of omics were produced using 24 platforms across 32 labs, and all are available through the Quartet Data Portal (J. Yang et al., 2023). Some additional omics datatypes and their corresponding repositories are described in Table 13.1.

TABLE 13.1
Major Omics Databases and Their Key Features

Database/Organizer	Data Type	Coverage	Key Features	Reference
The Cancer Genome Atlas (TCGA) *NIH & NCI*	RNA-Seq, DNA-Seq, miRNA, mutations, CNVs, DNA methylation, protein (RPPA), clinical data	>22,000 patients; 33 cancer types	Comprehensive genomic and clinical cancer data; freely available via GDC and UCSC Xena portals	Tomczak et al. (2015)
Cancer Cell Line Encyclopedia (CCLE) *Broad Institute & Novartis*	Gene expression, mutation, CNVs, proteomics, drug response	>1,000 human cancer cell lines	Integrates genomic and drug response data; used for biomarker discovery and drug sensitivity research; linked with DepMap and GDSC	Nusinow et al. (2020)
Omics Discovery Index (OmicsDI) *EMBL-EBI*	Genomics, transcriptomics, proteomics, metabolomics	Multi-species (human, model & non-model organisms)	Aggregates datasets from repositories (PRIDE, GEO, ArrayExpress, MetaboLights); follows FAIR data principles	Dass et al. (2020)
METABRIC (Molecular Taxonomy of Breast Cancer International Consortium) *Canada & UK*	Gene expression, SNPs, CNVs, clinical & survival data	>2,000 breast cancer samples	Defines ten molecular subtypes of breast cancer; identifies subgroups and therapeutic targets; accessible via cBioPortal	Guido & Gomez-Fernandez (2020)
TARGET (Therapeutically Applicable Research to Generate Effective Treatments) *U.S. National Cancer Institute*	Gene expression, miRNA, CNVs, mutations	Pediatric cancers (ALL, AML, neuroblastoma, Wilms tumor)	Improves understanding of childhood cancers; supports development of targeted therapies; data available via GDC	Gou[illegible] et al. (2021)
International Cancer Genome Consortium (ICGC/ICGC-ARGO) *Global Initiative (13+ Countries)*	Whole-genome, transcriptomic, methylation, CNVs, clinical data	>50 cancer types from diverse populations	ICGC catalogs cancer mutations; ICGC-ARGO links genomics to clinical outcomes for precision oncology	International Cancer Genome Consortium et al. (2010)

13.3 INTEGRATION STRATEGIES

13.3.1 Early Integration

Early integration is a method where all omics datasets (like genomics, transcriptomics, proteomics) are combined into one large table by placing their features side-by-side. This approach is popular because it is easy to use and allows Machine learning (ML) models to find relationships between different omics layers directly. It gives a combined view of all the data, helping models to find links between different omics types. This increases the number of features, but the number of samples remains the same. As a result, this combined data becomes large, complicated, noisy, and difficult for ML models to analyze. If one omics dataset has many more features than others, the model may focus more on that dataset and ignore the smaller ones. This method does not consider the special features of each omics type, which might lead to the finding of false patterns that are not biologically meaningful (Yadav et al., 2019; Zhang et al., 2024).

13.3.2 Mixed Integration

The mixed integration strategy first processes each omics dataset like genomics, proteomics, or metabolomics separately to make them simpler, smaller, and less noisy. This helps remove differences between datasets, such as data type or size. After that, the processed data from all omics are combined and analyzed using normal machine learning models. There are three main ways to transform the data (Table 13.2): kernel-based, graph-based, and deep learning (DL) methods (Picard et al., 2021; Saliba et al., 2024).

13.3.3 Intermediate Integration

Intermediate integration methods combine multi-omics datasets in a smarter way than just joining them (like in early integration), but without changing them into the same formats first. These methods generate shared features (common to all omics) and omics-specific features, which help to reduce the data's size and complexity for easier analysis. These methods are usually applied after preprocessing and feature selection, since raw omics data can be very different in scale and type. Some advanced tools like SLIDE, an improved version of mRMR (minimum Redundancy Maximum Relevance), can detect semi-shared patterns, which are relationships that appear in some omics but not all. A widely used method in this category is Non-negative Matrix Factorization (NMF) and its extensions, like Joint NMF, which identifies correlated modules across omics and Integrative NMF which helps to identify sample clusters and disease subtypes (Baião et al., 2025).

TABLE 13.2
Main Categories of Mixed Integration Strategies with Their Key Features and Applications

Integration Method	Description	Tools	Applications
Kernel-Based Methods	Measure similarity between samples using mathematical functions (kernels). Different kernel types capture different relationships among data (linear, Gaussian, polynomial).	SimpleMKL, SpicyMKL, SMO-MKL	Combine multiple omics datasets by finding optimal similarity kernels; improve classification and prediction accuracy.
Graph-Based Methods	Represent omics data as networks, where nodes are samples or genes and edges represent relationships or similarities. Networks from each omics type are fused for joint analysis.	Similarity Network Fusion (SNF), Affinity Network Fusion, Random Walk With Restart Fusion, iRANK, OmicsNet, DTINet, DeepDR, PIMKL, PAMOGK	Identify disease subtypes, predict outcomes, find key biomarkers, and explore molecular interactions.
Graph Neural Networks (GNNs)/ Graph Convolutional Networks (GCNs)	Advanced graph-based deep learning methods that analyze relationships and patterns directly from graph structures.	GNNs, GCNs	Predict side effects, classify samples, represent drugs or molecular structures.
Deep Learning (DL) Methods	Use multi-layer neural networks to combine or separately train omics datasets. Automatically learn hidden features and reduce noise.	Artificial Neural Networks (ANNs), Autoencoders, Restricted Boltzmann Machines (RBMs)	Feature extraction, denoising, and pattern recognition across omics layers for better biological interpretation.

13.3.4 Late Integration (Decision-Level Integration)

Late integration is the simplest and most straightforward method. It analyses each omics dataset separately using models designed for that specific omics type and then combines the predictions at the end. For example, one model might analyze gene expression, another might process copy number variation (CNV), and a third process clinical data. Their results are then averaged or merged to make a final prediction. Advanced versions like VCDN (View Correlation Discovery Network) not only merge predictions but also explore the relationships between omics results. A limitation of this approach is that it fails to capture interactions between different omics layers. The models never "talk to each other" during training, so they can't use complementary information shared across omics. Because of this, it's often treated as just a set of separate analyses, not true integration (Spooner et al., 2025).

13.3.5 Hierarchical Integration

Hierarchical integration reflects the natural flow of biological information, such as: DNA → RNA → Protein → Phenotype. It uses biological knowledge from databases and scientific literature to guide how omics layers are connected. For example, changes in DNA can affect gene expression, which affects protein levels, which then change the phenotype. Genotype-to-phenotype studies benefit from this approach. Omics layers are handled separately but in a biologically meaningful order, based on known regulatory paths. Several tools support hierarchical integration. iBAG combines methylation and gene expression to predict patient survival, while the Robust Network explores how copy number variations (CNVs) regulate gene expression. GRN Inference builds Gene Regulatory Networks (GRNs) using gene expression, CNVs, and methylation, and Fortelny and Bock develop a deep learning model that mirrors the logic of cellular regulation. Each node represents a gene or protein, and the model follows biological flow (expression → TFs → signaling → output). The result is easy to interpret by tracking which nodes are active (Wang et al., 2013).

13.4 TOOLS FOR INTEGRATIVE CLUSTERING AND SUBTYPE DISCOVERY IN MULTI-OMICS DATA

Many diseases, especially cancer, show a lot of variation between patients. This variation (called heterogeneity) happens due to differences in how the disease develops in each person, and can also be influenced by lifestyle and environmental factors. To understand a disease better and offer more targeted treatments, it's important to group patients into specific subtypes based on their biological (omics) data. Several tools are available that use multi-omics data to divide patients or samples into these meaningful subgroups based on molecular features. Table 13.3 provides an overview of key computational methods used for multi-omics data integration.

13.4.1 LRAcluster

LRAcluster is a method that helps to group disease samples into subtypes using a mathematical approach called low-rank approximation. It finds a simpler representation of complex multi-omics data (like mutations, methylation, gene expression, and CNVs) in fewer dimensions. LRAcluster is a fast and effective method for unsupervised classification of samples using multi-omics data (Wu et al., 2015).

13.4.2 PARADIGM (Bayesian-Based Tool)

PARADIGM is a tool that combines various types of omics data (like gene expression and copy number) from a single patient to estimate which biological pathways are active in their cells. It uses

TABLE 13.3
Computational Tools and Methods for Multi-Omics Data Integration

Tools	Principle	References
Joint non-negative matrix factorization (jNMF)/interactive NMF (intNMF)	Identifies both shared and unique patterns across different omics datasets. Handles data heterogeneity by separating common signals from dataset-specific ones, useful when data types vary	Mercadié et al. (2025); Yang & Michailidis (2016)
JDSNMF (Joint Deep Semi-Non-Negative Matrix Factorization)	Uses deep learning to find hidden biological patterns in multi-omics data. Works with both matched and unmatched datasets	Moon & Lee (2021)
MOFA/MOFA+	A Bayesian tool that reduces omics data into key features (shared and unique). Handles missing data well using probabilistic modeling	Argelaguet et al. (2020)
iCluster	Uses a statistical model assuming all omics data come from a shared hidden pattern. Combines clustering and feature selection	Athieniti & Spyrou (2023)
iClusterPlus	Improved iCluster that supports more data types and uses elastic-net/GLM. Adapts to different data distributions	Mo & Shen, n.d.
LRAcluster	Uses fast low-rank math to find hidden patterns efficiently, improving speed and accuracy	Wu et al. (2015)
moCluster	Uses a PCA-based method to find joint patterns. More stable than iClusterPlus and gives consistent results	Meng et al. (2016)
iClusterBayes	A fully Bayesian version of iCluster that adds uncertainty estimates to the clustering process	Mo et al. (2018)
JIVE/aJIVE	Joint and Individual Variation Explained (matrix decomposition into shared vs dataset-specific structure)	Cantini et al. (2021)
Integrated PCA (iPCA)	Identifies both shared and dataset-specific signals in multi-source data, giving a fuller picture	Tang & Allen (2021)
multiSLIDE	Builds on JIVE. Uses sparse and low-rank techniques to separate shared and unique components in omics datasets	Ghosh et al. (2021)
SNF (Similarity Network Fusion) (ab-SNF, NEMO)	Builds networks based on sample similarity for each omics layer, then combines them into one network for clustering. Graph-based approach	Cai et al. (2022)
DIABLO (mixOmics)	R package using multivariate, projection-based methods (PCA, ICA, PLS, rCCA, rGCCA, PLS-DA) for supervised and unsupervised data integration. Sparse variants with LASSO help select key biomarkers. Supervised multiblock PLS with correlated feature selection	Rohart et al. (2017); Singh et al. (2019)
moCluster	Uses sparse consensus PCA to detect shared patterns across omics layers, followed by K-means clustering. Faster and more stable than iCluster/iCluster+	Meng et al. (2016)
PFA (Pattern Fusion Analysis)	Uses PCA to extract local patterns from each omics dataset and aligns them into a shared low-dimensional space for integration. But cannot handle categorical data (e.g., mutations, SNPs); sensitive to dataset bias	Shi et al. (2017)
MFA (Multiple Factor Analysis)	MFA integrates numerical and categorical omics data into a common low-dimensional space. It first analyzes each omics dataset with PCA, then combines them. It also allows adding extra data (like biological processes) as supplementary variables	Grismer (2025)

a mathematical method called Bayesian factor graphs and information from curated biological databases to create a model. Each gene is represented as a network that includes its expression, activity, and interactions, allowing multiple data types to be combined. The result is a matrix showing pathway activity levels for each patient (Pulkstenis et al., 2017).

13.4.3 iCluster

iCluster is a tool used to group samples into clusters based on multiple data types, such as gene expression and DNA copy number. It works by building a shared model that finds hidden (latent) factors influencing all data types, while reducing complexity. It doesn't require labeled data (unsupervised learning) and uses a statistical approach called expectation-maximization to find the best grouping. However, iCluster can't handle both categorical (like mutation presence) and numerical data at once. iClusterPlus is an upgraded version of iCluster that can work with both types of data, numerical (like gene expression) and categorical (like mutation status). It uses generalized linear models and a method called LASSO regression to find key features (genes or markers) that define differences between subtypes. It identifies hidden variables (factors) that drive variation across the datasets. Although iClusterPlus is powerful, its main drawback is that it's computationally demanding and requires more effort to choose the best final model because of its complex statistical methods (Chalise et al., 2023).

13.4.4 Patient-Specific Data Fusion (PSDF)

PSDF is a method that combines CNV and gene expression data using a Bayesian model to classify patients into different subgroups. It checks whether the two data types (CNV and expression) agree for each sample. If they agree (concordant), the sample is considered fused and used in the analysis; if they disagree, the sample is left unfused, allowing for patient-specific grouping. The method also uses a statistical technique called Markov chain Monte Carlo (MCMC) to estimate how likely each sample is to be fused. Only features (genes) that are useful for grouping are selected, helping to remove noise (Gligorijevi et al., 2015).

13.4.5 Bayesian Consensus Clustering (BCC)

BCC is a method that combines results from different omics datasets (like gene expression, methylation, miRNA, and protein data) to create a consensus grouping of samples. It first builds clusters for each omics type, then merges them into an overall clustering that shows the common structure across all data sources. BCC uses a statistical model (finite Dirichlet mixture model) and can also estimate the best number of clusters (Lu & Lou, 2022).

13.4.6 Multiple Dataset Integration (MDI)

MDI integrates different types of omics data to identify meaningful clusters. It uses Dirichlet mixture models to group each data type (like gene expression or methylation) separately but also link the clusters together by shared features, such as specific genes. This helps to detect patterns where the same genes show similar behavior across multiple omics layers (Huang et al., 2017).

13.4.7 Similarity Network Fusion (SNF)

SNF is a network-based method for multi-omics data integration. It builds a separate similarity network for each data type (gene expression or DNA methylation). Then, fuses them into a single combined network using a message-passing technique, where each network updates itself to resemble others during multiple rounds (M. Yang et al., 2023).

13.4.8 PINSPlus

PINSPlus is a tool that finds disease subtypes from multi-omics data without needing labeled data (unsupervised). It works by repeatedly checking how often the same patients are grouped together when the data is slightly changed (perturbed), Different omics types are used, or different clustering methods are applied. PINSPlus builds graphs from each omics type where patients are nodes. These are merged into a similarity matrix, and subtypes are detected. It can also explore smaller subgroups within subtypes. The tool is customizable and fast enough to work with large datasets (Nguyen et al., 2019).

13.4.9 NEMO (Neighborhood-Based Multi-Omics Clustering)

NEMO is a simple, efficient clustering tool for integrating omics data. It starts by measuring how similar patients are (using Euclidean distance), builds similarity matrices for each omic layer, and then combines them into one. It uses spectral clustering to find disease subtypes. NEMO works even when some samples are missing some data types (partial datasets) (Heo et al., 2021; Rappoport & Shamir, 2019).

13.4.10 MCIA (Multiple Co-Inertia Analysis)

MCIA shows how multiple omics data types (genes, proteins, miRNAs) relate to each other across the **same set of samples**. It brings all datasets to a common scale and projects them into a **shared space**, making it easier to identify groups and shared features (Min & Long, 2020).

13.4.11 iNMF (Integrative Non-negative Matrix Factorization)

iNMF is an advanced version of the standard NMF technique, adapted to integrate data that contains both shared and unique signals across omics layers. It includes a tuning parameter **(λ)** to balance the integration (Yang & Michailidis, 2016).

13.4.12 JIVE (Joint and Individual Variation Explained)

JIVE separates what is shared between omics datasets and what is unique to each. It breaks data into shared signals, dataset-specific signals, and Noise. It uses permutation tests to determine which signals matter most (Lock et al., 2013).

13.4.13 web-rMKL

It is a web server that provides the integrative dimensionality reduction with subsequent clustering of samples based on data from multiple inputs. A machine learning method, rMKL-LPP, uses multiple kernels (mathematical similarity functions) and a dimensionality reduction method (LPP) to integrate different omics datasets. It automatically gives more importance to more informative data and avoids overfitting (Röder et al., 2019).

13.5 BIOLOGICAL INTERPRETATION AND VISUALIZATION TOOLS IN MULTI-OMICS INTEGRATION

A range of computational tools and platforms have been developed to support multi-omics data integration, visualization, and pathway analysis. These tools are used for pathway mapping, network construction, dimensionality reduction, and analyzing cross-omics relationships. Some visualization tools are summarized in Table 13.4.

TABLE 13.4
Commonly Used Visualization Tools in Multi-Omics Integration

Tool/Platform	Type	Function	Reference
Pathway Commons	Integrated Pathway Database	Aggregates pathway data from multiple sources (Reactome, BioPAX, etc.) for visualization and analysis	Rodchenkov et al. (2020)
PaintOmics 4, OmicsNet2.0	3D Multi-Omics Network Viewer	Builds 3D interactome networks using transcriptomics, proteomics, and metabolomics	Liu et al. (2022); Zhou et al. (2022)
ShinyOmics	Interactive Visualization Tool	Web-based tool for visualizing expression, network, and metadata for multi-omics datasets	Surujon & van Opijnen (2020)
Ingenuity Pathway Analysis (IPA)	Pathway Analysis Tool	Analyzes and visualizes molecular interactions, pathways, and networks from omics data	Gutierrez Reyes et al. (2024)
Galaxy, Galaxy-P	Open-source Analysis Platform	Enables multi-omics data analysis workflows with visualization and reproducibility	Boekel et al. (2015) McGowan et al. (2020)
Integrative Networks	Bipartite/Multiplex Networks	Represents cross-layer relationships (e.g., gene–protein, protein–phenotype)	Chierici et al. (2020)
Latent-Space Maps	Dimensionality Reduction	Projects multi-omics data into 2D/3D to reveal subgroups and hidden patterns	Otto et al. (2022)
Alluvial/Sankey Diagrams	Flow Visualization	Illustrates transitions and relationships across omics layers or timepoints	Otto et al. (2022); Yen et al. (2023)
CIRCOS	Circular Genomic Visualization	Shows inter-omics links (SNPs, expression, methylation, etc.) across chromosomes	Wilkinson & Friendly (2009); Lv et al. (2021); Pavlopoulos et al. (2018).
Cytoscape	Network Visualization Tool	Visualizes molecular interaction networks with omics data overlays	Shannon et al. (2003)
PaintOmics 3	Pathway & Expression Visualization	Maps transcriptomic, proteomic, and metabolomic data to biological pathways	Hernández-de-Diego et al. (2018)
MiBiOmics	network visualization	facilitates multi-omics data visualization, exploration, integration, and analysis	Zoppi et al. (2021)
OmicsAnalyst	interactive heatmaps	clustering, and dimensionality reduction	Zhou et al. (2021)
CFViSA	generating plots and customizing analyses	Customize visualizations and perform statistical tests on selected features across different experimental conditions	Wang et al. (2024)
InterTADs	analyze both gene expression and DNA methylation data	targets Topologically Associated Domains (TADs), regions where genes and regulatory elements interact closely	Tsagiopoulou et al. (2022)

13.6 CASE STUDIES

13.6.1 Multi-Omics Studies in Cancer Research

13.6.1.1 Chronic Lymphocytic Leukemia

Chronic lymphocytic leukemia (CLL) is a complex blood cancer. The development of high-throughput multi-omics technologies has greatly helped to understand CLL better. A tool called MOFA (Multi-Omics Factor Analysis) was used to study CLL patients by combining DNA methylation and gene expression data. MOFA identifies major patterns (called factors) across different omics layers. Some factors are common across multiple omics types, while others are unique to a single data type. These factors help in grouping patients into subtypes and can also be used to fill in missing data (Argelaguet et al., 2018; Tsagiopoulou & Gut, 2024; Lu et al., 2021).

13.6.1.2 Endometrial Cancer

Endometrial cancer is the most common gynecological malignancy in developed countries. Recent advances in multi-omics approaches have significantly contributed to the identification of novel molecular markers and therapeutic targets. Notably, *TMEFF2* and *RNF183* have been linked to estrogen receptor alpha (ERα) signaling. *ZBTB7A* is found to be downregulated in endometrial cancer. Other genes of interest include *HOXB9*, *ARL4C*, *LGR5*, *SST*, *ZNF558*, and *PTGDS*. Additionally, *BTG1* has emerged as a potential biomarker for endometrial cancer (An et al., 2025; Kumar et al., 2023; Boroń et al., 2022).

13.6.2 Multi-Omics Studies in Drug Resistance

13.6.2.1 BRAF Inhibition in Melanoma Cells

The B-Raf proto-oncogene, serine/threonine kinase (*BRAF*) is a key molecule involved in the mitogen-activated protein kinase (*MAPK*) pathway. Using a multi-omics approach, both genetic and non-genetic cause of resistance to BRAF inhibitors (BRAFi) was identified. Although BRAF inhibitors like vemurafenib and dabrafenib are used to treat melanoma, their effectiveness is reduced due to the development of resistance. TYRP1 protein, involved in melanin biosynthesis, was found to be overexpressed in vemurafenib and dabrafenib-resistant cell lines. Reactivation of the MAPK pathway may be responsible for BRAFi resistance, allowing cells to bypass their dependency on BRAF (Kim & Kim, 2025).

13.6.3 Multi-Omics in Metagenomics

Combining data from multiple omics approaches helps us better understand how the microbiome works. For example, when metagenomics (which tells us which genes are present) is combined with metatranscriptomics (which shows which genes are being used), scientists can calculate the ratio of transcripts to genes. This tells whether a gene is actively working or not. Metaproteomics when used with metagenomics, helps identify proteins using known gene sequences or to measure how much of each protein microbes produce. Metabolomics integrated with metagenomics helps find links between certain metabolites and the microbes that produce them. It can also show which specific groups of microbes make which metabolites. A study on the gut microbiome in people with type 1 diabetes (T1D) used a combination of metagenomics, metatranscriptomics, and metaproteomics (Heintz-Buschart et al., 2016; Kumar et al., 2024). Researchers found that some microbes changed their gene activity without changing in number. They also found human proteins in stool samples that matched changes in microbial activity. This study shows how combining different omics methods can give a clearer picture of how microbes interact with the human body (Zhang et al., 2019; Nair et al., 2024; Prajapati et al., 2024; James et al., 2024; Mishra et al., 2024).

13.7 CHALLENGES OF MULTI-OMICS DATA INTEGRATION

13.7.1 High-Dimensional Data Problem

In multi-omics studies, there are many features like genes or proteins in the samples. This makes it difficult for models to find useful patterns and often causes overfitting (when a model works well on training data but fails on new data). To solve this, researchers use feature selection (keeping only important variables) and feature extraction (reducing many features into a smaller set). Dimension reduction can be done by involving two main techniques: feature selection and feature extraction, Feature selection focuses on identifying the most important variables using methods like filters (CFS, mRMR, ReliefF), wrappers (SVM-RFE, Boruta), and embedded models (LASSO, Elastic Net), to simplify models and prevent overfitting and feature extraction, reducing many features into a smaller set using linear methods like PCA, MCIA, joint NMF, MOFA, and nonlinear techniques like t-SNE and autoencoders, preserving key patterns while reducing noise.

13.7.2 Data Heterogeneity & Missingness

In multi-omics integration, data often come from different platforms, formats and scales, which makes combining them difficult. Most tools need clean tables with features like genes and samples, but missing or incomplete data can reduce accuracy and give wrong results. To solve this, the data can be cleaned first using steps such as filtering, normalization, quality checks and filling missing values. Machine learning models such as linear models (LASSO, Elastic Net, TANDEM) are simple and easy to understand, while tree-based models (decision Trees, Random Forests can detect complex source patterns). Multi-task and multi–kernel learning combine data from different sources, and graph-based methods (SNF, PARADIGM, HeretoMed) or latent models (iClusterPlus, iClusterBayes) help find hidden links and subtypes in the data.

13.7.3 Class Imbalance

In many omics studies, data are often uneven—some groups (like rare mutations or specific binding sites) have very few samples compared to others. This imbalance can make machine learning models focus more on the larger group and give poor results for the smaller one. To solve this, researchers use methods like oversampling (e.g., SMOTE), undersampling, cost-sensitive learning, changing decision limits, and stratified cross-validation. It's also important to check model performance with measures that suit imbalanced data, such as Precision-Recall AUC, F1-score, and AUROC.

13.7.4 Scalability

In multi-omics studies, data with thousands or millions of features, when combined, become very large and complex. This is known as the scalability problem, and it can be solved by dimensionality reduction. Models can also be trained in small batches instead of all data at once and using parallel and distributed computing.

13.8 CONCLUSION

Multi-omics integration has become an important tool in biology and medicine. It combines data from different layers, such as proteomics, transcriptomics, metagenomics, and metabolomics and helps to better understand different diseases compared to single omics data analysis. Multi-omics can help to identify disease subtypes, discover biomarkers and improve treatments. When metagenomics is combined with multi-omics data, it helps to better understand the microbial communities, their behavior, interactions and environmental responses. This integration helps identify disease subtypes, discover reliable biomarkers, and guide the development of personalized therapeutic strategies. Metagenomics integration with multi-omics analyses enhances our understanding of microbial diversity, ecological interactions, and host–microbiome relationships that influence metabolism, immunity, and aging. Linking microbial community functions with host molecular pathways allows researchers to uncover causal mechanisms behind disease onset and progression. Such comprehensive analyses are paving the way for precision medicine, where diagnostic, prognostic, and therapeutic decisions are guided by system-level biological data. Continued advances in data integration, computational modeling, and machine learning will further enhance the interpretability and translational potential of multi-omics metagenomic research in the coming years.

REFERENCES

An, Y., Feng, Q., Jia, L., Sha, X., Zhang, T., Lu, L., Wang, R., & Bai, B. (2025). Present progress in biomarker discovery of endometrial cancer by multi-omics approaches. *Clinical Proteomics*, *22*(1), 15. https://doi.org/10.1186/s12014-025-09528-6

Argelaguet, R., Arnol, D., Bredikhin, D., Deloro, Y., Velten, B., Marioni, J. C., & Stegle, O. (2020). MOFA+: A statistical framework for comprehensive integration of multi-modal single-cell data. *Genome Biology*, *21*(1), 111. https://doi.org/10.1186/s13059-020-02015-1

Argelaguet, R., Velten, B., Arnol, D., Dietrich, S., Zenz, T., Marioni, J. C., Buettner, F., Huber, W., & Stegle, O. (2018). Multi-omics factor analysis—a framework for unsupervised integration of multi-omics data sets. *Molecular Systems Biology*, *14*(6), e8124. https://doi.org/10.15252/msb.20178124

Athieniti, E., & Spyrou, G. M. (2023). A guide to multi-omics data collection and integration for translational medicine. *Computational and Structural Biotechnology Journal*, *21*, 134–149. https://doi.org/10.1016/j.csbj.2022.11.050

Baião, A. R., Cai, Z., Poulos, R. C., Robinson, P. J., Reddel, R. R., Zhong, Q., Vinga, S., & Gonçalves, E. (2025). A technical review of multi-omics data integration methods: From classical statistical to deep generative approaches. *Briefings in Bioinformatics*, *26*(4), bbaf355. https://doi.org/10.1093/bib/bbaf355

Boekel, J., Chilton, J. M., Cooke, I. R., Horvatovich, P. L., Jagtap, P. D., Käll, L., Lehtiö, J., Lukasse, P., Moerland, P. D., & Griffin, T. J. (2015). Multi-omic data analysis using Galaxy. *Nature Biotechnology*, *33*(2), 137–139. https://doi.org/10.1038/nbt.3134

Boroń, D., Zmarzły, N., Wierzbik-Strońska, M., Rosińczuk, J., Mieszczański, P., & Grabarek, B. O. (2022). Recent multiomics approaches in endometrial cancer. *International Journal of Molecular Sciences*, *23*(3), 1237. https://doi.org/10.3390/ijms23031237

Cai, Z., Poulos, R. C., Liu, J., & Zhong, Q. (2022). Machine learning for multi-omics data integration in cancer. *iScience*, *25*(2), 103798. https://doi.org/10.1016/j.isci.2022.103798

Cantini, L., Zakeri, P., Hernandez, C., Naldi, A., Thieffry, D., Remy, E., & Baudot, A. (2021). Benchmarking joint multi-omics dimensionality reduction approaches for the study of cancer. *Nature Communications*, *12*(1), 124. https://doi.org/10.1038/s41467-020-20430-7

Chalise, P., Kwon, D., Fridley, B. L., & Mo, Q. (2023). Statistical methods for integrative clustering of multi-omics data. *Methods in Molecular Biology (Clifton, N.J.)*, *2629*, 73–93. https://doi.org/10.1007/978-1-0716-2986-4_5

Chierici, M., Bussola, N., Marcolini, A., Francescatto, M., Zandonà, A., Trastulla, L., Agostinelli, C., Jurman, G., & Furlanello, C. (2020). Integrative network fusion: A multi-omics approach in molecular profiling. *Frontiers in Oncology*, *10*, 1065. https://doi.org/10.3389/fonc.2020.01065

Clough, E., & Barrett, T. (2016). The gene expression omnibus database. *Methods in Molecular Biology (Clifton, N.J.)*, *1418*, 93–110. https://doi.org/10.1007/978-1-4939-3578-9_5

Dass, G., Vu, M.-T., Xu, P., Audain, E., Hitz, M.-P., Grüning, B. A., Hermjakob, H., & Perez-Riverol, Y. (2020). The omics discovery REST interface. *Nucleic Acids Research*, *48*(W1), W380–W384. https://doi.org/10.1093/nar/gkaa326

Ghosh, S., Datta, A., & Choi, H. (2021). MultiSLIDE is a web server for exploring connected elements of biological pathways in multi-omics data. *Nature Communications*, *12*(1), 2279. https://doi.org/10.1038/s41467-021-22650-x

Gligorijevi, V., Malod-Dognin, N., & Pr?ulj, N. (2015). Patient-specific data fusion for cancer stratification and personalised treatment. In *Biocomputing 2016* (pp. 321–332). WORLD SCIENTIFIC. https://doi.org/10.1142/9789814749411_0030

Gout, A. M., Arunachalam, S., Finkelstein, D. B., & Zhang, J. (2021). Data-driven approaches to advance research and clinical care for pediatric cancer. *Biochimica et Biophysica Acta (BBA)—Reviews on Cancer*, *1876*(1), 188571. https://doi.org/10.1016/j.bbcan.2021.188571

Grismer, L. L. (2025). Introducing multiple factor analysis (MFA) as a diagnostic taxonomic tool complementing principal component analysis (PCA). *ZooKeys*, *1248*, 93–109. https://doi.org/10.3897/zookeys.1248.159516

Guido, L. P., & Gomez-Fernandez, C. (2020). Advances in the molecular taxonomy of breast cancer. *Archives of Medical Research*, *51*(8), 777–783. https://doi.org/10.1016/j.arcmed.2020.08.001

Gutierrez Reyes, C. D., Alejo-Jacuinde, G., Perez Sanchez, B., Chavez Reyes, J., Onigbinde, S., Mogut, D., Hernández-Jasso, I., Calderón-Vallejo, D., Quintanar, J. L., & Mechref, Y. (2024). Multi omics applications in biological systems. *Current Issues in Molecular Biology*, *46*(6), 5777–5793. https://doi.org/10.3390/cimb46060345

Heintz-Buschart, A., May, P., Laczny, C. C., Lebrun, L. A., Bellora, C., Krishna, A., Wampach, L., Schneider, J. G., Hogan, A., De Beaufort, C., & Wilmes, P. (2016). Integrated multi-omics of the human gut microbiome in a case study of familial type 1 diabetes. *Nature Microbiology*, *2*(1), 16180. https://doi.org/10.1038/nmicrobiol.2016.180

Heo, Y. J., Hwa, C., Lee, G.-H., Park, J.-M., & An, J.-Y. (2021). Integrative multi-omics approaches in cancer research: From biological networks to clinical subtypes. *Molecules and Cells*, *44*(7), 433–443. https://doi.org/10.14348/molcells.2021.0042

Hernández-de-Diego, R., Tarazona, S., Martínez-Mira, C., Balzano-Nogueira, L., Furió-Tarí, P., Pappas, G. J., & Conesa, A. (2018). PaintOmics 3: A web resource for the pathway analysis and visualization of multi-omics data. *Nucleic Acids Research*, *46*(Web Server issue), W503–W509. https://doi.org/10.1093/nar/gky466

Huang, S., Chaudhary, K., & Garmire, L. X. (2017). More is better: Recent progress in multi-omics data integration methods. *Frontiers in Genetics*, *8*, 84. https://doi.org/10.3389/fgene.2017.00084

International Cancer Genome Consortium, Hudson, T. J., Anderson, W., Artez, A., Barker, A. D., Bell, C., Bernabé, R. R., Bhan, M. K., Calvo, F., Eerola, I., Gerhard, D. S., Guttmacher, A., Guyer, M., Hemsley, F. M., Jennings, J. L., Kerr, D., Klatt, P., Kolar, P., Kusada, J., . . . Yang, H. (2010). International network of cancer genome projects. *Nature*, *464*(7291), 993–998. https://doi.org/10.1038/nature08987

James, A. S., Adil, N. A., Goltz, D., Tangudu, D., Chaudhari, D. S., Shukla, R., Kumar, V., Kumar, A., Masternak, M. M., Holland, P., Labyak, C., Golden, A., Dangiolo, M., Arikawa, A. Y., Kociolek, J., Fraser, A., Williams, C., Agronin, M., Aymat, M., . . . Yadav, H. (2024). Abnormalities in gut virome signatures linked with cognitive impairment in older adults. *Gut Microbes*, *16*(1), 2431648. https://doi.org/10.1080/19490976.2024.2431648

Kim, H. K., & Kim, T. (2025). Integrating multi-omics in endometrial cancer: From molecular insights to clinical applications. *Cells*, *14*(17), 1404. https://doi.org/10.3390/cells14171404

Kumar, V., Pandey, A., Arora, A., Gautam, P., Bisht, D., Gupta, S., Chaurasia, A., & Sachan, M. (2023). Diagnostics and therapeutic potential of miR-205 and miR-34a in ovarian cancer management: A miRNA-target-based analysis. *DNA and Cell Biology*, *42*(3), 151–162. https://doi.org/10.1089/dna.2022.0487

Kumar, V., Sharma, J. K., Kumar, S., Gulia, J., Singh, N. P., & Santal, A. R. (2025). Metagenomic and proteomic approach for bioremediation of environmental pollutants. In M. P. Shah (Ed.), *Development in waste water treatment research and processes*. Elsevier. https://doi.org/10.1016/B978-0-443-13615-3.00020-0

Kumar, V., Shukla, R., Gangani, S., Shekhar, H., & Shukla, V. (2024). Metagenomics and microbiome analysis. In A. Chaudhary, S. K. Sethi, & A. Verma (Eds.), *Unraveling new frontiers and advances in bioinformatics* (pp. 77–92). Springer Nature, Singapore. https://doi.org/10.1007/978-981-97-7123-3_4

Liu, T., Salguero, P., Petek, M., Martinez-Mira, C., Balzano-Nogueira, L., Ramšak, Ž., McIntyre, L., Gruden, K., Tarazona, S., & Conesa, A. (2022). PaintOmics 4: New tools for the integrative analysis of multi-omics datasets supported by multiple pathway databases. *Nucleic Acids Research*, *50*(W1), W551–W559. https://doi.org/10.1093/nar/gkac352

Lock, E. F., Hoadley, K. A., Marron, J. S., & Nobel, A. B. (2013). Joint and individual variation explained (JIVE) for integrated analysis of multiple data types. *The Annals of Applied Statistics*, *7*(1), 523–542. https://doi.org/10.1214/12-AOAS597

Lu, J., Cannizzaro, E., Meier-Abt, F., Scheinost, S., Bruch, P.-M., Giles, H. A., Lütge, A., Hüllein, J., Wagner, L., Giacopelli, B., Nadeu, F., Delgado, J., Campo, E., Mangolini, M., Ringshausen, I., Böttcher, M., Mougiakakos, D., Jacobs, A., Bodenmiller, B., . . . Huber, W. (2021). Multi-omics reveals clinically relevant proliferative drive associated with mTOR-MYC-OXPHOS activity in chronic lymphocytic leukemia. *Nature Cancer*, *2*(8), 853–864. https://doi.org/10.1038/s43018-021-00216-6

Lu, Z., & Lou, W. (2022). Bayesian consensus clustering for multivariate longitudinal data. *Statistics in Medicine*, *41*(1), 108–127. https://doi.org/10.1002/sim.9225

Lv, Y., Huang, S., Zhang, T., & Gao, B. (2021). Application of multilayer network models in bioinformatics. *Frontiers in Genetics*, *12*, 664860. https://doi.org/10.3389/fgene.2021.664860

McGowan, T., Johnson, J. E., Kumar, P., Sajulga, R., Mehta, S., Jagtap, P. D., & Griffin, T. J. (2020). Multi-omics visualization platform: An extensible galaxy plug-in for multi-omics data visualization and exploration. *GigaScience*, *9*(4), giaa025. https://doi.org/10.1093/gigascience/giaa025

Meng, C., Helm, D., Frejno, M., & Kuster, B. (2016). moCluster: Identifying joint patterns across multiple omics data sets. *Journal of Proteome Research*, *15*(3), 755–765. https://doi.org/10.1021/acs.jproteome.5b00824

Mercadié, A., Gravier, É., Josse, G., Fournier, I., Viodé, C., Vialaneix, N., & Brouard, C. (2025). NMFProfiler: A multi-omics integration method for samples stratified in groups. *Bioinformatics*, *41*(2), btaf066. https://doi.org/10.1093/bioinformatics/btaf066

Min, E. J., & Long, Q. (2020). Sparse multiple co-Inertia analysis with application to integrative analysis of multi-omics data. *BMC Bioinformatics*, *21*(1), 141. https://doi.org/10.1186/s12859-020-3455-4

Mishra, S., Buddendorff, L., Hoover, J., Shukla, R., Kumar, V., Jain, S., & Yadav, H. (2024). Microbiome role in cognitive impairment: Miagb cohort insights. *Innovation in Aging*, *8*(Supplement_1), 811–811. https://doi.org/10.1093/geroni/igae098.2629

Mo, Q., & Shen, R. (2023). *iClusterPlus: Integrative clustering of multiple genomic data sets*. Bioconductor package vignette. Bioconductor. Available at: https://bioconductor.org/packages/iClusterPlus

Mo, Q., Shen, R., Guo, C., Vannucci, M., Chan, K. S., & Hilsenbeck, S. G. (2018). A fully Bayesian latent variable model for integrative clustering analysis of multi-type omics data. *Biostatistics*, *19*(1), 71–86. https://doi.org/10.1093/biostatistics/kxx017

Moon, S., & Lee, H. (2021). JDSNMF: Joint deep semi-non-negative matrix factorization for learning integrative representation of molecular signals in alzheimer's disease. *Journal of Personalized Medicine*, *11*(8), 686. https://doi.org/10.3390/jpm11080686

Nair, S. U., Lin, Y., Jain, S., Chaudhari, D. S., Shukla, R., Kumar, V., Mankowski, R. T., Meng, H., Buford, T. W., Kuchel, G. A., Yadav, H., & Consortium, M. (2024). Unique gut microbiome signatures associated with physical and cognitive function. *Alzheimer's & Dementia*, *20*(S2), e092044. https://doi.org/10.1002/alz.092044

Nguyen, H., Shrestha, S., Draghici, S., & Nguyen, T. (2019). PINSPlus: A tool for tumor subtype discovery in integrated genomic data. *Bioinformatics*, *35*(16), 2843–2846. https://doi.org/10.1093/bioinformatics/bty1049

Nusinow, D. P., Szpyt, J., Ghandi, M., Rose, C. M., McDonald, E. R., Kalocsay, M., Jané-Valbuena, J., Gelfand, E., Schweppe, D. K., Jedrychowski, M., Golji, J., Porter, D. A., Rejtar, T., Wang, Y. K., Kryukov, G. V., Stegmeier, F., Erickson, B. K., Garraway, L. A., Sellers, W. R., & Gygi, S. P. (2020). Quantitative proteomics of the cancer cell line encyclopedia. *Cell*, *180*(2), 387–402.e16. https://doi.org/10.1016/j.cell.2019.12.023

Otto, E., Culakova, E., Meng, S., Zhang, Z., Xu, H., Mohile, S., & Flannery, M. A. (2022). Overview of Sankey flow diagrams: Focusing on symptom trajectories in older adults with advanced cancer. *Journal of Geriatric Oncology*, *13*(5), 742–746. https://doi.org/10.1016/j.jgo.2021.12.017

Pavlopoulos, G. A., Kontou, P. I., Pavlopoulou, A., Bouyioukos, C., Markou, E., & Bagos, P. G. (2018). Bipartite graphs in systems biology and medicine: A survey of methods and applications. *GigaScience*, *7*(4), giy014. https://doi.org/10.1093/gigascience/giy014

Picard, M., Scott-Boyer, M.-P., Bodein, A., Périn, O., & Droit, A. (2021). Integration strategies of multi-omics data for machine learning analysis. *Computational and Structural Biotechnology Journal*, *19*, 3735–3746. https://doi.org/10.1016/j.csbj.2021.06.030

Prajapati, S. K., Kumar, V., Shukla, R., Szekeres, C., Ma, Y., Jain, S., & Yadav, H. (2024). Microbiome-immune interplay in aging brains: A pilot investigation from the MiaGB cohort. *Alzheimer's & Dementia*, *20*(S2), e090734. https://doi.org/10.1002/alz.090734

Pulkstenis, E., Patra, K., & Zhang, J. (2017). A Bayesian paradigm for decision-making in proof-of-concept trials. *Journal of Biopharmaceutical Statistics*, *27*(3), 442–456. https://doi.org/10.1080/10543406.2017.1289947

Rappoport, N., & Shamir, R. (2019). NEMO: Cancer subtyping by integration of partial multi-omic data. *Bioinformatics*, *35*(18), 3348–3356. https://doi.org/10.1093/bioinformatics/btz058

Rodchenkov, I., Babur, O., Luna, A., Aksoy, B. A., Wong, J. V., Fong, D., Franz, M., Siper, M. C., Cheung, M., Wrana, M., Mistry, H., Mosier, L., Dlin, J., Wen, Q., O'Callaghan, C., Li, W., Elder, G., Smith, P. T., Dallago, C., . . . Sander, C. (2020). Pathway commons 2019 update: Integration, analysis and exploration of pathway data. *Nucleic Acids Research*, *48*(D1), D489–D497. https://doi.org/10.1093/nar/gkz946

Röder, B., Kersten, N., Herr, M., Speicher, N. K., & Pfeifer, N. (2019). Web-rMKL: A web server for dimensionality reduction and sample clustering of multi-view data based on unsupervised multiple kernel learning. *Nucleic Acids Research*, *47*(W1), W605–W609. https://doi.org/10.1093/nar/gkz422

Rohart, F., Gautier, B., Singh, A., & Lê Cao, K.-A. (2017). MixOmics: An R package for 'omics feature selection and multiple data integration. *PLoS Computational Biology*, *13*(11), e1005752. https://doi.org/10.1371/journal.pcbi.1005752

Saliba, A., Du, Y., Feng, T., & Garmire, L. (2024). Multi-omics integration in nephrology: Advances, challenges, and future directions. *Seminars in Nephrology*, *44*(6), 151584. https://doi.org/10.1016/j.semnephrol.2025.151584

Shannon, P., Markiel, A., Ozier, O., Baliga, N. S., Wang, J. T., Ramage, D., Amin, N., Schwikowski, B., & Ideker, T. (2003). Cytoscape: A software environment for integrated models of biomolecular interaction networks. *Genome Research*, *13*(11), 2498–2504. https://doi.org/10.1101/gr.1239303

Shi, Q., Zhang, C., Peng, M., Yu, X., Zeng, T., Liu, J., & Chen, L. (2017). Pattern fusion analysis by adaptive alignment of multiple heterogeneous omics data. *Bioinformatics (Oxford, England)*, *33*(17), 2706–2714. https://doi.org/10.1093/bioinformatics/btx176

Shukla, S., Singh, B., & Mani, A. (2024). Machine learning and artificial intelligence in bioinformatics. In A. Chaudhary, S. K. Sethi, & A. Verma (Eds.), *Unraveling new frontiers and advances in bioinformatics* (pp. 133–142). Springer Nature, Singapore. https://doi.org/10.1007/978-981-97-7123-3_7

Singh, A., Shannon, C. P., Gautier, B., Rohart, F., Vacher, M., Tebbutt, S. J., & Lê Cao, K.-A. (2019). DIABLO: An integrative approach for identifying key molecular drivers from multi-omics assays. *Bioinformatics (Oxford, England)*, *35*(17), 3055–3062. https://doi.org/10.1093/bioinformatics/bty1054

Spooner, A., Moridani, M. K., Toplis, B., Behary, J., Safarchi, A., Maher, S., Vafaee, F., Zekry, A., & Sowmya, A. (2025). Benchmarking ensemble machine learning algorithms for multi-class, multi-omics data integration in clinical outcome prediction. *Briefings in Bioinformatics*, *26*(2), bbaf116. https://doi.org/10.1093/bib/bbaf116

Surujon, D., & van Opijnen, T. (2020). ShinyOmics: Collaborative exploration of omics-data. *BMC Bioinformatics*, *21*(1), 22. https://doi.org/10.1186/s12859-020-3360-x

Tang, T. M., & Allen, G. I. (2021). Integrated principal components analysis. *J Mach Learn Res*, *22*(1), 198:8953–198:9023.

Tomczak, K., Czerwińska, P., & Wiznerowicz, M. (2015). The cancer genome atlas (TCGA): An immeasurable source of knowledge. *Contemporary Oncology (Poznan, Poland)*, *19*(1A), A68–77. https://doi.org/10.5114/wo.2014.47136

Tsagiopoulou, M., & Gut, I. G. (2024). Machine learning and multi-omics data in chronic lymphocytic leukemia: The future of precision medicine? *Frontiers in Genetics*, *14*, 1304661. https://doi.org/10.3389/fgene.2023.1304661

Tsagiopoulou, M., Pechlivanis, N., Maniou, M. C., & Psomopoulos, F. (2022). InterTADs: Integration of multi-omics data on topologically associated domains, application to chronic lymphocytic leukemia. *NAR Genomics and Bioinformatics*, *4*(1), lqab121. https://doi.org/10.1093/nargab/lqab121

Wang, N., Li, Y., Han, S., Zhang, Y., Yang, J., Yin, Z., Deng, C., Liu, Z., Wu, Y., Wu, W., Xue, W., Yang, T., Xu, Y., Shen, Q., Jiang, G., & Wei, Z. (2024). CFViSA: A comprehensive and free platform for visualization and statistics in omics-data. *Computers in Biology and Medicine*, *171*, 108206. https://doi.org/10.1016/j.compbiomed.2024.108206

Wang, W., Baladandayuthapani, V., Morris, J. S., Broom, B. M., Manyam, G., & Do, K.-A. (2013). iBAG: Integrative Bayesian analysis of high-dimensional multiplatform genomics data. *Bioinformatics*, *29*(2), 149–159. https://doi.org/10.1093/bioinformatics/bts655

Wilkinson, L., & Friendly, M. (2009). The history of the cluster heat map. *The American Statistician*, *63*(2), 179–184. https://doi.org/10.1198/tas.2009.0033

Wu, D., Wang, D., Zhang, M. Q., & Gu, J. (2015). Fast dimension reduction and integrative clustering of multi-omics data using low-rank approximation: Application to cancer molecular classification. *BMC Genomics*, *16*(1), 1022. https://doi.org/10.1186/s12864-015-2223-8

Yadav, B. S., Yadav, A. K., Singh, S., Singh, N. K., & Mani, A. (2019). Methods in metagenomics and environmental biotechnology. In K. M. Gothandam, S. Ranjan, N. Dasgupta, & E. Lichtfouse (Eds.), *Nanoscience and biotechnology for environmental applications* (Vol. 22, pp. 85–113). Springer International Publishing. https://doi.org/10.1007/978-3-319-97922-9_4

Yang, J., Liu, Y., Shang, J., Chen, Q., Chen, Q., Ren, L., Zhang, N., Yu, Y., Li, Z., Song, Y., Yang, S., Scherer, A., Tong, W., Hong, H., Xiao, W., Shi, L., & Zheng, Y. (2023). The quartet data portal: Integration of community-wide resources for multiomics quality control. *Genome Biology*, *24*(1), 245. https://doi.org/10.1186/s13059-023-03091-9

Yang, M., Matan-Lithwick, S., Wang, Y., De Jager, P. L., Bennett, D. A., & Felsky, D. (2023). Multi-omic integration via similarity network fusion to detect molecular subtypes of ageing. *Brain Communications*, *5*(2), fcad110. https://doi.org/10.1093/braincomms/fcad110

Yang, Z., & Michailidis, G. (2016). A non-negative matrix factorization method for detecting modules in heterogeneous omics multi-modal data. *Bioinformatics*, *32*(1), 1–8. https://doi.org/10.1093/bioinformatics/btv544

Yen, P.-T., Chien, T.-W., Chou, W., & Tsai, K.-T. (2023). Using the Alluvial diagram to display variable characteristics for COVID-19 patients and research achievements on the topic of COVID-19, epidemiology, pathogenesis, and vaccine (CEPV): Bibliometric analysis. *Medicine*, *102*(25), e33873. https://doi.org/10.1097/MD.0000000000033873

Zhang, W., Mou, M., Hu, W., Lu, M., Zhang, H., Zhang, H., Luo, Y., Xu, H., Tao, L., Dai, H., Gao, J., & Zhu, F. (2024). MOINER: A novel multiomics early integration framework for biomedical classification and biomarker discovery. *Journal of Chemical Information and Modeling*, *64*(7), 2720–2732. https://doi.org/10.1021/acs.jcim.4c00013

Zhang, X., Li, L., Butcher, J., Stintzi, A., & Figeys, D. (2019). Advancing functional and translational microbiome research using meta-omics approaches. *Microbiome*, *7*, 154. https://doi.org/10.1186/s40168-019-0767-6

Zheng, Y., Liu, Y., Yang, J., Dong, L., Zhang, R., Tian, S., Yu, Y., Ren, L., Hou, W., Zhu, F., Mai, Y., Han, J., Zhang, L., Jiang, H., Lin, L., Lou, J., Li, R., Lin, J., Liu, H., . . . Shi, L. (2024). Multi-omics data integration using ratio-based quantitative profiling with Quartet reference materials. *Nature Biotechnology*, *42*(7), 1133–1149. https://doi.org/10.1038/s41587-023-01934-1

Zhou, G., Ewald, J., & Xia, J. (2021). OmicsAnalyst: A comprehensive web-based platform for visual analytics of multi-omics data. *Nucleic Acids Research*, *49*(W1), W476–W482. https://doi.org/10.1093/nar/gkab394

Zhou, G., Pang, Z., Lu, Y., Ewald, J., & Xia, J. (2022). OmicsNet 2.0: A web-based platform for multi-omics integration and network visual analytics. *Nucleic Acids Research*, *50*(W1), W527–W533. https://doi.org/10.1093/nar/gkac376

Zoppi, J., Guillaume, J.-F., Neunlist, M., & Chaffron, S. (2021). MiBiOmics: An interactive web application for multi-omics data exploration and integration. *BMC Bioinformatics*, *22*(1), 6. https://doi.org/10.1186/s12859-020-03921-8

14 Challenges, Ethics, and Policy in Metagenomics

Shreyasi Majumdar[1], *Puneet K. Samaiya*[2], *and Santosh Kumar Prajapati*[3]*
[1]Department of Pharmaceutical Technology, School of Health and Medical Sciences, Adamas University, Kolkata, India
[2]Department of Pharmacy, Shri G. S. Institute of Technology and Science, Indore, India
[3]Center of Excellence in Aging and Brain Repair, Department of Neurosurgery, Brain and Spine, University of South Florida, Morsani College of Medicine, Tampa, Florida, USA
*Corresponding author

14.1 INTRODUCTION

Scientists now study microbial communities in natural and clinical and industrial environments through the new field of metagenomics (Schloss & Handelsman, 2005). The method bypasses traditional microbiology limitations because it enables direct sequencing of environmental and host-derived genetic material without requiring laboratory cultivation of more than 99% of microbial species (Schloss & Handelsman, 2005; Kumar, Shukla, et al., 2024). The new analytical techniques have revealed that microbial diversity exceeds initial expectations while demonstrating their ability to perform multiple functions. This approach has transformed our understanding of ecosystems and disease mechanisms and biotechnological applications.

The study of gut microbiome through metagenomics has provided essential knowledge about its role in the development of inflammatory bowel disease and colorectal cancer and metabolic disorders (Lepage et al., 2013; Ma et al., 2021; Xu et al., 2022; Prajapati et al., 2024; Nair et al., 2024; James et al., 2024; Sarfraz et al., 2024). Soil metagenomics research enables scientists to discover microbial functions that enhance nutrient cycling and plant development and disease resistance which leads to environmentally friendly agricultural practices (Jagadesh et al., 2024; Nwachukwu & Babalola, 2022; Kumar et al., 2025). Research conducted through metagenomics in marine and terrestrial environments and harsh ecosystems has expanded the tree of life and demonstrated climate change impacts on microbial communities and guided international conservation strategies (Cavicchioli et al., 2019; Jansson & Hofmockel, 2020).

The field has made progress, but scientists and engineers need to solve multiple significant technical and scientific challenges. The vast quantity of sequencing data creates substantial difficulties for data storage and analysis and results reproduction. The use of different DNA extraction methods and sequencing equipment and bioinformatics tools generate systematic errors which reduce the ability to compare research results (Forry et al., 2024; Poussin et al., 2018; Kumar, Gangani, et al., 2024). The conversion of metagenomic data into practical medical or agricultural applications remains challenging because these results show inconsistent performance across different human subjects and geographic locations. The ethical and policies of metagenomic research present equal importance to scientific challenges. The storage of human microbiome

DOI: 10.1201/9781003670155-14

databases with host information creates privacy risks because it enables potential host identification (Chuong et al., 2017; Dupras & Bunnik, 2021). The collection of justice. The Nagoya Protocol's genetic resource sovereignty framework faces similar challenges as researchers debate the proper management samples from indigenous communities and resource-poor areas without proper benefit sharing creates ethical dilemmas about fairness and of metagenomic data (Buck & Hamilton, 2011; Muller, 2015). The dual-use nature of metagenomic data creates risks because it enables unauthorized pathogen reconstruction and biotechnological exploitation (Berger, 2021; Wang et al., 2025).

Therefore, the goal of this chapter is to give a short but critical summary of these connected problems. In particular, it talks about: (1) the biggest scientific and technological problems in metagenomics, (2) the moral problems with privacy, fairness, and responsible usage, and (3) the policy frameworks that need to be in place to make sure that everything is clear, reproducible, and fair. The chapter stresses that the future of metagenomics rests not only on new technologies but also on the creation of strong ethical and governance frameworks that make sure that everyone in society may benefit from it in a fair and responsible way (See Graphical abstract).

14.2 SCIENTIFIC AND TECHNICAL CHALLENGES

Metagenomics is revolutionary, yet many of its most exciting discoveries are still restricted by scientific and technical challenges. These problems are generally connected: huge sequencing datasets need advanced computer systems; sampling and extraction biases affect how results are interpreted; and translational barriers stop discoveries from being used in medicine, agriculture, or ecology. This part looks at three main types of problems: the amount and complexity of the data, how to understand it and avoid bias, and how to put it into practice.

14.2.1 Data Volume and Complexity

14.2.1.1 Explosion of Sequencing Data

High-throughput sequencing technology advancements during the past decade have produced terabases of sequencing data for each research study. The Human Microbiome Project (HMP) and Earth Microbiome Project (EMP) demonstrated broad microbial diversity through their large-scale initiatives which produced massive datasets that scientists use in their current research (Gilbert et al., 2014; McDonald et al., 2015). The massive scale of these projects exposed a fundamental problem which involves handling and processing enormous amounts of data.

The size of shotgun metagenomic datasets ranges from 5 to 20 gigabases per sample while research studies with hundreds or thousands of samples reach petabyte levels (Quince et al., 2017). The analysis of clinical microbiome consortia through disease-related microbial shift studies generates datasets that amount to terabytes in size (Aggarwal et al., 2022; Gupta et al., 2025). The storage of these datasets requires both adequate storage solutions and efficient scalable computational systems.

14.2.1.2 Storage and Computational Infrastructure

The storage requirements for raw reads form only a small part of the complete challenge. The total storage requirements for preprocessing and assembly and annotation and statistical analysis operations grow between 2 to 5 times larger than the original dataset size. The assembly and binning of complex microbial communities require advanced computing systems to function properly. The construction of large soil metagenomes using MEGAHIT or metaSPAdes software requires more than 250 GB of RAM and multiple CPU cores according to (Li et al., 2016; Navaux et al., 2023). The lack of institutional cluster access creates difficulties for laboratories with limited resources.

The three major cloud platforms AWS and Google Cloud and Microsoft Azure help researchers solve their computational problems (Mathur, 2024). The ability to scale up operations exists in these platforms but users must pay high costs when working with large datasets. The process of moving terabytes of data between cloud servers faces major delays because most institutions lack sufficient bandwidth for real-time operations.

14.2.1.3 Reproducibility and Versioning

The process of achieving study reproducibility becomes challenging when researchers handle large datasets. The same dataset produces different results when different software tool versions are used because these tools receive continuous updates. The same HMP data produced different taxonomic patterns when analyzed with modern classifiers compared to previous studies (Bolyen et al., 2019; Lozupone et al., 2013). The growth of reference genomes leads to better classification accuracy but makes it impossible to perform direct comparisons between different time periods. The combination of Nextflow and Snakemake workflow management tools with Docker and Singularity containerization technologies provides solutions for creating reproducible pipelines and standardized environments according to research (Di Tommaso et al., 2017). The adoption of these solutions remains inconsistent because smaller laboratories with limited computational expertise show low acceptance rates.

14.2.1.4 Big Data Integration

The combination of multi-omics data from metabolomics and metaproteomics and metatranscriptomics produces datasets that surpass the complexity of individual metagenomic datasets. The analysis of multi-modal data requires new computational methods which include advanced machine learning models that can identify relationships between different data types (Bileschi et al., 2022). The full potential of microbiome research remains limited because researchers lack these analytical methods.

14.2.2 Interpretation and Bias

14.2.2.1 Unequal Representation of Microbial Diversity

The growing use of sequencing methods has created an imbalance in database representation of microbial diversity. The majority of collected microbiome samples originate from human gut microbiomes but deep subterranean ecosystems and seas and deserts and polar regions remain poorly represented. The worldwide understanding of microbial diversity and function becomes skewed because of this biased representation (Haryono et al., 2022). The Global Ocean Sampling mission expanded our knowledge of microbial diversity but similar extensive research efforts have not been conducted on terrestrial ecosystems. The existing gap between different ecosystems prevents us from making general ecological or evolutionary conclusions.

14.2.2.2 Biases in Sampling and DNA Extraction

The extraction and sampling process in microbiome research creates methodological biases which affect community profile accuracy and make biological interpretation more complex. The diversity of microbial populations across different body locations and tissue depths and environmental habitats results in sampling bias. The selection of limited or specific sampling methods leads to missing essential microbial populations which results in incomplete and unreliable research outcomes. The patterns of diversity and abundance in the gut and oral cavity undergo substantial changes because of geographical differences between these areas (Parfrey & Knight, 2012; Pollock et al., 2018). The correct characterization of microbial communities requires sampling methods that understand local site differences. Extraction bias represents a major confounding factor in microbiome research. The combination of Gram-positive bacteria and spore-formers and fungus remains underrepresented during extraction because their cell wall structures and lytic resistance levels exceed the effectiveness of current enzymatic (lysozyme, mutanolysin) and mechanical (bead-beating) extraction methods. The extraction process fails to detect certain species properly which leads to biased community composition results. The combination of bead-beating with enzymatic lysis produces the highest DNA yield and microbial diversity according to studies which demonstrated that DNA extraction methods significantly affect results (Costea et al., 2017; Karst et al., 2021; Sinha et al., 2017).

The four biological samples of blood, placenta, cerebrospinal fluid and air represent low biomass specimens which become easily contaminated through laboratory procedures and chemical substances. The contaminants present in sequencing results can be mistaken for actual microbial signals because they dominate the results (Eisenhofer et al., 2019; Salter et al., 2014). The discovery of placental microbiota linked to pollution contamination represents one of the many disputes that arose from these artifacts (De Goffau et al., 2019). The solution to this problem requires scientists to use bioinformatic filtering pipelines together with proper reagent decontamination and negative control implementation (Weiss et al., 2019). The microbiome literature shows that methodological errors between studies create contradictory results which hinders both study reproducibility and makes it hard to compare findings between different studies. The interpretation of results depends on standardized procedures and strict technological controls and open reporting methods to minimize artifacts.

14.2.2.3 Database Incompleteness and Annotation Errors

Greengenes and NCBI RefSeq: The majority of microbial species remain unsequenced and uncultured which makes metagenomic readings often difficult to interpret. Research indicates that environmental samples contain interpretation of taxonomic and functional data heavily relies on reference databases which include UniProt, SILVA, 30–50% of unidentified sequences which scientists have named the "dark matter" of microbiome studies (Nissimov, 2017). The process of annotation errors creates additional uncertainty in the analysis. The spread of incorrect genome assignments through reference databases leads to flawed ecological and functional interpretations of the data (Gao et al., 2020).

14.2.2.4 Impact on Downstream Analyses

The way scientists interpret data leads to major consequences. Scientists who study microbiomes between sick and healthy groups may mistake procedural differences in DNA extraction and sequencing technology for disease-related changes. Similarly, if databases have duplicate or incorrectly annotated information, functional conclusions about metabolic pathways or antibiotic resistance may be exaggerated. Standardized validation is necessary since benchmarking studies reveal that even popular classifiers like Kraken2 and MetaPhlAn provide different results under different circumstances (Edwin et al., 2024; Kobras et al., 2021).

14.2.3 Translation to Practice

14.2.3.1 From Discovery to Application

Metagenomics has yielded thousands of correlations between microbial genes or taxa and environmental changes, agricultural output, or human health effects. A major challenge still exists, nevertheless, in converting these correlations into useful biomarkers or treatments. Microbiome correlations are usually context-dependent and heterogeneous, in contrast to human genomes, where certain polymorphisms are frequently associated with distinct phenotypic outcomes. In a particular study, for example, microbial biomarkers associated with colorectal cancer frequently do not replicate in different cohorts or geographic groups (Avuthu & Guda, 2022; Dai et al., 2018). Similarly, crop yield predictions derived from soil microbes could not be applicable in other climates or with various farming methods.

14.2.3.2 Limited Validation across Populations

The lack of cross-population validation makes it difficult to translate microbiome research into clinical diagnosis. The makeup of the microbiome is extremely sensitive to location, environment, lifestyle, and nutrition, leading to context-specific indicators that frequently don't generalize (Gupta et al., 2017; Yatsunenko et al., 2012). **Small sample sizes** in many studies further reduce statistical power and increase the risk of false discoveries (Sinha et al., 2017). Further complicating replication is the fact that different sequencing techniques and bioinformatic pipelines provide contradictory findings among investigations (Knight et al., 2018; Pollock et al., 2018). The regulatory standards for repeatability and validation across separate cohorts are thus not met by many microbiome-based signatures. Current microbiome discoveries seldom ever meet the standardized, cross-validated biomarker requirements that agencies like the FDA stress are necessary before authorizing therapeutic applications (Integrative, 2014).

14.2.3.3 Functional Uncertainty

Functional interpretation presents another difficulty. Although metagenomics may identify a community's genetic potential, it cannot identify the genes that are being expressed. Therefore, without substantiating transcriptomic, proteomic, or metabolomic data, many anticipated activities remain conjectural (Queiroz et al., 2019). This makes it more difficult to relate genetic information to phenotype or the results of interventions.

14.2.3.3.1 Regulatory and Clinical Integration

The implementation of metagenomic signatures into clinical and environmental and agricultural decision systems requires both strong regulatory systems and scientific evidence validation. The clinical use of genetic data faces ongoing challenges regarding patient privacy protection and data security and ethical genetic data utilization (Arellano et al., 2018; Manolio et al., 2013). The implementation of agricultural microbial genes that indicate antibiotic resistance poses significant risks to biosecurity and biosafety which demands strict regulatory oversight (Fao, 2019). The Nagoya Protocol serves as an international framework which establishes rules for microbial genetic resource ownership and access and benefit distribution in environmental management (Kamau et al., 2010; Morgera et al., 2014). The lack of well-defined policies for metagenomic discovery translation will result in delayed or uncoordinated or unfair outcomes which will damage both public trust and scientific integrity.

14.3 ETHICAL CONSIDERATIONS

The field of metagenomics now faces unavoidable ethical challenges because its scientific progress has reached a point where it directly impacts agricultural practices and health care and environmental conservation. In contrast to conventional microbiology, metagenomics generates extensive datasets that frequently include information related to humans, come from a variety of ecological and cultural contexts, and may be used in ways that go beyond the original researchers' goals. Therefore, there are three primary areas of ethical dilemmas in metagenomics: privacy of the human microbiome, benefit sharing and equity, and dual-use issues (see Figure 14.1).

14.3.1 Human Microbiome Privacy

The human microbiome is becoming more widely acknowledged as a biometric identifier with implications for data security and privacy. Individual microbiome profiles can be utilized to re-identify individuals or infer sensitive health-related features, despite the fact that microbiome data are technically different *3.1.1. Microbiome as a Personal Identifier* from host genetic data, as multiple studies have shown.

For example, strain-level heterogeneity in the gut microbiome serves as a stable "fingerprint," allowing people to be re-identified across samples taken months apart, as demonstrated by Franzosa

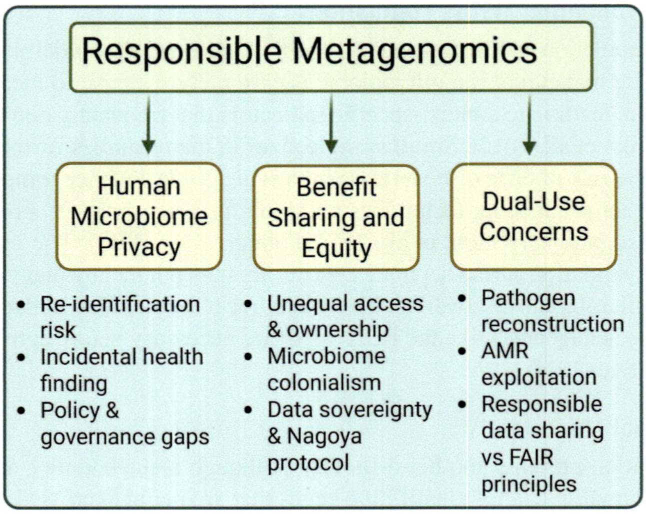

FIGURE 14.1 Schematic representation of the three core ethical domains in metagenomics-human microbiome privacy, benefit sharing and equity, and dual-use concerns-highlighting their key issues and their central role in guiding responsible metagenomics research.

et al. (2015). According to Oh et al. (2016), there are enough strain-specific characteristics in the skin and oral microbiomes to allow for individual identification within a cohort (Oh et al., 2016). This ability to re-identify calls into question the widely held belief that microbiome data are automatically anonymized. To ensure responsible data sharing and to preserve public confidence in microbiome research, microbiome datasets now require parallel protections, much like genomic datasets did when it became evident that re-identification was possible from seemingly de-identified SNP arrays (Gymrek et al., 2013).

14.3.1.1 Incidental Findings and Health Inference

Human-associated metagenomic datasets may unintentionally disclose private health information in addition to identifying problems. Gut microbial profiles and diseases including obesity, type 2 diabetes, inflammatory bowel disease, and colorectal cancer have been linked to more and more strong correlations (Fan & Pedersen, 2021; Wirbel et al., 2019). For instance, Thomas et al. (2019) showed that microbial indicators may accurately stratify both healthy controls and those with colorectal cancer (Thomas et al., 2019). Similar to worries already present in genetic medicine, such accidental results increase the possibility of stigmatization, discrimination, or insurance prejudice if communicated or abused (Presidential Commission for the Study of Bioethical Issues, 2012). The microbiome contains encoded social information because it affects how people eat and live and where they reside and their economic status (Gupta et al., 2017; Yatsunenko et al., 2012). The incorrect interpretation of these patterns creates risks for both prejudice development and collective harm. The presentation of microbial differences as natural deficiencies instead of structural and environmental inequalities leads to unfair discrimination against disadvantaged groups (Rhodes et al., 2013). The combination of these problems requires data governance systems and ethical frameworks which protect both social implications of microbiome research and personal privacy rights.

14.3.1.2 Policy and Governance Gaps

The current privacy laws contain major regulatory deficiencies because they were designed for genetic and clinical data yet fail to protect microbiome information according to Nebeker and Harlow et al. (Nebeker et al., 2017). As ethical frameworks start to take shape, there is continuous discussion on whether microbiome profiles, given their proven ability to be re-identified and to infer health information, can be considered individually identifiable health information (Lynch &

Pedersen, 2016). In response, consent procedures in microbiome research are focusing more on possible dangers, such as unintentional health discoveries and re-identification (Malin et al., 2013). Nonetheless, there are still differences in practice between nations and organizations. In multinational partnerships where microbiome data are extensively shared, the lack of worldwide harmonization in consent norms, data governance, and participant rights puts participants at risk for unequal protections.

14.3.2 Benefit Sharing and Equity

14.3.2.1 Unequal Access and Ownership

Metagenomics research now faces rising ethical challenges regarding ownership rights and distribution of benefits and data accessibility. The communities has become common in scientific research (Warbrick et al., 2023). The scientific value of these samples creates disputes about who collection of microbiome samples from biodiverse habitats and unique health and lifestyle settings of low-income and indigenous should control the resulting data and which parties should benefit from additional research findings (Kowal, 2013). The use of biological resources without proper compensation in biodiversity and genomics research has led to the term "biopiracy" in ongoing discussions (Shiva, 2016). Research studies about indigenous and rural gut microbiomes have generated knowledge which scientists later turned into commercial products including probiotics and diagnostic tools and medical treatments. The commercialization of these products fails to provide any substantial benefits or development opportunities to the communities that supplied the original biological materials (Kamau et al., 2010; Reardon, 2019). The implementation of Nagoya Protocol Access and Benefit-Sharing mechanisms represents a necessary solution to address these problems in microbiome research governance. The absence of protective measures makes it probable the continuation of extractive research practices.

14.3.2.2 Microbiome Colonialism

High-income country researchers have established "microbiome colonialism" as a term which describes their practice of extracting biological resources and microbiome data from low- and middle-income countries (LMICs) without proper collaboration or capacity development or fair authorship recognition (Bezuidenhout & Chakauya, 2018; Rogers et al., 2019). The unequal distribution of scientific recognition and medical benefits from biological samples collected from marginalized groups threatens to intensify existing health disparities between developed and developing regions (Ng et al., 2015). The worldwide collection of stool samples to study gut microbiota serves as an illustration of this practice. The development of thorough reference databases depends on samples from under-represented populations in Africa, Asia, and Latin America, which have frequently been the focus of extensive efforts (Yatsunenko et al., 2012). The exclusion of local scientists and people from study design, data analysis, authorship, and benefit-sharing, however, raises ethical questions that are similar to those raised by previous discussions in genomes and biodiversity research (Arif et al., 2025). To address these disparities, real international collaborations are needed, with a focus on equitable benefit distribution, fair credit distribution, and capacity building. Without these changes, research on the microbiome runs the risk of replicating extractive scientific methods under a different name.

14.3.2.3 Data Sovereignty and the Nagoya Protocol

Data sovereignty is increasingly being used to define issues of ownership and equity in microbiome research. Countries and communities who provide biological resources should participate in the advantages of their usage, according to the Nagoya Protocol on Access and Benefit Sharing (ABS), an international agreement under the Convention on Biological Diversity (CBD) (Overmann & Scholz, 2017). Although the Protocol has been extensively used for genetic resources in genomics and biodiversity research, its applicability to microbiome data is still up for question, especially when datasets are exchanged internationally and stored in open-access repositories (Lyall, 2007).

In order to overcome these obstacles, microbiome research ethics best practices increasingly require:

- Participation of the community in the planning and execution of studies,
- Context-sensitive and transparent consent procedures, and
- Developing local ability to guarantee that scientific partnerships yield enduring advantages in underserved or resource-constrained environments.

The implementation of these measures will protect microbiome science from worsening existing inequalities while creating fair distribution of benefits and worldwide scientific participation and research trust.

14.3.3 Dual-Use Concerns

14.3.3.1 Synthetic Biology and Pathogen Reconstruction

Potentially harmful viral, plasmid, or mobile genetic element sequences are commonly found in metagenomic datasets. Even while these sequences are crucial for comprehending host-pathogen interactions, microbial ecology, and antibiotic resistance, they also highlight dual-use problems, or the potential for useful scientific resources to be misused for detrimental ends. It is now theoretically possible to recreate bacterial or viral genomes from scratch thanks to recent developments in DNA synthesis and genome assembly. The availability of public metagenomic data has generated concerns because it could be exploited for synthetic biology applications or biological warfare purposes (Diggans & Leproust, 2019; Koblentz, 2017). The analysis of human sewage and wastewater and environmental samples through virome datasets has revealed complete genomic sequences of important human diseases. The addition of such data enables scientific progress but simultaneously creates new possibilities for accidental or harmful applications (Diggans & Leproust, 2019; Koblentz, 2017). The need for proper data governance becomes evident because scientists must establish strict data release protocols and ethical oversight and worldwide standards which protect open science while ensuring biosecurity.

14.3.3.2 Antimicrobial Resistance (AMR) Exploitation

Metagenomics serves as a method to study the resistome which represents antimicrobial resistance genes that exist within microbial communities. The identification of AMR genes through mapping serves as a critical requirement for public health monitoring and surveillance operations (Bengtsson-Palme et al., 2018). The potential exists for new resistance genes and mobile genetic elements to be misused through business exploitation for profit or criminal activities to evade current medical treatments (Nesme & Simonet, 2015). The dual-use problem in genomics mirrors the discussion about open data-sharing benefits versus unanticipated applications which also occurs in this context (Diggans & Leproust, 2019).

14.3.3.3 Governance and Responsible Data Sharing

The storage of sensitive metagenomic data including new antibiotic resistance genes and pathogenic viral sequences should occur in restricted-access databases to minimize dual-use risks according to multiple international initiatives (Aubry, 2019; Diggans & Leproust, 2019). The FAIR principles (Findable, Accessible, Interoperable, Reusable) which the microbiome research community supports for data sharing face challenges from these security measures despite their biosecurity benefits (Wilkinson et al., 2016). The ethical challenge of metagenomics remains to achieve proper transparency and accountability balance.

14.4 POLICY AND REGULATORY CHALLENGES

The fast development of metagenomic research has created substantial discovery opportunities but it has also generated complex legal and policy challenges. The data-intensive nature of metagenomics differs from traditional microbiology because it generates large datasets that move across borders for public repository storage and ecological and agricultural and medical applications. The worldwide data-driven approach requires governance systems which protect openness and fairness while ensuring responsible use of data. The current legal frameworks fail to keep up with the rapid pace of scientific progress. The three connected domains of concern include data governance and intellectual property management and access control and international and environmental policy frameworks (see Figure 14.2).

14.4.1 Data Governance

14.4.1.1 Need for Harmonized Global Policies

The field of metagenomics operates as a worldwide collaborative scientific discipline. The Earth Microbiome Project and Human Microbiome Project used open science principles to demonstrate the value of FAIR data standards which ensure data accessibility and reusability and findability and interoperability (Wilkinson et al., 2016). The genomic guidelines Consortium (GSC) developed the Minimum Information about any (x) Sequence (MIxS) guidelines which establish fundamental data criteria for repositories to achieve interoperability through standardization of sample types and collection methods and sequencing platforms. The existing guidelines do not achieve complete data

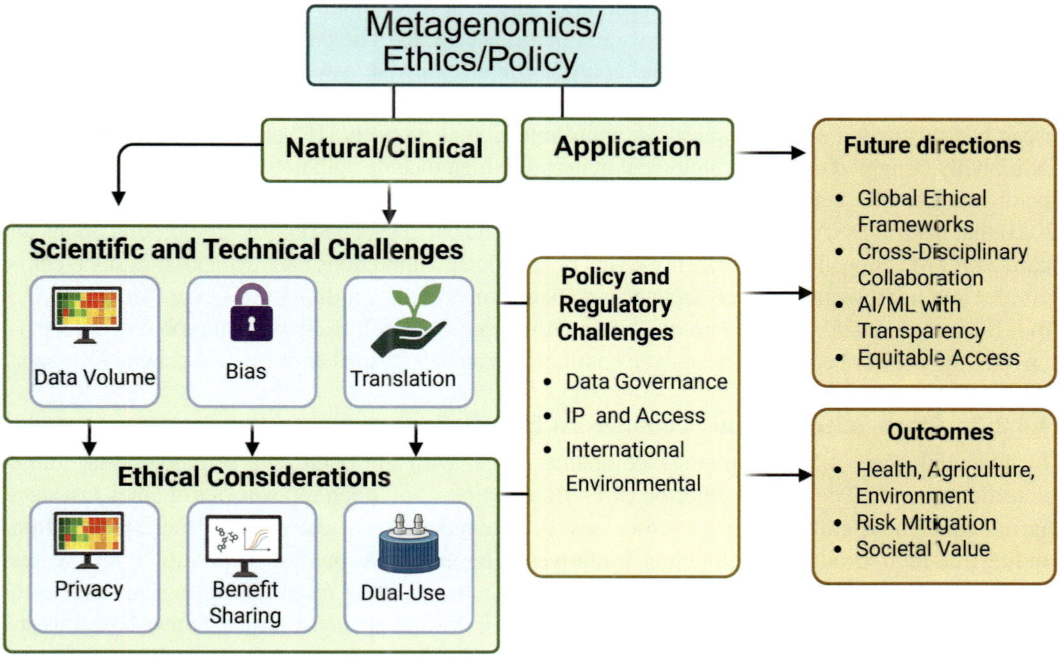

FIGURE 14.2 Overview of ethical, scientific, and policy dimensions in metagenomics research and applications. The framework illustrates the interplay between natural/clinical and application-oriented aspects of metagenomics, highlighting key scientific and technical challenges (data volume, bias, and translational barriers), ethical considerations (privacy, benefit sharing, and dual-use risks), and policy and regulatory challenges (data governance, intellectual property, and international environmental concerns). Addressing these challenges through future directions-including global ethical frameworks, cross-disciplinary collaboration, transparent AI/ML integration, and equitable access-will promote positive outcomes across health, agriculture, and environmental sectors while enhancing societal value and risk mitigation.

harmonization (Yilmaz et al., 2011). The NCBI Sequence Read Archive and EBI-ENA and MG-RAST repositories operate with different permission systems and metadata standards and submission protocols. Researchers face difficulties when trying to meet multiple repository requirements simultaneously because this creates inefficiencies that block both meta-analysis and study comparison.

14.4.1.2 Patchwork of Repository and Consent Practices

The main challenge stems from the need for data ownership and informed permission. The policies for participant consent in human-associated microbiome studies vary between countries and institutions when research takes place in clinical environments. The laws of different governments require researchers to obtain consent that matches their specific research goals but other jurisdictions allow researchers to obtain broad consent for any future data utilization. The different consent policies between nations create barriers for both data reuse and international scientific collaboration. The need for tiered-access systems becomes more critical because of privacy concerns (Section 3.1). The implementation of restricted access to sensitive data is required for security but excessive limitations could violate the FAIR principles. The implementation of governance frameworks requires a balance between data protection and openness to preserve scientific value of data while respecting human rights and cultural traditions.

14.4.2 Intellectual Property and Access

14.4.2.1 Patentability of Microbiome-Derived Products

The commercial industry now uses metagenomic data to develop modified microorganisms and probiotics and prebiotics and diagnostic tools and pharmaceuticals. The development of microbial consortia for medical use and stool sample analysis for disease risk assessment through microbiome diagnostics represents two examples of business applications. The new developments in microbiome research have created essential questions about intellectual property (IP) protection. Patent law has traditionally struggled to protect biological materials which include both natural sequences and living organisms. The U.S. Supreme Court ruled in Association for Molecular Pathology v. Myriad Genetics (2013) that natural DNA sequences cannot receive patents but cDNA and engineered modifications do qualify for protection. The patentability status of functional annotations and artificial microbial communities and microbiome-derived sequences remains uncertain according to this case (Morgan et al., 2014; Bisht et al., 2025). The trademark ability of diagnostic algorithms that use microbiome information remains unclear because they analyze community patterns instead of individual genetic elements.

14.4.2.2 Open Science versus Commercial Exploitation

The financial goals of biotechnology companies clash with the open data principles that guide scientific research. The implementation of FAIR principles and open repositories enables fast data sharing which accelerates discovery processes. Private businesses use exclusive practices to obtain funding for their product development initiatives. The excessive pursuit of privatization creates knowledge silos which block scientists from verifying or extending research findings according to critics. The path to clinical practice for translational innovations such as microbiome-based treatments requires commercialization pathways to succeed. The main regulatory challenge exists in finding an appropriate equilibrium between safeguarding intellectual property rights and maintaining open scientific practices.

14.4.3 International and Environmental Policy

14.4.3.1 Nagoya Protocol and Genetic Resource Access

The Nagoya Protocol on ABS stands as the most significant international agreement for metagenomics because it received approval under the Convention on Biological Diversity. The Nagoya

Protocol establishes that nations maintain full authority over their biological resources while they must distribute advantages from resource utilization equitably. The Nagoya Protocol started with physical biological samples but now extends its authority to cover metagenomic data together with digital sequence information (DSI). The extension of the Nagoya Protocol to sequence data from ecosystems creates a dispute about whether nations should control access to their ecosystem data even when these data exist in international databases. The implementation of strict access controls faces opposition because it could block scientific partnerships and slow down essential research during health crises but supporters argue that such measures are essential for achieving fairness.

14.4.3.2 Metagenomics in Environmental Monitoring

Metagenomics technology finds increasing adoption as a monitoring system for environmental purposes and policy enforcement. The research of Larkin et al. (2025) shows that ocean microbiome surveys help scientists track climate-induced changes in marine ecosystems (Larkin et al., 2025), and soil metagenomics helps scientists understand carbon storage and sustainable farming practices (Lefeuvre et al., 2019). The bioindicator system of metagenomics has gained recognition from policymakers who see its potential for climate change agreements and conservation frameworks and biodiversity treaties. The regulatory framework has not yet adapted to this new technology. Standardization of collection methods and sequencing and analytical techniques must be established to establish metagenomics as a reliable policy instrument for environmental impact assessments and protected area performance monitoring.

14.5 FUTURE DIRECTIONS AND RECOMMENDATIONS

Metagenomics continues to evolve through enhanced complexity while expanding its reach across the world and gaining greater social importance. The future development of metagenomics depends on how society along with policymakers and scientists handle the ethical and legal and equity matters discussed in previous sections. The following proposals establish essential objectives to steer metagenomics toward its proper development.

14.5.1 Development of Global Ethical Frameworks for Microbiome Research

Metagenomic research generates ethical problems which extend past national borders because they involve privacy concerns and dual-use risks and benefit distribution issues. The current methods operate independently from one another because they use different consent procedures and they do not follow ABS principles consistently and they fail to assess dual-use risks. A worldwide ethical framework stands as an urgent necessity to address these matters. The new framework needs to handle microbiome data-specific challenges while building upon existing standards that include MIxS recommendations and FAIR principles (Wilkinson et al., 2016; Yilmaz et al., 2011). The privacy measures for metagenomic datasets need to surpass traditional genomic privacy standards because these datasets might contain accidental health information from their host organisms. The Nagoya Protocol's ABS guidelines should be integrated into international frameworks to achieve fair benefit sharing but without creating excessive restrictions on DSI that would block open scientific collaboration (Greiber, 2012). Such frameworks need to emerge from collaborative dialogue between scientists and ethicists and politicians and affected communities instead of being forced by top-down authorities.

14.5.2 Stronger Cross-Disciplinary Collaboration

The analysis of metagenomic data requires interdisciplinary collaboration because of its complex nature and its wide-ranging social implications. The independent work of microbiologists and bioinformaticians fails to address the broader issues related to dual-use applications and governance and equity. Research that combines social responsibility with scientific validity requires essential collaboration between ethicists and legal scholars and politicians and social scientists. The process requires equal involvement from the community. Research participants who supply samples should be treated as research partners instead of being viewed as data providers. The participatory research models which genetics and public health fields now adopt should guide similar initiatives in microbiome studies. The approach will create transparent results while building trust between researchers and participants who will jointly own the findings. The collaborative methods will help solve immediate policy challenges by enabling the integration of metagenomics into environmental governance systems and the development of IP protection mechanisms that support open science practices.

14.5.3 AI/ML Integration with Transparency and Explainability

Artificial intelligence (AI) and machine learning (ML) enable scientists to analyze patterns and forecast functions and model ecosystems at scales which were previously impossible thus revolutionizing metagenomics. Deep learning methods outperform traditional homology-based approaches by delivering better results for gene function prediction (Bileschi et al., 2022). The analysis shows that ML models show potential for disease risk prediction through microbiome composition analysis (Marcos-Zambrano et al., 2021). The new developments create multiple obstacles for researchers to overcome. The inability of black-box models to provide transparent predictions makes it difficult to validate their output. The need for explainable models exists in clinical and policy contexts because it builds trust and enables proper model deployment. The development of AI/ML methods which balance interpretability with predictive accuracy should be the main focus of future research. The development of accurate predictions depends on obtaining diverse high-quality training data which must be free from biases. The achievement of global data-sharing protocols will become essential for ensuring that predictions reflect diverse human experiences and environmental settings.

14.5.4 Promoting Equitable Access to Metagenomic Technologies and Benefits

The last recommendation involves distributing metagenomics advantages across the world in a fair manner. The current distribution of sequencing infrastructure and computing power and data analytic capabilities between high-income nations and low- and middle-income nations creates an outside partnership dependency for the latter. The current inequality between nations threatens to maintain the "microbiome colonialism" situation. The following steps will help address this issue: The establishment of training programs and local sequencing facilities and high-performance computing systems requires funding support. The development of open-source bioinformatics tools and portable sequencing technology (such as ONT MinION) at affordable prices should be implemented. The establishment of fair partnerships in international collaborations should include equal authorship rights and proper recognition and fair distribution of benefits. The process of translating research results into practical applications represents a critical aspect of making scientific knowledge accessible to all. The development of microbiome-based diagnostics and probiotics and therapies needs to focus on affordability and accessibility to prevent worsening global health disparities.

14.6 CONCLUSION

The study of microbial diversity through metagenomics has evolved into a transformative field which delivers substantial impacts on agricultural practices and health care and biotechnology and environmental sustainability. The field will advance further when scientists address essential scientific and ethical and policy matters. The sector requires solutions to handle big data management and analytical bias reduction while protecting privacy and ensuring fair distribution of benefits to vulnerable groups and preventing data misuse. The development of global trust requires standardized protocols and equal access to technology and ethical frameworks that guide responsible innovation practices. Metagenomics supports scientific discovery through interdisciplinary collaboration and open computational development which respects social norms. The future of metagenomics depends on scientists who will unlock microbial potential while using this knowledge to create sustainable benefits for ecosystems and human health and sustainable development.

REFERENCES

Aggarwal, N., Kitano, S., Puah, G. R. Y., Kittelmann, S., Hwang, I. Y., & Chang, M. W. (2022). Microbiome and human health: Current understanding, engineering, and enabling technologies. *Chemical Reviews*, *123*(1), 31–72.

Arellano, A. M., Dai, W., Wang, S., Jiang, X., & Ohno-Machado, L. (2018). Privacy policy and technology in biomedical data science. *Annual Review of Biomedical Data Science*, *1*(1), 115–129.

Arif, S. J., Graham, S. P., Abdill, R. J., & Blekhman, R. (2025). Analyzing human gut microbiome data from global populations: Challenges and resources. *Trends in Microbiology*, *33*(11), 1212–1223.

Aubry, S. (2019). The future of digital sequence information for plant genetic resources for food and agriculture. *Frontiers in Plant Science*, *10*, 1046.

Avuthu, N., & Guda, C. (2022). Meta-analysis of altered gut microbiota reveals microbial and metabolic biomarkers for colorectal cancer. *Microbiology Spectrum*, *10*(4), e00013–00022.

Bengtsson-Palme, J., Kristiansson, E., & Larsson, D. J. (2018). Environmental factors influencing the development and spread of antibiotic resistance. *FEMS Microbiology Reviews*, *42*(1), fux053.

Berger, K. M. (2021). Technological advances that test the dual-use research of concern model. In *Applied biosecurity: Global health, biodefense, and developing technologies* (pp. 133–160). Springer.

Bezuidenhout, L., & Chakauya, E. (2018). Hidden concerns of sharing research data by low/middle-income country scientists. *Global Bioethics*, *29*(1), 39–54.

Bileschi, M. L., Belanger, D., Bryant, D. H., Sanderson, T., Carter, B., Sculley, D., Bateman, A., DePristo, M. A., & Colwell, L. J. (2022). Using deep learning to annotate the protein universe. *Nature Biotechnology*, *40*(6), 932–937.

Bisht, D., Kumar, V., Gupta, S., & Sachan, M. (2025). Navigating circulating tumour cells and tumour microemboli: Where we stand in bridging their molecular dynamics with clinical translation. *Biochimica et Biophysica Acta (BBA)—Reviews on Cancer*, *1880*(4), 189384. https://doi.org/10.1016/j.bbcan.2025.189384

Bolyen, E., Rideout, J. R., Dillon, M. R., Bokulich, N. A., Abnet, C. C., Al-Ghalith, G. A., Alexander, H., Alm, E. J., Arumugam, M., & Asnicar, F. (2019). Reproducible, interactive, scalable and extensible microbiome data science using QIIME 2. *Nature Biotechnology*, *37*(8), 852–857.

Buck, M., & Hamilton, C. (2011). The Nagoya Protocol on access to genetic resources and the fair and equitable sharing of benefits arising from their utilization to the convention on biological diversity. *Review of European Community & International Environmental Law*, *20*(1), 47–61.

Cavicchioli, R., Ripple, W. J., Timmis, K. N., Azam, F., Bakken, L. R., Baylis, M., Behrenfeld, M. J., Boetius, A., Boyd, P. W., & Classen, A. T. (2019). Scientists' warning to humanity: Microorganisms and climate change. *Nature Reviews Microbiology*, *17*(9), 569–586.

Chuong, K. H., Hwang, D. M., Tullis, D. E., Waters, V. J., Yau, Y. C., Guttman, D. S., & O'Doherty, K. C. (2017). Navigating social and ethical challenges of biobanking for human microbiome research. *BMC Medical Ethics*, *18*(1), 1.

Costea, P. I., Coelho, L. P., Sunagawa, S., Munch, R., Huerta-Cepas, J., Forslund, K., Hildebrand, F., Kushugulova, A., Zeller, G., & Bork, P. (2017). Subspecies in the global human gut microbiome. *Molecular Systems Biology*, *13*(12), 960.

Dai, Z., Coker, O. O., Nakatsu, G., Wu, W. K., Zhao, L., Chen, Z., Chan, F. K., Kristiansen, K., Sung, J. J., & Wong, S. H. (2018). Multi-cohort analysis of colorectal cancer metagenome identified altered bacteria across populations and universal bacterial markers. *Microbiome*, *6*(1), 70.

De Goffau, M. C., Lager, S., Sovio, U., Gaccioli, F., Cook, E., Peacock, S. J., Parkhill, J., Charnock-Jones, D. S., & Smith, G. C. (2019). Human placenta has no microbiome but can contain potential pathogens. *Nature*, *572*(7769), 329–334.

Di Tommaso, P., Chatzou, M., Floden, E. W., Barja, P. P., Palumbo, E., & Notredame, C. (2017). Nextflow enables reproducible computational workflows. *Nature Biotechnology*, *35*(4), 316–319.

Diggans, J., & Leproust, E. (2019). Next steps for access to safe, secure DNA synthesis. *Frontiers in Bioengineering and Biotechnology*, *7*, 86.

Dulanto Chiang, A., & Dekker, J. P. (2020). From the pipeline to the bedside: Advances and challenges in clinical metagenomics. *The Journal of Infectious Diseases*, *221*(Supplement_3), S331–S340.
Dupras, C., & Bunnik, E. M. (2021). Toward a framework for assessing privacy risks in multi-omic research and databases. *The American Journal of Bioethics*, *21*(12), 46–64.
Edwin, N. R., Fitzpatrick, A. H., Brennan, F., Abram, F., & O'Sullivan, O. (2024). An in-depth evaluation of metagenomic classifiers for soil microbiomes. *Environmental Microbiome*, *19*(1), 19.
Eisenhofer, R., Minich, J. J., Marotz, C., Cooper, A., Knight, R., & Weyrich, L. S. (2019). Contamination in low microbial biomass microbiome studies: Issues and recommendations. *Trends in Microbiology*, *27*(2), 105–117.
Fan, Y., & Pedersen, O. (2021). Gut microbiota in human metabolic health and disease. *Nature Reviews Microbiology*, *19*(1), 55–71.
FAO. (2019). *The state of food and agriculture 2019: Moving forward on food loss and waste reduction*. UN.
Forry, S. P., Servetas, S. L., Kralj, J. G., Soh, K., Hadjithomas, M., Cano, R., Carlin, M., Amorim, M. G. D., Auch, B., & Bakker, M. G. (2024). Variability and bias in microbiome metagenomic sequencing: An interlaboratory study comparing experimental protocols. *Scientific Reports*, *14*(1), 9785.
Franzosa, E. A., Huang, K., Meadow, J. F., Gevers, D., Lemon, K. P., Bohannan, B. J., & Huttenhower, C. (2015). Identifying personal microbiomes using metagenomic codes. *Proceedings of the National Academy of Sciences*, *112*(22), E2930–E2938.
Gao, Y., Cao, D., Zhu, J., Feng, H., Luo, X., Liu, S., Yan, X.-X., Zhang, X., & Gao, P. (2020). Structural insights into assembly, operation and inhibition of a type I restriction–modification system. *Nature Microbiology*, *5*(9), 1107–1118.
Gilbert, J. A., Jansson, J. K., & Knight, R. (2014). The earth microbiome project: Successes and aspirations. *BMC Biology*, *12*(1), 69.
Greiber, T. (2012). *An explanatory guide to the Nagoya protocol on access and benefit-sharing*. IUCN.
Gupta, A., Pahal, S., Kumar, V., & Chaudhary, A. (2025). Role of monoclonal antibodies in alzheimer's disease: Successes, failures, and future directions. *Neurological Sciences*. https://doi.org/10.1007/s10072-025-08456-5
Gupta, V. K., Paul, S., & Dutta, C. (2017). Geography, ethnicity or subsistence-specific variations in human microbiome composition and diversity. *Frontiers in Microbiology*, *8*, 1162.
Gymrek, M., McGuire, A. L., Golan, D., Halperin, E., & Erlich, Y. (2013). Identifying personal genomes by surname inference. *Science*, *339*(6117), 321–324.
Haryono, M. A., Law, Y. Y., Arumugam, K., Liew, L. C.-W., Nguyen, T. Q. N., Drautz-Moses, D. I., Schuster, S. C., Wuertz, S., & Williams, R. B. (2022). Recovery of high quality metagenome-assembled genomes from full-scale activated sludge microbial communities in a tropical climate using longitudinal metagenome sampling. *Frontiers in Microbiology*, *13*, 869135.
Integrative, H. (2014). The integrative human microbiome project: Dynamic analysis of microbiome-host omics profiles during periods of human health and disease. *Cell Host & Microbe*, *16*(3), 276–289.
Jagadesh, M., Dash, M., Kumari, A., Singh, S. K., Verma, K. K., Kumar, P., Bhatt, R., & Sharma, S. K. (2024). Revealing the hidden world of soil microbes: Metagenomic insights into plant, bacteria, and fungi interactions for sustainable agriculture and ecosystem restoration. *Microbiological Research*, *285*, 127764.
James, A. S., Adil, N. A., Goltz, D., Tangudu, D., Chaudhari, D. S., Shukla, R., Kumar, V., Kumar, A., Masternak, M. M., Holland, P., Labyak, C., Golden, A., Dangiolo, M., Arikawa, A. Y., Kociolek, J., Fraser, A., Williams, C., Agronin, M., Aymat, M., . . . Yadav, H. (2024). Abnormalities in Gut Virome signatures linked with cognitive impairment in older adults. *Gut Microbes*, *16*(1), 2431648. https://doi.org/10.1080/19490976.2024.2431648
Jansson, J. K., & Hofmockel, K. S. (2020). Soil microbiomes and climate change. *Nature Reviews Microbiology*, *18*(1), 35–46.
Kamau, E. C., Fedder, B., & Winter, G. (2010). The Nagoya protocol on access to genetic resources and benefit sharing: What is new and what are the implications for provider and user countries and the scientific community. *Law Env't Dev J*, *6*, 246.
Karst, S. M., Ziels, R. M., Kirkegaard, R. H., Sørensen, E. A., McDonald, D., Zhu, Q., Knight, R., & Albertsen, M. (2021). High-accuracy long-read amplicon sequences using unique molecular identifiers with Nanopore or PacBio sequencing. *Nature Methods*, *18*(2), 165–169.
Knight, R., Vrbanac, A., Taylor, B. C., Aksenov, A., Callewaert, C., Debelius, J., Gonzalez, A., Kosciolek, T., McCall, L.-I., & McDonald, D. (2018). Best practices for analysing microbiomes. *Nature Reviews Microbiology*, *16*(7), 410–422.
Koblentz, G. D. (2017). The de novo synthesis of horsepox virus: Implications for biosecurity and recommendations for preventing the reemergence of smallpox. *Health Security*, *15*(6), 620–628.

Kobras, C. M., Fenton, A. K., & Sheppard, S. K. (2021). Next-generation microbiology: From comparative genomics to gene function. *Genome Biology*, *22*(1), 123.

Kowal, E. (2013). Orphan DNA: Indigenous samples, ethical biovalue and postcolonial science. *Social Studies of Science*, *43*(4), 577–597.

Kumar, V., Gangani, S., Shukla, R., Prajapati, S. K., Shekhar, H., & Shukla, V. (2024). Next-generation sequencing and genomic data analysis. In A. Chaudhary, S. K. Sethi, & A. Verma (Eds.). *Unraveling new frontiers and advances in bioinformatics* (pp. 49–76). Springer Nature, Singapore. https://doi.org/10.1007/978-981-97-7123-3_3

Kumar, V., Sharma, J. K., Kumar, S., Gulia, J., Singh, N. P., & Santal, A. R. (2025). Metagenomic and proteomic approach for bioremediation of environmental pollutants. In M. P. Shah (Ed.), *Development in waste water treatment research and processes*. Elsevier. https://doi.org/10.1016/B978-0-443-13615-3.00020-0

Kumar, V., Shukla, R., Gangani, S., Shekhar, H., & Shukla, V. (2024). Metagenomics and microbiome analysis. In A. Chaudhary, S. K. Sethi, & A. Verma (Eds.), *Unraveling new frontiers and advances in bioinformatics* (pp. 77–92). Springer Nature, Singapore. https://doi.org/10.1007/978-981-97-7123-3_4

Larkin, A. A., Brock, M. L., Fagan, A. J., Moreno, A. R., Gerace, S. D., Lees, L. E., Suarez, S. A., Eloe-Fadrosh, E. A., & Martiny, A. C. (2025). Climate-driven succession in marine microbiome biodiversity and biogeochemical function. *Nature Communications*, *16*(1), 3926.

Lefeuvre, P., Martin, D. P., Elena, S. F., Shepherd, D. N., Roumagnac, P., & Varsani, A. (2019). Evolution and ecology of plant viruses. *Nature Reviews Microbiology*, *17*(10), 632–644.

Lepage, P., Leclerc, M. C., Joossens, M., Mondot, S., Blottière, H. M., Raes, J., Ehrlich, D., & Doré, J. (2013). A metagenomic insight into our gut's microbiome. *Gut*, *62*(1), 146–158.

Li, D., Luo, R., Liu, C.-M., Leung, C.-M., Ting, H.-F., Sadakane, K., Yamashita, H., & Lam, T.-W. (2016). MEGAHIT v1. 0: A fast and scalable metagenome assembler driven by advanced methodologies and community practices. *Methods*, *102*, 3–11.

Lozupone, C. A., Stombaugh, J., Gonzalez, A., Ackermann, G., Wendel, D., Vázquez-Baeza, Y., Jansson, J. K., Gordon, J. I., & Knight, R. (2013). Meta-analyses of studies of the human microbiota. *Genome Research*, *23*(10), 1704–1714.

Lyall, C. (2007). Governing genomics: New governance tools for new technologies? *Technology Analysis & Strategic Management*, *19*(3), 369–386.

Lynch, S. V., & Pedersen, O. (2016). The human intestinal microbiome in health and disease. *New England Journal of Medicine*, *375*(24), 2369–2379.

Ma, Y., Zhang, Y., Jiang, H., Xiang, S., Zhao, Y., Xiao, M., Du, F., Ji, H., Kaboli, P. J., & Wu, X. (2021). Metagenome analysis of intestinal bacteria in healthy people, patients with inflammatory bowel disease and colorectal cancer. *Frontiers in Cellular and Infection Microbiology*, *11*, 599734.

Malin, B. A., Emam, K. E., & O'Keefe, C. M. (2013). Biomedical data privacy: Problems, perspectives, and recent advances. *Journal of the American Medical Informatics Association*, *20*(1), 2–6.

Manolio, T. A., Chisholm, R. L., Ozenberger, B., Roden, D. M., Williams, M. S., Wilson, R., Bick, D., Bottinger, E. P., Brilliant, M. H., & Eng, C. (2013). Implementing genomic medicine in the clinic: The future is here. *Genetics in Medicine*, *15*(4), 258–267.

Marcos-Zambrano, L. J., Karaduzovic-Hadziabdic, K., Loncar Turukalo, T., Przymus, P., Trajkovik, V., Aasmets, O., Berland, M., Gruca, A., Hasic, J., & Hron, K. (2021). Applications of machine learning in human microbiome studies: A review on feature selection, biomarker identification, disease prediction and treatment. *Frontiers in Microbiology*, *12*, 634511.

Mathur, P. (2024). Cloud computing infrastructure, platforms, and software for scientific research. *High Performance Computing in Biomimetics: Modeling, Architecture and Applications*, 89–127.

McDonald, D., Birmingham, A., & Knight, R. (2015). Context and the human microbiome. *Microbiome*, *3*(1), 52.

Morgan, J. L., McNamara, J. T., & Zimmer, J. (2014). Mechanism of activation of bacterial cellulose synthase by cyclic di-GMP. *Nature Structural & Molecular Biology*, *21*(5), 489–496.

Morgera, E., Tsioumani, E., & Buck, M. (2014). *Unraveling the nagoya protocol: A commentary on the nagoya protocol on access and benefit-sharing to the convention on biological diversity*. Brill.

Muller, M. R. (2015). *Genetic resources as natural information: Implications for the convention on biological diversity and Nagoya protocol*. Routledge.

Nair, S. U., Lin, Y., Jain, S., Chaudhari, D. S., Shukla, R., Kumar, V., Mankowski, R. T., Meng, H., Buford, T. W., Kuchel, G. A., Yadav, H., & Consortium, M. (2024). Unique gut microbiome signatures associated with physical and cognitive function. *Alzheimer's & Dementia*, *20*(S2), e092044. https://doi.org/10.1002/alz.092044

Navaux, P. O. A., Lorenzon, A. F., & da Silva Serpa, M. (2023). Challenges in high-performance computing. *Journal of the Brazilian Computer Society*, *29*(1), 51–62.

Nebeker, C., Harlow, J., Espinoza Giacinto, R., Orozco-Linares, R., Bloss, C. S., & Weibel, N. (2017). Ethical and regulatory challenges of research using pervasive sensing and other emerging technologies: IRB perspectives. *AJOB Empirical Bioethics*, *8*(4), 266–276.

Nesme, J., & Simonet, P. (2015). The soil resistome: A critical review on antibiotic resistance origins, ecology and dissemination potential in telluric bacteria. *Environmental Microbiology*, *17*(4), 913–930.

Ng, L. C., Hanlon, C., Yimer, G., Henderson, D. C., & Fekadu, A. (2015). Ethics in global health research: The need for balance. *The Lancet Global Health*, *3*(9), e516–e517.

Nissimov, J. (2017). A communal catalogue reveals Earth's multiscale microbial diversity. *Nature*, *551*, 457–463.

Nwachukwu, B. C., & Babalola, O. O. (2022). Metagenomics: A tool for exploring key microbiome with the potentials for improving sustainable agriculture. *Frontiers in Sustainable Food Systems*, *6*, 886987.

Oh, J., Byrd, A. L., Park, M., Kong, H. H., & Segre, J. A. (2016). Temporal stability of the human skin microbiome. *Cell*, *165*(4), 854–866.

Overmann, J., & Scholz, A. H. (2017). Microbiological research under the Nagoya protocol: Facts and fiction. *Trends in Microbiology*, *25*(2), 85–88.

Parfrey, L., & Knight, R. (2012). Spatial and temporal variability of the human microbiota. *Clinical Microbiology and Infection*, *18*, 5–7.

Pollock, J., Glendinning, L., Wisedchanwet, T., & Watson, M. (2018). The madness of microbiome: Attempting to find consensus "best practice" for 16S microbiome studies. *Applied and Environmental Microbiology*, *84*(7), e02627–02617.

Poussin, C., Sierro, N., Boué, S., Battey, J., Scotti, E., Belcastro, V., Peitsch, M. C., Ivanov, N. V., & Hoeng, J. (2018). Interrogating the microbiome: Experimental and computational considerations in support of study reproducibility. *Drug Discovery Today*, *23*(9), 1644–1657.

Prajapati, S. K., Kumar, V., Shukla, R., Szekeres, C., Ma, Y., Jain, S., & Yadav, H. (2024). Microbiome-immune iterplay in aging brains: A pilot investigation from the MiaGB cohort. *Alzheimer's & Dementia*, *20*(S2), e090734. https://doi.org/10.1002/alz.090734

Queiroz, R. M., Smith, T., Villanueva, E., Marti-Solano, M., Monti, M., Pizzinga, M., Mirea, D.-M., Ramakrishna, M., Harvey, R. F., & Dezi, V. (2019). Comprehensive identification of RNA–protein interactions in any organism using orthogonal organic phase separation (OOPS). *Nature Biotechnology*, *37*(2), 169–178.

Quince, C., Walker, A. W., Simpson, J. T., Loman, N. J., & Segata, N. (2017). Shotgun metagenomics, from sampling to analysis. *Nature Biotechnology*, *35*(9), 833–844.

Reardon, J. (2019). *The postgenomic condition: Ethics, justice, and knowledge after the genome*. University of Chicago Press.

Rhodes, R., Gligorov, N., & Schwab, A. P. (2013). *The human microbiome: Ethical, legal and social concerns*. Oxford University Press.

Rogers, G. B., Ward, J., Brown, A., & Wesselingh, S. L. (2019). Inclusivity and equity in human microbiome research. *The Lancet*, *393*(10173), 728–729.

Salter, S. J., Cox, M. J., Turek, E. M., Calus, S. T., Cookson, W. O., Moffatt, M. F., Turner, P., Parkhill, J., Loman, N. J., & Walker, A. W. (2014). Reagent and laboratory contamination can critically impact sequence-based microbiome analyses. *BMC Biology*, *12*(1), 87.

Sarfraz, H., Czerniecki, B. J., Stankiewicz, K., Jaglal, M. V., Jain, S., Yadav, H., Liu, M., Shukla, R., Kumar, V., Alhomsi, M., Hathout, M., Tuz Zahra, F., Darwin, A., Joshi, U., Sen, A., Muzaffar, J., & Bari, S. (2024). Evaluating gut microbiome as a biomarker of pathological complete response in patients with early locally advanced triple negative breast cancer (LA TNBC): A pilot study. *Journal of Clinical Oncology*, *42*(Supplement_16), 593–593. https://doi.org/10.1200/JCO.2024.42.16_suppl.593

Schloss, P. D., & Handelsman, J. (2005). Metagenomics for studying unculturable microorganisms: Cutting the Gordian knot. *Genome Biology*, *6*(8), 229.

Shiva, V. (2016). *Biopiracy: The plunder of nature and knowledge*. North Atlantic Books.

Sinha, R., Abu-Ali, G., Vogtmann, E., Fodor, A. A., Ren, B., Amir, A., Schwager, E., Crabtree, J., Ma, S., & Consortium, M. Q. C. P. (2017). Assessment of variation in microbial community amplicon sequencing by the microbiome quality control (MBQC) project consortium. *Nature Biotechnology*, *35*(11), 1077–1086.

Thomas, A. M., Manghi, P., Asnicar, F., Pasolli, E., Armanini, F., Zolfo, M., Beghini, F., Manara, S., Karcher, N., & Pozzi, C. (2019). Metagenomic analysis of colorectal cancer datasets identifies cross-cohort microbial diagnostic signatures and a link with choline degradation. *Nature Medicine*, *25*(4), 667–678.

Wang, X., Huang, M., Shao, H., Li, K., & Zhai, X. (2025). Ethical reflection of Chinese scientists on the dual-use concerns of emerging medical biotechnology. *Journal of Medical Ethics*, *51*(10), 662–667.

Warbrick, I., Heke, D., & Breed, M. (2023). Indigenous knowledge and the microbiome—bridging the disconnect between colonized places, peoples, and the unseen influences that shape our health and well-being. *Msystems*, *8*(1), e00875–00822.

Weiss, D. J., Lucas, T. C., Nguyen, M., Nandi, A. K., Bisanzio, D., Battle, K. E., Cameron, E., Twohig, K. A., Pfeffer, D. A., & Rozier, J. A. (2019). Mapping the global prevalence, incidence, and mortality of Plasmodium falciparum, 2000–17: A spatial and temporal modelling study. *The Lancet*, *394*(10195), 322–331.

Wilkinson, M. D., Dumontier, M., Aalbersberg, I. J., Appleton, G., Axton, M., Baak, A., Blomberg, N., Boiten, J.-W., da Silva Santos, L. B., & Bourne, P. E. (2016). The FAIR guiding principles for scientific data management and stewardship. *Scientific Data*, *3*(1), 1–9.

Wirbel, J., Pyl, P. T., Kartal, E., Zych, K., Kashani, A., Milanese, A., Fleck, J. S., Voigt, A. Y., Palleja, A., & Ponnudurai, R. (2019). Meta-analysis of fecal metagenomes reveals global microbial signatures that are specific for colorectal cancer. *Nature Medicine*, *25*(4), 679–689.

Xu, X., Ocansey, D. K. W., Hang, S., Wang, B., Amoah, S., Yi, C., Zhang, X., Liu, L., & Mao, F. (2022). The gut metagenomics and metabolomics signature in patients with inflammatory bowel disease. *Gut Pathogens*, *14*(1), 26.

Yatsunenko, T., Rey, F. E., Manary, M. J., Trehan, I., Dominguez-Bello, M. G., Contreras, M., Magris, M., Hidalgo, G., Baldassano, R. N., & Anokhin, A. P. (2012). Human gut microbiome viewed across age and geography. *Nature*, *486*(7402), 222–227.

Yilmaz, P., Kottmann, R., Field, D., Knight, R., Cole, J. R., Amaral-Zettler, L., Gilbert, J. A., Karsch-Mizrachi, I., Johnston, A., & Cochrane, G. (2011). Minimum information about a marker gene sequence (MIMARKS) and minimum information about any (x) sequence (MIxS) specifications. *Nature Biotechnology*, *29*(5), 415–420.

15 The Future of Metagenomics

Vanshika Rustagi[1#], Shradheya R. R. Gupta[1#], Monu Kaushik[2], and Indrakant Kumar Singh[1,3]*

[1]*Molecular Biology & Health Informatics Lab, Dr. B. R. Ambedkar Centre for Biomedical Research, University of Delhi, Delhi, India*

[2]*Indian Institute of Technology, Kanpur, India*

[3]*Delhi School of Public Health, Institute of Eminence, University of Delhi, Delhi, India*

[#]*Equal Contribution*

[*]*Corresponding Author*

15.1 INTRODUCTION

The term "meta" (beyond) and "genomics" (study of genomes), when combined together, form the word metagenomics. The term metagenomics refers to the analysis of the collective genomes of microbial communities in a given environment comprehensively, extending beyond the study of individual organisms (see Figure 15.1) (Panda et al., 2025). The field is culture independent; this means that it studies the genetic material recovered directly from environmental or host samples, and by this, the field has fundamentally altered because microbial communities are now investigated as a whole (Garza & Dutilh, 2015). In its initial days, metagenomic approaches were mainly descriptive; this means that their main focus was on generating inventories of microbial taxa through 16S rRNA sequencing or shotgun sequencing (Peterson et al., 2021). There is no doubt that these surveys provided unmatched insights into microbial diversity, but the information offered was limited to the functional capacity of microbial communities. The current evolved field is markedly different. Metagenomics is advancing at a faster pace from descriptive catalogues to predictive and functional frameworks that reveal proper microorganisms' roles, interactions, and dynamics within complex ecosystems (Alves et al., 2018; Kumar, Shukla et al., 2024).

A convergence of technologies and conceptual advances is driving the evolution of metagenomics. First, improvements in next-generation sequencing (NGS) platforms and the advent of long-read sequencing technologies (e.g., PacBio HiFi, Oxford Nanopore) are helping researchers assemble genomes more accurately at strain-level resolution (Gonzalez et al., 2025; Han et al., 2024; Wang 王博 et al., 2025). Second, advancements in bioinformatics pipelines, cloud computing, and artificial intelligence (AI) allow the integration of large datasets and the development of accurate predictive models that can simulate microbial interactions and functional potential (Alsulimani et al., 2024; Peterson et al., 2021). Third, the rise of multi-omics strategies, such as meta-transcriptomics, -proteomics, and -metabolomics, provides a holistic perspective to researchers by linking genetic potential to actual activity and ecological function (Yang et al., 2025). Together, these innovations are redefining what metagenomics can achieve.

The implications of the metagenomics go beyond academic curiosity. In the case of clinical diagnostics, metagenomics is transforming rapid pathogen identification, surveillance of antimicrobial resistance (AMR), and the discovery of biomarkers for diseases ranging from infectious to chronic conditions (Chiu & Miller, 2019; Napit et al., 2025; Kumar, Gangani, et al., 2024).

DOI: 10.1201/9781003670155-15

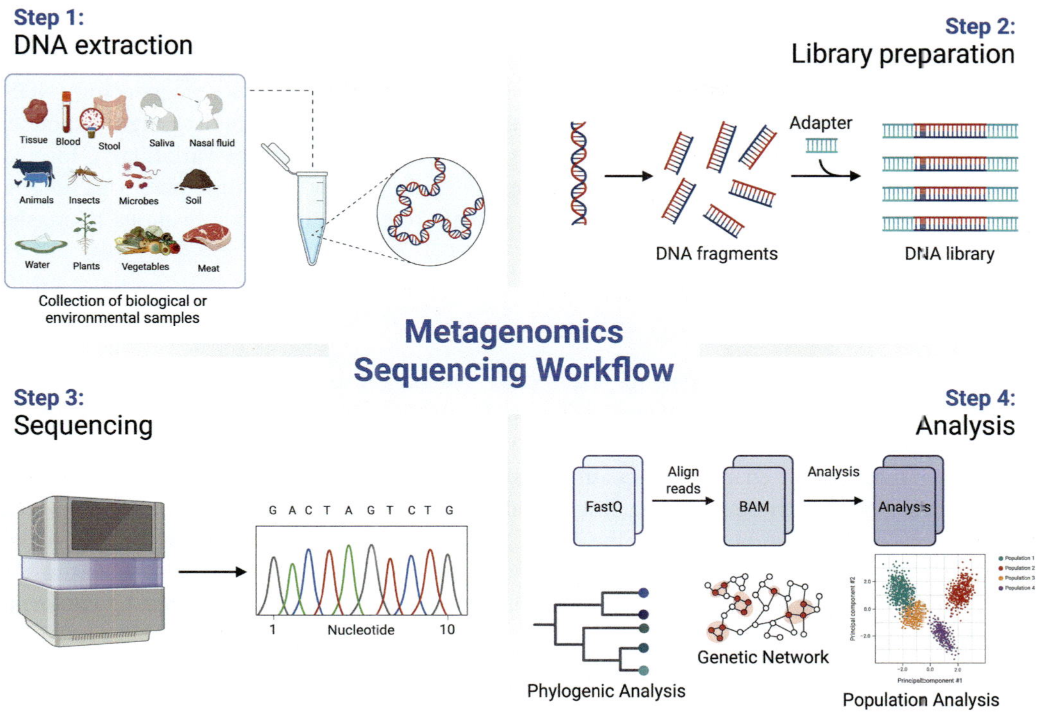

FIGURE 15.1 Step wise metagenomics sequencing workflow, DNA extraction from biological or environmental samples, followed by library preparation, sequencing, and analysis using computers. The approach enables phylogenetic profiling, genetic network construction, and population-level insights into microbial communities.

In the field of precision medicine, the ability to identify, detect, and profile host-associated microbiomes at high resolution offers new field to personalized therapies, microbiome-targeted interventions, and predictive risk assessments (Kashyap et al., 2017; Prajapati et al., 2024; Mishra et al., 2024; Nair et al., 2024; James et al., 2024; Sarfraz et al., 2024). In agriculture, soil and plant microbiome studies enhance nutrient cycling, crop resilience, and the biocontrol of pathogens (Wani et al., 2025). Similarly, in environmental science, metagenomic analyses are central to monitoring biodiversity, understanding microbial contributions to climate-related processes, such as carbon sequestration and methane emissions, and predicting ecosystem responses to anthropogenic stressors (Das et al., 2025; Panda et al., 2025; Kumar et al., 2025).

The field of metagenomics is likely to move toward real-time, portable, and clinically validated applications. Its integration with ML-driven predictive analytics and synthetic biology tools will enable researchers and companies to observe microbial communities and design, manipulate, and optimize them for desired outcomes for the development of society as a whole. This shift brings metagenomics to the heart of a scientific revolution, with implications for human health, food security, environmental sustainability, and biotechnology innovation.

In this chapter, we have explored the likely future of metagenomics, highlighting its emerging trends, technological breakthroughs, translatable opportunities, and ethical considerations that will shape its trajectory in diagnostics and beyond.

15.2 EMERGING SEQUENCING TECHNOLOGIES

The future of the field is tied to developing sequencing platforms that can deliver greater accuracy, longer read lengths, much faster turnaround times, and reduced costs (Table 15.1) (Purushothaman et al., 2022). Early metagenomic studies relied on short-read next-generation sequencing (NGS) platforms, such as Illumina. These technologies usually struggle with assembling highly complex microbial genomes, resolving repetitive elements, and detecting structural variations (Howe & Chain, 2015). However, recent advances in long-read sequencing and real-time platforms address these limitations, opening new frontiers in metagenomic research and its applications (Gonzalez et al., 2025).

15.2.1 Long-Read and Real-Time Platforms

Technologies for long-read sequencing, specifically Pacific Biosciences (PacBio) HiFi sequencing and Oxford Nanopore Technologies (ONT), are transforming metagenomic research. They are capable of generating reads that span thousands to tens of thousands of base pairs (Udaondo et al., 2021). These read lengths enable high-quality genome assemblies that can resolve complete microbial genomes, including those from rare, low-abundance, or previously unculturable taxa. They also provide strain-level resolution, allowing the distinction of related strains within a community, which is essential for monitoring microbial dynamics, pathogenic variants, and horizontal gene transfer events. Additionally, long-read platforms allow the detection of structural variations by capturing

TABLE 15.1
Sequencing Technologies in Metagenomics. Short-Reads Aid Profiling, Long-Reads/Hybrids Improve Assemblies, and Real-Time Methods Enable Rapid, Portable Diagnostics

Technology	Platform Examples	Key Advantages	Limitations	Relevance to Metagenomics
Sanger sequencing (first generation, ~1977)	ABI capillary sequencers	Gold standard, high accuracy (Crossley et al., 2020)	Low throughput, costly, not suitable for large-scale studies	Early microbial sequencing, foundational but largely replaced
Short-read NGS (Illumina, ~2005 onward)	Illumina HiSeq, NovaSeq	High accuracy, high throughput, cost-effective (Slatko et al., 2018)	Short reads hinder assembly of complex genomes, poor in repetitive/structural regions	Taxonomic profiling, abundance estimation, gene catalogues
Long-read sequencing (PacBio/ONT, ~2010 onward)	PacBio HiFi, ONT MinION/ PromethION	Long reads resolve complex/ repetitive regions, better genome assembly, and structural variation detection (Espinosa et al., 2024)	Higher error rates (though HiFi improves this), higher cost	Enables near-complete microbial genomes, strain-level resolution
Hybrid sequencing (~2015 onward)	Illumina + PacBio/ONT	Combines high accuracy of short reads with long-read scaffolding (Miller et al., 2017)	More complex workflows, higher cost of integration	Produces high-quality assemblies and functional annotations
Real-time sequencing (~2016 onward)	Oxford Nanopore portable devices	Rapid turnaround, portable devices, in-field applications (P. Chen et al., 2023)	Error correction needed, variable throughput	On-site diagnostics, environmental monitoring, outbreak tracking

genomic elements, such as plasmids, transposons, and phages, which are usually fragmented or missed in short-read assemblies (Begum et al., 2021).

PacBio's HiFi technology combines long-read lengths with high base accuracy, which makes it well suited for producing reference-quality genomes from metagenomic samples (Benoit et al., 2024). On the other hand, ONT platforms have unique advantages in terms of portability and real-time analysis (Logsdon et al., 2020). MinION-like devices have enabled sequencing outside traditional laboratory settings, ranging from clinical bedside pathogen detection to on-site environmental monitoring in remotely situated field locations (Bloemen et al., 2023).

ONT real-time capabilities are transformative. Data can be analyzed simultaneously with sequencing, allowing rapid identification of pathogens during outbreaks, detection of antimicrobial resistance genes in clinical samples, and monitoring of microbial communities in agriculture and aquaculture (Taxt et al., 2020). This speed and mobility make ONT-based real-time sequencing highly attractive for diagnostics, biosecurity, and environmental surveillance (Barrero et al., 2017).

Challenges remain, as long-read sequencing platforms often require higher DNA input, which can be limiting for specific low-biomass samples, and exhibit increased error rates, although these are improving and remain higher than those of short-read platforms (De Coster et al., 2021). To balance accuracy with assembly continuity, hybrid approaches that integrate short- and long-read sequencing are increasingly used, offering robust solutions for reconstructing complex metagenomes (Goussarov et al., 2024).

15.2.2 Ultra-High Throughput

Illumina NovaSeq, BGI/MGI DNBSEQ, and Element Biosciences AVITI platforms deliver billions of reads per run, driving the shift from small-scale exploratory studies to population-level and global-scale microbial community profiling (Eisenstein, 2023; Kim et al., 2021; Liu et al., 2024). This increases the throughput and dramatically reduces per-base sequencing costs. These platforms are democratizing access to metagenomics and broadening its applications across research, medicine, agriculture, and ecology.

The ability of ultra-high-throughput sequencing is to generate deep coverage of microbial communities (Caporaso et al., 2012). This helps improve the detection of low-abundance organisms and enhances the resolution of strain-level variation, rare biosynthetic gene clusters, and antimicrobial resistance determinants (Goltsman et al., 2018; Nicholls et al., 2019).Sensitivity at this level is critical in clinical diagnostics, where detecting a rare and clinically relevant pathogen or resistance gene can directly impact patient outcomes.

The scaling of high throughput allows for comparative metagenomics at population scales. Extensive cohort studies that involve thousands of individuals can be undertaken to examine how host-associated microbiomes influence disease risk, treatment responses, and health trajectories. Similarly, in the fields of agriculture and environmental science, ultra-high-throughput sequencing enables large-scale soil, plant, and aquatic microbiome surveys, uncovering microbial patterns that inform sustainable practices and ecosystem management (Loos et al., 2024).

Reductions in sequencing costs have made global microbial diversity initiatives possible. Projects such as the Earth Microbiome Project and the Human Microbiome Project largely rely on high-throughput sequencing. These efforts help capture microbial diversity across continents and habitats (Gilbert et al., 2014). They are creating reference databases that will serve as benchmarks for diagnostics, evolutionary studies, and biotechnological exploration.

These throughput sequencing platforms present challenges, despite their revolutionary impact. Voluminous data require robust bioinformatics pipelines. To analyze the data, researchers require substantial storage infrastructure and standardized frameworks to ensure that generated results are reproducible and interoperable (Loos et al., 2024). The reliance of this technology on short-read sequencing remains dominant in ultra-high-throughput platforms, limiting assembly quality, fragmenting genomes, and obscuring complex structural features (Dohm et al., 2008). Hybrid approaches integrating short- and long-read sequencing are increasingly adopted to mitigate these limitations.

Continued improvements in throughput, accuracy, and cost efficiency are pushing metagenomics closer to real-time, population-scale surveillance systems. This allows microbial communities to be tracked across space and time, creating predictive frameworks for infectious disease outbreaks, antimicrobial resistance dissemination, agricultural productivity, and ecosystem resilience on a truly global scale.

15.3 INTEGRATION WITH MULTI-OMICS

Metagenomics coming frontier lies in its integration with other omics technologies which includes meta-transcriptomics, meta-proteomics, metabolomics, and host genomics (Fernandez-Carro et al., 2025; Gupta et al., 2025). While metagenomics provides insights into the genetic potential of microbial communities, it does not reveal which genes are actively expressed, which proteins are produced, or which metabolites are synthesized. By combining layers of biological information, researchers can now link microbial community composition to function and interaction, providing a systems-level perspective of microbiome dynamics. This shift from descriptive to functional and mechanistic frameworks will be central to the future of metagenomics.

15.3.1 Beyond Taxonomy

Traditionally, metagenomic studies have cataloged microbial diversity, answering the fundamental question, "Who is present in a given environment?" However, microbial community structure alone cannot fully explain ecological functions, disease processes, or interactions between hosts and microbes. The future of metagenomics lies in moving far beyond taxonomy to establish links between microbial identity and functional activity (National Research Council, Board on Life Sciences, Committee on Metagenomics, & Functional Applications, 2007).

One promising direction is the integration of metagenomics with metabolomics, which enables the characterization of microbial metabolite production (Puig-Castellví et al., 2023). Microbial metabolites, including short-chain fatty acids, bile acids, and secondary metabolites, play important roles in shaping host physiology, modulating immune responses, and influencing disease states (Masse & Lu, 2023). By linking genomic potential with metabolite profiles, researchers can predict how microbial communities contribute to health or pathology and identify novel biomarkers and therapeutic targets.

Combining metagenomics with meta-transcriptomics and meta-proteomics provides insights into gene expression and protein activity within microbial ecosystems (Dubey et al., 2020). These approaches can reveal pathways that are actively used under specific conditions. They also help researchers understand how microbes respond to environmental stresses and how host-associated microbiomes adapt during disease progression or treatment (Martin et al., 2014). One example is the study of host–microbe interactions that integrate immune profiling with microbiome transcriptomics, which is uncovering mechanistic pathways by which microbial communities influence inflammation, cancer progression, or metabolic disorders.

In environmental microbiology, this functional perspective is equally transformative. Metagenomics links community composition to biogeochemical processes, such as carbon cycling, nitrogen fixation, and methane production, providing predictive insights into ecosystem resilience and responses to climate change (Parab et al., 2025). Such functional understanding goes beyond biodiversity surveys and helps design strategies for ecosystem restoration, pollution mitigation, and sustainable agriculture.

Ultimately, researchers require advanced computational modeling and network analysis to move beyond taxonomy. Integrative frameworks coupled with machine learning and systems biology are being developed to connect microbial genomes to phenotypes, host responses, and environmental outputs (Qu et al., 2019). This predictive modeling capacity allows metagenomics not only to describe microbial communities but also to forecast their behavior, enabling translational applications in diagnostics, therapeutics, and environmental management.

15.3.2 Systems Biology Approaches

Integrating metagenomics with other techniques, such as transcriptomics, proteomics, and metabolomics, makes a transformative shift in the research field from descriptive microbiome research to systems biology. The merged sciences help researchers learn about microbial community dynamics as interconnected networks rather than static species assemblages (Yang et al., 2025). Systems biology approaches aim to reconstruct these networks, revealing individual microbial functions, their interactions, emergent properties, and contributions to host or environmental stability.

At the core of systems biology lies the understanding of microbial community behavior—that it cannot be fully explained by studying genomes alone and in isolation. Meta-transcriptomics reveals which genes are active and expressed under specific conditions, offering insights into regulatory responses to diet, antibiotics, or environmental stressors (Ota et al., 2024). Meta-proteomics extends this view by identifying protein structures that mediate biochemical processes, thereby linking gene expression to enzymatic activity (Shrestha et al., 2021). Metabolomics provides the final functional readout by characterizing the small molecules produced and exchanged within microbial networks and between microbes and their hosts.

When combined with metagenomic data, these layers enable researchers to construct systems-level models that describe microbial community interactions with one another, hosts, and their environment. Such models capture dynamic feedback loops—for instance, how microbial metabolites regulate host immune pathways, shaping microbial composition and creating complex bidirectional interactions (Rooks & Garrett, 2016).

In clinical research, systems biology approaches are yielding breakthroughs. Gut microbiome multi-omics studies have revealed that intricate microbial–host networks are linked to obesity, inflammatory bowel disease, diabetes, and cancer (Su et al., 2025). By integrating microbial gene expression, metabolite profiles, and host transcriptomics, researchers have identified mechanistic pathways and biomarkers. These are invisible to metagenomics alone (Johnson et al., 2016). These insights are laying the foundation for precision microbiome-based therapies.

In environmental microbiology, models designed at the system level can predict microbial contributions to biogeochemical cycles, such as carbon sequestration, methane emissions, and nitrogen cycling (American Society for Microbiology, 2023). By modeling microbial interactions within soil and marine ecosystems, researchers can now assess ecosystem resilience and forecast responses to climate change, pollution, or land-use alterations (Mesa-Marín et al., 2019).

The demand for computational power in systems biology is substantial. This power is used to build accurate predictive models and requires advances in bioinformatics, network inference algorithms, and machine learning tools. These tools are capable of integrating heterogeneous datasets across multiple scales. Emerging techniques, such as genome-scale metabolic models (GEMs) and agent-based modeling, allow simulation of microbial interactions under different conditions, offering powerful tools for hypothesis generation and testing (Lee et al., 2025).

By revealing how microbial communities work as integrated, adaptive systems, systems biology approaches elevate metagenomics from a descriptive science to a predictive and actionable discipline. This transition is required to translate microbiome research into real-world applications in healthcare, agriculture, biotechnology, and environmental stewardship.

15.4 SINGLE-CELL AND SPATIAL METAGENOMICS

Traditionally, metagenomics uses DNA pooling from microbial communities within a sample to generate composite profiles of microbes (Kumar Awasthi et al., 2020). This technique is powerful,

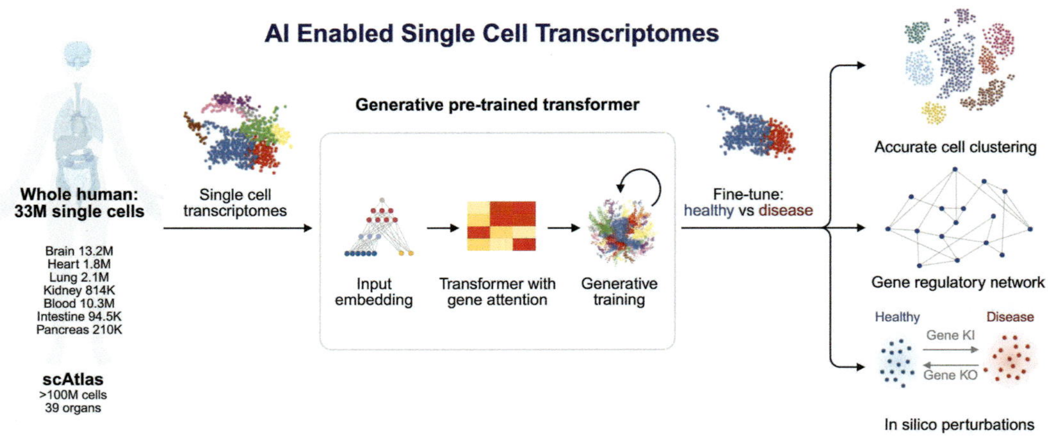

FIGURE 15.2 A transformer-based deep learning framework integrates single-cell transcriptomes from multiple human organs to model gene expression. The approach supports accurate cell clustering, gene regulatory network inference, and in silico perturbations, distinguishing healthy and disease states.

but the approach obscures the contributions of individual cells and overlooks spatial organization, both of which are critical for understanding microbial ecology (see Figure 15.2). New advances in single-cell and spatial metagenomics techniques are overcoming these limitations, providing high and accurate resolution into the heterogeneity of microbial communities and the micro-niches in which they operate (J. Chen et al., 2023).

In single-cell metagenomics, the genome of individual microbial cells is isolated and sequenced using techniques such as fluorescence-activated cell sorting (FACS), microfluidics, or laser capture microdissection, depending on the researcher's needs and the study (Yilmaz & Singh, 2012). Using this approach, genome recovery from unculturable organisms that are difficult to detect in bulk metagenomes becomes feasible. It helps researchers gain strain-level resolution and capture rare taxa and genetic variability within populations. It also provides insights into horizontal gene transfer, plasmid carriage, and functional heterogeneity across cells of the same species (Bedhomme et al., 2017). This technique help researchers to reconstruct genomes cell by cell, researchers can link specific organisms to ecological functions and phenotypes that would otherwise be masked in community-level sequencing. Low-abundance microbes, sometimes called the "microbial dark matter," can now be studied using single-cell approaches, as these microbes play important but hidden roles in ecosystems (Rinke, 2018).

While single-cell sequencing captures heterogeneity, it does not provide information on the spatial organization of microbial communities. Spatial metagenomics combines sequencing with imaging and in situ molecular techniques to map microbes within their native environments (Lee et al., 2025). Methods such as spatial transcriptomics, fluorescence in situ hybridization (FISH)-based approaches, and imaging mass spectrometry allow researchers to visualize how microbes are arranged and interact with one another and their hosts (Gu et al., 2022).

A spatial perspective is valuable and needed for understanding microbial niches in the gut, oral cavity, skin, soil, and plant rhizosphere (Lee et al., 2025). Microbe–host interactions, including proximity to epithelial cells, immune cells, or disease sites, are influenced by spatial organization. Spatial arrangement often dictates metabolic cross-feeding, syntrophy, and competition within communities (Martinez-Rabert et al., 2022).

In the human gut, spatial mapping has revealed structured microbial consortia in which specific taxa cluster in micro-niches that shape nutrient metabolism and immune modulation. In soils and marine sediments, this approach has uncovered microscale oxygen, pH, or nutrient gradients that define ecological niches for distinct microbial groups (Yang et al., 2023).

Both single-cell and spatial metagenomics are transforming our understanding of microbial ecology by providing direct links between genotype, phenotype, and habitat context. Both approaches are particularly powerful when combined with other multi-omics and systems biology frameworks. This enables scientists to develop models that capture microbial community composition and function, as well as their organization in space and time (Pinu et al., 2019). However, challenges remain, including the complexity of isolating cells or preserving spatial integrity, high costs, and intensive data analysis requirements for each sample. As technologies continue to mature, single-cell and spatial metagenomics will become central techniques in advancing microbiome research toward precision diagnostics, targeted therapeutics, and ecosystem engineering.

15.5 COMPUTATIONAL AND AI INNOVATIONS

The exponential growth of metagenomic data, fueled by advances in sequencing technologies, has made computational innovation crucial. Modern microbiome studies routinely generate terabytes of raw sequence data that must be processed, assembled, annotated, and interpreted (Navgire et al., 2022). Standard tools are insufficient for such scale and complexity, which has led to the development of next-generation bioinformatics pipelines, including machine learning methods and AI-driven frameworks. These computational innovations enable researchers to extract meaningful biological insights from massive, heterogeneous datasets, ultimately transforming metagenomics into a predictive and translational science.

15.5.1 Advanced Bioinformatics

Currently, metagenomics faces various challenges, among which one is the reconstruction of microbial genomes from highly complex community datasets. This is because DNA often consists of fragments from hundreds or even thousands of organisms, many of which are closely related (Howe & Chain, 2015). In such cases, advanced bioinformatics assembly algorithms, such as MEGAHIT, metaSPAdes, and Hi-C-based scaffolding methods, are useful, as they allow the generation of longer, more accurate contigs and, in some cases, complete microbial genomes from metagenomic data (Navgire et al., 2022). Binning approaches, like MetaBAT, CONCOCT, and MaxBin, use sequence composition, coverage, and linkage information to cluster contigs into metagenome-assembled genomes (MAGs), providing strain-level resolution (Navgire et al., 2022). Complementing these, annotation pipelines such as Prokka, DRAM, and eggNOG-mapper automate the functional characterization of genes and metabolic pathways, building connections between community composition and ecological or clinical functions (Cantalapiedra et al., 2021; Shaffer et al., 2024).

These improvements allow researchers to move beyond fragmentary snapshots toward high-resolution reconstructions of microbial community structure and function, supporting the discovery of novel enzymes, biosynthetic pathways, and microbial interactions.

With the increase in automation of bioinformatics pipelines, features such as user-friendliness and cloud-based access have reduced barriers for non-specialists. Frameworks like QIIME 2, MG-RAST, and Kraken2 streamline workflows for taxonomic classification, diversity analysis, and functional annotation (Kibegwa et al., 2020). Using cloud platforms such as Google Cloud and AWS, global collaboration of reproducible pipelines is possible (Gnimpieba et al., 2024). These advances in clinical and environmental applications, are rapid and leads to accurate data interpretation. For example, automated bioinformatics pipelines can quickly identify pathogens, resistance genes, or functional biomarkers in hospital settings. At the same time, in environmental monitoring, they can track microbial responses to pollution or climate change at the ecosystem level.

Despite significant progress, much work remains, as metagenomic datasets are noisy, heterogeneous, and biased by DNA extraction, sequencing depth, and reference database limitations (Nearing et al., 2021). This provides a path for new algorithms to handle novel organisms not represented in current databases and to integrate multi-omics data for a systems-level understanding. The future lies in the fusion of advanced bioinformatics with AI and ML, which will provide powerful predictive models that allow researchers to simulate microbial interactions, forecast ecosystem dynamics, and design microbiome-based interventions with accuracy (Shahid, 2025 (Shahid, 2025).

15.5.2 Machine Learning Applications

The growing size and complexity of metagenomic datasets mean that traditional statistical methods are restricted in capturing nonlinear relationships, high dimensionality, and hidden patterns within microbial communities (Aplakidou et al., 2024). Using ML and deep learning (DL) can overcome these challenges, providing predictive and translational insights from microbiome data (Table 15.2).

TABLE 15.2
Comparative Summary of Machine Learning Methods Used in Metagenomics, Highlighting Their Key Characteristics, Strengths, Limitations, and Applications

Method	Key Characteristics	Advantages	Limitations	Applications
Random Forest (RF) (Wassan et al., 2023)	Ensemble of decision trees with bagging	Robust to noise, interpretable feature importance	Can overfit with noisy features, less effective with very high-dimensional sparse data	Biomarker discovery, taxonomic classification, feature selection
Support Vector Machines (SVMs) (Liu et al., 2013)	Separates data using hyperplanes; kernel trick	Accurate with small/medium datasets, works well in high-dimensional spaces	Computationally expensive for large datasets, kernel selection critical	Disease classification, gene expression analysis, AMR prediction
Artificial Neural Networks (ANNs) (Herazo-Alvarez et al., 2025)	Layers of interconnected neurons	Captures complex relationships, flexible	Needs large data, risk of overfitting, less interpretable	Toxicity prediction, biological activity prediction, functional genomics
Convolutional Neural Networks (CNNs) (Al-Ajlan & El Allali, 2019)	Uses convolutions for spatial/structured data	Detects local motifs/patterns, scalable	High computational cost, less useful for tabular data	Protein/DNA motif detection, metagenomic read classification, biomedical imaging
Recurrent Neural Networks (RNNs)/LSTMs (Sherstinsky, 2020)	Sequence-aware neural networks	Models long-term dependencies	Training instability (vanishing/exploding gradients), slow	RNA-seq dynamics, temporal omics, clinical outcome prediction
Graph Neural Networks (GNNs) (Herazo-Alvarez et al., 2025)	Learns from graph-structured data (nodes/edge)	Captures relationships and network effects	Computationally intensive, complex training	Protein–protein interaction networks, drug–target interactions, microbiome networks
Autoencoders (AEs)/Variational Aes (Lindez et al., 2023)	Dimensionality reduction via encoding-decoding	Good for denoising, latent feature learning	Interpretability limited	Omics data compression, anomaly detection, genome/metabolome representations
Transformers (Wang et al., 2025)	Attention-based models for long-range dependencies	State-of-the-art performance, scalable	Requires huge data and compute, interpretability	Protein language models (AlphaFold), large-scale genomics, functional annotation

ML models have the capability to process thousands of microbial features, including taxa, genes, pathways, and metabolites. These profiles help identify patterns that are difficult to detect using conventional methods. In metagenomics, these applications are diverse and impactful. ML models are used to predict community shifts by forecasting how microbial populations respond to disease, diet, drugs, or environmental stressors. For example including models trained on gut microbiome data that can early diagnose type 2 diabetes, inflammatory bowel disease, and colorectal cancer (Caballero-María et al., 2025; Gibbons et al., 2022). It helps to explore biomarker discovery, where feature selection algorithms such as random forests and support vector machines identify microbial taxa, gene signatures, or metabolites that serve as biomarkers for diagnosis, prognosis, or treatment, particularly in complex diseases where single biomarkers are insufficient (Acharjee et al., 2020). ML approaches are used to explore therapeutic targets by mapping microbial features to host phenotypes, supporting the development of potential microbiome-based products such as probiotics, prebiotics, or microbiome-modulating drugs (Preidis & Versalovic, 2009). In environmental science, ML helps monitor ecosystems by predicting microbial responses to climate change and nutrient cycling, guiding ecosystem restoration (Panda et al., 2025; Madigan et al., 2025).

Advances in deep learning architectures have further enhanced the capacity to model complex microbiome data. DL methods are used to model microbial interaction networks, predict functional gene expression, and simulate community dynamics under perturbation (Lee et al., 2020).

ML is not only descriptive but also predictive and prescriptive. When combined with experimental validation, these models can be used to design personalized microbiome therapies, forecast antibiotic resistance evolution, and identify early warning signals of ecosystem collapse (Marcos-Zambrano et al., 2021). Although these advancements exist, some ML models still suffer from overfitting, limited interpretability, and bias due to uneven training datasets (Hassija et al., 2023; Kumar et al., 2023). Microbiome data are inherently noisy due to variability introduced by sample collection, sequencing protocols, and batch effects (Zhou et al., 2023; Bisht et al., 2025). Researchers are currently developing interpretable AI (XAI) methods and standardized pipelines that promote robustness, reproducibility, and clinical reliability (Band et al., 2023). In the future, machine learning will remain central to metagenomics. As datasets grow through global microbiome initiatives and real-time sequencing, ML-powered frameworks will be capable of predicting, manipulating, and optimizing microbial communities across health, agriculture, and environmental systems.

15.5.3 Cloud and Federated Frameworks

As the complexity of datasets increases, the demand for infrastructure also shifts. This leads to the necessity of techniques that can manage massive datasets, heterogeneous workflows, and international collaborations. Traditional local computing resources usually have restricted capacity for such demands, leading to the adoption of new cloud-based platforms and emerging federated analysis frameworks. These changes are essential for enabling scalable, collaborative, and secure microbiome research across the globe.

Cloud-based analysis platforms provide flexible, on-demand computational power and storage capacity making processing of large metagenomics datasets easier (Belmann et al., 2025). Amazon Web Services (AWS), Google Cloud, and Microsoft Azure support scalable pipelines for sequence assembly, genome binning, and functional annotation (Ahmed et al., 2021). Platforms like MG-RAST, QIIME 2 and Galaxy on cloud servers are community-driven. These lower technical barriers by offering user-friendly interfaces and reproducible workflows (Batut et al., 2018; Meyer et al., 2019; Kumar et al., 2019). Cloud computing also provides real-time data sharing and collaboration, that helps researchers across globe to access the same datasets, tools, and results without the need for local high-performance computing systems.

Particularly in clinical and microbiome studies, issues such as data privacy and sovereignty emerge, which are easily addressed by federated frameworks (Casaletto et al., 2023). These frameworks allow researchers to perform ML and statistical modeling across distributed datasets without transferring raw data to a central location. Models can be trained locally on institutional datasets, and only aggregated results or model parameters are shared. By enabling large-scale comparative analyses across cohorts and populations, these frameworks help preserve patient confidentiality, ensure regulatory compliance (e.g., GDPR, HIPAA), and maintain institutional data ownership (Pati et al., 2024).

For global health and diagnostics, federated frameworks are important tools, as they allow combining microbiome datasets across hospitals, countries, or continents, enhancing biomarker discovery and predictive modeling (Brisimi et al., 2018). Federated approaches expand the reach of metagenomics into precision medicine and public health surveillance by enabling cross-cohort comparisons without centralizing sensitive patient data. Cloud-based workflows can incur high costs, require specialized expertise, and raise concerns about data security. Federated frameworks, while promising, face technical issues in standardizing metadata and harmonizing workflows to ensure reproducibility across heterogeneous datasets (Rujano et al., 2024). But with maturity, these frameworks will enable global collaboration, leading to ethical and computationally efficient metagenomic research.

In the future, the combination of cloud platforms, federated learning, and AI-driven pipelines will be beneficial, as these will enable population-scale, real-time metagenomics, transforming microbiome science into a globally connected discipline with direct clinical, agricultural, and environmental applications.

15.6 TRANSLATIONAL APPLICATIONS

Metagenomics has progressed from basic research into applied domains, where it can directly impact human health, agriculture, biotechnology, and environmental sustainability. By translating genomic and multi-omics insights into actionable outcomes, metagenomics has emerged as a cornerstone of diagnostics, precision medicine, and public health surveillance (Afshinnekoo et al., 2017). Among which clinical diagnostics is one the most immediate and transformative area.

15.6.1 Diagnostics

Metagenomics offers a drastic shift in diagnostics by promoting the culture-independent identification of pathogens, antimicrobial resistance (AMR) genes, and microbial community signatures linked to disease (Liu et al., 2023). Unlike older diagnostic methods, which rely on culturing microbes or targeted molecular assays such as PCR or qPCR, metagenomics provides an unbiased, comprehensive snapshot of all microbial DNA or RNA present in a sample (Chiu & Miller, 2019). This approach has many advantages, including pathogen detection using shotgun metagenomic sequencing, which can detect both known and novel pathogens without prior assumptions (Leung et al., 2020). During outbreaks of infectious diseases, metagenomics has been used to identify causative agents and track their spread (Forbes et al., 2017). Metagenomic pipelines are used to identify Antimicrobial resistance (AMR) genes and plasmids directly from clinical samples, enabling clinicians to tailor antibiotic therapies and monitor resistance evolution (Olsen & Riber, 2025). Beyond infectious disease, metagenomic profiling is also used to identify microbial consortia associated with inflammatory bowel disease, colorectal cancer, obesity, diabetes, and even neurological disorders (Rujano et al., 2024).

A drastic innovation in this area is the rise of portable sequencing platforms, especially Oxford Nanopore's MinION and Flongle devices (Akamatsu et al., 2024). These tools allow clinicians to perform real-time sequencing at the bedside or in field hospitals, bypassing centralized laboratories. Portable sequencing has been deployed for Ebola and COVID-19 outbreak surveillance, and in hospital settings, it enables rapid turnaround for diagnosing sepsis, respiratory infections, and drug-resistant pathogens (Quick et al., 2016). Metagenomic diagnostics also offer value in non-traditional clinical contexts. For example, prenatal and neonatal care can benefit from detecting pathogenic microbes in amniotic fluid or neonatal intensive care units before they cause severe infections (Agudelo-Perez et al., 2023). In the field of Oncology, profiling the tumor microbiome to identify microbial biomarkers linked to cancer initiation, progression, or therapy response (Zhang et al., 2025). Monitoring the microbiome of organ donors and recipients to reduce infection risk and improve transplant outcomes (Schreiber et al., 2019).

Despite these advances, significant challenges remain before metagenomic diagnostics can be routinely implemented in clinical practice. Key barriers include time, cost, regulatory approval, clinical validation, and data interpretation (Greninger, 2018). In particular, distinguishing between colonization and actual infection remains a concern. Moreover, the lack of standardized workflows and reference databases complicates comparisons and reproducibility. Advancements in automated bioinformatics pipelines, AI-driven pathogen identification, and cloud-based clinical decision support tools that helps to integrate metagenomic diagnostics into routine healthcare (Ren et al., 2025). The combination of portable sequencing and real-time analytics can transform hospitals, field clinics, and even home testing into rapid, precision diagnostic hubs.

15.6.2 Therapeutics and Precision Medicine

Apart from diagnostics, metagenomics shapes the development of therapeutics that target the microbiome. By revealing the structure, function, and dynamics of microbial communities at the individual level, metagenomics enables the design of personalized therapies unique patient's microbiome profile (see Figure 15.3) (Sun et al., 2025). This shift represents a key of precision medicine, where treatments are informed not only by host genetics and lifestyle but also by the microbial ecosystem that influences health and disease.

One promising therapeutic avenue involves engineering microbial consortia to restore or enhance beneficial functions. Instead of traditional single-strain probiotics, synthetic communities comprising carefully selected and engineered microbes can be designed to perform targeted functions, such as producing anti-inflammatory metabolites, degrading toxins, or outcompeting pathogens

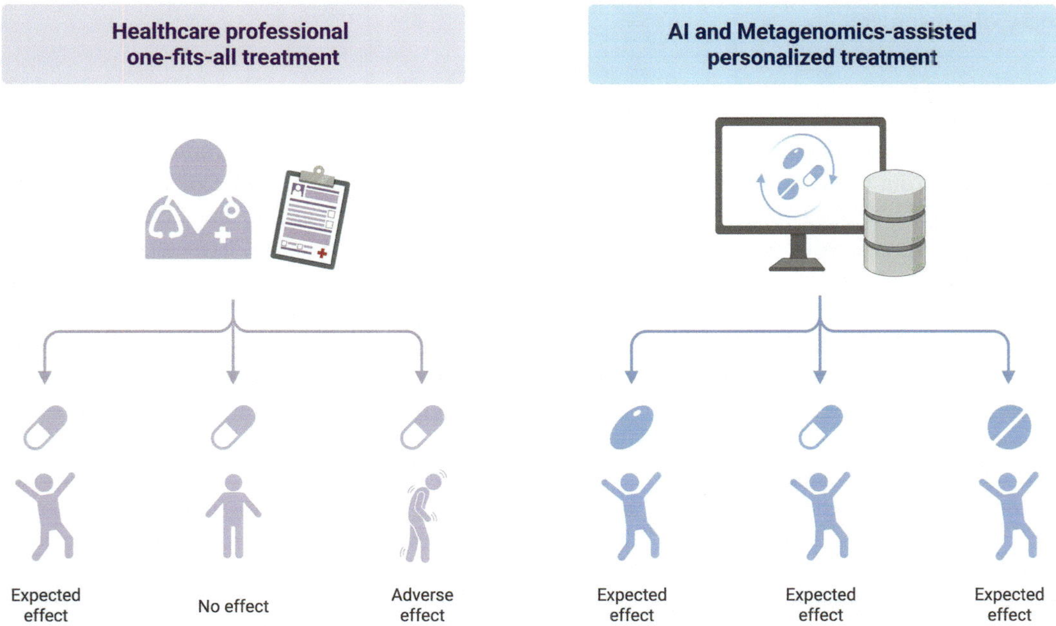

FIGURE 15.3 A transformer-based deep learning framework integrates single-cell transcriptomes from multiple human organs to model gene expression. The approach supports accurate cell clustering, gene regulatory network inference, and in silico perturbations, distinguishing healthy and disease states.

(Dosoky et al., 2020). For example, microbial consortia engineered to produce short-chain fatty acids (SCFAs) are being investigated as therapies for inflammatory bowel disease and metabolic disorders (Sankarganesh et al., 2025). Advances in synthetic biology are accelerating this field. Technologies have emerged that can engineer microbes with synthetic circuits that can sense disease biomarkers and respond by producing therapeutic molecules, offering the potential for medicines that dynamically adapt to the host environment. Metagenomics plays a critical role here by identifying microbial genes, pathways, and interactions that inform the design of these engineered pools.

Traditional probiotics and prebiotics have variable effects on different individuals due to differences in host genetics, diet, and microbiome composition (Petschow et al., 2013). With the help of Metagenomic profiling developing personalized probiotics and prebiotics matching to the patient's existing microbiome is easier. For example, if a patient's gut lacks distinct butyrate-producing taxa, the intervention aims to restore these functions through targeted microbial supplementation or dietary strategies (Zhu et al., 2021). Similarly, fecal microbiota transplantation (FMT), traditionally used for treating recurrent *Clostridioides difficile* infections, can be refined using metagenomic analyses to ensure donor–recipient compatibility, helping reduce risks and improve therapeutic outcomes (Baunwall et al., 2020). Future developments may replace whole FMT with distinct microbial cocktails, guided by metagenomic insights. Metagenomic data also facilitate the discovery of novel therapeutic targets by linking microbial genes and metabolites to disease progression. For example, specific microbial pathways have been implicated in drug metabolism, influencing efficacy and toxicity (Li et al., 2016). By profiling patient microbiomes, clinicians could anticipate drug–microbiome interactions, enabling delivery of personalized treatments.

Despite promising advancements, the field faces challenges that include regulatory approval, standardization of manufacturing processes, safety concerns, and interindividual variability. Long-term studies are required to evaluate the stability and efficacy of engineered and personalized probiotics, particularly in studies involving genetically modified organisms, where ethical considerations are required. However, as sequencing costs fall and computational models improve, metagenomics will increasingly enable precision therapeutics for microbial, host, and environmental factors (Ariaeenejad et al., 2024). This integration will make microbiome-based therapies a key of personalized medicine, with demand in infectious diseases, oncology, autoimmune conditions, and metabolic disorders.

15.6.3 Agriculture and Environmental Management

Metagenomics has drastically influenced healthcare, agriculture, and environmental management by offering powerful tools to manage food security, sustainability, and ecological resilience (Jagadesh et al., 2024). This is achieved by providing insights that can be translated into sustainable farming practices, biodiversity conservation, and bioremediation strategies. Characterizing the taxonomic and functional diversity of microbial communities in soil, crops, animals, and ecosystems has significantly influenced the area (Jagadesh et al., 2024).

Earth's most diverse microbial ecosystems, nutrient cycling, plant health, and agricultural productivity are derived from soil (Chen et al., 2024). Using metagenomic profiling, beneficial taxa involved in these processes (Jain et al., 2024). These insights allow the design of microbial inoculants and biofertilizers that reduce dependence on synthetic agrochemicals (Alori & Babalola, 2018). For example, metagenomics-mediated processes can promote colonization by plant growth-promoting rhizobacteria (PGPR) or arbuscular mycorrhizal fungi, improving nutrient uptake and crop yields (Bigatton et al., 2024). Likewise, monitoring soil microbiomes under different conditions, such as organic farming, crop rotation, or no-till agriculture, helps identify strategies that maintain soil fertility and microbial resilience (Suman et al., 2022).

Using metagenomics one can explore about crop and livestock by exploring host-associated microbiomes (Zhuang et al., 2024). Microbes on plant's leaves and roots play critical roles in pathogen defense and stress tolerance (Singh et al., 2024). In livestock, gut microbiome analysis can unwind microbial contributions to feed efficiency, disease resistance, and methane emissions (Badhan et al., 2025). Further using these analysis development of tailored probiotics, dietary interventions, and microbial management practices can be made.

Beyond agriculture, metagenomics is becoming valuable for ecosystem monitoring and conservation biology (Rout et al., 2025). By studying microbes in soil, freshwater, marine environments, and even air, analysts can track responses to pollution, habitat loss, and climate change (American Society for Microbiology, 2022). Fluctuations in microbial diversity often serve as indicators of ecosystem degradation. In biodiversity conservation, metagenomic surveys provide microbial data for endangered habitats and species, supporting conservation strategies that account for both microbial and macro-organismal variations (Stumpf et al., 2016). With advances in sequencing environmental DNA (eDNA), rare or invasive species can be detected by analyzing microbial traces in water or soil samples (Sahu et al., 2023).

Metagenomics also paves the way for discovery of microbes and enzymes capable of degrading pollutants such as hydrocarbons, plastics, pesticides, and heavy metals (Rout et al., 2025). These results enhance bioremediation strategies that harness natural microbial processes to detoxify contaminated environments (Borthakur et al., 2022). Besides, the role of microbial communities in carbon sequestration, methane cycling, and greenhouse gas regulation positions metagenomics as a vital tool for understanding and mitigating climate change (Iqbal et al., 2025).

Metagenomics still faces some limitations. Scaling laboratory findings to field conditions, ensuring microbial inoculant stability, and balancing productivity with ecological sustainability require further research. Additionally, robust regulatory frameworks are needed to guide the application of engineered microbes in open environments. Integrating metagenomics with remote sensing, precision agriculture technologies, and ecological modeling will provide real-time monitoring of agroecosystems and natural environments (Aderele et al., 2025; Ambaru et al., 2025). Such unification provide assurance for achieving sustainable agriculture, climate resilience, and ecosystem restoration in the current era.

15.7 FUTURE CHALLENGES AND DIRECTIONS

Although there has been remarkable progress in the field of metagenomics, it still faces several challenges. These challenges must be addressed before fully realizing its potential across medicine, agriculture, and environmental management. One major limitation is the lack of standardized methods. Variations in sample collection, DNA extraction, sequencing strategies, and bioinformatic pipelines lead to inconsistencies across studies, making it difficult to compare results among laboratories and integrate findings into meta-analyses. Therefore, developing internationally recognized standards and benchmarking datasets is essential to ensure reproducibility and build confidence in metagenomic applications, particularly in clinical and regulatory contexts.

Another challenge lies in the limitations of reference databases. The majority of data come from easily available and well-cultured microbes, but the vast majority of microbial diversity, sometimes called microbial "dark matter," remains uncharacterized (Li et al., 2023). This gap hampers precise taxonomic classification and functional annotation. Expanding databases to include genomes from rare, unculturable, and environment-specific taxa, viruses, and mobile genetic elements is crucial for advancing the field. Approaches such as single-cell genomics, metagenome-assembled genomes, and large-scale sequencing initiatives hold promise for filling this gap.

Additionally, the field needs to transform from descriptive surveys of microbial communities to developing predictive models that connect microbial composition and function with health outcomes, agricultural productivity, and ecosystem resilience. This shift requires integrating metagenomic data with other multi-omic approaches, longitudinal studies, and experimental validation. Advanced computational tools play an important role in uncovering hidden patterns and building models capable of predicting microbial community effects on disease, diet, environmental stress, or interventions (Marcos-Zambrano et al., 2021).

The growth of metagenomic research brings challenges for data management, ethics, and security, as high-throughput sequencing generates massive datasets that require robust storage, analysis, and sharing platforms (Bhuvaneshwar et al., 2015). Cloud-based and federated frameworks offer solutions but raise concerns about privacy, intellectual property, and equitable access, especially involving human-associated microbiomes (Zhang et al., 2023). Establishing clear ethical guidelines will be important to ensure that data sharing promotes collaboration while protecting sensitive information.

Translating metagenomic discoveries into practical applications remains a challenge. In healthcare, diagnostic tools based on metagenomics must demonstrate clinical utility and cost-effectiveness before routine implementation. In agriculture and environmental management, microbial inoculants, engineered communities, or bioremediation strategies must be validated for safety, stability, and ecological impact (Ibrahim et al., 2025; Trabelsi & Mhamdi, 2013). Overcoming these barriers requires strong collaborations among researchers, clinicians, policymakers, and industry stakeholders.

Microbial community science is complex and interdisciplinary, demanding collaborative approaches. No single discipline can fully explain the interactions between microbes, hosts, and environments. Progress relies on teamwork that brings together microbiologists, ecologists, clinicians, computational scientists, engineers, and data ethicists. International initiatives and data-sharing networks are particularly important for addressing global challenges such as antimicrobial resistance, climate change, food insecurity, and emerging infectious diseases (Proctor et al., 2025).

In summary, addressing these challenges through standardization, database expansion, predictive modeling, ethical data governance, translational pipelines, and cross-disciplinary collaboration will determine the trajectory of metagenomics in the coming decades. If these challenges are overcome, metagenomics will deepen our understanding of microbial communities and provide actionable insights for advancing precision medicine, sustainable agriculture, environmental stewardship, and global health.

15.8 CONCLUSION

Metagenomics has transformed how microbial communities are studied. It enables the classification of microbial diversity and the extraction of predictive, functional, and actionable insights. By combining sequencing platforms, integrative multi-omics approaches, and computational frameworks, our understanding of how microbes shape health, disease, agriculture, and the environment has been re-evaluated. By uniting exploration with application, metagenomics has evolved into a transformative tool for innovation, transitioning from a discovery-driven science.

Metagenomics lays the foundation for healthcare, particularly for next-generation diagnostics, personalized therapeutics, and precision medicine approaches that tailor interventions to the specific microbial signatures of individuals. In agriculture, it provides strategies for enhancing soil fertility, improving crop yields, and promoting sustainable farming through microbiome-informed management. In environmental science, metagenomics serves as both a monitoring system and a solution, enabling early detection of ecosystem disturbances and the development of microbial strategies for conservation and bioremediation.

The future of metagenomics involves addressing challenges related to methodological standardization, database expansion, predictive modeling, ethical data governance, and translational barriers. Overcoming these hurdles will require interdisciplinary collaboration and global initiatives that bring together microbiologists, clinicians, ecologists, computational scientists, and policymakers.

Metagenomics is not just a technological advancement; it represents a biological shift. It can transform diagnostics, medicine, agriculture, and ecosystem sustainability by revealing the hidden world of microbial communities and linking their dynamics to human and planetary health. As sequencing, computational, and translational innovations converge, the field holds the promise of becoming a foundational pillar of the life sciences in the 21st century.

15.9 CREDIT AUTHORSHIP CONTRIBUTION STATEMENT

V.R.: Conceptualization, Literature Review, Original draft writing and Chapter preparation. S.R.R.G.: Conceptualization, Figure preparation, Original draft writing and Chapter preparation. M.K.: Original draft preparation. I.K.S.: Conceptualization, Reviewing, Editing, Supervision and Chapter approval.

ACKNOWLEDGMENT

V.R. acknowledges the CSIR for the CSIR SRF-Direct Fellowship. S.R.R.G. acknowledges the Indian Council of Medical Research (ICMR), Delhi, India (Grant No. 2020 – 1353,33/14/2023/RD/BMS) for Fellowship. The authors acknowledge the University of Delhi for providing access to https://BioRender.com, which was used to prepare all figures. Further, authors acknowledge the use of grammar and language enhancement tools, to enhance the overall quality of the text. However, the content and ideas presented in this work are entirely original and were not generated by the tools mentioned above.

CONFLICT OF INTEREST

The authors declare that the work was conducted without any commercial or financial relationships that could be construed as a potential conflict of interest.

REFERENCES

Acharjee, A., Larkman, J., Xu, Y., Cardoso, V. R., & Gkoutos, G. V. (2020). A random forest based biomarker discovery and power analysis framework for diagnostics research. *BMC Med Genomics*, *13*(1), 178. https://doi.org/10.1186/s12920-020-00826-6

Aderele, M. O., Srivastava, A. K., Butterbach-Bahl, K., & Rahimi, J. (2025). Integrating machine learning with agroecosystem modelling: Current state and future challenges. *European Journal of Agronomy*, *168*, 127610. https://doi.org/10.1016/j.eja.2025.127610

Afshinnekoo, E., Chou, C., Alexander, N., Ahsanuddin, S., Schuetz, A. N., & Mason, C. E. (2017). Precision metagenomics: Rapid metagenomic analyses for infectious disease diagnostics and public health surveillance. *J Biomol Tech*, *28*(1), 40–45. https://doi.org/10.7171/jbt.17-2801-007

Agudelo-Perez, S., Fernandez-Sarmiento, J., Rivera Leon, D., & Pelaez, R. G. (2023). Metagenomics by next-generation sequencing (mNGS) in the etiological characterization of neonatal and pediatric sepsis: A systematic review. *Front Pediatr*, *11*, 1011723. https://doi.org/10.3389/fped.2023.1011723

Ahmed, Z., Renart, E. G., & Zeeshan, S. (2021). Genomics pipelines to investigate susceptibility in whole genome and exome sequenced data for variant discovery, annotation, prediction and genotyping. *PeerJ*, *9*, e11724. https://doi.org/10.7717/peerj.11724

Akamatsu, S., Mitsuhashi, S., Soga, K., Mizukami, H., Shiraishi, M., Frith, M. C., & Yamano, Y. (2024). Targeted nanopore sequencing using the Flongle device to identify mitochondrial DNA variants. *Sci Rep*, *14*(1), 25161. https://doi.org/10.1038/s41598-024-75749-8

Al-Ajlan, A., & El Allali, A. (2019). CNN-MGP: Convolutional neural networks for metagenomics gene prediction. *Interdiscip Sci*, *11*(4), 628-635. https://doi.org/10.1007/s12539-018-0313-4

Alori, E. T., & Babalola, O. O. (2018). Microbial inoculants for improving crop quality and human health in Africa. *Front Microbiol*, *9*, 2213. https://doi.org/10.3389/fmicb.2018.02213

Alsulimani, A., Akhter, N., Jameela, F., Ashgar, R. I., Jawed, A., Hassani, M. A., & Dar, S. A. (2024). The impact of artificial intelligence on microbial diagnosis. *Microorganisms*, *12*(6). https://doi.org/10.3390/microorganisms12061051

Alves, L. F., Westmann, C. A., Lovate, G. L., de Siqueira, G. M. V., Borelli, T. C., & Guazzaroni, M. E. (2018). Metagenomic approaches for understanding new concepts in microbial science. *Int J Genomics*, *2018*, 2312987. https://doi.org/10.1155/2018/2312987

Ambaru, B., Manvitha, R., & Madas, R. (2025). Synergistic integration of remote sensing and soil metagenomics data: Advancing precision agriculture through interdisciplinary approaches [opinion]. *Frontiers in Sustainable Food Systems*, *8*. https://doi.org/10.3389/fsufs.2024.1499973

American Society for Microbiology. (2022). Microbes and climate change—science, people & impacts: Report on an American academy of microbiology. In *Microbes and climate change—science, people & impacts: Report on an American academy of microbiology virtual colloquium held on November 5, 2021*. https://doi.org/10.1128/AAMCol.Nov.2021

American Society for Microbiology. (2023). The role of microbes in mediating methane emissions. In *The role of microbes in mediating methane emissions*. https://doi.org/10.1128/AAMCol.Nov.2023

Aplakidou, E., Vergoulidis, N., Chasapi, M., Venetsianou, N. K., Kokoli, M., Panagiotopoulou, E., Iliopoulos, I., Karatzas, E., Pafilis, E., Georgakopoulos-Soares, I., Kyrpides, N. C., Pavlopoulos, G. A., & Baltoumas, F. A. (2024). Visualizing metagenomic and metatranscriptomic data: A comprehensive review. *Comput Struct Biotechnol J*, *23*, 2011–2033. https://doi.org/10.1016/j.csbj.2024.04.060

Ariaeenejad, S., Gharechahi, J., Foroozandeh Shahraki, M., Fallah Atanaki, F., Han, J. L., Ding, X. Z., Hildebrand, F., Bahram, M., Kavousi, K., & Hosseini Salekdeh, G. (2024). Precision enzyme discovery through targeted mining of metagenomic data. *Nat Prod Bioprospect*, *14*(1), 7. https://doi.org/10.1007/s13659-023-00426-8

Badhan, A., Wang, Y., Terry, S., Gruninger, R., Guan, L. L., & McAllister, T. A. (2025). Invited review: Interplay of rumen microbiome and the cattle host in modulating feed efficiency and methane emissions. *J Dairy Sci*, *108*(6), 5489–5501. https://doi.org/10.3168/jds.2024-26063

Band, S., Yarahmadi, A., Hsu, C.-C., Biyari, M., Sookhak, M., Ameri, R., Dehzangi, I., Chronopoulos, A. T., & Liang, H.-W. (2023). Application of explainable artificial intelligence in medical health: A systematic review of interpretability methods. *Informatics in Medicine Unlocked*, *40*, 101286. https://doi.org/10.1016/j.imu.2023.101286

Barrero, R. A., Napier, K. R., Cunnington, J., Liefting, L., Keenan, S., Frampton, R. A., Szabo, T., Bulman, S., Hunter, A., Ward, L., Whattam, M., & Bellgard, M. I. (2017). An internet-based bioinformatics toolkit for plant biosecurity diagnosis and surveillance of viruses and viroids. *BMC Bioinformatics*, *18*(1), 26. https://doi.org/10.1186/s12859-016-1428-4

Batut, B., Gravouil, K., Defois, C., Hiltemann, S., Brugere, J. F., Peyretaillade, E., & Peyret, P. (2018). ASaiM: A Galaxy-based framework to analyze microbiota data. *Gigascience*, *7*(6). https://doi.org/10.1093/gigascience/giy057

Baunwall, S. M. D., Lee, M. M., Eriksen, M. K., Mullish, B. H., Marchesi, J. R., Dahlerup, J. F., & Hvas, C. L. (2020). Faecal microbiota transplantation for recurrent Clostridioides difficile infection: An updated systematic review and meta-analysis. *EClinicalMedicine*, *29–30*, 100642. https://doi.org/10.1016/j.eclinm.2020.100642

Bedhomme, S., Perez Pantoja, D., & Bravo, I. G. (2017). Plasmid and clonal interference during post horizontal gene transfer evolution. *Mol Ecol*, *26*(7), 1832–1847. https://doi.org/10.1111/mec.14056

Begum, G., Albanna, A., Bankapur, A., Nassir, N., Tambi, R., Berdiev, B. K., Akter, H., Karuvantevida, N., Kellam, B., Alhashmi, D., Sung, W. W. L., Thiruvahindrapuram, B., Alsheikh-Ali, A., Scherer, S. W., & Uddin, M. (2021). Long-read sequencing improves the detection of structural variations impacting complex non-coding elements of the genome. *Int J Mol Sci*, *22*(4). https://doi.org/10.3390/ijms22042060

Belmann, P., Osterholz, B., Kleinbolting, N., Puhler, A., Schluter, A., & Sczyrba, A. (2025). Metagenomics-toolkit: The flexible and efficient cloud-based metagenomics workflow featuring machine learning-enabled resource allocation. *NAR Genom Bioinform*, *7*(3), lqaf093. https://doi.org/10.1093/nargab/lqaf093

Benoit, G., Raguideau, S., James, R., Phillippy, A. M., Chikhi, R., & Quince, C. (2024). High-quality metagenome assembly from long accurate reads with metaMDBG. *Nat Biotechnol*, *42*(9), 1378–1383. https://doi.org/10.1038/s41587-023-01983-6

Bhuvaneshwar, K., Sulakhe, D., Gauba, R., Rodriguez, A., Madduri, R., Dave, U., Lacinski, L., Foster, I., Gusev, Y., & Madhavan, S. (2015). A case study for cloud based high throughput analysis of NGS data using the globus genomics system. *Comput Struct Biotechnol J*, *13*, 64–74. https://doi.org/10.1016/j.csbj.2014.11.001

Bigatton, E. D., Verdenelli, R. A., Haro, R. J., Ayoub, I., Barbero, F. M., Martin, M. P., Dubini, L. E., Jorrin Novo, J. V., Lucini, E. I., & Castillejo, M. A. (2024). Metagenomic analysis to assess the impact of plant growth-promoting rhizobacteria on peanut (Arachis hypogaea L.) crop production and soil enzymes and microbial diversity. *J Agric Food Chem*, *72*(40), 22385–22397. https://doi.org/10.1021/acs.jafc.4c05687

Bisht, D., Kumar, V., Gupta, S., & Sachan, M. (2025). Navigating circulating tumour cells and tumour microemboli: Where we stand in bridging their molecular dynamics with clinical translation. *Biochimica et Biophysica Acta (BBA)—Reviews on Cancer*, *1880*(4), 189384. https://doi.org/10.1016/j.bbcan.2025.189384

Bloemen, B., Gand, M., Vanneste, K., Marchal, K., Roosens, N. H. C., & De Keersmaecker, S. C. J. (2023). Development of a portable on-site applicable metagenomic data generation workflow for enhanced pathogen and antimicrobial resistance surveillance. *Sci Rep*, *13*(1), 19656. https://doi.org/10.1038/s41598-023-46771-z

Borthakur, D., Rani, M., Das, K., Shah, M. P., Sharma, B. K., & Kumar, A. (2022). Bioremediation: An alternative approach for detoxification of polymers from the contaminated environment. *Lett Appl Microbiol*, *75*(4), 744–758. https://doi.org/10.1111/lam.13616

Brisimi, T. S., Chen, R., Mela, T., Olshevsky, A., Paschalidis, I. C., & Shi, W. (2018). Federated learning of predictive models from federated electronic health records. *Int J Med Inform*, *112*, 59–67. https://doi.org/10.1016/j.ijmedinf.2018.01.007

Caballero-María, P., Caballero-Villarraso, J., Arenas-Montes, J., Díaz-Cáceres, A., Castañeda-Nieto, S., Alcalá-Díaz, J. F., Delgado-Lista, J., Rodríguez-Cantalejo, F., Pérez-Martínez, P., López-Miranda, J., & Camargo, A. (2025). Deep learning model approach to predict diabetes type 2 based on clinical, biochemical, and gut microbiota profiles. *Applied Sciences*, *15*(4), 2228. https://www.mdpi.com/2076-3417/15/4/2228

Cantalapiedra, C. P., Hernandez-Plaza, A., Letunic, I., Bork, P., & Huerta-Cepas, J. (2021). eggNOG-mapper v2: Functional annotation, orthology assignments, and domain prediction at the metagenomic scale. *Mol Biol Evol*, *38*(12), 5825–5829. https://doi.org/10.1093/molbev/msab293

Caporaso, J. G., Lauber, C. L., Walters, W. A., Berg-Lyons, D., Huntley, J., Fierer, N., Owens, S. M., Betley, J., Fraser, L., Bauer, M., Gormley, N., Gilbert, J. A., Smith, G., & Knight, R. (2012). Ultra-high-throughput microbial community analysis on the Illumina HiSeq and MiSeq platforms. *ISME J*, *6*(8), 1621–1624. https://doi.org/10.1038/ismej.2012.8

Casaletto, J., Bernier, A., McDougall, R., & Cline, M. S. (2023). Federated analysis for privacy-preserving data sharing: A technical and legal primer. *Annu Rev Genomics Hum Genet*, *24*, 347–368. https://doi.org/10.1146/annurev-genom-110122-084756

Chen, J., Wang, Y., & Ko, J. (2023). Single-cell and spatially resolved omics: Advances and limitations. *J Pharm Anal*, *13*(8), 833–835. https://doi.org/10.1016/j.jpha.2023.07.002

Chen, P., Sun, Z., Wang, J., Liu, X., Bai, Y., Chen, J., Liu, A., Qiao, F., Chen, Y., Yuan, C., Sha, J., Zhang, J., Xu, L. Q., & Li, J. (2023). Portable nanopore-sequencing technology: Trends in development and applications. *Front Microbiol*, *14*, 1043967. https://doi.org/10.3389/fmicb.2023.1043967

Chen, Q., Song, Y., An, Y., Lu, Y., & Zhong, G. (2024). Soil microorganisms: Their role in enhancing crop nutrition and health. *Diversity*, *16*(12), 734. https://www.mdpi.com/1424-2818/16/12/734

Chiu, C. Y., & Miller, S. A. (2019). Clinical metagenomics. *Nat Rev Genet*, *20*(6), 341–355. https://doi.org/10.1038/s41576-019-0113-7

Crossley, B. M., Bai, J., Glaser, A., Maes, R., Porter, E., Killian, M. L., Clement, T., & Toohey-Kurth, K. (2020). Guidelines for Sanger sequencing and molecular assay monitoring. *J Vet Diagn Invest*, *32*(6), 767–775. https://doi.org/10.1177/1040638720905833

Das, B. K., Gadnayak, A., Chakraborty, H. J., Pradhan, S. P., Raut, S. S., & Das, S. K. (2025). Exploring microbial players for metagenomic profiling of carbon cycling bacteria in sundarban mangrove soils. *Sci Rep*, *15*(1), 4784. https://doi.org/10.1038/s41598-025-89418-x

De Coster, W., Weissensteiner, M. H., & Sedlazeck, F. J. (2021). Towards population-scale long-read sequencing. *Nat Rev Genet*, *22*(9), 572–587. https://doi.org/10.1038/s41576-021-00367-3

Dohm, J. C., Lottaz, C., Borodina, T., & Himmelbauer, H. (2008). Substantial biases in ultra-short read data sets from high-throughput DNA sequencing. *Nucleic Acids Res*, *36*(16), e105. https://doi.org/10.1093/nar/gkn425

Dosoky, N. S., May-Zhang, L. S., & Davies, S. S. (2020). Engineering the gut microbiota to treat chronic diseases. *Appl Microbiol Biotechnol*, *104*(18), 7657–7671. https://doi.org/10.1007/s00253-020-10771-0

Dubey, R. K., Tripathi, V., Prabha, R., Chaurasia, R., Singh, D. P., Rao, C. S., El-Keblawy, A., & Abhilash, P. C. (2020). Metatranscriptomics and metaproteomics for microbial communities profiling. In *Unravelling the soil microbiome* (pp. 51–60). Springer International Publishing. https://doi.org/10.1007/978-3-030-15516-2_5

Eisenstein, M. (2023). Innovative technologies crowd the short-read sequencing market. *Nature*, *614*(7949), 798–800. https://doi.org/10.1038/d41586-023-00512-4

Espinosa, E., Bautista, R., Larrosa, R., & Plata, O. (2024). Advancements in long-read genome sequencing technologies and algorithms. *Genomics*, *116*(3), 110842. https://doi.org/10.1016/j.ygeno.2024.110842

Fernandez-Carro, E., Letsiou, S., Tsironi, S., Chaniotis, D., Ciriza, J., & Beloukas, A. (2025). Alternatives integrating omics approaches for the advancement of human skin models: A focus on metagenomics, metatranscriptomics, and metaproteomics. *Microorganisms*, *13*(8). https://doi.org/10.3390/microorganisms13081771

Forbes, J. D., Knox, N. C., Ronholm, J., Pagotto, F., & Reimer, A. (2017). Metagenomics: The next culture-independent game changer. *Front Microbiol*, *8*, 1069. https://doi.org/10.3389/fmicb.2017.01069

Garza, D. R., & Dutilh, B. E. (2015). From cultured to uncultured genome sequences: Metagenomics and modeling microbial ecosystems. *Cell Mol Life Sci*, *72*(22), 4287–4308. https://doi.org/10.1007/s00018-015-2004-1

Gibbons, S. M., Gurry, T., Lampe, J. W., Chakrabarti, A., Dam, V., Everard, A., Goas, A., Gross, G., Kleerebezem, M., Lane, J., Maukonen, J., Penna, A. L. B., Pot, B., Valdes, A. M., Walton, G., Weiss, A., Zanzer, Y. C., Venlet, N. V., & Miani, M. (2022). Perspective: leveraging the gut microbiota to predict personalized responses to dietary, prebiotic, and probiotic interventions. *Adv Nutr*, *13*(5), 1450–1461. https://doi.org/10.1093/advances/nmac075

Gilbert, J. A., Jansson, J. K., & Knight, R. (2014). The earth microbiome project: Successes and aspirations. *BMC Biol*, *12*, 69. https://doi.org/10.1186/s12915-014-0069-1

Gnimpieba, E. Z., Hartman, T. W., Do, T., Zylla, J., Aryal, S., Haas, S. J., Agany, D. D. M., Gurung, B. D. S., Doe, V., Yosufzai, Z., Pan, D., Campbell, R., Huber, V. C., Sani, R., Gadhamshetty, V., & Lushbough, C. (2024). Biofilm marker discovery with cloud-based dockerized metagenomics analysis of microbial communities. *Brief Bioinform*, *25*(Supplement_1). https://doi.org/10.1093/bib/bbae429

Goltsman, D. S. A., Sun, C. L., Proctor, D. M., DiGiulio, D. B., Robaczewska, A., Thomas, B. C., Shaw, G. M., Stevenson, D. K., Holmes, S. P., Banfield, J. F., & Relman, D. A. (2018). Metagenomic analysis with strain-level resolution reveals fine-scale variation in the human pregnancy microbiome. *Genome Res*, *28*(10), 1467–1480. https://doi.org/10.1101/gr.236000.118

Gonzalez, A., Fullaondo, A., & Odriozola, A. (2025). Why are long-read sequencing methods revolutionizing microbiome analysis? *Microorganisms*, *13*(8). https://doi.org/10.3390/microorganisms13081861

Goussarov, G., Mysara, M., Cleenwerck, I., Claesen, J., Leys, N., Vandamme, P., & Van Houdt, R. (2024). Benchmarking short-, long- and hybrid-read assemblers for metagenome sequencing of complex microbial communities. *Microbiology (Reading)*, *170*(6). https://doi.org/10.1099/mic.0.001469

Greninger, A. L. (2018). The challenge of diagnostic metagenomics. *Expert Rev Mol Diagn*, *18*(7), 605–615. https://doi.org/10.1080/14737159.2018.1487292

Gu, J., Wang, H., Zhang, M., Xiong, Y., Yang, L., Ren, B., & Huang, R. (2022). Application of fluorescence in situ hybridization (FISH) in oral microbial detection. *Pathogens*, *11*(12). https://doi.org/10.3390/pathogens11121450

Gupta, A., Pahal, S., Kumar, V., & Chaudhary, A. (2025). Role of monoclonal antibodies in alzheimer's disease: Successes, failures, and future directions. *Neurological Sciences*. https://doi.org/10.1007/s10072-025-08456-5

Han, Y., He, J., Li, M., Peng, Y., Jiang, H., Zhao, J., Li, Y., & Deng, F. (2024). Unlocking the potential of metagenomics with the PacBio high-fidelity sequencing technology. *Microorganisms*, *12*(12), 2482. https://doi.org/10.3390/microorganisms12122482

Hassija, V., Chamola, V., Mahapatra, A., Singal, A., Goel, D., Huang, K., Scardapane, S., Spinelli, I., Mahmud, M., & Hussain, A. (2023). Interpreting black-box models: A review on explainable artificial intelligence. *Cognitive Computation*, *16*(1), 45–74. https://doi.org/10.1007/s12559-023-10179-8

Herazo-Alvarez, J., Mora, M., Cuadros-Orellana, S., Vilches-Ponce, K., & Hernandez-Garcia, R. (2025). A review of neural networks for metagenomic binning. *Brief Bioinform*, *26*(2). https://doi.org/10.1093/bib/bbaf065

Howe, A., & Chain, P. S. (2015). Challenges and opportunities in understanding microbial communities with metagenome assembly (accompanied by IPython Notebook tutorial). *Front Microbiol*, *6*, 678. https://doi.org/10.3389/fmicb.2015.00678

Ibrahim, A., Oginga, B., Zhang, Y., Ling, W., Tang, L., Elatafi, E., Abady, M., & Gao, Y. (2025). Bioremediation of soils with emerging organic contaminants using immobilized microorganisms. *Environmental Technology & Innovation*, *40*, 104345. https://doi.org/10.1016/j.eti.2025.104345

Iqbal, S., Begum, F., Nguchu, B. A., Claver, U. P., & Shaw, P. (2025). The invisible architects: Microbial communities and their transformative role in soil health and global climate changes. *Environ Microbiome*, *20*(1), 36. https://doi.org/10.1186/s40793-025-00694-6

Jagadesh, M., Dash, M., Kumari, A., Singh, S. K., Verma, K. K., Kumar, P., Bhatt, R., & Sharma, S. K. (2024). Revealing the hidden world of soil microbes: Metagenomic insights into plant, bacteria, and fungi interactions for sustainable agriculture and ecosystem restoration. *Microbiol Res*, *285*, 127764. https://doi.org/10.1016/j.micres.2024.127764

Jain, A., Sarsaiya, S., Singh, R., Gong, Q., Wu, Q., & Shi, J. (2024). Omics approaches in understanding the benefits of plant-microbe interactions. *Front Microbiol*, *15*, 1391059. https://doi.org/10.3389/fmicb.2024.1391059

James, A. S., Adil, N. A., Goltz, D., Tangudu, D., Chaudhari, D. S., Shukla, R., Kumar, V., Kumar, A., Masternak, M. M., Holland, P., Labyak, C., Golden, A., Dangiolo, M., Arikawa, A. Y., Kociolek, J., Fraser, A., Williams, C., Agronin, M., Aymat, M., . . . Yadav, H. (2024). Abnormalities in gut virome signatures linked with cognitive impairment in older adults. *Gut Microbes*, *16*(1), 2431648. https://doi.org/10.1080/19490976.2024.2431648

Johnson, C. H., Ivanisevic, J., & Siuzdak, G. (2016). Metabolomics: Beyond biomarkers and towards mechanisms. *Nat Rev Mol Cell Biol*, *17*(7), 451–459. https://doi.org/10.1038/nrm.2016.25

Kashyap, P. C., Chia, N., Nelson, H., Segal, E., & Elinav, E. (2017). Microbiome at the frontier of personalized medicine. *Mayo Clin Proc*, *92*(12), 1855–1864. https://doi.org/10.1016/j.mayocp.2017.10.004

Kibegwa, F. M., Bett, R. C., Gachuiri, C. K., Stomeo, F., & Mujibi, F. D. (2020). A comparison of two DNA metagenomic bioinformatic pipelines while evaluating the microbial diversity in feces of Tanzanian small holder dairy cattle. *Biomed Res Int*, *2020*, 2348560. https://doi.org/10.1155/2020/2348560

Kim, H. M., Jeon, S., Chung, O., Jun, J. H., Kim, H. S., Blazyte, A., Lee, H. Y., Yu, Y., Cho, Y. S., Bolser, D. M., & Bhak, J. (2021). Comparative analysis of 7 short-read sequencing platforms using the Korean reference genome: MGI and illumina sequencing benchmark for whole-genome sequencing. *Gigascience*, *10*(3). https://doi.org/10.1093/gigascience/giab014

Kumar, V., Gangani, S., Shukla, R., Prajapati, S. K., Shekhar, H., & Shukla, V. (2024). Next-generation sequencing and genomic data analysis. In A. Chaudhary, S. K. Sethi, & A. Verma (Eds.), *Unraveling new frontiers and advances in bioinformatics* (pp. 49–76). Springer Nature. https://doi.org/10.1007/978-981-97-7123-3_3

Kumar, V., Pandey, A., Arora, A., Gautam, P., Bisht, D., Gupta, S., Chaurasia, A., & Sachan, M. (2023). Diagnostics and therapeutic potential of miR-205 and miR-34a in ovarian cancer management: A miRNA-target-based analysis. *DNA and Cell Biology*, *42*(3), 151–162. https://doi.org/10.1089/dna.2022.0487

Kumar, V., Sharma, J. K., Kumar, S., Gulia, J., Singh, N. P., & Santal, A. R. (2025). Metagenomic and proteomic approach for bioremediation of environmental pollutants. In M. P. Shah (Ed.), *Development in waste water treatment research and processes*. Elsevier. https://doi.org/10.1016/B978-0-443-13615-3.00020-0

Kumar, V., Shukla, R., Gangani, S., Shekhar, H., & Shukla, V. (2024). Metagenomics and microbiome analysis. In A. Chaudhary, S. K. Sethi, & A. Verma (Eds.), *Unraveling new frontiers and advances in bioinformatics* (pp. 77–92). Springer Nature, Singapore. https://doi.org/10.1007/978-981-97-7123-3_4

Kumar, V., Singh, A., Gautam, P., & Sachan, M. (2019). Techniques/tools to study epigenetic biomarkers in human cancer detection. In S. Paul (Ed.), *Biomedical engineering and its applications in healthcare*. Springer, Singapore. https://doi.org/10.1007/978-981-13-3705-5_14

Kumar Awasthi, M., Ravindran, B., Sarsaiya, S., Chen, H., Wainaina, S., Singh, E., Liu, T., Kumar, S., Pandey, A., Singh, L., & Zhang, Z. (2020). Metagenomics for taxonomy profiling: Tools and approaches. *Bioengineered*, *11*(1), 356–374. https://doi.org/10.1080/21655979.2020.1736238

Lee, J. Y., Sadler, N. C., Egbert, R. G., Anderton, C. R., Hofmockel, K. S., Jansson, J. K., & Song, H. S. (2020). Deep learning predicts microbial interactions from self-organized spatiotemporal patterns. *Comput Struct Biotechnol J*, *18*, 1259–1269. https://doi.org/10.1016/j.csbj.2020.05.023

Lee, Y. Q., Choi, Y. M., Park, S. Y., Kim, S. K., Lee, M., Kim, D., Koduru, L., Lakshmanan, M., Jung, S., Kim, M. J., Choe, Y. H., & Lee, D. Y. (2025). Genome-scale metabolic model-guided systematic framework for designing customized live biotherapeutic products. *NPJ Syst Biol Appl*, *11*(1), 73. https://doi.org/10.1038/s41540-025-00555-5

Leung, C. M., Li, D., Xin, Y., Law, W. C., Zhang, Y., Ting, H. F., Luo, R., & Lam, T. W. (2020). MegaPath: Sensitive and rapid pathogen detection using metagenomic NGS data. *BMC Genomics*, *21*(Supplement_6), 500. https://doi.org/10.1186/s12864-020-06875-6

Li, H., He, J., & Jia, W. (2016). The influence of gut microbiota on drug metabolism and toxicity. *Expert Opin Drug Metab Toxicol*, *12*(1), 31–40. https://doi.org/10.1517/17425255.2016.1121234

Li, S., Lian, W. H., Han, J. R., Ali, M., Lin, Z. L., Liu, Y. H., Li, L., Zhang, D. Y., Jiang, X. Z., Li, W. J., & Dong, L. (2023). Capturing the microbial dark matter in desert soils using culturomics-based metagenomics and high-resolution analysis. *NPJ Biofilms Microbiomes*, *9*(1), 67. https://doi.org/10.1038/s41522-023-00439-8

Lindez, P. P., Johansen, J., Kutuzova, S., Sigurdsson, A. I., Nissen, J. N., & Rasmussen, S. (2023). Adversarial and variational autoencoders improve metagenomic binning. *Commun Biol*, *6*(1), 1073. https://doi.org/10.1038/s42003-023-05452-3

Liu, S., Obert, C., Yu, Y. P., Zhao, J., Ren, B. G., Liu, J. J., Wiseman, K., Krajacich, B. J., Wang, W., Metcalfe, K., Smith, M., Ben-Yehezkel, T., & Luo, J. H. (2024). Utility analyses of AVITI sequencing chemistry. *BMC Genomics*, *25*(1), 778. https://doi.org/10.1186/s12864-024-10686-4

Liu, Y., Guo, J., Hu, G., & Zhu, H. (2013). Gene prediction in metagenomic fragments based on the SVM algorithm. *BMC Bioinformatics*, *14*(Supplement_5), S12. https://doi.org/10.1186/1471-2105-14-S5-S12

Liu, Y., Xu, Y., Xu, X., Chen, X., Chen, H., Zhang, J., Ma, J., Zhang, W., Zhang, R., & Chen, J. (2023). Metagenomic identification of pathogens and antimicrobial-resistant genes in bacterial positive blood cultures by nanopore sequencing. *Front Cell Infect Microbiol*, *13*, 1283094. https://doi.org/10.3389/fcimb.2023.1283094

Logsdon, G. A., Vollger, M. R., & Eichler, E. E. (2020). Long-read human genome sequencing and its applications. *Nat Rev Genet*, *21*(10), 597-614. https://doi.org/10.1038/s41576-020-0236-x

Loos, D., Filho, A., Dutilh, B. E., Barber, A. E., & Panagiotou, G. (2024). A global survey of host, aquatic, and soil microbiomes reveals shared abundance and genomic features between bacterial and fungal generalists. *Cell Rep*, *43*(4), 114046. https://doi.org/10.1016/j.celrep.2024.114046

Madigan, K. E., Maffei, V. J., Shukla, R., Kumar, V., Yadav, H., & Weinberg, R. B. (2025). Fecal microbiome richness and evenness are reduced and the relative abundance of firmicutes species is increased in patients with intestinal methanogen overgrowth. *Gastroenterology*, *169*(Supplement_1), S-1263. https://doi.org/10.1016/S0016-5085(25)03787-4

Marcos-Zambrano, L. J., Karaduzovic-Hadziabdic, K., Loncar Turukalo, T., Przymus, P., Trajkovik, V., Aasmets, O., Berland, M., Gruca, A., Hasic, J., Hron, K., Klammsteiner, T., Kolev, M., Lahti, L., Lopes, M. B., Moreno, V., Naskinova, I., Org, E., Paciencia, I., Papoutsoglou, G., . . . Truu, J. (2021). Applications of machine learning in human microbiome studies: A review on feature selection, biomarker identification, disease prediction and treatment. *Front Microbiol*, *12*, 634511. https://doi.org/10.3389/fmicb.2021.634511

Martin, R., Miquel, S., Langella, P., & Bermudez-Humaran, L. G. (2014). The role of metagenomics in understanding the human microbiome in health and disease. *Virulence*, *5*(3), 413–423. https://doi.org/10.4161/viru.27864

Martinez-Rabert, E., van Amstel, C., Smith, C., Sloan, W. T., & Gonzalez-Cabaleiro, R. (2022). Environmental and ecological controls of the spatial distribution of microbial populations in aggregates. *PLoS Comput Biol*, *18*(12), e1010807. https://doi.org/10.1371/journal.pcbi.1010807

Masse, K. E., & Lu, V. B. (2023). Short-chain fatty acids, secondary bile acids and indoles: Gut microbial metabolites with effects on enteroendocrine cell function and their potential as therapies for metabolic disease. *Front Endocrinol (Lausanne)*, *14*, 1169624. https://doi.org/10.3389/fendo.2023.1169624

Mesa-Marín, J., Pérez-Romero, J. A., Mateos-Naranjo, E., Bernabeu-Meana, M., Pajuelo, E., Rodríguez-Llorente, I. D., & Redondo-Gómez, S. (2019). Effect of plant growth-promoting Rhizobacteria on Salicornia ramosissima seed germination under salinity, CO2 and temperature stress. *Agronomy*, *9*(10), 655. https://www.mdpi.com/2073-4395/9/10/655

Meyer, F., Bagchi, S., Chaterji, S., Gerlach, W., Grama, A., Harrison, T., Paczian, T., Trimble, W. L., & Wilke, A. (2019). MG-RAST version 4-lessons learned from a decade of low-budget ultra-high-throughput metagenome analysis. *Brief Bioinform*, *20*(4), 1151–1159. https://doi.org/10.1093/bib/bbx105

Miller, J. R., Zhou, P., Mudge, J., Gurtowski, J., Lee, H., Ramaraj, T., Walenz, B. P., Liu, J., Stupar, R. M., Denny, R., Song, L., Singh, N., Maron, L. G., McCouch, S. R., McCombie, W. R., Schatz, M. C., Tiffin, P., Young, N. D., & Silverstein, K. A. T. (2017). Hybrid assembly with long and short reads improves discovery of gene family expansions. *BMC Genomics*, *18*(1), 541. https://doi.org/10.1186/s12864-017-3927-8

Mishra, S., Buddendorff, L., Hoover, J., Shukla, R., Kumar, V., Jain, S., & Yadav, H. (2024). Microbiome role in cognitive impairment: MiaGB cohort insights. *Innovation in Aging*, *8*(Supplement_1), 811–811. https://doi.org/10.1093/geroni/igae098.2629

Nair, S. U., Lin, Y., Jain, S., Chaudhari, D. S., Shukla, R., Kumar, V., Mankowski, R. T., Meng, H., Buford, T. W., Kuchel, G. A., Yadav, H., & Consortium, M. (2024). Unique gut microbiome signatures associated with physical and cognitive function. *Alzheimer's & Dementia*, *20*(S2), e092044. https://doi.org/10.1002/alz.092044

Napit, R., Gurung, A., Poudel, A., Chaudhary, A., Manandhar, P., Sharma, A. N., Raut, S., Pradhan, S. M., Joshi, J., Poyet, M., Groussin, M., Rajbhandari, R. M., & Karmacharya, D. B. (2025). Metagenomic analysis of human, animal, and environmental samples identifies potential emerging pathogens, profiles antibiotic resistance genes, and reveals horizontal gene transfer dynamics. *Sci Rep*, *15*(1), 12156. https://doi.org/10.1038/s41598-025-90777-8

National Research Council, Board on Life Sciences, Committee on Metagenomics, & Functional Applications. (2007). *Challenges and functional applications. The new science of metagenomics: Revealing the secrets of our microbial planet*. https://doi.org/https://www.ncbi.nlm.nih.gov/books/NBK54011/

Navgire, G. S., Goel, N., Sawhney, G., Sharma, M., Kaushik, P., Mohanta, Y. K., Mohanta, T. K., & Al-Harrasi, A. (2022). Analysis and interpretation of metagenomics data: An approach. *Biol Proced Online*, *24*(1), 18. https://doi.org/10.1186/s12575-022-00179-7

Nearing, J. T., Comeau, A. M., & Langille, M. G. I. (2021). Identifying biases and their potential solutions in human microbiome studies. *Microbiome*, *9*(1), 113. https://doi.org/10.1186/s40168-021-01059-0

Nicholls, S. M., Quick, J. C., Tang, S., & Loman, N. J. (2019). Ultra-deep, long-read nanopore sequencing of mock microbial community standards. *Gigascience*, *8*(5). https://doi.org/10.1093/gigascience/giz043

Olsen, N. S., & Riber, L. (2025). Metagenomics as a transformative tool for antibiotic resistance surveillance: Highlighting the impact of mobile genetic elements with a focus on the complex role of phages. *Antibiotics (Basel)*, *14*(3). https://doi.org/10.3390/antibiotics14030296

Ota, Y., Chen, F., Prah, I., Mahazu, S., Watanabe, K., Kinoshita, T., Gu, Y., Nukui, Y., & Saito, R. (2024). Metatranscriptomic analysis reveals actively expressed antimicrobial-resistant genes and their hosts in hospital wastewater. *Antibiotics (Basel)*, *13*(12). https://doi.org/10.3390/antibiotics13121122

Panda, A., Rout, S. S., Dey, S., Parida, C. K., Jena, R., Dhar, S., Tiwari, B., Singh, R. K., & Singh, A. K. (2025). Metagenomics and its application in environmental monitoring. In A. K. Rout, R. K. Singh, A. K. Shukla, & B. K. Behera (Eds.), *Advances in omics technologies* (pp. 1–37). Springer Nature, Singapore. https://doi.org/10.1007/978-981-95-0285-1_1

Parab, A. S., Ghose, M., & Manohar, C. S. (2025). Metagenomic reconstruction of microbial genomes and biogeochemical pathways: Insights into carbon and nitrogen flux dynamics in the eastern Arabian Sea. *Mar Environ Res*, *210*, 107292. https://doi.org/10.1016/j.marenvres.2025.107292

Pati, S., Kumar, S., Varma, A., Edwards, B., Lu, C., Qu, L., Wang, J. J., Lakshminarayanan, A., Wang, S., Sheller, M. J., Chang, K., Singh, P., Rubin, D. L., Kalpathy-Cramer, J., & Bakas, S. (2024). Privacy preservation for federated learning in health care. *Patterns*, *5*(7), 100974. https://doi.org/10.1016/j.patter.2024.100974

Peterson, D., Bonham, K. S., Rowland, S., Pattanayak, C. W., Consortium, R., & Klepac-Ceraj, V. (2021). Comparative analysis of 16S rRNA gene and metagenome sequencing in pediatric gut microbiomes [original research]. *Front Microbiol*, *12*, 670336. https://doi.org/10.3389/fmicb.2021.670336

Petschow, B., Dore, J., Hibberd, P., Dinan, T., Reid, G., Blaser, M., Cani, P. D., Degnan, F. H., Foster, J., Gibson, G., Hutton, J., Klaenhammer, T. R., Ley, R., Nieuwdorp, M., Pot, B., Relman, D., Serazin, A., & Sanders, M. E. (2013). Probiotics, prebiotics, and the host microbiome: The science of translation. *Ann N Y Acad Sci*, *1306*(1), 1–17. https://doi.org/10.1111/nyas.12303

Pinu, F. R., Beale, D. J., Paten, A. M., Kouremenos, K., Swarup, S., Schirra, H. J., & Wishart, D. (2019). Systems biology and multi-omics integration: Viewpoints from the metabolomics research community. *Metabolites*, *9*(4). https://doi.org/10.3390/metabo9040076

Prajapati, S. K., Kumar, V., Shukla, R., Szekeres, C., Ma, Y., Jain, S., & Yadav, H. (2024). Microbiome-immune interplay in aging brains: A pilot investigation from the MiaGB cohort. *Alzheimer's & Dementia*, *20*(S2), e090734. https://doi.org/10.1002/alz.090734

Preidis, G. A., & Versalovic, J. (2009). Targeting the human microbiome with antibiotics, probiotics, and prebiotics: Gastroenterology enters the metagenomics era. *Gastroenterology*, *136*(6), 2015–2031. https://doi.org/10.1053/j.gastro.2009.01.072

Proctor, L. M., Dore, J., Maguin, E., Wannerberger, K., & Vincent, C. (2025). A global partnership to advance the microbiome in human, plant, animal, and planetary health. *Lancet Microbe*, *6*(8), 101120. https://doi.org/10.1016/j.lanmic.2025.101120

Puig-Castellví, F., Pacheco-Tapia, R., Deslande, M., Jia, M., Andrikopoulos, P., Chechi, K., Bonnefond, A., Froguel, P., & Dumas, M.-E. (2023). Advances in the integration of metabolomics and metagenomics for human gut microbiome and their clinical applications. *TrAC Trends in Analytical Chemistry*, *167*, 117248. https://doi.org/https://doi.org/10.1016/j.trac.2023.117248

Purushothaman, S., Meola, M., & Egli, A. (2022). Combination of whole genome sequencing and metagenomics for microbiological diagnostics. *Int J Mol Sci*, *23*(17). https://doi.org/10.3390/ijms23179834

Qu, K., Guo, F., Liu, X., Lin, Y., & Zou, Q. (2019). Application of machine learning in microbiology. *Front Microbiol*, *10*, 827. https://doi.org/10.3389/fmicb.2019.00827

Quick, J., Loman, N. J., Duraffour, S., Simpson, J. T., Severi, E., Cowley, L., Bore, J. A., Koundouno, R., Dudas, G., Mikhail, A., Ouedraogo, N., Afrough, B., Bah, A., Baum, J. H., Becker-Ziaja, B., Boettcher, J. P., Cabeza-Cabrerizo, M., Camino-Sanchez, A., Carter, L. L., . . . Carroll, M. W. (2016). Real-time, portable genome sequencing for Ebola surveillance. *Nature*, *530*(7589), 228–232. https://doi.org/10.1038/nature16996

Ren, F., Wei, J., Chen, Q., Hu, M., Yu, L., Mi, J., Zhou, X., Qin, D., Wu, J., & Wu, A. (2025). Artificial intelligence-driven multi-omics approaches in Alzheimer's disease: Progress, challenges, and future directions. *Acta Pharmaceutica Sinica B*. https://doi.org/10.1016/j.apsb.2025.07.030

Rinke, C. (2018). Single-cell genomics of microbial dark matter. *Methods Mol Biol*, *1849*, 99–111. https://doi.org/10.1007/978-1-4939-8728-3_7

Rooks, M. G., & Garrett, W. S. (2016). Gut microbiota, metabolites and host immunity. *Nat Rev Immunol*, *16*(6), 341–352. https://doi.org/10.1038/nri.2016.42

Rout, A. K., Rout, S. S., Panda, A., Tripathy, P. S., Kumar, N., Parida, S. N., Dey, S., Dash, S. S., Behera, B. K., & Pandey, P. K. (2025). Potential applications and future prospects of metagenomics in aquatic ecosystems. *Gene*, *967*, 149720. https://doi.org/10.1016/j.gene.2025.149720

Rujano, M. A., Boiten, J. W., Ohmann, C., Canham, S., Contrino, S., David, R., Ewbank, J., Filippone, C., Connellan, C., Custers, I., van Nuland, R., Mayrhofer, M. T., Holub, P., Alvarez, E. G., Bacry, E., Hughes, N., Freeberg, M. A., Schaffhauser, B., Wagener, H., . . . Panagiotopoulou, M. (2024). Sharing sensitive data in life sciences: An overview of centralized and federated approaches. *Brief Bioinform*, *25*(4). https://doi.org/10.1093/bib/bbae262

Sahu, A., Kumar, N., Pal Singh, C., & Singh, M. (2023). Environmental DNA (eDNA): Powerful technique for biodiversity conservation. *Journal for Nature Conservation*, *71*, 126325. https://doi.org/10.1016/j.jnc.2022.126325

Sankarganesh, P., Bhunia, A., Ganesh Kumar, A., Babu, A. S., Gopukumar, S. T., & Lokesh, E. (2025). Short-chain fatty acids (SCFAs) in gut health: Implications for drug metabolism and therapeutics. *Medicine in Microecology*, *25*, 100139. https://doi.org/https://doi.org/10.1016/j.medmic.2025.100139

Sarfraz, H., Czerniecki, B. J., Stankiewicz, K., Jaglal, M. V., Jain, S., Yadav, H., Liu, M., Shukla, R., Kumar, V., Alhomsi, M., Hathout, M., Tuz Zahra, F., Darwin, A., Joshi, U., Sen, A., Muzaffar, J., & Bari, S. (2024). Evaluating gut microbiome as a biomarker of pathological complete response in patients with early locally advanced triple negative breast cancer (LA TNBC): A pilot study. *Journal of Clinical Oncology*, *42*(Supplement_16), 593–593. https://doi.org/10.1200/JCO.2024.42.16_suppl.593

Schreiber, P. W., Kufner, V., Hubel, K., Schmutz, S., Zagordi, O., Kaur, A., Bayard, C., Greiner, M., Zbinden, A., Capaul, R., Boni, J., Hirsch, H. H., Mueller, T. F., Mueller, N. J., Trkola, A., & Huber, M. (2019). Metagenomic virome sequencing in living donor and recipient kidney transplant pairs revealed JC polyomavirus transmission. *Clin Infect Dis*, *69*(6), 987–994. https://doi.org/10.1093/cid/ciy1018

Shaffer, M., Borton, M. A., Bolduc, B., Faria, J. P., Flynn, R. M., Ghadermazi, P., Edirisinghe, J. N., Wood-Charlson, E. M., Miller, C. S., Chan, S. H. J., Sullivan, M. B., Henry, C. S., & Wrighton, K. C. (2024). Correction to: Kb_DRAM: Annotation and metabolic profiling of genomes with DRAM in KBase. *Bioinformatics*, *40*(5). https://doi.org/10.1093/bioinformatics/btae325

Shahid, U. (2025). Microbiome-guided precision medicine: Mechanistic insights, multi-omics integration, and translational horizons. *The Journal of Precision Medicine: Health and Disease*, *3*. https://doi.org/10.1016/j.premed.2025.100018

Sherstinsky, A. (2020). Fundamentals of recurrent neural network (RNN) and long short-term memory (LSTM) network. *Physica D: Nonlinear Phenomena*, *404*, 132306. https://doi.org/https://doi.org/10.1016/j.physd.2019.132306

Shrestha, H. K., Appidi, M. R., Villalobos Solis, M. I., Wang, J., Carper, D. L., Burdick, L., Pelletier, D. A., Doktycz, M. J., Hettich, R. L., & Abraham, P. E. (2021). Metaproteomics reveals insights into microbial structure, interactions, and dynamic regulation in defined communities as they respond to environmental disturbance. *BMC Microbiol*, *21*(1), 308. https://doi.org/10.1186/s12866-021-02370-4

Singh, D. P., Maurya, S., Satnami, L., Renu, Prabha, R., Sarma, B. K., & Rai, N. (2024). Roots of resistance: Unraveling microbiome-driven plant immunity. *Plant Stress*, *14*, 100661. https://doi.org/10.1016/j.stress.2024.100661

Slatko, B. E., Gardner, A. F., & Ausubel, F. M. (2018). Overview of next-generation sequencing technologies. *Curr Protoc Mol Biol*, *122*(1), e59. https://doi.org/10.1002/cpmb.59

Stumpf, R. M., Gomez, A., Amato, K. R., Yeoman, C. J., Polk, J. D., Wilson, B. A., Nelson, K. E., White, B. A., & Leigh, S. R. (2016). Microbiomes, metagenomics, and primate conservation: New strategies, tools, and applications. *Biological Conservation*, *199*, 56–66. https://doi.org/10.1016/j.biocon.2016.03.035

Su, F., Su, M., Wei, W., Wu, J., Chen, L., Sun, X., Liu, M., Sun, S., Mao, R., Bourgonje, A. R., & Hu, S. (2025). Integrating multi-omics data to reveal the host-microbiota interactome in inflammatory bowel disease. *Gut Microbes*, *17*(1), 2476570. https://doi.org/10.1080/19490976.2025.2476570

Suman, J., Rakshit, A., Ogireddy, S. D., Singh, S., Gupta, C., & Chandrakala, J. (2022). Microbiome as a key player in sustainable agriculture and human health [review]. *Frontiers in Soil Science*, *2*. https://doi.org/10.3389/fsoil.2022.821589

Sun, X., Qiu, Q., Wang, F., Wu, J., Bian, D., & Huo, H. (2025). Metagenomics reveals microbial community metabolic and functional changes in electroplating washing wastewater treatment systems under ultralong operating mode. *Journal of Environmental Chemical Engineering*, *13*(4), 117374. https://doi.org/10.1016/j.jece.2025.117374

Taxt, A. M., Avershina, E., Frye, S. A., Naseer, U., & Ahmad, R. (2020). Rapid identification of pathogens, antibiotic resistance genes and plasmids in blood cultures by nanopore sequencing. *Sci Rep*, *10*(1), 7622. https://doi.org/10.1038/s41598-020-64616-x

Trabelsi, D., & Mhamdi, R. (2013). Microbial inoculants and their impact on soil microbial communities: A review. *Biomed Res Int*, *2013*, 863240. https://doi.org/10.1155/2013/863240

Udaondo, Z., Sittikankaew, K., Uengwetwanit, T., Wongsurawat, T., Sonthirod, C., Jenjaroenpun, P., Pootakham, W., Karoonuthaisiri, N., & Nookaew, I. (2021). Comparative analysis of PacBio and Oxford nanopore sequencing technologies for transcriptomic landscape identification of penaeus monodon. *Life (Basel)*, *11*(8). https://doi.org/10.3390/life11080862

Wang 王博, B., Jia 贾鹏, P., Gao 高胜寒, S., Zhao 赵焕焕, H., Zheng 郑高洋, G., Xu 许林峰, L., & Ye 叶凯, K. (2025). Long and accurate: How HiFi sequencing is transforming genomics. *Genomics Proteomics Bioinformatics*, *23*(1). https://doi.org/10.1093/gpbjnl/qzaf003

Wang, N., Wu, M., Gu, W., Dai, C., Shao, Z., & Subbalakshmi, K. P. (2025). MSFT-transformer: A multistage fusion tabular transformer for disease prediction using metagenomic data. *Brief Bioinform*, *26*(3). https://doi.org/10.1093/bib/bbaf217

Wani, A. K., Qadir, F., Elboughdiri, N., Rahayu, F., Saefudin, Pranowo, D., Martasari, C., Kosmiatin, M., Suhara, C., Sudaryono, T., Prayogo, Y., Yadav, K. K., Muzammil, K., Eltayeb, L. B., Alreshidi, M. A., & Singh, R. (2025). Metagenomics and plant-microbe symbioses: Microbial community dynamics, functional roles in carbon sequestration, nitrogen transformation, sulfur and phosphorus mobilization for sustainable soil health. *Biotechnol Adv*, *82*, 108580. https://doi.org/10.1016/j.biotechadv.2025.108580

Wassan, J. T., Wang, H., & Zheng, H. (2023). Developing a new phylogeny-driven random forest model for functional metagenomics. *IEEE Trans Nanobioscience*, *22*(4), 763–770. https://doi.org/10.1109/TNB.2023.3283462

Yang, S. Y., Han, S. M., Lee, J. Y., Kim, K. S., Lee, J. E., & Lee, D. W. (2025). Advancing gut microbiome research: The shift from metagenomics to multi-omics and future perspectives. *J Microbiol Biotechnol*, *35*, e2412001. https://doi.org/10.4014/jmb.2412.12001

Yang, Y., Rivera Perez, C. A., Richter-Heitmann, T., Nimzyk, R., Friedrich, M. W., & Reich, M. (2023). Effects of oxygen availability on mycobenthic communities of marine coastal sediments. *Sci Rep*, *13*(1), 15218. https://doi.org/10.1038/s41598-023-42329-1

Yilmaz, S., & Singh, A. K. (2012). Single cell genome sequencing. *Curr Opin Biotechnol*, *23*(3), 437–443. https://doi.org/10.1016/j.copbio.2011.11.018

Zhang, C., Pan, Z., & Hou, C. (2023). Marketing data security and privacy protection based on federated gamma in cloud computing environment. *International Journal of Intelligent Networks*, *4*, 261–271. https://doi.org/10.1016/j.ijin.2023.09.003

Zhang, W., Xiang, Y., Ren, H., Liu, Y., Wang, Q., Ran, M., Zhou, W., Tian, L., Zheng, X., Qiao, C., Liu, Y., & Yan, M. (2025). The tumor microbiome in cancer progression: Mechanisms and therapeutic potential. *Mol Cancer*, *24*(1), 195. https://doi.org/10.1186/s12943-025-02403-w

Zhou, R., Ng, S. K., Sung, J. J. Y., Goh, W. W. B., & Wong, S. H. (2023). Data pre-processing for analyzing microbiome data—a mini review. *Comput Struct Biotechnol J*, *21*, 4804–4815. https://doi.org/10.1016/j.csbj.2023.10.001

Zhu, L. B., Zhang, Y. C., Huang, H. H., & Lin, J. (2021). Prospects for clinical applications of butyrate-producing bacteria. *World J Clin Pediatr*, *10*(5), 84–92. https://doi.org/10.5409/wjcp.v10.i5.84

Zhuang, Y., Liu, S., Xiao, J., Chen, T., Gao, D., Xu, Y., Jiang, W., Wang, J., Hou, G., Li, S., Zhao, X., Huang, Y., Li, S., Zhang, S., Li, M., Wang, W., Li, S., & Cao, Z. (2024). Metagenomics reveals the characteristics and potential spread of microbiomes and virulence factor genes in the dairy cattle production system. *J Hazard Mater*, *480*, 136005. https://doi.org/10.1016/j.jhazmat.2024.136005